ASQUEZ

HYSICIENS

DEZ

ABORATOIRES

SURREALISM, ART AND MODERN SCIENCE

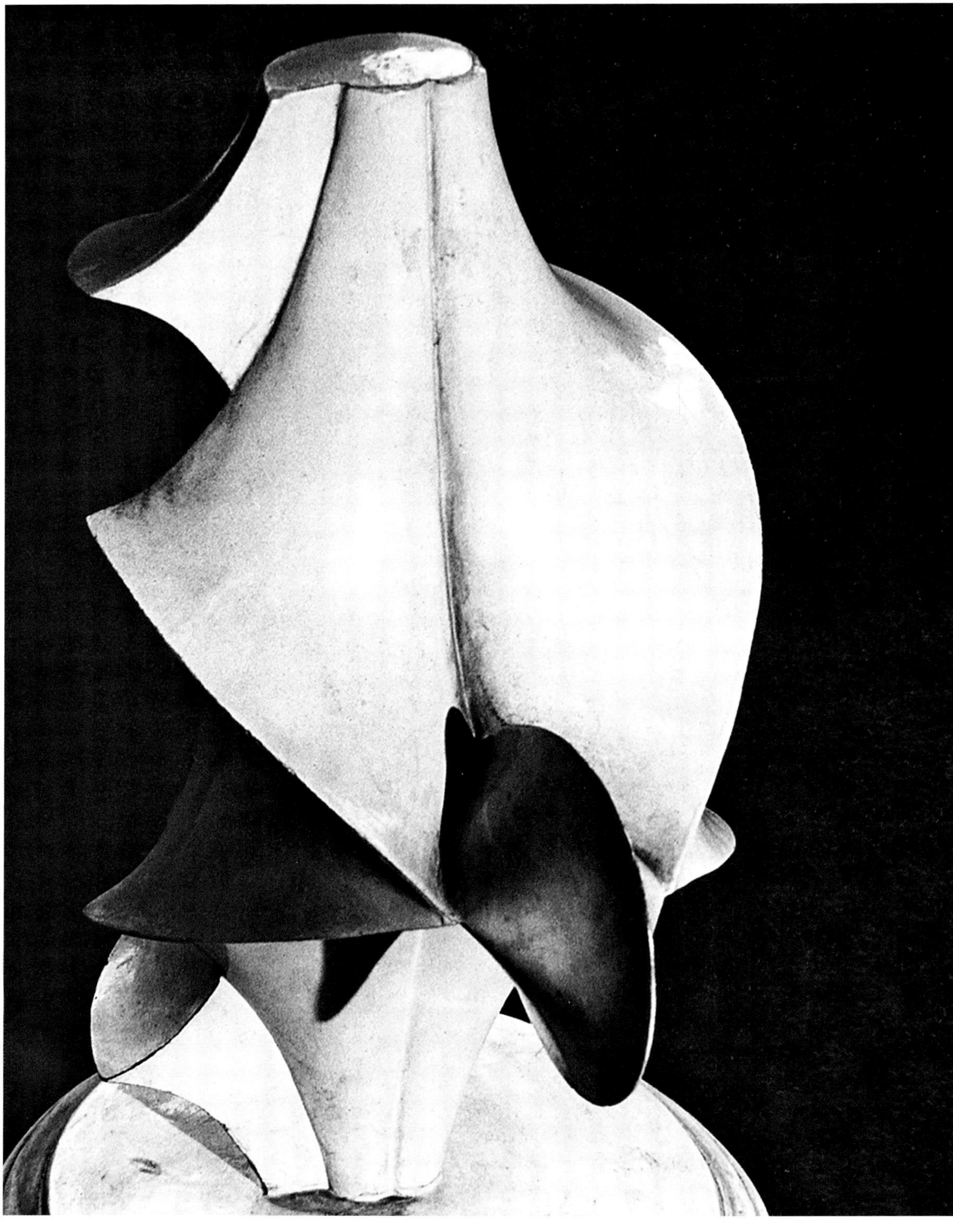

Gavin Parkinson

SURREALISM, ART AND MODERN SCIENCE

Relativity, Quantum Mechanics, Epistemology

YALE UNIVERSITY PRESS NEW HAVEN AND LONDON

Printed in China

Designed by Paul Sloman

Library of Congress Cataloging-in-Publication Data

Parkinson, Gavin.
Surrealism, art, and modern science : relativity, quantum mechanics, epistemology / Gavin Parkinson.
p. cm.
Includes bibliographical references and index.
ISBN 978-0-300-09887-7 (cl : alk. paper)
1. Surrealism–France. 2. Arts, Modern–France–20th century. 3. Physics and the arts–France–History–20th century.
4. Physics–Philosophy–History–20th century. I. Title.
NX549.A1P377 2007
709.04'063–dc22

2007025368

ENDPAPERS
The Surrealist group, *Démasquez les Physiciens, Videz les laboratoires* (18 February 1958).

FRONTISPIECE
Man Ray, *Surface à courbure constant negative d'ennepar, dérivée de la pseudo-sphère* (1934–36). Photograph, 29.5 x 24 cm. Man Ray Trust.

CONTENTS

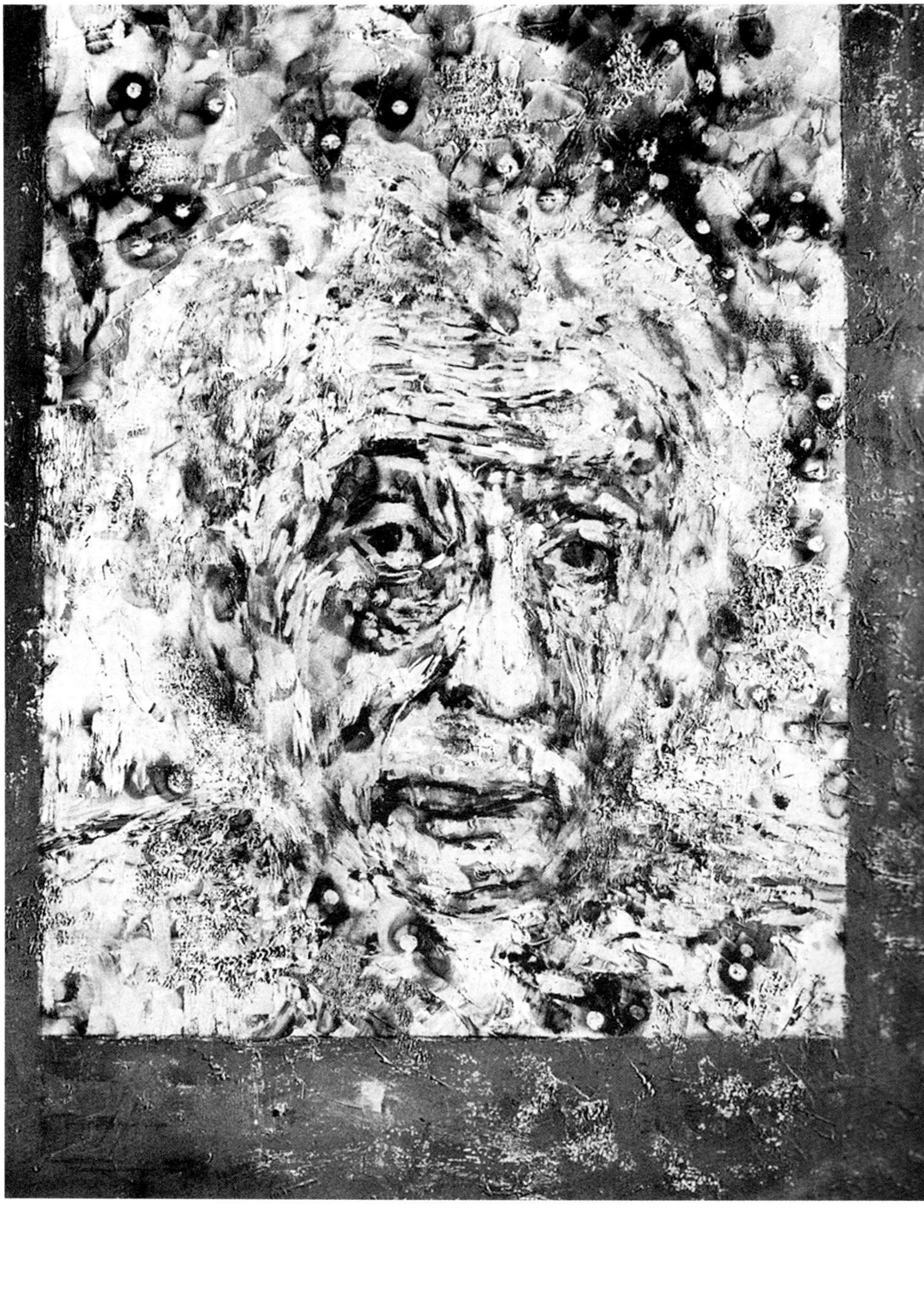

ACKNOWLEDGEMENTS

Wolfgang Paalen, *Portrait of Albert Einstein* (1940). Oil on canvas. Dimensions and present whereabouts unknown.

This began life as a Ph.D. thesis in 1997 at the Courtauld Institute of Art, London. It was completed in 2000, thanks largely to a grant from what was then the British Academy's Arts and Humanities Research Board. Without research leave, I worked on turning it into a book when I could, whilst trying to earn a living here and there as a university lecturer. Over a period of ten years, a project like this benefits from various kinds of support from individuals and institutions, frequently unquantifiable, and often difficult even to express satisfactorily. The best I can do is to list here colleagues, friends and family who helped in some way to make my book what it is, and I hope they will understand the depth of my gratitude.

In its earliest stages, my research was helped along by comments and suggestions made by Jacqueline Chénieux-Gendron, Anthony Hill, Arthur I. Miller, Colin Sanderson, Martica Sawin and Amy Winter. For their nurturing advice, I am grateful to staff in the Department of Art History and Archaeology at the University of Manchester, at the School of History of Art, Film & Visual Media at Birkbeck College, University of London, at the Courtauld Institute, and at the Department of the History of Art and Visual Culture at the University of Oxford: Caroline Arscott, Shulamith Behr, Victoria Brown, Suzy Butters, Andrew Causey, Christine Hawkins, Geraldine Johnson, Martin Kemp, Marius Kwint, David Lomas, Robert Maniura, Livia Morgan, Mignon Nixon, Gervase Rosser, Simon Shaw-Miller, Julian Stallabrass, Sarah Wilson and Rachel Woodruff. At Yale University Press, I received expert guidance from Gillian Malpass and Paul Sloman.

For technical, intellectual and/or emotional support, I thank Dawn Ades, Tanya Barrett, Lee Beard, Lesley Bell, Kiki Benzon, Joanne Bernstein, Fae Brauer, Ian, Karen, Rachel and Katie Brooks, Fiona Candlin, Jacqueline Cockburn, Tom Corby, David Cunningham, Anna Dezeuze, Marc Domènech, James Elkins, Zavier Ellis, Simon Elmer, Jo Episcopo, Morgan Falconer, Briony Fer, Krzysztof Fijalkowski, Claudine Frank, John Frusciante, David Getsy, Bob Gillan, Rebecca Gilpin, Kate Horbury, Jolyon Howarth, Nancy Ireson, Gareth Jenkins, Gillian and Michael Keber, Julia Kelly, Elliott H. King, Alex Klar, Samantha Lackey, Alexandra Lembert, Robert Linsley, Hero Lotti, Karl Lucas, Alyce Mahon, Amna Malik, Joel Marshall, la famiglia Mazzoli, Robert McNab, Jordana Mendelson, Charles Miller, Sarah Monks, Joanne Morra, Jennifer Mundy,

Andreas Neufert, Mariska Nietzman, Edward Parkinson, Simone Perks, Richard Plant, Katie Price, Peg Rawes, Michael Richardson, Donna Roberts, Christian Roy, Marquard Smith, Louise Sorensen, Michael R. Taylor, Rebecca Virag, Jonathan Wallis, Michael Whitworth, Kelly Wilder, Laurence Wolff, Jon Wood, Olivia Woodhouse, Christian Zimmerman, Rita Zuccari and, especially, Alessandra Montioni and Greta Parkinson who let me get on with it.

Deepest gratitude goes to Steven Harris who read the original thesis and offered innumerable suggestions for its transformation into a book; Linda Dalrymple Henderson who has been a great personal friend and professional guide over the last five years or so; and most of all I thank my supervisor and mentor from the beginning of this project, Christopher Green, whose patient and sustained commentary, tactful criticism and skilful guidance have been both inspiration and model.

ψ

Thou foster-child of silence and slow time,
Sylvan historian, who canst thus express
A flowery tale more sweetly than our rhyme

I dedicate this first book to Jane and Faruque Hussein.

Castel Ritaldi, Italia
vigilia di Natale, 2006

INTRODUCTION

Scholarship on Surrealism has refined and nuanced its reception of the movement largely with reference to theories and histories of psychoanalysis, medicine, ethnology, avant-garde and modernist art and literature, Romanticism, Symbolism, Hegelianism and Marxism. Over the last few years, an understanding has begun to take hold that Surrealism's enthusiasm for modern physics was another defining feature of its theory and practice between the two wars. The purpose of this book is to offer the first comprehensive history, analysis and interpretation of that unlikely sounding juncture (figs 1 and 2).

From its beginnings in the carnage and pessimism of the First World War, Surrealism was fuelled by a sense of historical authority that drew mainly on French and German literary and intellectual traditions. Given this, the shortage of literature on the Surrealists' reception of the new physics is unsurprising. Yet Surrealism emerged in France during probably the most momentous period in the history of physics, and the same decade that saw it forge its identity through André Breton and Philippe Soupault's *Magnetic Fields* of 1919 – the first text written using the key Surrealist technique of 'automatism' – and Breton's two manifestos (1924 and 1929–30) was traversed by the popular reception of the Theory of Relativity and the revolutionary discoveries of quantum theory.

I do not mean to say this historical coincidence has gone completely unnoticed. Writers on friendly terms with Surrealism such as Michel Carrouges are only too happy to cite it as a means of giving authority to Surrealism's 'scientific' or 'experimental' credentials:

> It has not been stressed sufficiently that Surrealism was born in the era of the great revolution in the physical sciences and mathematics brought about by Einstein, Niels Bohr, Riemann, Heisenberg, de Broglie, and many others.[1]

If anything, my complaint about casual remarks like this (I give other similar examples in the chapters which follow) is that they are made about Surrealism too often. As with the other cases I point to, Carrouges' intention is to bolster Surrealism's standing as 'une science d'un type nouveau, irremplaçable' – the kind of underexplored, featherweight judgement I will be at pains to avoid.[2]

True, Carrouges hits on something when he relates quotations by philosopher of science Gaston Bachelard and physicist Louis de Broglie to Surrealism, the latter one of the central figures of quantum mechanics.[3] However, his typical, broadly drawn suggestion that Surrealist poetry and the mathematical language of quantum physics share a point of contact in a symbolism free of naturalistic representation is, again, pure guesswork without a supporting theory of knowledge. Every post-Foucauldian historian recognises the value of situating objects and texts in a broad epistemic field, a move of which Breton himself was mindful as we shall see. But speculative parallels between Surrealism and modern physics hastily made by Surrealists and commentators on the movement (Carrouges was both) nearly always lack, in my view, respectable historical foundations.

In their place, I am concerned with discovering empirical connections and with sometimes using those to propose epistemological links between science and the arts and humanities. Although Bachelard and de Broglie are important to my discussion too, I consider them alongside Surrealism in what might be called a historio-epistemological framework. This is not primarily a work of epistemology, then, nor an attempt to evaluate the differences or similarities between art and science. When I do ponder that relationship – which is unavoidable, and usually takes place here along the lines of creativity and method – it is always embedded in my main task, which is to give a history of the reception of modern physics by the Surrealists.

Carrouges does not broach the question of Surrealist art and physics; indeed, the subject has barely been treated beyond the few pages of Linda Dalrymple Henderson's landmark study *The Fourth Dimension and Non-Euclidean Geometry in Modern Art*.[4] Exhaustive and pioneering in the depth of its research, the reach of its ambition and the originality of its inquiry, Henderson's book was an important point of reference for my own. However, where Henderson reviews the impact on modern art of both non-Euclidean geometries and the popular, idealist, sometimes mystical spaces of higher dimensions, I decided to limit myself solely to an examination of the reception of new developments in the physical sciences by the Surrealists and their friends. Because of the subject matter I was dealing with, I often came into contact with 'pataphysics, alchemy and the (non-Einsteinian) fourth dimension within the context of Surrealism during the course of my research. But I decided to deal with those only in the forms in which they emerged in the main line of my inquiry; that is, strictly through their interplay with Relativity and quanta. A study of Surrealism and 'pseudo-physics' would give that wide-ranging subject the attention it deserves. Henderson touches on the question of the fourth dimension and even non-Euclidean geometry in Surrealist art but the nature of her analysis and its primary focus precludes extended reflection in her book.[5]

Apart from Carrouges' speculative chapter on Surrealism and science and Henderson's brief passages at the end of her account of modern art and science where the emphasis is placed mainly on the art of the previous avant-garde generation, only one other study bears mention here, Virginia Parrott Williams' little-known *Surrealism, Quantum Philosophy, and World War I*.[6] Although the title suggests similar material to my own, Williams is entirely concerned with ascertaining the existence of a 'war

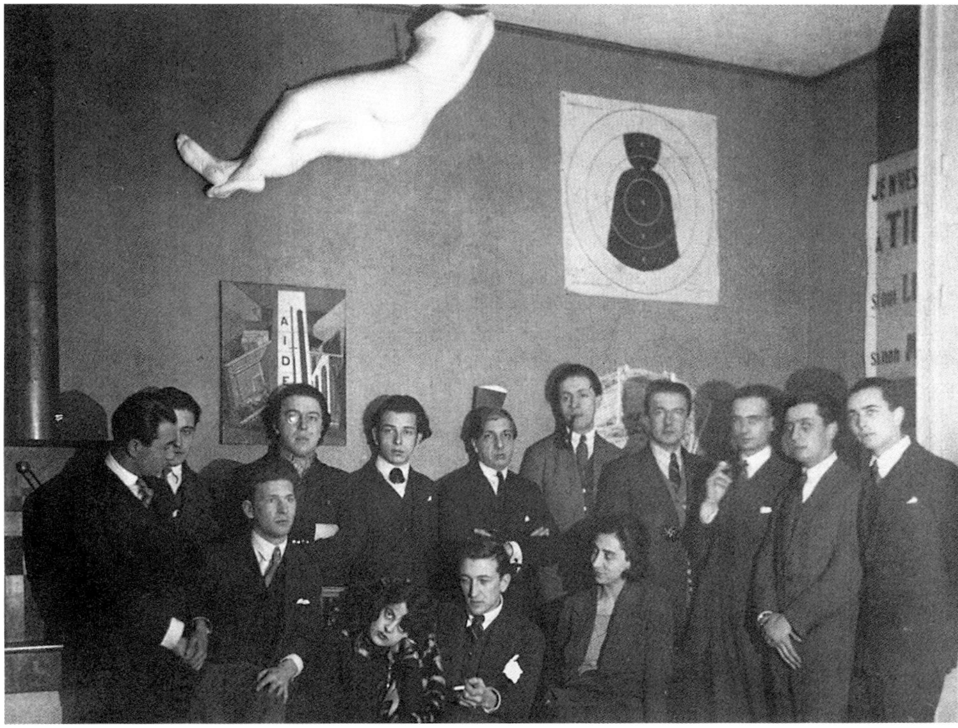

1 Man Ray, photograph of the Surrealists at the *Bureau de recherches surréalistes*, Paris, 1924.

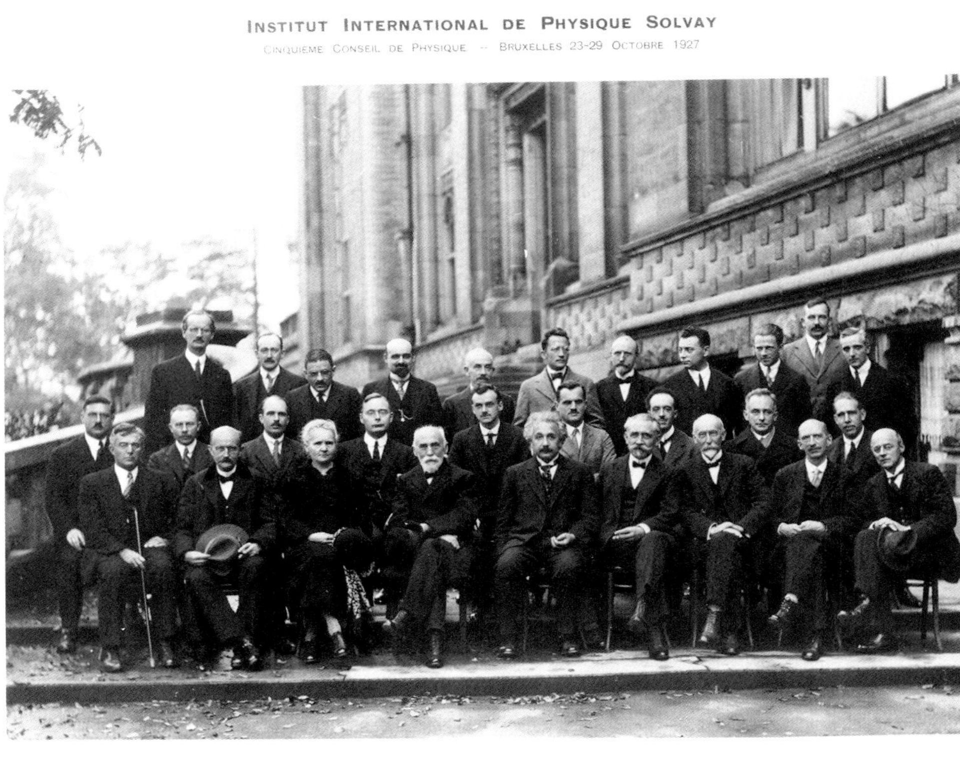

2 Photograph of physicists at fifth Solvay Conference, Brussels, October 1927.

generation' – for which she takes the Surrealists and quantum physicists as representatives – whose work can be analysed as a response to the horror and depravity of the First World War.[7] Her investigation refrains entirely from demonstrating any interest on the part of the Surrealists in the concepts, theory or epistemology of twentieth-century physics, which is the object of my study.

Although the implications for knowledge of Relativity and quantum mechanics were first recognised by the handful of physicists mainly responsible for the formulation of those theories, the epistemology that emerged from them was developed and made known to a larger public by philosophers of science, historians of science and philosophers. Surrealism has generally been seen as figuring an *écart absolu* from contemporary philosophy (notwithstanding the part it played in the rise of Hegelianism in France in the 1930s). But through its contextualisation within this broader and expanding field of scientific and philosophic thought it becomes clearer how much it shared with what Breton would have perceived as the *nouveaux frissons* running through intellectual life at the time, not just in Paris but across Europe. That is to say, individuals and 'schools' rarely associated with Surrealism are brought to the fore by a study like this one, extending its reach out of the humanities towards science. My approach places Surrealism at the crossroads of physics, philosophy, psychoanalysis and politics, requiring an examination of the Surrealists' position within a broad band of French and European thought of the 1920s and 1930s.

This might seem like an obvious step, but it has rarely been taken. True, there have been successful studies of Surrealism and psychoanalysis, Surrealism and politics, Surrealism and Hegelianism or 'Surrealism and its enemies' (Purism, late Cubism, abstraction, 'dissident Surrealism'), but these isolate Surrealism, seal it within an avant-garde, or highlight antagonisms. The alternative historical contextualisation of Surrealism has been a *linear* one, placing it in a direct line from Romanticism, the Baroque and/or Symbolism. This second scholarly tendency originates in the numerous lists of precursors that litter Breton's writings, intended to secure a distinguished yet alternative lineage for Surrealism and confirm the validity of its undertakings. It has been encouraged by the Surrealists' insistence on their independence and on publishing their work as a group, usually in their own journals, leading to the notion of the movement as a reservoir of thought isolated, once again, from French philosophy between the wars.

Although I was looking specifically at Surrealism's view of modern physics and seeking to map that reception, I was constantly led to wonder where Surrealism could be situated *laterally*. What did this purportedly cut-off group share with contemporary thought in France? What links were forged between the avant-garde and academia? How does Surrealism's 'philosophie de la concrète' of the 1930s compare with similar currents emanating from Europe?

Misunderstandings, prejudices and inept appropriations occur frequently in the windy narrative recounting the supposedly perennial antipathy between the humanities and science, covered and exacerbated by C. P. Snow's seminal lecture of 1959, 'The Two Cultures'.[8] Snow argues that the two ways of organising and comprehending the world developed by the humanities and science respectively had become increasingly distinct, while the two domains themselves had become more and more antagonistic towards one another. Dramatised particularly since the beginning of the twentieth century, 'this cultural divide,' he insists, 'is not just an English phenomenon: it exists all over the Western world'.[9] A novelist whose early training was in science, Snow had a larger, admirable objective, which was to convince his audience that the polarisation of knowledge in the West around two principal areas of intellectual activity – a dualistic model he would later regret and one that today is rejected as inadequate – had led to practical difficulties in resolving problems on the social scale, of hunger, disease and poverty.

In spite of remarks made for the sake of even-handedness about scientists who were uninterested in Dickens, Snow's main target is spelled out in the title of the second section of his four-part lecture: 'Intellectuals as Natural Luddites'. By this he means to refer chiefly to literary figures opposed to advances in science and technology, who idealised the era preceding the Industrial Revolution. He says this attitude led many in the arts and humanities to reject basic science and to misuse and abuse its ideas:

> The clashing point of two subjects, two disciplines, two cultures . . . ought to produce creative chances . . . The chances are there now. But they are there, as it were, in a vacuum, because those in the two cultures can't talk to each other. It is bizarre how very little of twentieth-century science has been assimilated into twentieth-century art. Now and then one used to find poets conscientiously using scientific expressions, and getting them wrong – there was a time when 'refraction' kept cropping up in verse in a mystifying fashion, and when 'polarised light' was used as though writers were under the illusion that it was a specially admirable kind of light.[10]

Problems in education had led to and been worsened by this 'great misunderstanding', and social inequality in the West and the underdevelopment (in Western terms) of poorer nations had been the price.

While I acknowledge the continued, pressing relevance of many of Snow's social questions, I hope to reveal a more complex situation in France in the 1920s, 1930s and 1940s. Against a background of intellectual eclecticism and mature interdisciplinary debate in that period, I present a variety of responses from within and around Surrealism to the epistemological crisis set in train by the physics of Relativity and quanta. While there were certainly some who made gestures towards scientific terminology of the kind thought inappropriate by Snow (though I would question whether poetic and artistic appropriation of scientific terms and concepts is necessarily evil), there were others who deployed ideas from the new physics and its philosophy in ways that would have met with Snow's approval. This was not a stable situation but one modified by broader political, social, economic and cultural changes, and I am indebted to the social history of art (as all art history is today) for enabling a socio-political account of Surrealism's engagement with modern physics. It allows me to build a set of arguments around Surrealism's approval, opposition, use and abuse of the language and discoveries of physics in contrast with the oversimplified model offered by Snow of a steadily increasing mutual incomprehension over sixty years, for which he gives no historical reasons.

Given my subject matter, I should also look here at how the latest manifestation of the debate about the breach that supposedly divides the humanities and science impacts on my book. The story of the 'Sokal Hoax' is now well known.[11] In 1996, an article by Alan Sokal, Professor of Physics at New York University, was accepted for inclusion in the American literary journal *Social Text* and was carried in its spring–summer issue.[12] Using writings of contemporary, mainly French thinkers which appropriate ideas from science and mathematics, Sokal's text takes a radical position, arguing that all '"eternal" physical laws' are in fact socially constructed.[13] Continuing in a relativistic and pluralistic vein, he declares:

> It has become increasingly apparent that physical 'reality,' no less than social 'reality,' is at bottom a social and linguistic construct; that scientific 'knowledge,' far from being objective, reflects and encodes the dominant ideologies and power relations of the culture that produced it; that the truth claims of science are inherently theory-laden and self-referential; and consequently, that the discourse of the scientific community, for all its undeniable value, cannot assert a privileged epistemological status with respect to counter-hegemonic narratives emanating from dissident or marginalized communities.[14]

Under the twin banner of relativism and postmodernism, Sokal argues 'the *P* of Euclid and the *G* of Newton, formerly thought to be constant and universal, are now perceived in their ineluctable historicity.'[15]

When Sokal revealed that the essay was a hoax, packed with genuine though imprecise quotations and a spurious line of argument, the fun and games began in earnest. The sense of a direct hit on a pretentious and deserving constituency was compounded when it was revealed that Sokal was by no means the species of caricatural right-wing critic who could easily be ignored but, in his own words, 'an unabashed Old Leftist' intent on exposing what he perceived as the left's retreat from rationalism, objectivity and value judgement into a labyrinth of limitless relativism.[16]

Sokal's rather smug follow up volume co-written with Jean Bricmont, *Intellectual Impostures*, gives over whole chapters to alleged abuses of mathematics and physics by a roster of French megabrains, including Jacques Lacan, Julia Kristeva, Luce Irigaray, Jean Baudrillard, Gilles Deleuze and Félix Guattari. The point of Sokal/Bricmont's close readings, of course, is to bring to the fore a level of balderdash, obscurantism and plain ignorance on the part of what they term 'postmodernist' authors (and their followers) in their redundant forays into the precise sciences. However, a lack of historical perspective weakens Sokal/Bricmont's argument and introduces part of mine:

> what is new and curious about postmodernism is that it is an anti-rationalist form of thought that has seduced part of the left . . . For most of the past two centuries,

> the left has been identified with science and against obscurantism, believing that rational thought and the fearless analysis of objective reality . . . are incisive tools for combating the mystifications promoted by the powerful – not to mention desirable human ends in their own right. And yet, over the past two decades, a large number of 'progressive' or 'leftist' academic humanists and social scientists . . . have turned away from this Enlightenment legacy.[17]

Contrary to this assertion I wish to demonstrate that well before Sokal/Bricmont's postmodernists, the Surrealists took a sceptical view of the logic of the Enlightenment, or 'rationalisme fermé' as they labelled it. Moreover, they took this term from Gaston Bachelard, a perfect example of Sokal/Bricmont's 'academic humanist', who coined the expression and formulated its 'arational' characteristics in 1934 while Professor of Philosophy at the University of Dijon. As borne out by the Surrealists' alliance with Bachelard following the formulation of quantum mechanics, in the 1930s dispute over rationalism the warring parties for and against did not separate sharply along the party lines of science and the humanities, nor those of academia and the avant-garde:

> the 1930s were characterised by a new intensification of the crisis of rationalism, which, increasingly menaced, tried nevertheless to maintain its position with the foundation of the Union rationaliste by [Albert] Bayet, [Marcel] Boll, and [Paul] Langevin, and with the publication of *Ages de l'intelligence* and *Progrès de la conscience dans la philosophie occidentale* by Léon Brunschvicg. These fits and starts would not protect rationalism, in spite of its continuing hold over most university posts, from increasing attack. After several years of silent meditation, Bergson was to publish in 1932 his last great book, *The Two Sources of Morality and Religion*, which widened his audience further. The progress of quantum physics, led by Louis de Broglie in France, would encourage scientists and philosophers to question again the notions of causality and determinism. Inspired by Freudian research into the unconscious, philosophic thought widened by opening itself up to more subtle exploration of the mysterious margins and profound zones of consciousness with, for example, the works of Bachelard elaborating a method of total knowledge and an explanatory critique of the symbolism of images.[18]

Evidently, the emphasis placed by Sokal/Bricmont on American academic fashion caused them to overlook the general dispute over rationalism that took place in France in the 1920s and 1930s, which had its roots in Bergson's philosophy. My book offers a kind of cross-section through that story, though it deserves a comprehensive and critical historical study of its own which will add to our understanding of Surrealism's formation and intellectual development.

Although it does not offer a critique of the complex and subtle discourse on rationalism that took place in 1930s France, the Sokal/Bricmont polemic is still germane to one area of my study: the nature of the Surrealists' appropriation of ideas emanating from physics. As culpable of the 'abuse' of science as today's postmodernists (if you choose to look at it that way) are the suspiciously neat historical parallels between science and poetry drawn up by Breton; the extension of questions surrounding scientific concepts to the social sphere by René Crevel, Tristan Tzara and Pierre Mabille; the festive, stellar orgy gleaned from modern astrophysics by Georges Bataille; pictorial improvisations on non-Euclidean geometry by Matta and the free and easy use of features of Relativity by Salvador Dalí (which even in their own day drew the ire of some of Dalí's contemporaries).

However, my main purpose here is not really to point out where the Surrealists 'misused' or ran into difficulties with the technical details of the new physics. While this is a critical account on the whole which indicates some of the Surrealists' theoretical inconsistencies along the way, I want to show primarily how the 1930s alliance with Bachelard both sanctioned and refocused the direction taken by Surrealism, equipping it with a language emphasising the *coupure* that lay between it and previous forms of (Enlightenment) knowledge. Central to my discussion, Bachelard is not mentioned by Sokal/Bricmont, yet I show he occupied a crucial dual position: between the humanities and science in the thirties, and at the dawn of modern French thought. Recalling Snow (and quoting him), Sokal declares that the two cultures 'are probably farther apart in mentality than at any time in the past fifty years' (Snow's estimate in 1959 was sixty years).[19] I will offer a more complex and conciliatory view, showing an engagement throughout Surrealism with Relativity and quantum mechanics between the wars followed by a rejection of nuclear physics in the era of the Cold War and subsequent suppression of that earlier history.

ψ

This book is roughly chronological. My first chapter lays a path from the conceptual languages and implications for knowledge of the Theory of Relativity and quantum mechanics to Surrealism. I set this process in motion with an overview of the historical and theoretical development of both theories. I establish the means by which the new physics, together with the challenge it posed to a philosophical tradition still in thrall to Cartesianism, was communicated to French-speaking intellectuals, writers and artists in the 1920s and 1930s. This entails a discussion of physics in France up to the beginning of the 1930s and a brief survey of philosophy reviews and popular science publications of the period to give a taste of how closely placed were research and theory in modern physics, philosophy and psychology. Although the first two-thirds of this chapter were a nightmare to write and will be the most difficult section to read, it was the most intellectually rewarding chapter for me to research and I think it will be the one my non-scientific readers get the most out of, provided they do not approach it as a hurdle.

The second chapter is concerned principally with Surrealism's chief theorist André Breton, beginning with an outline of what the origins of Surrealism owed to the Theory of Relativity. Shared by many in the early 1920s, Breton's interest in Relativity soon waned though he took a positive view of physics again in the 1930s, even regarding it as the field of research with the most to offer Surrealism. I account for this renewed enthusiasm by means of a close comparison between Bachelard's writings and Breton's 1936 essay 'Crisis of the Object'. Since no other tract by Breton shares the same warmth towards scientific endeavour or a comparable anxiety to seek out ground held in common with a separate and, apparently, hostile body of theory to that of poetry, 'Crisis of the Object' works well as a case study.

These first two chapters, and to a certain extent the third, are heavy on facts and are meant as a direct response to the high-pitched squeal of a question that forms an invisible arc over this book. How could Surrealism, which from its earliest days owed its existence, identity and continuing vitality to an anti-positivist engagement with the world, take an interest in a field of theory and experiment that traditionally based its knowledge on such a strictly empirical mode of investigation? No doubt the thought of Surrealism usefully brandishing ideas from physics seems at first about as likely to scholars and students of the movement as a badger playing a violin (or a more suitable Surrealist variant). For that reason, this first part of my book is meant to make sense of that initially unaccustomed juxtaposition.

By contrast, the literature on Surrealism and politics is becoming familiar. Over the last few years, English-language scholarship on the movement has converged on the Popular Front period and the Surrealists' perennially fraught and occasionally farcical relationship with Communism and the Parti Communiste français. The centrality of politics to the culture of Surrealism has never been more clear and for that reason it is important to give some space to the political and social context for Surrealism's exploration of modern physics. Left-wing physicists Jean Perrin and Paul Langevin come to the fore as significant figures in the political landscape at this time with various kinds of contact with Surrealism. But my principal case study in this third chapter is the one-number-only journal *Inquisitions* in which a group of former Surrealists attempted to triangulate their residual Surrealism with the politics of the Popular Front and the astonishing progress of physics. Bachelard contributed to *Inquisitions*, and I use this to extend my discussion of his intellectual relationship with Surrealism, here in the political context. Surrealism shifted its footing before the beginning of the Second World War, losing confidence in the capacity of Communism to bring about fundamental change in the world and ultimately placing greater faith in alchemy, magic and the occult, yet I show these areas of study were often linked by the Surrealists to recent advances in physics.

I shift subtexts from alchemy to mysticism for the next chapter to focus on Georges Bataille, who was never a member of the Surrealist group but remained to a greater or lesser extent grudgingly in its orbit for the period I cover. Demonstrating their interest in modern physics at many levels of theory and practice, the Surrealists frequently had recourse to what the French term the *vulgarisation*, especially those of the English astrophysicist Arthur Eddington, enjoyed by an international audience for their combination of wit and rigour. Bataille's text 'Corps célestes' (1938) draws upon Eddington's sketch of contemporary celestial mechanics in his *Rotation of the Galaxy* (1930) to generate a peculiarly Bataillean vision of a festive dance of the stars. Extravagantly inimical to Eddington's sober, 'Quakerish' account of the behaviour of heavenly bodies,

Bataille conceives an orgiastic, Dionysian spectacle, refracted through a reading of Nietzsche, sacred ritual and sacrifice. While my aim here is to examine two radically different and irreconcilable theories of constellation, I also want to constellate Eddington and Bataille themselves *through* this difference – against a backdrop of mysticism – in a configuration that is conciliatory and complementary.

Parallels are easily made between Bataille's early texts and the 1930s writings on art by critic Carl Einstein and historian of religion Henri-Charles Puech, which insist that a revolution in knowledge had been brought about by Relativity, quantum mechanics and psychoanalysis. In my fifth chapter I argue that the antiocularcentrism promulgated by Carl Einstein and Puech together with their proto-postmodern recital of an external and internal fragmentation, dispersion or dislocation of descriptive possibilities, offers a theoretical framework in which Matta's painting of the 1940s can be situated. I am mainly concerned here with the interest held by particle physics and quantum mechanics for artists, and Wolfgang Paalen offers the most obvious monographic case study to extend this investigation. Accordingly, I examine his art against his major 'manifesto' statement 'Art and Science' (1942). Finally and briefly (and probably unexpectedly) in this chapter, I turn to Max Ernst to show how Surrealism can, indeed, be related to positivism in the light of quantum mechanics. Breton made a close connection between his own poetic theory and the then fashionable Theory of Relativity in 1921 when he wrote about Ernst's collages. Subsequently, Ernst drew upon the recent findings of quantum theory as they were set out in the science journal *Die Naturwissenschaften* when he discussed his own artistic practice and theory in the context of Surrealism in 1934.

I make this argument about positivism with reference in part to the art and writings of Salvador Dalí, which were shaped by the innovations and peculiarities introduced into physics by Relativity from the moment he joined the Surrealists and began formulating his method of 'paranoia-criticism' in 1929. In my final chapter, I show how his writings of the 1930s blend the concepts and 'narratives' of physics and psychoanalysis. Both Dalí and the writer Roger Caillois endured equivocal couplings with Surrealism and I contrast their work produced inside and outside its confines to show how the terrain of modern physics and the fourth dimension became contested ground within the Surrealist milieu. I round out the trajectory of my book with a coda sketching the Surrealists' changing attitude towards modern physics and its philosophy in the nuclear age. While Dalí enthused over atomic physics and quantum mechanics with all the infatuated zeal of a religious convert in the 1950s, an antagonistic attitude was struck by the Surrealists towards physics in the era of the Cold War. Breton's own 1952 account of Surrealist painting, physics and epistemology was hatched in that climate. To the opinions he expresses there, the following chapters can be considered a response.

1

MODERN PHYSICS AND ITS PHILOSOPHY

The Reception in France

In this first chapter I outline some of the early history and conceptual detail of the Theory of Relativity and quantum mechanics (QM) and their reception and interpretation by philosophers and writers in France between the wars. My first aim here is to give a clear account of the 'new' physics for the non-specialist audience of this book. I take this to be mainly art historians and intellectual historians, but also artists, cultural historians, perhaps historians of science and French culture, and maybe even the odd scientist. Secondly, I want to reconstruct part of the narrative of popular, critical and intellectual discourse on the new physics in France within which I can go on to situate Surrealism throughout my book.[1] Accordingly, I examine the identity, content and tone of some key physics journals, philosophy reviews and popular science books to give a sense of the interpretation of Relativity and QM and the (changing) position of physics as an activity in France in the 1920s and 1930s. This allows me to contextualise without further explanation in the chapters that follow the diverse uses of physics – philosophical, political, theoretical, analogical, metaphorical, poetical, pictorial – by individuals in and around the Surrealist group in Paris.

Relativity and QM are best viewed as supplementing and inflecting classical, Newtonian mechanics rather than simply replacing it. Taken together, the Theory of Relativity and quantum theory were worked on and brought to fruition between 1900 and 1927. Their findings were processed by philosophers and became known to a larger public from 1919. This interval of 1919–27 corresponds closely with the period during which Surrealism initially made its presence felt, going on to gain influence, reputation and visibility throughout Europe in the lead-up to the Second World War for an epistemological radicalism that I compare to that of the new physics. Although they were obviously located in different milieux, concerned with unlike phenomena, dealt with divergent material and formulated different goals, both the physicists and the Surrealists comprised international communities that were research-driven and experimental in their orientation. Sharing an obligation to develop forms of rationalism that could force the mind onto new paths of discovery by fundamentally questioning earlier languages, concepts and bases of knowledge, modern physics and Surrealism disclose historical and epistemological resemblance. The conceptual accounts I give here of Relativity and QM are shaped and summarised with the rest of the book and my imagined reader in mind. However, the section on QM is more detailed and historical in acknowledgement of the complexity and magnitude of its achievement. Often described as the greatest achievement in the history of Western thought, quantum theory is difficult, but I will forgo the customary routine of telling my readers they can skip it and still understand the rest of the book. It is worth spending some time over.

The Theory of Relativity

Forever associated with the name of Albert Einstein (1879–1955), the Theory of Relativity emerged out of a history of investigation into the velocity and propagation of light going back as far as Galileo (fig. 3).[2] The young Einstein began to reflect on the most profoundly accepted theories of physics whilst enrolled at the Swiss Federal Polytechnic (Eidgenössische Technische Hochschule) in Zürich between 1895 and 1900. Increasingly detached from the pure mathematics to which he had been devoted over the preceding years, he discovered and began to nurture an ability 'to scent out that which was able to lead to fundamentals and to turn aside from everything else, from the multitude of things which clutter up the mind and divert it from the essential'.[3]

The governing fields of knowledge in physics at the close of the nineteenth century were Newtonian mechanics and Maxwell's equations explaining the workings of electricity and magnetism. The weakness of Greek (Aristotelian) physics had been its lack of corroboration by experiment, leading to erroneous conclusions as to the comparative movement of bodies (the difference in speed between the fall of a stone and a piece of cloth, for instance). This confusion carried to astronomy, and the most insightful findings of ancient astronomy, those of the second century Egyptian Ptolemy, were still accepted at the time of the Renaissance. The great leap forward into the modern age taken by Nicolaus Copernicus (1473–1543) at the beginning of the sixteenth century was to assume the sun was at rest and the planets revolved around it. However, it was Johannes Kepler (1571–1630) who later in the century broke with the Greek notion of the uniform circular motion of the planets in favour of elliptical orbits. In 1632, Galileo (1564–1642) demonstrated that contrary to Aristotelian physics, all objects fell with equal acceleration, no matter what their mass. He also proposed a new understanding of the dual forces at work on bodies in motion – of inertial propulsion and gravity – without however, recognising the significance of their relationship.

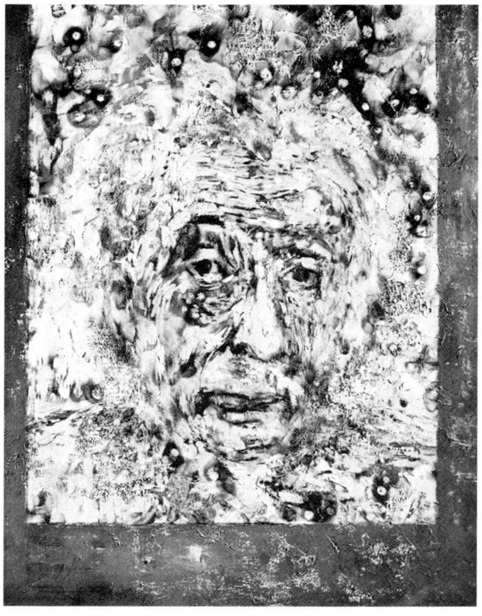

3 Wolfgang Paalen, *Portrait of Albert Einstein* (1940). Oil on canvas. Dimensions and present whereabouts unknown.

This was left to Isaac Newton (1643–1727), who systematised, quantified and generalised Galileo's findings, and whose

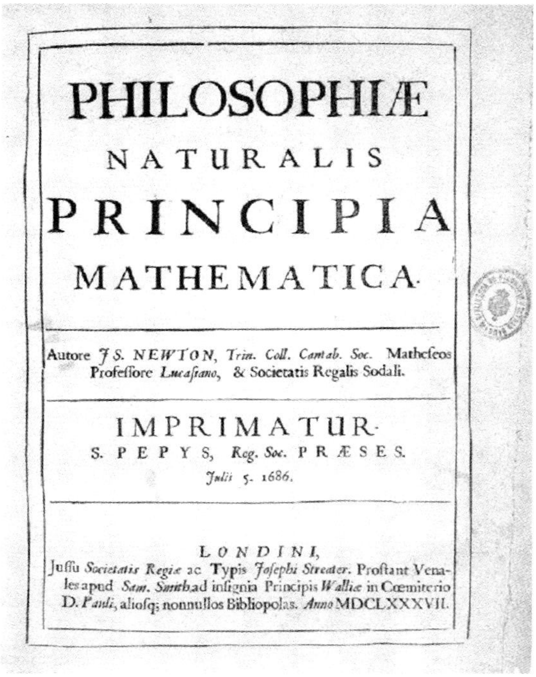

PHILOSOPHIÆ
NATURALIS
PRINCIPIA
MATHEMATICA.

Autore J S. NEWTON, *Trin. Coll. Cantab. Soc.* Matheseos Professore *Lucasiano*, & Societatis Regalis Sodali.

IMPRIMATUR.
S. PEPYS, *Reg. Soc.* PRÆSES.
Julii 5. 1686.

LONDINI,
Jussu *Societatis Regiæ* ac Typis *Josephi Streater*. Prostant Venales apud *Sam. Smith* ad insignia Principis *Walliæ* in Cœmiterio D. *Pauli*, aliosq; nonnullos Bibliopolas. *Anno* MDCLXXXVII.

4 Isaac Newton, *Philosophiae Naturalis Principia Mathematica* (1687).

Philosophiae Naturalis Principia Mathematica (1687) remained unchallenged until the end of the nineteenth century (fig. 4). Newton broke up and analysed orbital motions (where Galileo had only looked at orbital motions as a whole) and through his invention of differential calculus was able to give the first detailed accounts (through differential equations) of the change in behaviour of objects at any point along their orbit. He also gave mathematical expression to the force of gravity and through these findings ultimately demonstrated the relationship between an orbit and its initial conditions (velocity). Both Kepler's and Galileo's observations were given theoretical explanation and quantitative form. But they were also given predictive power, for Newtonian mechanics asserts that once initial conditions and forces are calculated and related, it is possible in theory to determine the future conditions of all particles, and therefore the universe should be viewed as existing on a fixed and calculable course. The determinism of Newtonian mechanics is usually summarised by historians of science and philosophy through the words of the French physicist and mathematician Pierre Simon de Laplace, who wrote in his *Méchanique Céleste* (1825):

> We must envisage the present state of the universe as the effect of its previous state, and as the cause of that which will follow. An intelligence that could know, at a given instant, all the forces governing the natural world, and the respective positions of the entities which compose it, if in addition it was great enough to analyse all this information, would be able to embrace in a single formula the movements of the largest bodies in the universe and those of the lightest atom: nothing would be uncertain for it, and the future like the past, would be present to its observation.[4]

These events are played out in a universe in which time and space are absolute, different and separate. The extraordinary success of Newton's physics in explaining the behaviour of moving bodies from particles to planets led some physicists to an understanding that physics had reached near-completion by the end of the nineteenth century.

The sense of being close to a unified theory was strengthened by the remarkable successes of the equations of James Clerk Maxwell (1831–1879) who died in the year Einstein was born. Defining the nature of and relationship between electricity and magnetism, Maxwell's equations were built on the discovery of electromagnetic induction (the production of electrical currents by magnets) by the experimental physicist Michael Faraday (1791–1867), and Faraday's concept of electromagnetic fields (the electrical field of influence produced by magnets and observable through the movement of iron filings and compass needles later illustrated by the Surrealist artist Matta). Maxwell was an incredibly precocious mathematician and was able to formulate equations quantifying the 'lines of force' observed by Faraday. Maxwell predicted the propagation of electromagnetic energy in a vacuum to his sceptical contemporaries, and this was achieved after Maxwell's death by Heinrich Hertz in 1888. By then it had long been accepted that light had a measurable speed, and Hertz was able to confirm experimentally that the Maxwell waves he created with special oscillators travelled at the speed of light.

But if light were now to be regarded as an electromagnetic phenomenon which could carry in a vacuum, Newtonian mechanics demanded some kind of material support that it could vibrate *through*. As increasingly refined experiments determined the speed of light with greater exactitude, scientists focused their attention on the medium that allowed the distribution (propagation) of light waves. A ubiquitous, stationary ether had traditionally offered scientists – including Maxwell himself – a speculative object against which to set (and preserve) absolute time and space, and therefore motion. This now seemed indispensable. The failure of the Michelson/Morley experiments of 1881–87 to discover an 'ether wind' (to which the earth was subject due to its rotation) through the registration of variation in the intensity (or velocity) of light did not act as a deterrent to this idea. For over the years following this famous experiment it was argued controversially that the measuring devices used had been subject to the 'Fitzgerald-Lorentz Contraction', a notion originally intended to explain the contraction of objects under the force of the ether (which would later be recycled by Einstein as a consequence of the Special Theory of Relativity and, I argue, would also present Salvador Dalí with a new artistic vocabulary). Shrinking in the direction of the motion of the earth, it was claimed that the arm of Michelson/Morley's interferometer moving against the ether failed to register the expected evidence of the effects of the ether on light speed. The idea of a fixed ether as the tangible substance carrying vibrating light waves was thus retained.

In the wake of the findings of Hendrik Lorentz and the writings of Ernst Mach and Henri Poincaré, the young Einstein became increasingly critical of classical mechanics and electrodynamics, despairing of 'the possibility of discovering the true laws by means of constructive efforts based on known facts'.[5] He now carried out a thought experiment from within the framework of Newtonian mechanics, which nevertheless uncovered a paradox that contradicted classical physics. Imagining how light might be experienced by someone travelling at light speed, he already had the seed of Special Relativity in 1895:

> After ten years of reflection such a principle resulted from a paradox upon which I had already hit at the age of sixteen: If I pursue a beam of light with the veloc-

ity *c* (velocity of light in a vacuum), I should observe such a beam of light as a spatially oscillatory electromagnetic field at rest. However, there seems to be no such thing, whether on the basis of experience or according to Maxwell's equations. From the very beginning it appeared to me intuitively clear that, judged from the standpoint of such an observer, everything would have to happen according to the same laws as for an observer who, relative to the earth, was at rest. For how, otherwise, should the first observer know, i.e., be able to determine, that he is in a state of fast uniform motion?[6]

Two options presented themselves: either an observer travelling at the speed of light would view a light wave as being at rest (contradicting Maxwell), or an observer is unable to travel at the speed of light (contradicting Newton).

Einstein claims first in his 1905 Special Relativity paper 'On the Electrodynamics of Moving Bodies' that the laws of nature are consistent from one 'frame of reference' to another and only relative motion between these is detectable (fig. 5). Obviously, for our purposes on earth we can *assume* a still object (our planet) upon and against which we can measure out speeds, so Newtonian fixity and relativity are appropriate for our daily needs. But thought experiments like this one led Einstein to assert the meaninglessness of absolute rest or absolute motion. Although Newton himself was aware of this problem, he satisfied himself with a theological compromise appropriate to his Christian mysticism, establishing the consciousness of God as his ultimate frame of reference. This religious background disappeared behind the immense achievements of Newtonian mechanics, yet it became visible again in a modern age also sceptical of the immobile ether demanded by Hertz's confirmation of the existence of Maxwell waves. Einstein's Relativity does not exactly *disprove* the existence of the ether, but argues for the futility of searching for it by indicating that everything in our universe is in motion, and an absolute, fixed ether is superfluous and undetectable, given that only relative motion is measurable.[7]

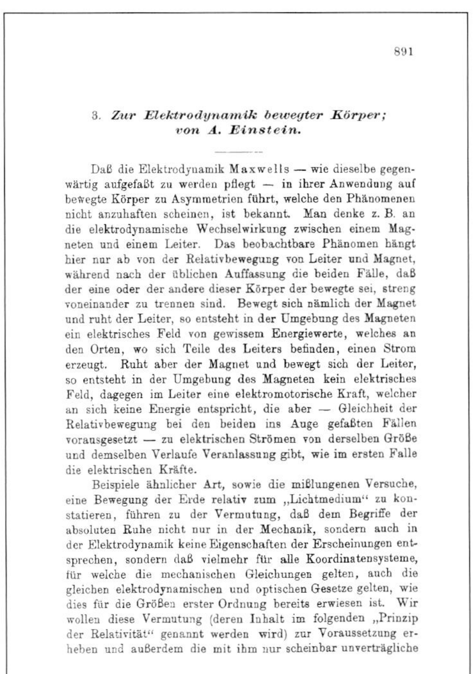

891

3. ***Zur Elektrodynamik bewegter Körper; von A. Einstein.***

Daß die Elektrodynamik Maxwells — wie dieselbe gegenwärtig aufgefaßt zu werden pflegt — in ihrer Anwendung auf bewegte Körper zu Asymmetrien führt, welche den Phänomenen nicht anzuhaften scheinen, ist bekannt. Man denke z. B. an die elektrodynamische Wechselwirkung zwischen einem Magneten und einem Leiter. Das beobachtbare Phänomen hängt hier nur ab von der Relativbewegung von Leiter und Magnet, während nach der üblichen Auffassung die beiden Fälle, daß der eine oder der andere dieser Körper der bewegte sei, streng voneinander zu trennen sind. Bewegt sich nämlich der Magnet und ruht der Leiter, so entsteht in der Umgebung des Magneten ein elektrisches Feld von gewissem Energiewerte, welches an den Orten, wo sich Teile des Leiters befinden, einen Strom erzeugt. Ruht aber der Magnet und bewegt sich der Leiter, so entsteht in der Umgebung des Magneten kein elektrisches Feld, dagegen im Leiter eine elektromotorische Kraft, welcher an sich keine Energie entspricht, die aber — Gleichheit der Relativbewegung bei den beiden ins Auge gefaßten Fällen vorausgesetzt — zu elektrischen Strömen von derselben Größe und demselben Verlaufe Veranlassung gibt, wie im ersten Falle die elektrischen Kräfte.

Beispiele ähnlicher Art, sowie die mißlungenen Versuche, eine Bewegung der Erde relativ zum „Lichtmedium" zu konstatieren, führen zu der Vermutung, daß dem Begriffe der absoluten Ruhe nicht nur in der Mechanik, sondern auch in der Elektrodynamik keine Eigenschaften der Erscheinungen entsprechen, sondern daß vielmehr für alle Koordinatensysteme, für welche die mechanischen Gleichungen gelten, auch die gleichen elektrodynamischen und optischen Gesetze gelten, wie dies für die Größen erster Ordnung bereits erwiesen ist. Wir wollen diese Vermutung (deren Inhalt im folgenden „Prinzip der Relativität" genannt werden wird) zur Voraussetzung erheben und außerdem die mit ihm nur scheinbar unverträgliche

5 Albert Einstein, 'Zur Elektrodynamik bewegter Körper,' *Annalen der Physik*, vol. 17, no. 10, 26 September 1905, 891-921.

Secondly, Einstein states in the great 1905 paper that light is propagated at a definite velocity, *c*, irrespective of the state of motion of its source. This postulate leads to the counterintuitive conclusion that if a star and the earth are approaching each other at a combined velocity of, say, 100 000 miles per second, or moving away from each other at that speed, the light leaving the star will still travel in our direction at light speed (186 000 miles per second) *not* at a rate of the sum of the two quantities (286 000 miles per second in one example, 86 000 miles per second in the other). If we imagine such behaviour on the Newtonian level of experience (as Einstein would for clarification and elaboration), we must assume that if a man on a train travelling at thirty miles per hour throws a ball at forty miles per hour relative to the train in the direction in which it is moving, the ball will *not* travel at a combined speed of seventy miles an hour relative to the ground, but at a speed of forty miles per hour. Although it seems to defy common sense, Einstein viewed the constant velocity of light relative to an observer as a basic law of the universe.

From these two revolutionary postulates, Einstein was able to develop equations explaining certain phenomena with greater precision than had been possible with Newtonian mechanics and also to make remarkable predictions later corroborated by experiment. An essential component in this was Einstein's generalisation of the Lorentz transformation equations (from electromagnetic forces only to all matter), devised by Lorentz in 1895 to account for the results gained by Michelson and Morley. These nineteenth-century equations, formulated entirely within the framework of Newtonian mechanics, were now dusted down by Einstein and used to relate the velocity, distance, time, mass, length and so on of events relative to each other, proposing and explaining the behaviour of time and space in a manner incompatible with classical physics.

The Special Theory of Relativity verifies the Fitzgerald-Lorentz Contraction, which describes the contraction of bodies in the direction of motion as observed externally to the system (but not noticeable within the system itself), though this is invisible to the naked eye, of course, and barely even measurable on the human scale of events. It also demonstrates that the rate of temporal flow varies between positions in space and with the velocity of bodies in motion ('time dilation', which abolished the notion of absolute time). Before Einstein, time was thought of as a river flowing in one direction and shared by one and all. The sheer strangeness of time is one of the characteristic discoveries of Relativity, which tells us that two events that are simultaneous for one observer are not simultaneous for other observers. Single events may even be experienced in a different order by observers (without one or the other seeing them in the 'right' order, absolute time now rendered invalid by Relativity) depending on each observer's position and velocity relative to the others. Relativity's weirdest conclusions are reached with the slowing down of time in a moving frame of reference as compared to another at 'rest', meaning that if one twin is sent out on a journey to a distant star at a velocity close to the speed of light while the other stays on earth, the first will return home younger than the second. While Relativity's findings would bemuse some physicists at first, from 1919 they provided a fund of material for cartoonists, novelists, Surrealists and the like.

The Special Theory of Relativity accounts for an object's increased mass at a higher velocity and shows how large a quantity of energy is equivalent to a small mass given that the additional energy that accrues to the additional mass of an object increasing in velocity is equal to the increase in mass multiplied by the velocity of light squared. From this, Einstein deduced that in nuclear processes there will be a quantity of energy

(E) associable with *any* mass (m) that can be found when the latter is multiplied by the square of the velocity of light (c^2): $E = mc^2$.

Special Relativity was largely the fruit of Einstein's labour at the Swiss Federal Polytechnic and his reflections after leaving there in 1900. Since his departure, his academic career had stalled, though his first publication appeared in the *Annalen der Physik* in 1901. Although his job search was hampered for fairly straightforward reasons (he had not yet received his doctorate), historians like to dramatise Einstein's accomplishments further (as though they needed it) by noting that at the time the Special Relativity paper appeared in 1905, preceded in June and July of that year by two other pioneering publications on light quanta and Brownian motion, he was employed as an Expert Third Class at the Swiss Patent Office in Bern.[8] Einstein had to continue in this humble-sounding post for a couple more years (apart from a promotion to Expert Second Class in 1906) while the implications of Relativity seeped through to the physics community. This process was furthered in 1908, when Einstein's former Professor of Mathematics in Zürich, Hermann Minkowski, discovered an elegant mathematical formalism and sophisticated language for Einstein's theory, enabling the translation of the Lorentz transformation equations of Relativity into a four-dimensional Euclidean space. Initially Einstein's suspicion of the notion of the fourth dimension led him to take the view that 'no statement can be more banal than that our familiar world is a four-dimensional time-space continuum.'[9] However, later he came to value Minkowski's 1908 intervention. In that year, with a growing understanding among physicists of his achievements, Einstein was made a *privatdozent* at Bern University, then swiftly appointed to the Chair of Theoretical Physics at Zurich University in 1909 (on the same salary he had lately received at the Patent Office).

By then, Einstein had realised that the Special Theory of Relativity could not be his last word on the matter and he was well on his way to extending his findings about constant relative velocities to bodies moving with velocities at variance with each other, enshrined in the General Theory of Relativity of 1916 (fig. 6).[10] Usually regarded as his masterpiece, this is Einstein's theory of gravitation, based on the notion of the 'principle of equivalence', which according to one commentator states that 'the effects of a uniform constant acceleration on an observer or his measuring instruments are indistinguishable from – that is, equivalent to – the observer's being at rest but acted on by a uniform field of gravitation'.[11] That is to say, Einstein realised a geometrical form might be available for motion due to a gravitational field of force alone, intuiting 'the laws of geometry in a space where there exists a gravitational field are different from those in a space which is "of free forces" in the old sense'.[12] Einstein made a basic postulate of the principle of equivalence, and derived from it the profound conclusions of the General Theory of Relativity.

There were three of these, constituting the first substantiated modification to Newtonian mechanics. Although Newton had charted the elliptical orbits of planets around the sun with sufficient accuracy to satisfy astronomers for more than two centuries, Einstein's theory shows these ellipses are not stationary (one orbit mapped exactly over the previous one) but displaced, so that they fan out like petals on a flower (to exaggerate these minute alterations). The so-called 'perihelion advance' of Mercury – the point on its orbit closest to the sun which alters with each orbit – is tiny, but the easiest in our solar system to measure. Where the failure of Newton's law of universal gravitation to explain the displacement of Mercury's orbit had long puzzled astronomers, the General Theory now calculated it exactly. Secondly, Einstein's theory predicted another idiosyncrasy of time in total disagreement with the notion of absolute time: that it moves slower when affected by a gravitational field than when free of the effects of gravity. The stronger the gravity, the slower time passes, so a clock on the sun will run slower than an exactly synchronised one on the earth (Einstein predicted one second on the sun would be equal to 1.000002 seconds on the earth). This effect was demonstrated by the comparison of the frequencies of the vibrations of atoms on the sun with those of the same atoms on earth. General Relativity correctly predicts the shift of the frequencies of the sun's light to the red end of the visible spectrum as an outcome of its powerful gravitational field, because red light has a smaller frequency (due to slower atomic vibrations).

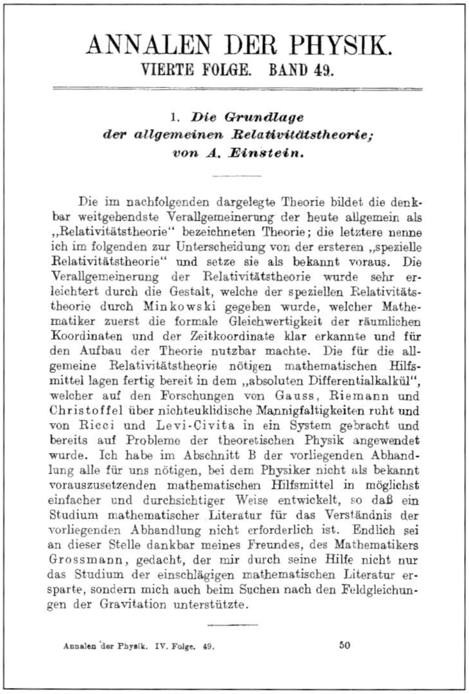

ANNALEN DER PHYSIK.

VIERTE FOLGE. BAND 49.

1. ***Die Grundlage der allgemeinen Relativitätstheorie; von A. Einstein.***

Die im nachfolgenden dargelegte Theorie bildet die denkbar weitgehendste Verallgemeinerung der heute allgemein als „Relativitätstheorie" bezeichneten Theorie; die letztere nenne ich im folgenden zur Unterscheidung von der ersteren „spezielle Relativitätstheorie" und setze sie als bekannt voraus. Die Verallgemeinerung der Relativitätstheorie wurde sehr erleichtert durch die Gestalt, welche der speziellen Relativitätstheorie durch Minkowski gegeben wurde, welcher Mathematiker zuerst die formale Gleichwertigkeit der räumlichen Koordinaten und der Zeitkoordinate klar erkannte und für den Aufbau der Theorie nutzbar machte. Die für die allgemeine Relativitätstheorie nötigen mathematischen Hilfsmittel lagen fertig bereit in dem „absoluten Differentialkalkül", welcher auf den Forschungen von Gauss, Riemann und Christoffel über nichteuklidische Mannigfaltigkeiten ruht und von Ricci und Levi-Civita in ein System gebracht und bereits auf Probleme der theoretischen Physik angewendet wurde. Ich habe im Abschnitt B der vorliegenden Abhandlung alle für uns nötigen, bei dem Physiker nicht als bekannt vorauszusetzenden mathematischen Hilfsmittel in möglichst einfacher und durchsichtiger Weise entwickelt, so daß ein Studium mathematischer Literatur für das Verständnis der vorliegenden Abhandlung nicht erforderlich ist. Endlich sei an dieser Stelle dankbar meines Freundes, des Mathematikers Grossmann, gedacht, der mir durch seine Hilfe nicht nur das Studium der einschlägigen mathematischen Literatur ersparte, sondern mich auch beim Suchen nach den Feldgleichungen der Gravitation unterstützte.

Annalen der Physik. IV. Folge. 49. 50

6 Albert Einstein, 'Die Grundlage der allgemeinen Relativitätstheorie,' *Annalen der Physik*, vol. 49, no. 7, 11 May 1916, 769–822.

But General Relativity predicted that a body with sufficient gravitational attraction could modify not only time but also space. In that case, 'flat' Euclidean geometry would do as a measure for spaces where gravitational attraction was at a minimum (away from large bodies like planets), but the non-Euclidean geometries discovered in the nineteenth century were now necessary to measure the spaces 'distorted' by bodies like the sun, whose gravitational pull is 330 000 times that of the earth. Visible evidence for the curving of space can be gleaned from the behaviour of light under the influence of gravity, which Einstein concluded would travel in geodesic lines. These are straight lines in curved space, similar to the shape of lines of latitude and longitude on the surface of the earth, traced in the paintings of Matta as a kind of Surrealist alternative to the straight line, 'Copernican' perspective of the Renaissance. According to General Relativity, such is the sun's mass that it curves space-time, deflecting the path of light along one of these geodesics. Einstein mentioned that the apparent shift in location of the stars by the distortion of the path of light travelling from them by the sun (predicted by him as 1.74 seconds of arc) could be measured during a solar eclipse.

The technical difficulty of the General Theory of Relativity (grossly simplified in my account) meant that its reception was subdued initially and treated by much of its spe-

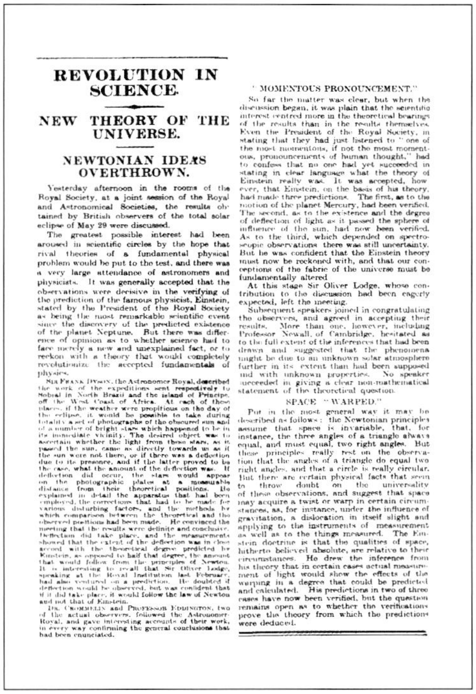

REVOLUTION IN SCIENCE.

NEW THEORY OF THE UNIVERSE.

NEWTONIAN IDEAS OVERTHROWN.

Yesterday afternoon in the rooms of the Royal Society, at a joint session of the Royal and Astronomical Societies, the results obtained by British observers of the total solar eclipse of May 29 were discussed.

The greatest possible interest had been aroused in scientific circles by the hope that rival theories of a fundamental physical problem would be put to the test, and there was a very large attendance of astronomers and physicists. It was generally accepted that the observations were decisive in the verifying of the prediction of the famous physicist, Einstein, stated by the President of the Royal Society as being the most remarkable scientific event since the discovery of the predicted existence of the planet Neptune. But there was difference of opinion as to whether science had to face merely a new and unexplained fact, or to reckon with a theory that would completely revolutionize the accepted fundamentals of physics.

SIR FRANK DYSON, the Astronomer Royal, described the work of the expeditions sent respectively to Sobral in North Brazil and the island of Principe, off the West Coast of Africa. At each of these places, if the weather were propitious on the day of the eclipse, it would be possible to take during totality a set of photographs of the obscured sun and of a number of bright stars which happened to be in its immediate vicinity. The desired object was to ascertain whether the light from these stars, as it passed the sun, came as directly towards us as if the sun were not there, or if there was a deflection due to its presence, and if the latter proved to be the case, what the amount of the deflection was. If deflection did occur, the stars would appear on the photographic plates at a measurable distance from their theoretical positions. He explained in detail the apparatus that had been employed, the corrections that had to be made for various disturbing factors, and the methods by which comparison between the theoretical and the observed positions had been made. He convinced the meeting that the results were definite and conclusive. Deflection did take place, and the measurements showed that the extent of the deflection was in close accord with the theoretical degree predicted by Einstein, as opposed to half that degree, the amount that would follow from the principles of Newton. It is interesting to recall that Sir Oliver Lodge, speaking at the Royal Institution last February, had also ventured on a prediction. He doubted if deflection would be observed, but was confident that if it did take place, it would follow the law of Newton and not that of Einstein.

DR. CROMMELIN and PROFESSOR EDDINGTON, two of the actual observers, followed the Astronomer-Royal, and gave interesting accounts of their work, in every way confirming the general conclusions that had been enunciated.

"MOMENTOUS PRONOUNCEMENT."

So far the matter was clear, but when the discussion began, it was plain that the scientific interest centred more in the theoretical bearings of the results than in the results themselves. Even the President of the Royal Society, in stating that they had just listened to "one of the most momentous, if not the most momentous, pronouncements of human thought," had to confess that no one had yet succeeded in stating in clear language what the theory of Einstein really was. It was accepted, however, that Einstein, on the basis of his theory, had made three predictions. The first, as to the motion of the planet Mercury, had been verified. The second, as to the existence and the degree of deflection of light as it passed the sphere of influence of the sun, had now been verified. As to the third, which depended on spectroscopic observations there was still uncertainty. But he was confident that the Einstein theory must now be reckoned with, and that our conceptions of the fabric of the universe must be fundamentally altered

At this stage Sir Oliver Lodge, whose contribution to the discussion had been eagerly expected, left the meeting.

Subsequent speakers joined in congratulating the observers, and agreed in accepting their results. More than one, however, including Professor Newall, of Cambridge, hesitated as to the full extent of the inferences that had been drawn and suggested that the phenomena might be due to an unknown solar atmosphere further in its extent than had been supposed and with unknown properties. No speaker succeeded in giving a clear non-mathematical statement of the theoretical question.

SPACE "WARPED."

Put in the most general way it may be described as follows: the Newtonian principles assume that space is invariable, that, for instance, the three angles of a triangle always equal, and must equal, two right angles. But these principles really rest on the observation that the angles of a triangle do equal two right angles, and that a circle is really circular. But there are certain physical facts that seem to throw doubt on the universality of these observations, and suggest that space may acquire a twist or warp in certain circumstances, as, for instance, under the influence of gravitation, a dislocation in itself slight and applying to the instruments of measurement as well as to the things measured. The Einstein doctrine is that the qualities of space, hitherto believed absolute, are relative to their circumstances. He drew the inference from his theory that in certain cases actual measurement of light would show the effects of the warping in a degree that could be predicted and calculated. His predictions in two of three cases have now been verified, but the question remains open as to whether the verifications prove the theory from which the predictions were deduced.

7 'Revolution in Science,' *The Times*, 7 November 1919, 12.

cialised audience with scepticism. The English theoretical physicist Arthur Eddington had been interested in Relativity since 1911, but by 1916 he had become one of its main advocates and was anxious to verify its findings. With A. C. Crommelin, assistant at the Royal Observatory at Greenwich, Eddington led the now-legendary British expedition, setting out in February 1919 to observe the solar eclipse visible in May from West Africa and Brazil.[13] Eddington hoped the eclipse would prove one of the three main effects of Relativity predicted by Einstein: that the path of a ray of light would be deflected when passing in sufficiently close proximity to a large body like the sun. In autumn 1919 the results gained from the expedition agreed with Einstein's (or were near enough to show his theory was closer than the prediction of Newton's law), securing Relativity's institutional respectability, catapulting Einstein onto the front pages, and guaranteeing his instant mystique (fig. 7).

Countering common sense assumptions about the behaviour of time, space and matter, the Theory of Relativity holds a peculiar position of importance in histories of physics, thought and culture. Yet quantum theory was even more revolutionary, stirring much greater passions within physics and having a greater impact on the material culture that surrounded it. While it received a muted reception in the wider world in comparison with Einstein's theory, the set of interactions, equations and concepts covered by the terms 'quantum mechanics' and 'quantum theory' constitutes the most successful model yet devised to explain the fundamental structure and behaviour of the physical world.

Quantum Mechanics and Quantum Theory

Predating Einstein's 1905 *annus mirabilis*, the origins of quantum theory can be traced back to the beginning of the twentieth century, though the theory reached maturity a few years after Relativity.[14] In 1900, Max Planck (1858–1947), Professor of Physics in Berlin, made a troubling, groundbreaking discovery about the distribution of energy (fig. 8). Through a series of experiments with black body radiation (a body absorbing all of the radiation that falls on it and emitting completely all of that radiation), Planck found that energy emission and absorption took place not smoothly and continuously as dictated by the view of nature inherited from the classical physics of Newton and

Maxwell, but discontinuously in pulses or packets of energy: that is, in discrete measurable units or 'quanta'.[15] He showed that the amount of energy in each quantum was directly proportional to the frequency of the radiation (frequency measures the number of wave crests that pass a given point in a second). This proportion is an irreducible property of the natural world and is found by multiplying the light frequency with the constant of nature now known as 'Planck's constant' or *h*.[16] According to Max Jammer's near-definitive account, Planck seems at first to have been unsure whether *h* was a universal property or 'merely a mathematical device'.[17] As part of a generation that subscribed unanimously to the principles of classical physics, Planck was deeply concerned at his inability to adapt it to those tenets, while at the same time he dimly perceived his discovery might be on a par with those of Newton.

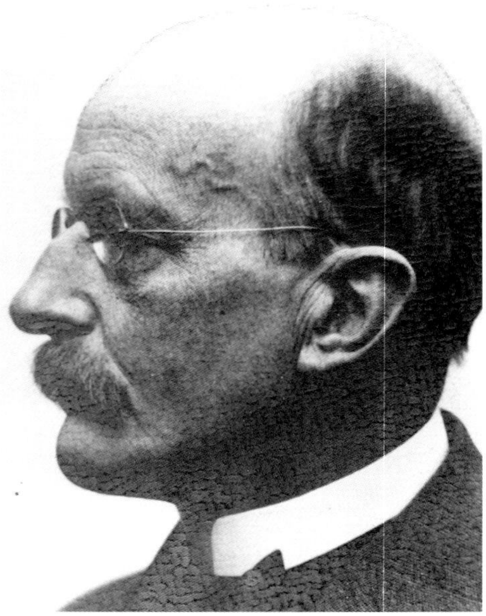

8 Max Planck, 1921. Courtesy Niels Bohr archive.

In the broad (oversimplified) historical paraphrase carried by popular books on QM, Planck's momentous discovery led initially to the quantisation of light in 1905 by Einstein, disclosed through his examination of the photoelectric effect. This is the phenomenon whereby electrons are dislodged from metal placed under a beam of light by the transfer of energy from the radiation to the electrons in the metal. Classical theory presented light as composed of waves, decreeing that a higher light intensity would displace more electrons and that the frequency of the light would play no role in the result of the experiment. In experimental practice, however, quite the contrary was often discovered to be the case: an intense beam of light with a low frequency might eject no electrons, whilst a weak light with a higher frequency might cause electrons to escape from the metal.

Pondering the problem of the photoelectric effect with Planck's recent work in mind, Einstein proposed that light was composed not of waves as taught by classical electromagnetic theory but quanta (later termed 'photons' in this context). Electrons were sufficiently activated to leave the metal only when quanta struck them and transferred to them their energy. The relation between intensity and frequency of radiation and the behaviour of electrons was explained by Einstein's approach to the photoelectric effect through Planck's theory, published as 'On a Heuristic Viewpoint Concerning the Production and Transformation of Light.'[18] A higher frequency of the radiation led to proportionally more energy in the quanta in line with Planck's constant, whereas a higher intensity of the beam only constituted more light quanta and not more energy (that is, greater intensity of radiation – more photons – can lead to *more* electrons escaping, but

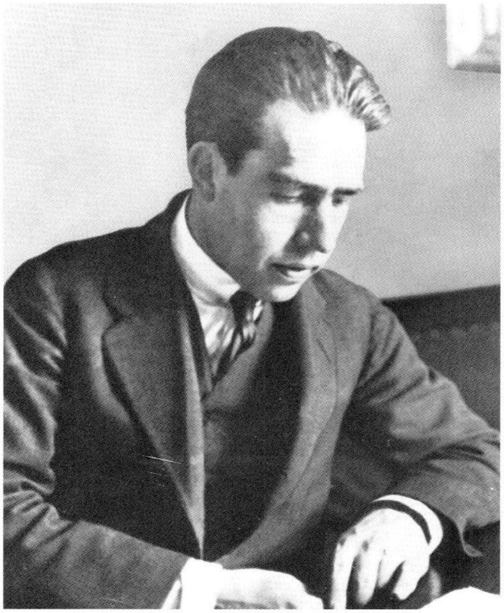

9 Niels Bohr, 1916. Courtesy Niels Bohr Archive.

value of frequency of radiation decides *whether or not* electrons are displaced). Like Planck, Einstein was bothered by the resistance of the 'old' theory to the 'new', later writing: 'My attempts . . . to adapt the theoretical foundation of physics to this knowledge failed completely.'[19] He goes on to express perfectly the instant revolution in physics and sense of an epistemological crisis: 'It was as if the ground had been pulled out from under one, with no firm foundation to be seen anywhere, upon which one could have built.'[20]

The third step in these early years of quantum theory was to remodel the structure of the atom, taken initially by the Danish physicist Niels Bohr (1885–1962) whilst working with Ernest Rutherford in Manchester (fig. 9). The classical, tentative 'plum pudding' model of the atom devised by J. J. Thomson and Lord Kelvin, in which the negative charge of the electrons (the plums) was balanced by the positive charge distributed throughout the rest of the atom (the pudding), had sufficed in the absence of a reasonable alternative until 1909. In that year, Rutherford and two younger colleagues Hans Geiger and Ernest Marsden found to their astonishment that tiny, positively charged alpha-particles fired at a thin gold film often passed through as expected but sometimes *rebounded*. How could a bullet ricochet off a pudding? Rutherford had to assume that the positive charge was not dispersed throughout the atom but formed a solid core at its centre, and that the alpha particles had made contact with this charge and bounced back (since like charges repel each other, of course). From the unexpected results of this experiment, Rutherford conceived the 'solar system' model of the atom in 1911, consisting of a positive nucleus orbited by negative electrons. Though it explained observed effects, this model also played havoc with previous classical theory, which dictated that the changes in direction of circling electrons would cause them to lose some of their energy through radiation and spiral towards the nucleus. In addition to this disastrous instability in the atom, when classical electromagnetic theory was applied to this new model it ruled that the steady pattern of radiating energy given off by these orbiting electrons contradicted the more complex and discrete record of spectral frequencies of light. These sharp spectral lines had been thought to be random until their regularity was discovered in 1885 through the mathematical formulation of the Swiss school teacher Johann Jakob Balmer. The Balmer formula predicted the frequencies of the four visible spectral lines of hydrogen with such accuracy that it strongly suggested the discovery of a physical law, the true nature of which would remain beyond physicists for several years until the advent of QM.

Aware of these glaring inconsistencies in the theory of the atom, Bohr patched things up temporarily, hypothesising from Planck's intuition of the discontinuous nature of radiation that activity within the atom was similarly characterised by 'breaks'.[21] Just as absorption and emission of radiation was parcelled out in quanta, so, Bohr claimed, electrons could not move in just any conceivable orbits but were located on allowed orbits or 'stationary states' of specific value, corresponding with energy states between which they could jump as the atom absorbed or emitted chunks of energy or quanta. He was able to predict these stable radii at precise distances around the nucleus, asserting '*the angular momentum* [rotational motion] *of every electron round the centre of its orbit will in the permanent state of the system be equal to* [the universal value] $h/2\pi$, where h is Planck's constant.'[22] Whilst giving the atom back its stability, what became known as the 'quantum jumps' between stable orbits in Bohr's model also accommodated and mirrored the sharp spectral lines predicted by Balmer's formula. This is because Bohr found that all the possible permutations of leaps by the hydrogen electron added up to and were in agreement with the number of spectral lines assigned to it by Balmer. In this way the more speculative classical theory, which had not fixed the electron to predetermined radii, was superseded and replaced by another theory driven by Planck's quanta. The budding theory of quanta explained and incorporated into its model the simple and mathematically consistent results gained by Balmer, the fundamental significance of which, for atomic theory, had tantalised and baffled physicists up to that day.

Einstein's daring revival of Newton's corpuscular composition of light, together with Bohr's equally bold assumption of quantum behaviour in subatomic processes and its subsequent sensational experimental agreement with observation of spectral lines, were, in retrospect, serious blows to classical physics. Many were bothered by the implications of quanta though and continued to ignore both Planck's discovery and Einstein's and Bohr's extensions of his work, which together now constitute the 'old quantum theory'. Physics would for a few years be confused and hampered by the ill fit between the emerging quantum theory of nature and the deeply ingrained, 'respectable' classical physics of Newton and Maxwell. Planck expressed reservations about the further cultivation of the theory, and in exchanges with his colleagues even Einstein advised caution over his theory of light, concerned by its lack of congruence with experimentally established wave theory and puzzled at the continued experimental verification of the wave formation of light.

The confusion initiated by quanta over whether light conformed to a wave or particle description was furthered in 1923. In that year, the American physicist Arthur Compton (1892–1962) examined in his laboratory at Washington University the behaviour of X-rays when fired at electrons, discovering that the X-rays ricocheted off the electrons and scattered at prescribable angles as though they were particles, losing energy and changing to a lower frequency as they did so. This could be explained by recourse to Planck again, who had discovered that energy was directly proportional to frequency, yet it brought about a profound paradox. On the one hand, as high frequency electromagnetic radiation (light), X-rays were thought by everyone (except Ein-

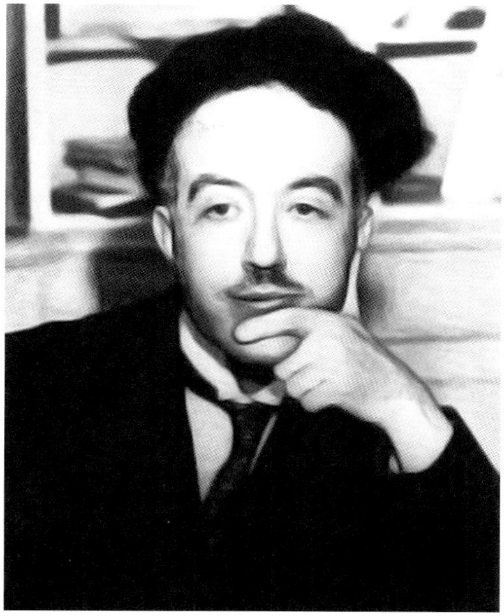

10 Louis de Broglie, c.1924.

stein) to be *wave*-like, and they should not therefore be deflected at angles measurable in agreement with the laws of *particle* physics; nor should X-rays change their frequency through collision with electrons; rather, in accordance with their wave properties, they should be absorbed and emitted as energy by electrons whilst retaining the same frequency. On the other hand, Compton partly recorded this particle-like behaviour (the loss of energy by the photons through their collision with the electrons) by measuring the change in frequency of the X-rays, yet it is waves, not particles, that have frequencies.[23] What became known as the Compton effect – that is, the demonstration of the quantised nature of light through the interpretation of the angles at which the X-rays scatter, the transfer of energy to the electron, and the change of frequency in the radiation – could be measured only by assuming both particle and wave descriptions. Compton concluded that the fit between theory and experiment left 'little doubt that the scattering of X-rays is a quantum phenomenon'.[24]

While bewilderment reigned about the constitution of the physical world there was a sign of greater acceptance of their ideas by the scientific establishment in the presentation of a Nobel Prize each to Planck, Einstein and Bohr for their work on quanta in 1918, 1921 and 1922 respectively. In the undertow of the First World War the tide was about to turn and swamp the classical worldview in the most dramatic period in the history of physics, bringing quantum theory to completion through a set of truly revolutionary ideas. Perhaps predictably, because they were unimpeded by the habits encouraged by the old physics, mainly young men were responsible for these radical revisions to the understanding of the physical world. This 'boys' physics' (*Knabenphysik*) as it was once termed by one of its architects, Wolfgang Pauli, took place largely between 1924 and 1927.

Compton's results had inevitably deepened the ongoing, productive crisis felt within the physics community, which was exacerbated further when the French physicist Louis de Broglie (1892–1987) extended the still-obscure particle/wave duality to all matter in his doctoral thesis examined at the Sorbonne on 25 November 1924 (fig. 10). On the examining committee for de Broglie's doctorate were two of France's greatest physicists, both of whom would later have a brush with Surrealism. They were Jean Perrin (1870–1942), who would win the Nobel Prize in 1926 for his groundbreaking work on atoms, and Paul Langevin (1872–1946), de Broglie's supervisor, whose research centred on the molecular structure of gases and the theory of magnetism, and who was a close friend of Einstein. When Einstein read the advance copy of de Broglie's doctorate sent to him by Langevin, he immediately understood its significance.[25]

Over the previous two years de Broglie had published articles on quanta in the leading French physics journal *Comptes Rendus* and in the lesser-read *Journal de Physique*.[26] These served as the basis for his doctoral thesis, which reached a larger audience when it appeared early in 1925 in another of the principal French physics periodicals, *Annales de Physique*.[27] Referring to Einstein's perplexed initiation of serious consideration of particle/wave duality in his work on the photoelectric effect, de Broglie proposes a generalisation of Einstein's findings (which were themselves a generalisation of Planck's turn-of-the-century discovery of quanta) to all of the physical world. Intuiting, around the time Compton showed waves to be particles, that particles might therefore be waves, de Broglie translated this particle/wave duality into mathematical relationships that drew upon the equations of both Planck and Einstein. In addition, however, de Broglie grasped the idea that *electrons*, which had never been conceived since their discovery by J. J. Thomson in 1897 as anything other than consummate particles with all of the attendant behaviour, could be thought of as *waves*. This was so insofar as the relationship between their curious quantum-jumping and the stable orbits they occupied and leapt between inside Bohr's atom could easily be imagined instead as a vibrating 'standing wave' quantised as a whole number of wavelengths (which had to 'fit'), extending around the circumference of the atomic nucleus. The frequency of the wave is equal to the energy state of the atom previously identified with the quantised emission and absorption of energy by the electron-as-particle.

One of the great unifying discoveries of physics, de Broglie's theory took into account all of the discoveries around quanta so far, and consequently had a powerful impact on physics throughout the 1920s. The 'idea-image' he presented to rationalise the contradictory notions of waves with angular momentum and particles possessed of frequency was that of a 'pilot' wave guiding each particle. The wave and its particle were mathematically related by de Broglie through the multiplication of momentum and wavelength (wavelength measures the distance between a wave and the one immediately following) to give Planck's constant ($p\lambda = h$).

Although the examiners of his doctorate were convinced by the mathematics, they were more sceptical about the physical reality of the scheme. However, de Broglie could not only predict the wave-like behaviour of electrons mathematically, but was ready with a suggestion as to how these effects might be observed. Diffraction patterns can be detected only when a hole is used that is smaller than the wavelength of the wave directed through it, and de Broglie proposed, therefore, that a beam of electrons (which have a very small wavelength) interacting with a crystal (because the atoms of crystals are separated by unusually tiny gaps) would present the 'electron waves' that his theory and mathematics predicted. In fact, such an experiment had already been carried out, though its results had been misunderstood (not surprisingly, given the 'non-existence' of the oddity of electron waves before de Broglie). Two separate experiments were subsequently completed in 1927 by Clinton Davisson (who had earlier misinterpreted the same effect) in America and George Paget Thomson in England, proving de Broglie correct and paving the way for his Nobel Prize of 1929 and the joint award of the Prize to Davisson and Thomson in 1937. Historians of science like to point out that George Thomson won the Nobel Prize for demonstrating the electron is a wave while his father J. J. Thomson won it in 1906 for showing the electron was a particle. Both were correct, and the paradox is very satisfying in the context of quantum theory, the discoveries of which made the world seem increasingly unreal from 1924.

Written, examined and published in Paris at exactly the moment André Breton wrote in the *Manifesto of Surrealism*, '[i]magination alone offers me some intimation of what *can be* . . .', de Broglie's was a pioneering piece of work that definitively marked the passing of one description of the world and confirmed the potential existence of a new one.[28] From the moment of its appearance, its author became a key figure in French science, and he consequently has singular importance in this book as philosopher, epistemologist and populariser of the new physics in France in the 1920s and 1930s, not least in the art and writings of the Surrealist artist Wolfgang Paalen, whose pictorial language stems directly from his reading of de Broglie.

In spite of all of the trailblazing work done on quanta, physics itself seemed something of a mess in early 1925. Rather than a new, sustainable, coherent, shared theory of quanta, physicists were faced with either patched up aggregates of classical and quantum theory or local breakthroughs of apparent significance. Although it had achieved a stunning success in deriving the Balmer formula (the correct values for hydrogen line spectra) from a model based on the radii of stable electron orbits, Bohr's model of the atom, which held such promise in 1913, seemed increasingly like an unsatisfactory example of the former.

One of its problems was that the structure of Bohr's atom used whole numbers (integers) to describe its energy transitions (jumping electrons). These single quantum numbers assigned to each orbit, mirrored in the number of spectral lines, could not account for the more refined results emerging from advanced spectroscopy. Now a much greater splitting of spectral lines (even in hydrogen) could be observed than Bohr's use of the quantised integer n (called the 'principal quantum number') was able to allow. This problem had been partially resolved during the First World War by Arnold Sommerfeld (1868–1951) in Munich, who approached the question as though the spectral lines were describing *elliptical* orbits (Bohr had allowed for circular electron orbits only). As with Bohr's stable orbital radii for electrons, only a *limited* number of values could be ascribed to the shapes of the orbits, and to this variable Sommerfeld assigned a second quantum number k.

To the delight of Einstein and Bohr, Sommerfeld advanced atomic theory further in 1916 with a quantum explanation of the Zeeman Effect. In 1896, the Dutch physicist Pieter Zeeman (Nobel Prize winner in 1902) found that extra spectral lines were produced by excited atoms placed in a magnetic field, though a physical description of these observed effects – the exact details of what was going on inside the atom – had not been forthcoming at the time. A respectable and complete atomic theory had to offer a solution for the phenomenon of the Zeeman Effect, and Sommerfeld's second contribution provided this by proposing that this multiplication of spectral lines indicated variations in the orientation (direction) of the electron's orbit caused by the magnetism. As with the size (n) and shape (k) of the electron's orbit, its direction (the 'magnetic quantum number' m) could not be arbitrary but was quantised in units of $h/2\pi$.

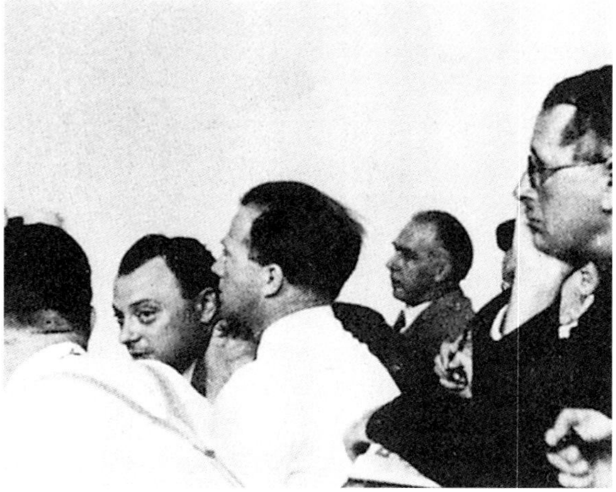

11 Photograph of lecture at Institute for Theoretical Physics, Copenhagen, around 1930 with Wolfgang Pauli facing the camera second from the left. Pascual Jordan is on the far left, Werner Heisenberg is in the centre foreground, Niels Bohr is in the background in the centre.

Now Bohr had three quantum numbers assigned to the behaviour of the electron, which could be combined to give a refined account of transitions in the energy level of the atom and the splitting of spectral lines, yet they still proved insufficient. Although the Bohr–Sommerfeld orbits were remarkably successful in explaining the Zeeman Effect, the limits of their effectiveness were exposed by experiments with magnetism that produced even more spectral lines – results referred to by confounded physicists as the Anomalous Zeeman Effect (AZE).

The task of discovering a solution to the AZE was a major preoccupation in the early 1920s for the young Viennese theoretical physicist Wolfgang Pauli (1900–1958; fig. 11). As a student at the University of Munich, Pauli was already a precocious mathematician. From there he went to Göttingen in 1921 as Max Born's assistant, then to Hamburg, and from there to Copenhagen in 1922–23 to work with Bohr where he wrestled with the AZE. Eventually in the autumn of 1924, Pauli suggested that a fourth quantum number with a '"classically non-describable two-valuedness"' should be assigned to the electron to account for the AZE's proliferating spectral lines.[29]

This 'new quantum theoretic property' proposed by Pauli was soon identified with the electron's extra angular momentum caused by its rotation on its axis or 'spin.'[30] The concept of spin was not Pauli's – it was first mooted by the American physicist Ralph Kronig in January 1925 – and like several of his colleagues Pauli rejected it initially because the revolving electron had no stability in the context of the Theory of Relativity, on which he had written brilliantly and at length. On the other hand, spin taken in its literal sense – of an electron spinning on its axis – was a misleading image too indebted to classical physics. By the time spin was proposed in 1925, the work of Einstein, Compton and de Broglie had created deep suspicion about such concrete imagery, and spin made no sense now that all three had shown 'particles' and 'waves' to be synonymous. Although Kronig's observations went unpublished in the face of these inconsistencies, a kind of inevitability accrued to the concept of spin: ultimately, it gained a currency in the physics community after it was tentatively touted again as 'Rotation des Elektrons' in October 1925, this time by George Uhlenbeck and Samuel Goudsmit of the Institute for Theoretical Physics in Leyden, as a short notice in the pages of Arnold Berliner's interdisciplinary journal *Die Naturwissenschaften*, which counted the Surrealist artist Max Ernst among its readers.[31]

Like the electron's two-valuedness, spin has a quantum mechanical and mathematical 'reality' only and it was on those terms that it was later accepted by Pauli. In its two potential directions of rotation (clockwise – spin up, and anticlockwise – spin down), it describes conveniently the two-valuedness he required for his fourth quantum number (*s*) to explain the AZE. However, the angular momentum of the supposedly spinning electron is called spin 1/2, being only half of the quantised angular momentum unit $h/2\pi$ arrived at by Bohr in 1913, meaning that if this 'spin' were taken literally the electron would get back to its starting point only by revolving on its axis twice! Such was the imperfect coalition of classical pictures and quantum interactions, and such, too, was the potential for poetic appropriation of ideas like spin, soon to be noted by ex-Surrealist Roger Caillois.

Spin, the fourth quantum number, relates directly to Pauli's major contribution to quantum theory, the Exclusion Principle or Pauli Principle that would win him the Nobel Prize in 1945 for its resolution of hitherto unexplained behaviour within Bohr's atom.[32] Physicists had yet to explain why in the case of atoms containing more than one electron – like lithium for example, which has three and potassium, which has nineteen – all of the electrons did not fall into the ground state (the lowest level of energy available to an atom where the electron occupies the 'shell' or orbit closest to the nucleus). Using the periodic table of elements designed in the 1890s by Dmitri Mendeleev and 'like' chemicals as a guide, Bohr intuited that electrons arranged themselves systematically so that a maximum of two were allowed on the inner shell, eight on the next, eighteen on the one after that, then thirty-two, and so on. That is why, he argued, in spite of their differences, hydrogen with one electron, lithium with three, and sodium with eleven display like properties: because each has a 'free' electron on its outer orbit (since the inner shell of lithium can only contain two and the inner two shells of sodium contain two and eight respectively). Remarkably, Bohr could show by 1921 that the behaviour of electrons he had conjectured and described could explain convincingly the structure of the periodic table. His success in offering such a simple rationale, from the field of theoretical physics, for a core problem in chemistry (the arrangement of the elements in their positions of increasing atomic number in the periodic table) received a luminous compliment from Einstein, who saw Bohr's idea as 'the highest form of musicality in the sphere of thought'.[33] However, as Pauli later reported, when Bohr gave a series of lectures on these matters in Göttingen in 1922 (where Pauli met Bohr for the first time), the question of 'why all electrons of an atom in its ground state were not bound in the innermost shell' remained unanswered.[34]

The task of responding to this fundamental problem was met in 1925 by the Exclusion Principle. Once Pauli added the fourth, two-valued number for the property of spin to the other three quantum numbers (size, shape and orientation of the electron's orbit), he was able to say that each electron possessed a set of four quantum numbers exclusive to it, a state demanding that the next electron go to a higher energy state. In this way, the shells of an atom fill to their maximum (two on the inner ring, then eight on the next, and so on) from lowest to highest, and the uniqueness of each electron creates the 'stacking up' process that prevents them from all falling into the ground state.[35]

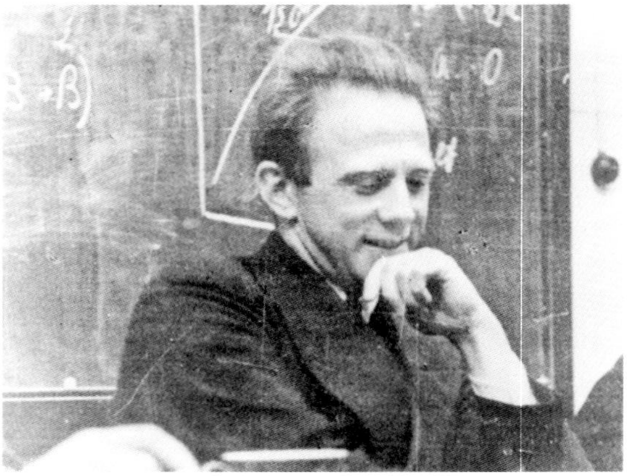

12 Werner Heisenberg, Leipzig, 1927. Courtesy Niels Bohr Archive.

After a quarter of a century of initially hesitant then increasingly intense work on quanta, events accelerated from this point, bringing to fruition no less than three different yet related theories formulated over a year from the middle of 1925. The German theoretical physicist Werner Heisenberg (1901–1976) is often given pre-eminence by scientists and historians of science, just as he is frequently identified as the presiding boy genius of QM by those in thrall to its heroic narrative (fig. 12). Heisenberg read physics in Munich under Sommerfeld from 1920 where he met Pauli. He witnessed the same Bohr lectures in Göttingen that Pauli saw in 1922, already critical of the application of classical language and pictures to the explanation of quantum processes. In the course of an important conversation with Bohr after one of the Göttingen lectures – during which it became clear to Bohr that the twenty-year-old Heisenberg would master the problems of QM – he discovered that Bohr felt the same way.

This was the ongoing problem in physics for the duration of what Heisenberg later termed 'the preparatory years' of QM (1922–24): its continuing obligation to utilise classical laws alongside the muddled, incomplete, yet irrefutable glimmers (given all that had recently been learned of the atom) of quantum logic in the face of physicists' awareness of their mutual epistemological incompatibility.[36] At Göttingen in the winter of 1922–23, Heisenberg was part of a small seminar group led by the German Max Born (1882–1970), who had recently become Professor of Theoretical Physics at the university, attended by a handful of others including another gifted 'boy physicist', the German Pascual Jordan (1902–1980). From these exchanges, Heisenberg became convinced that Bohr's theory had not received precise confirmation from experimental data because it was hampered by its baggage of imagery, language and concepts. He concluded that he would arrive at an overarching quantum theory only by dispensing with the imagery of electrons and all other 'perceptual' aids and models. By the middle of 1925, while many problems seemed insoluble in physics, Heisenberg was working with Born in Göttingen and Bohr at the Institute for Theoretical Physics in Copenhagen, yet he was increasingly eager to depart from the imagery to which they had reluctantly subscribed. In a letter written around that time he even reproached Pauli for falling into the 'old' classical language: '[m]y entire meagre efforts,' he complains, 'go toward killing off and suitably replacing the concept of the orbital paths that one cannot observe.'[37]

The clearest, most economical rendition of Heisenberg's insistent empiricism is given by Paul Dirac:

> Heisenberg's idea was that one should try to construct a theory in terms of quantities which are provided by experiment, rather than building it up, as people had done previously, from an atomic model which involved many quantities which could not be observed. By this brilliant idea Heisenberg really started a new philosophy, a philosophy that physics – physical theory – should keep close to the experimentally obtained data and should not depart into the use of quantities which are only very remotely connected with observation.[38]

Heisenberg, then, made his major discoveries following his sweeping rejection of the concepts of 'electrons' with 'orbits' now viewed as misleading descriptive analogies attached to actual events observed experimentally. Instead, he opted to attend to the associations between energy states given numerically, developing in the summer of 1925 a tabular mathematics that allowed the description of pairs of states of the atom with a finesse unachievable using ordinary numbers. In May and June, Heisenberg derived a formula for comprehensively linking the energy states and quantum numbers of atoms with the frequencies and intensities of light spectra, allowing the calculation of probabilities from initial conditions.

Although the mathematical consistency of his results suggested he had achieved something crucial, Heisenberg was troubled by an oddity of the calculations whereby the results depended on the order in which the multiplication was executed; that is to say, he had invented a form of mathematics that was non-commutative (where 2 x 3 does *not* equal 3 x 2). He completed his paper titled 'On a Quantum Theoretical Interpretation of Kinematical and Mechanical Relations', still feeling unsure about the non-commutation curiosity, and when he went to Cambridge to lecture in July he left it with Born who approved it and sent the historic paper on its way for publication.[39] Born soon realised that Heisenberg's algebra converted into a fairly obscure form of mathematics called matrix calculus, remembered from his student days by Born because of its odd non-commutative properties, which ultimately gave Heisenberg's quantum theory its name of 'matrix mechanics'.

Heisenberg and his subsequent admirers have heroised his achievement along the lines of the usual rhetoric of the 'lone genius', closely connecting his breakthrough to a brief sojourn he made in May 1925, seeking refuge from his chronic hay fever on the rocky island of Helgoland, a dot in the North Sea off the coast of Germany. However, he had sought support, advice and guidance from Bohr and Pauli along the way, and now he turned to Born and his assistant Jordan in Göttingen. In September, they had developed Heisenberg's work into what Max Jammer calls 'the first rigorous formulation of matrix mechanics',[40] publishing the key relation $pq - qp = h/2\pi i$ (QM's equivalent to Relativity's $E = mc^2$), in which p and q are matrices representing non-commutative quantum variables, respectively momentum and position.[41] Now Heisenberg worked with them until November to give these results greater mathematical rigour, sharpening the focus on the key matrix equation (which the Born–Jordan paper calls the 'exact quantum condition'),[42] and discovering with excitement the general applicability of the new mechanics through their success in deriving from it the same results as those achieved by Newtonian mechanics.[43] Now legendary and widely read, Jammer calls their paper, completed in mid-November 1925 and published in February 1926, 'the first comprehensive exposition of the foundations of modern quantum mechanics in its matrix formulation.'[44]

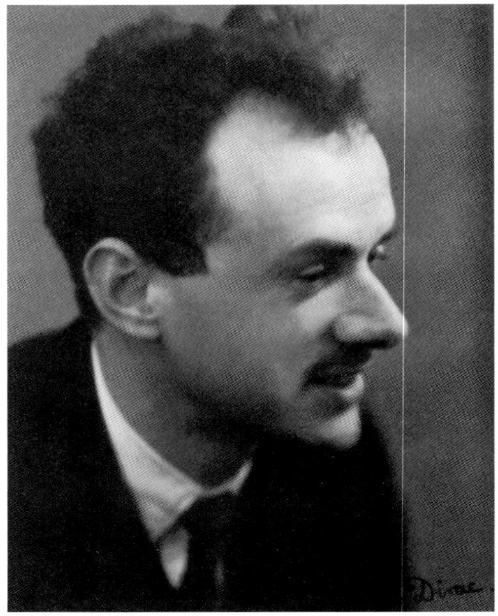

13 Paul Dirac, 1938. Courtesy Niels Bohr Archive.

In spite of their complex and unusual mathematics, the importance of the results gained by the Göttingen physicists gained limited recognition from a small audience straight away, most profoundly from the young Englishman Paul Dirac (1902–1984) (fig. 13). Dirac was destined to create the most complete version of QM, though he had aspired initially to engineering and mathematics before gaining a research grant in 1923 to study for a physics doctorate at St. John's College in Cambridge, where he first heard about QM. During his July lectures in Cambridge, Heisenberg had not discussed publicly his most recent discoveries, but following private conversations he sent a copy of the first, then-unpublished breakthrough paper to Ralph Fowler, Dirac's doctoral supervisor. It quickly made its way to Dirac himself who soon saw its value in resolving longstanding problems associated with the 'old' quantum theory.

While Born, Heisenberg and Jordan collaborated on their fundamental paper, Dirac worked alone. He was struck and impressed by the non-commuting physical quantities of Heisenberg's matrices, yet unlike Heisenberg he felt comfortable with non-commutation, regarding it as the core of the new theory. Accordingly he set about concerning himself with 'how the ordinary dynamics which people had been using till then should be modified to include it'.[45] For the task of adapting classical mechanics to non-commutation, Dirac was helped by his knowledge of the Bohr–Sommerfeld orbits and their connection with a form of nineteenth-century mechanics called 'Hamiltonian equations', developed as a means of calculating the orbits of bodies like planets. As Dirac remembered it, 'Hamilton's form of dynamics was just the form which was most suitable for bringing in non-commutation', providing the means by which he could engineer a fit between quantum and classical ideas within a few weeks.[46]

Giving prominence to non-commutation where the parallel work of the 'Göttingen three' concentrated chiefly on experimental results (viewing non-commutation as secondary and symptomatic), Dirac produced an extremely refined schema using what he termed 'quantum algebra' in his first major statement on the subject. Declaring, '[*a*]*ll*

14 Erwin Schrödinger. Courtesy Niels Bohr Archive.

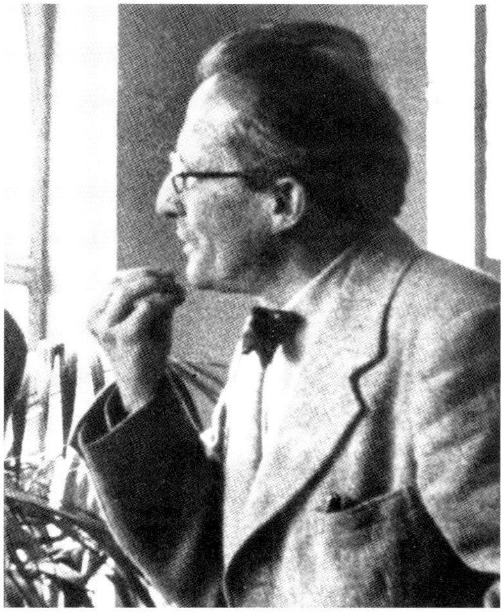

the information supplied by the classical theory can . . . be made use of in the new theory', Dirac presented a simple, elegant formalism that included Heisenberg's matrix mechanics, constituting a step forward from the fundamental yet raw QM of Born, Heisenberg and Jordan.[47] Nevertheless, on reading Dirac's results in his paper of December 1925, Heisenberg was unstinting in his praise, warmly communicating his congratulations to Dirac and declaring that his accomplishment was equally profound and subtler than the labours of the Göttingen/Copenhagen contingency. From here, Dirac became one of the leading figures of QM, going on to publish a string of classic papers up to July 1926 (put together as his doctoral thesis) and beyond, consolidating an international reputation and winning the Nobel Price in 1933.[48]

By the end of 1925, two versions of quantum mechanics existed, matrix mechanics and quantum algebra, which, in spite of their abstract mathematical formalism, set out vaguely wedded to the notion of the transition of quantum states of an electron-as-particle. Yet Einstein's 1905 theory of light quanta and Compton's experimental observations of X-rays in 1923 had demonstrated the perils of tying quantum theory too closely to one or other description of the physical world, particle or wave. Crucially de Broglie's theory had emphasised the (ambiguous) ontology of electron waves in 1924–25, a reality that received approving comment from Einstein in an important paper of 1925, seemingly making incomplete any theory (like those of Heisenberg and Dirac) that did not give notice to the wave structure of matter.[49]

Einstein's remark caught the eye of the Viennese physicist Erwin Schrödinger (1887-1961), who would share the 1933 Nobel Prize with Dirac for creating the third comprehensive version of QM from January 1926, in the wake of that of Heisenberg, Born, and Jordan, and that of Dirac (fig. 14). Schrödinger detected a mathematical relationship between de Broglie's findings and his own work, which he indicated in a letter to Einstein of November 1925, referring enthusiastically to de Broglie's 'ingenious thesis'.[50] For its resolute fidelity to the wave concept extended from de Broglie's insightful work – and, some argued, retrograde return to the classical picture of the physical world – Schrödinger's theory was titled 'wave mechanics'.

Schrödinger was of a generation somewhere between the architects of the old quantum theory and the new. By the time he wrote his monumental papers of 1926, he was twenty years into his life in science (which had begun in Vienna), had been Professor of Physics in Zürich for five years and was approaching the middle of his career.

His age probably helps to account for his attraction to de Broglie's more 'traditional' model of the atom, leading him in 1925 to attempt to adapt de Broglie's equation to electron waves moving in one of the Bohr orbits ('standing waves', where de Broglie had dealt only with travelling electron waves). Initially Schrödinger's new wave equation failed even to get a result with the spectrum of the hydrogen atom that agreed with Bohr's energy levels and quantum numbers. Yet he found later that this was because he had not allowed for spin, which was still a little-known property until it was introduced to a wide audience by Uhlenbeck and Goudsmit in *Die Naturwissenschaften* in November of that year.

Schrödinger was subsequently asked to organise a colloquium in Zürich on de Broglie's theory, and on returning to his earlier ruminations he found that if he removed the relativistic motion of the electron (spin) to give an approximate value in certain of his calculations for situations where spin was not a major factor, his equation was accurate. Essentially Schrödinger discovered the relation whereby the energy values and quantum numbers for the hydrogen atom could be derived as eigenvalues of three-dimensional matter waves (eigenvalues are the energies of a standing wave in distinct modes of vibration). The famous equation – which may be written $\Delta\psi + (8\pi^2 m/h^2)(E - V)\psi = 0$ – duly came to light in the first of Schrödinger's six fundamental contributions to QM, entitled 'Quantization as a Problem of Eigenvalues'.[51] There Schrödinger notes his indebtedness to de Broglie, and goes on to supplement the wave equations by imagining an elegant, vibrational activity within the atom, where a *continuous* process of energy transitions replaces the *discontinuous* quantum jumps of Bohr and his associates.[52]

Such remarks and those that swiftly followed in Schrödinger's further contributions to quantum theory guaranteed that his equation created waves of a different kind in Göttingen. Years later, Heisenberg recorded in mild terms his surprise at the sudden appearance at the beginning of 1926 of wave mechanics, but this initial reaction soon gave way to anger as Schrödinger's work gained powerful converts, seduced by the way it seemed to do away with the troublesome discontinuity of the QM of Bohr/Heisenberg.[53] In a footnote to his so-called 'equivalence' paper or 'third communication', Schrödinger had already accredited the patrimony of his work to de Broglie and Einstein rather than Heisenberg, brusquely recording how he had been 'discouraged, if not repelled, by what appeared to me as very difficult methods of transcendental algebra' pioneered by Heisenberg, 'and by the want of visualisibility [*Anschaulichkeit*]'.[54] The second point became a crucial one over which Heisenberg's views were equally uncompromising, leading to an increasingly heated polemic between Zürich and Göttingen over the visualisibility of subatomic processes. While Schrödinger found visualisation desirable and possible, Heisenberg in a letter to Pauli of 8 June 1926 used the word 'repulsive' to express his feelings of disgust at 'the physical portion of the Schrödinger theory', then in a famous passage went further: 'What Schrödinger writes about the visualisibility of his theory "is probably not quite right," in other words it's crap [*Mist*].'[55] Such were the passionate differences raised – even though mathematical equivalence had been reached – between the two main versions of QM over the interpretation of physical reality.

Schrödinger remained intransigent in the face of Pauli's insistence to him upon discontinuity in quantum behaviour in May 1926. However, a face-to-face encounter with Heisenberg during a lecture he gave on the new wave theory in Munich in July – where Heisenberg posed the obvious-sounding question of how Schrödinger's *continuum* physics could account for well-established *quantised* processes like the results gained by Planck from black body radiation, Einstein's interpretation of the photoelectric effect, or Compton scattering – gave him plenty to think about. These exchanges culminated in Schrödinger's visit to Copenhagen at the end of September for what turned into (literally) fevered discussions about QM with Bohr, some of which were witnessed and presumably joined by Heisenberg. Predictably, these turned on seemingly irresolvable differences about the physical constitution of matter and particularly the reality or not of discontinuity. Schrödinger began to realise that the irrefutable experimental existence of quantum jumps threatened the unity, clarity and completeness of his wave mechanics, whilst Bohr and Heisenberg could not ignore the unparalleled problem-solving success of Schrödinger's equation.

Conflict seemed to be the order of the day at the end of 1926 as a general dissatisfaction settled on the warring parties about the means to unite the various successful theories and conceive a satisfactory fit between laboratory experience and equations describing atomic processes. However, these exchanges would lead to the formulation of the two most controversial, non-classical and philosophically profound aspects of QM, effectively completing it and rounding out this history: Heisenberg's Uncertainty Principle (HUP) and Bohr's Complementarity, which together make up the Copenhagen Interpretation.

From October 1926 to February 1927, further extensive discussions were undertaken in Göttingen as to the usability of the concepts inherited from classical physics in the era of quantum interactions. Stirred into action by the reality of quantum leaps, the successes harvested by Schrödinger's wave equation and the unbridgeable gulf between the concepts deployed by Schrödinger and Bohr respectively in their Copenhagen conversations, Heisenberg moved in February and March 1927 to account for the gap between experimental results and mathematical apparatus in the formulation of the now well-known (and widely abused) uncertainty or indeterminacy relations, in his paper entitled 'On the Intuitive Content of the Quantum-Theoretical Kinematics and Mechanics' (fig. 15).[56]

The inconsistency between the unviewable (*anschaulich*) behaviour of atomic processes and the equations of the new mechanics demanded that the old Newtonian imagery and concepts of momentum, position, mass and orbit, along with more recent concepts forced on physics by Newtonian habits like 'spin', were now misleading rather than useful. New patterns of behaviour had to be inferred from the mathematics to describe laboratory observations now, and these had to be accepted no matter how strange they sounded. For example, at the quantum level the concepts paired in the key

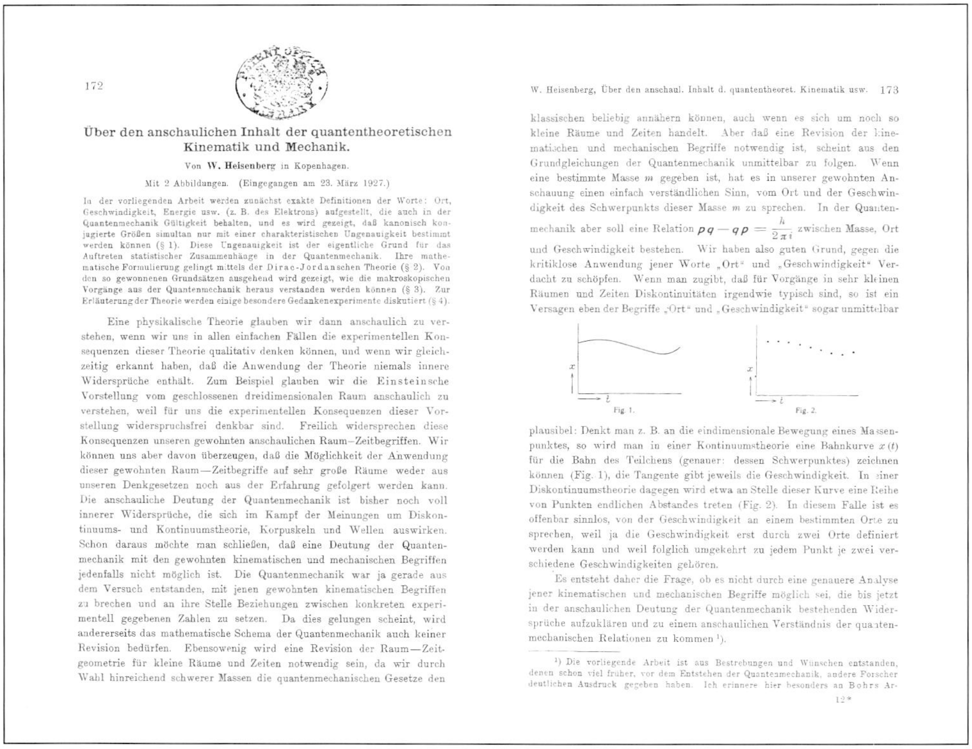

172

Über den anschaulichen Inhalt der quantentheoretischen Kinematik und Mechanik.

Von **W. Heisenberg** in Kopenhagen.

Mit 2 Abbildungen. (Eingegangen am 23. März 1927.)

In der vorliegenden Arbeit werden zunächst exakte Definitionen der Worte: Ort, Geschwindigkeit, Energie usw. (z. B. des Elektrons) aufgestellt, die auch in der Quantenmechanik Gültigkeit behalten, und es wird gezeigt, daß kanonisch konjugierte Größen simultan nur mit einer charakteristischen Ungenauigkeit bestimmt werden können (§ 1). Diese Ungenauigkeit ist der eigentliche Grund für das Auftreten statistischer Zusammenhänge in der Quantenmechanik. Ihre mathematische Formulierung gelingt mittels der Dirac-Jordanschen Theorie (§ 2). Von den so gewonnenen Grundsätzen ausgehend wird gezeigt, wie die makroskopischen Vorgänge aus der Quantenmechanik heraus verstanden werden können (§ 3). Zur Erläuterung der Theorie werden einige besondere Gedankenexperimente diskutiert (§ 4).

Eine physikalische Theorie glauben wir dann anschaulich zu verstehen, wenn wir uns in allen einfachen Fällen die experimentellen Konsequenzen dieser Theorie qualitativ denken können, und wenn wir gleichzeitig erkannt haben, daß die Anwendung der Theorie niemals innere Widersprüche enthält. Zum Beispiel glauben wir die Einsteinsche Vorstellung vom geschlossenen dreidimensionalen Raum anschaulich zu verstehen, weil für uns die experimentellen Konsequenzen dieser Vorstellung widerspruchsfrei denkbar sind. Freilich widersprechen diese Konsequenzen unseren gewohnten anschaulichen Raum–Zeitbegriffen. Wir können uns aber davon überzeugen, daß die Möglichkeit der Anwendung dieser gewohnten Raum—Zeitbegriffe auf sehr große Räume weder aus unseren Denkgesetzen noch aus der Erfahrung gefolgert werden kann. Die anschauliche Deutung der Quantenmechanik ist bisher noch voll innerer Widersprüche, die sich im Kampf der Meinungen um Diskontinuums- und Kontinuumstheorie, Korpuskeln und Wellen auswirken. Schon daraus möchte man schließen, daß eine Deutung der Quantenmechanik mit den gewohnten kinematischen und mechanischen Begriffen jedenfalls nicht möglich ist. Die Quantenmechanik war ja gerade aus dem Versuch entstanden, mit jenen gewohnten kinematischen Begriffen zu brechen und an ihre Stelle Beziehungen zwischen konkreten experimentell gegebenen Zahlen zu setzen. Da dies gelungen scheint, wird andererseits das mathematische Schema der Quantenmechanik auch keiner Revision bedürfen. Ebensowenig wird eine Revision der Raum—Zeitgeometrie für kleine Räume und Zeiten notwendig sein, da wir durch Wahl hinreichend schwerer Massen die quantenmechanischen Gesetze den

W. Heisenberg, Über den anschaul. Inhalt d. quantentheoret. Kinematik usw. 173

klassischen beliebig annähern können, auch wenn es sich um noch so kleine Räume und Zeiten handelt. Aber daß eine Revision der kinematischen und mechanischen Begriffe notwendig ist, scheint aus den Grundgleichungen der Quantenmechanik unmittelbar zu folgen. Wenn eine bestimmte Masse m gegeben ist, hat es in unserer gewohnten Anschauung einen einfach verständlichen Sinn, vom Ort und der Geschwindigkeit des Schwerpunkts dieser Masse m zu sprechen. In der Quantenmechanik aber soll eine Relation $pq - qp = \frac{h}{2\pi i}$ zwischen Masse, Ort und Geschwindigkeit bestehen. Wir haben also guten Grund, gegen die kritiklose Anwendung jener Worte „Ort" und „Geschwindigkeit" Verdacht zu schöpfen. Wenn man zugibt, daß für Vorgänge in sehr kleinen Räumen und Zeiten Diskontinuitäten irgendwie typisch sind, so ist ein Versagen eben der Begriffe „Ort" und „Geschwindigkeit" sogar unmittelbar

Fig. 1.

Fig. 2.

plausibel: Denkt man z. B. an die eindimensionale Bewegung eines Massenpunktes, so wird man in einer Kontinuumstheorie eine Bahnkurve $x(t)$ für die Bahn des Teilchens (genauer: dessen Schwerpunktes) zeichnen können (Fig. 1), die Tangente gibt jeweils die Geschwindigkeit. In einer Diskontinuumstheorie dagegen wird etwa an Stelle dieser Kurve eine Reihe von Punkten endlichen Abstandes treten (Fig. 2). In diesem Falle ist es offenbar sinnlos, von der Geschwindigkeit an einem bestimmten Orte zu sprechen, weil ja die Geschwindigkeit erst durch zwei Orte definiert werden kann und weil folglich umgekehrt zu jedem Punkt je zwei verschiedene Geschwindigkeiten gehören.

Es entsteht daher die Frage, ob es nicht durch eine genauere Analyse jener kinematischen und mechanischen Begriffe möglich sei, die bis jetzt in der anschaulichen Deutung der Quantenmechanik bestehenden Widersprüche aufzuklären und zu einem anschaulichen Verständnis der quantenmechanischen Relationen zu kommen [1]).

[1]) Die vorliegende Arbeit ist aus Bestrebungen und Wünschen entstanden, denen schon viel früher, vor dem Entstehen der Quantenmechanik, andere Forscher deutlichen Ausdruck gegeben haben. Ich erinnere hier besonders an Bohrs Ar-

12*

15 Werner Heisenberg, 'Über den anschaulichen Inhalt der quantentheoretischen Kinematik und Mechanik', *Zeitschrift für Physik*, vol. 43, 31 May 1927, 172–98.

relation pq – qp = *h*/2πι – 'momentum' and 'position' – could no longer be applied innocently to an electron, but were linked to the very conditions of observation, so when either of those two properties was measured the result would include within it the effect of the light photons hitting the electron to enable observation. Because Heisenberg's mechanics developed from reciprocal relationships between states of the atom, all measurements were affected by the strange non-commutability of the two variables (2 x 3 does *not* equal 3 x 2): the greater the accuracy of the measurement of one property, the less accurate the measurement of its partner.

The implications of Heisenberg's Uncertainty Principle were vast for both physics and philosophy. If the location and velocity of a particle could not now be measured with precision, then the epistemology inherited from the Newtonian–Kantian causality – insisting on the universe as a teleological mechanism, the future state of which could in theory be predicted from measurement of the location and velocity of a particle and any outlying conditions – was now obsolete. Artfully ending his seminal paper with a revolutionary flourish, the twenty-five year old Heisenberg confidently brought an epoch in philosophy to a close with the conclusion that the 'invalidity of the law of causality is definitely established by quantum mechanics'.[57]

Away on a skiing trip in Norway whilst Heisenberg dreamed up the HUP, Bohr was annoyed on his return to find that Heisenberg had gone ahead prematurely with a paper that considered only particles and quantum discontinuity, and not waves, insisting at first that Heisenberg withdraw it. After a heated exchange and lengthy stand off, Heisenberg tagged a conciliatory paragraph to the end of the paper. In the meantime Bohr worked on a compromise between the Heisenberg and Schrödinger factions, now known as Bohr's Principle of Complementarity.[58] For Bohr in 1927, Heisenberg's was only the latest chapter in a story showing consistently that 'an independent reality in the ordinary physical sense can neither be ascribed to the phenomena nor to the agencies of observation' due to the arbitrariness of the concept of observation in its handling of 'the quantum postulate with its inherent "irrationality"' (an irrationality determined by the partially obsolete framework of the epistemology of classical physics).[59] This is the heart of what Surrealist fellow-traveller Gaston Bachelard would soon call the 'new scientific spirit', out of sync with classical rationalism and therefore aligned with Surrealism's epistemological iconoclasm. First aired at a conference at Lake Como in Italy on 16 September 1927, Bohr's Complementarity made the assertion that under different experimental conditions, alternative, 'mutually exclusive but collectively necessary' models of reality may be mobilised – such as the particle or wave descriptions – dependent upon the experimental approach favoured.[60] This new formulation was fully accepted by Heisenberg and Pauli, though less so by others. Like Bohr, Heisenberg was not embarrassed by the loss to physics of a general description of the behaviour and constitution of matter:

> The use of 'matter waves' is convenient, for example, when dealing with the radiation emitted by the atom. By means of its frequencies and intensities the radiation gives information about the oscillating charge distribution in the atom, and there the wave picture comes much nearer to the truth than the particle picture. Therefore, Bohr advocated the use of both pictures, which he called 'complementary' to each other. The two pictures are of course mutually exclusive, because a certain thing cannot at the same time be a particle (i.e., substance confined to a very small volume) and a wave (i.e., a field spread out over a large space), but the two complement each other. By playing with both pictures, by going from the one picture to the other and back again, we finally get the right impression of the strange kind of reality behind our atomic experiments. Bohr uses the concept of 'complementarity' at several places in the interpretation of quantum theory.[61]

Because the issue is never settled one way or the other, the description of nature as given by Bohr's Complementarity was seen by many to have lapsed into undecidability. Most significantly, Einstein regarded the theory of QM as incomplete and temporary pending a 'final' answer, and his rejection of quantum theory is now a well-known part of its history (and his biography). But Bohr's defence of Heisenberg's indeterminacy and argument for Complementarity held in the face of resistance by Einstein and others, and the so-called Copenhagen Interpretation of 1927 is still the orthodoxy in physics even though it has been challenged (unsuccessfully) up to the present day.

Even more revolutionary than Relativity, QM predicts with great accuracy not only experimental findings in physics but also those in chemistry and molecular biology. Its mathematics also have immense application for technology, enabling the development of microprocessors, lasers and fibre-optic cables. However, the subatomic world it describes goes against the grain of common sense. Throwing into question classical concepts such as 'mass', 'location', 'velocity' and even 'determinism', it demands a wholesale departure from the well-worn language and imagery of physics, now largely ill-suited to the description of behaviour and phenomena on the smallest scale. Quantum theory certainly lacked a 'defining moment' to bring it forcefully to the attention of a wide public, as Eddington's 1919 verification of Relativity had done for Einstein's theory, and it could not be embodied in a single, charismatic individual like Einstein, or reduced in the popular mind to a 'simple' sum/symbol/image like $E = mc^2$. Yet its failure to make the headlines did not prevent recognition of its importance by philosophers and writers in the late 1920s and throughout the 1930s, who were soon convinced that its dramatic paradoxes demanded an overhaul of epistemological propositions about the relationships between human beings and nature.

Rejecting or at least radically questioning classical reason, the new world of time, space and matter teased out by Relativity and QM met with some ambivalence on the part of French physicists, who had their own ideas about how science should be done. However, the disquiet caused in French physics and philosophy by the 'epistemological break' forced by modern physics with Newtonian, Cartesian and Kantian forms of reason was not shared by the Surrealists, and even helped lay Surrealism's roots in its formative years immediately following the First World War. The case of Relativity in France acts as a useful test case for highlighting some of these prejudices and defining the nature of Surrealism's receptivity to the new physics.

Physics, Philosophy and Relativity in France

At the beginning of the twentieth century, France was one of the four countries with the most physicists, alongside Germany, Britain and the United States. Its standing as a great scientific nation was recognised when the first International Congress of Physics met in Paris in 1900. But as Relativity and QM made their initial inroads in the first decade of the new century, French physics went into decline. One reason for this lay in France's stubborn disposition at the time towards experimental as opposed to theoretical physics (which Relativity and QM were perceived to be) that lasted into the 1930s. This was not simply a pragmatic decision determined by economic or practical limits at the institutional level, but was ingrained pedagogical dogma. More specifically, the traditional views espoused by the *grandes écoles* cultivated 'a conservative orthodoxy' immensely resistant to change (especially when dictated from abroad), carrying into and shaping practice in scientific institutions.[62] This conservatism extended to fundamental beliefs about physics, sometimes covering the experimental and theoretical. An example of this is the hardened belief at the time in the existence of an ether, a key component in the conceptual framework of French physics due to its irreducible significance as an underlying material support in Newtonian mechanics. Michel Paty summarises the situation for pre-Relativity physics in France:

> mechanics was all-powerful and affected a faultless orthodoxy that was spread by a system which taught physics as a rigid and solid body of doctrine from which all weak points had been eliminated or disposed of. Science was presented as a fully realized achievement, encased in certainty, organized around Newtonian categories and justified by a positivism that took on the colours of mechanism, or, in a lesser fashion, those of energetics.[63]

Wedded to longstanding beliefs at the institutional and pedagogical levels about how science should be carried out, the unyielding orientation of French physics both experimentally and theoretically towards the classical model suggested that Relativity was in for a lean time in France.

There is worse to come though, for the awkward reception given the new physics in France was additionally mired in longstanding philosophical and cultural proclivities, comparable to those that lay in the path of the introduction of psychoanalysis into that country.[64] Surveying France's retarded reception of Relativity, Michel Biezunski observes that the coalition of Cartesian mechanism and Comtian positivism 'constituted the "mental set" of generations of French physicists':

> It gave rise to the mechanistic point of view in science: every phenomenon can be reduced to a description in terms of elementary movements, and this it true not only for physics, but for chemistry and biology as well. The supremacy of what was called in France 'rational mechanics' – as a reference to Descartes' philosophy – had been made explicit by Auguste Comte in his classification of the sciences. In Comte's scheme mechanics was considered totally different from and superior to physics. It was a 'science rationnelle' in the sense that it established a connection between an imperfect reality and an ideal, mathematical model.[65]

Although forewarned by the insightful and far-reaching analysis of classical mechanics carried out by Henri Poincaré, the nation of Comte did not take easily to new models of the physical world such as that of Relativity, rightly seen by its philosophers as a fundamental threat to the familiar Newtonian–Cartesian–Kantian notions of absolute time, space and motion.[66] Until then these had been only cosily debated in academic philosophy circles by positivists, empiricists and rationalists. By contrast, Relativity questioned 'certain traditional norms and concepts which had been previously accepted without criticism; in the country of Descartes, at the top of the list was "common sense"'.[67]

There were exceptions to those philosophers propounding 'rational mechanics', of course, even in the universities. Among them was rationalist and philosopher of science Émile Meyerson, who charged that science did not conform to the simple observation of facts cherished by both French positivists and the followers of Ernst Mach. Dealing with unobservable phenomena, Relativity had recourse to physical concepts, while Heisenbergian QM did away with observation altogether, resorting instead in the first instance to an unforgiving mathematical positivism. In the late teens and early twenties, Meyerson's objections cut little ice with French philosophers – most of whom persisted with positivism – and still less with physicists, while his teachings were assimilated in a complex manner by the avant-garde.

In fact the anti-Cartesian, anti-Comtian convictions of a French physicist rather than a philosopher turned out to have greater implications for philosophy in the long run. Paul Langevin had begun teaching Relativity as early as 1906, counting Louis de Broglie among his students. Friends with Einstein from the moment of the inaugural Solvay Congress of 1911, Langevin published his first article on Relativity in *Scientia* that year, subsequently becoming its most vocal French supporter and performing a similar task for Relativity in France as Arthur Eddington carried out for it in Britain.[68] According to Paty, Langevin 'detected in the new notions of Relativity the unravelling of the positivist spirit (in the Comtian sense) which organised knowledge in a hierarchical manner around absolute and normative categories'.[69]

Not surprisingly, then, André Breton was unusually receptive to Relativity in the early 1920s. In the *Manifesto of Surrealism*, he aggressively expresses identical sentiments to Langevin's towards canonical philosophy in Surrealism's unequivocal rejection of the French heritage of Cartesian common sense and Comtian positivism:

> In this day and age logical methods are applicable only to solving problems of secondary interest. The absolute rationalism that is still in vogue allows us to consider only facts relating directly to our experience. Logical ends, on the contrary, escape us. It is pointless to add that experience itself has found itself increasingly circumscribed. It paces back and forth in a cage from which it is more and more difficult to make it emerge. It too leans for support on what is most immediately expedient, and it is protected by the sentinels of common sense.[70]

> The realistic attitude, inspired by positivism, from Saint Thomas Aquinas to Anatole France, clearly seems to me to be hostile to any intellectual or moral advancement. I loathe it, for it is made up of mediocrity, hate, and dull conceit. It is this attitude which today gives birth to these ridiculous books, these insulting plays. It constantly feeds on and derives strength from the newspapers and stultifies both science and art by assiduously flattering the lowest of tastes . . .[71]

Chauvinistic criticism of just about anything deemed anti-French or anti-academic virtually guaranteed it the favourable attention of the Surrealists in the years following the First World War, and Relativity was no exception.

The war had devastated a whole generation of potential French scientists (an estimated forty percent of Sorbonne students who went to fight were killed or wounded), and dramatically exacerbated hostility in France towards German culture and the

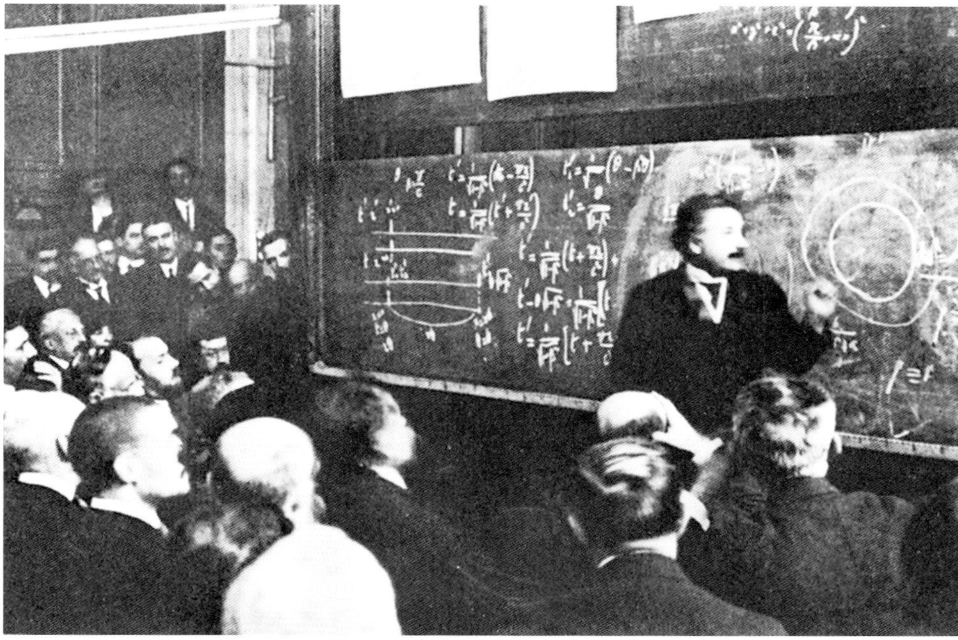

16 Photograph of Einstein lecturing at the Collège de France, Paris in 1922. Courtesy Niels Bohr Archive.

German language (also, inconveniently, the *lingua franca* of much of the new physics). In this light, Einstein's visit to Paris from 28 March to 10 April 1922 seems all the more remarkable (fig. 16). Breaking the boycott on German physicists dating from 1919, Langevin's invitation to Einstein to lecture on Relativity at the Collège de France was made during a time of crisis for French physics, still losing ground to other nations and slow to assimilate new discoveries from abroad. Einstein's presence caused a sensation in France as it did wherever he travelled during the period. This was manifested entertainingly in press articles, photographs and cartoons where Einstein was the main story in the forty-plus French newspapers, but also scandalously among French physicists.[72] Most did not fully accept Relativity in 1922, while many did not understand it or continued to resist outright the new 'German' (or 'Jewish') science on political, nationalist or racist grounds, as transcendental mathematics or metaphysics.[73] Even the mathematician Paul Painlevé, France's Prime Minister in 1917 and 1925, got in on the act, advising caution over its acceptance as late as 1922 in the hope of retaining at least a version of Newton's absolute time.

French physics was not exactly at a standstill, but views like those expressed over Relativity did not help its development, nor did indifference at government level. Jean Perrin's new Laboratoire de Chimie physique was established in the same year of Einstein's visit and Perrin was able to use his influence to pressure government to institute a new scientific organisation for funding research projects. But these were years of underfunding for French science. In 1924, the year of Breton's *Manifesto of Surrealism*, one of Einstein's French mediators, the mathematician Émile Borel, was successful in his new guise as member of the Chamber of Deputies in getting a tax levied on industry to support laboratory research, but this already meagre percentage was reduced in subsequent years. In 1926, the same year he won the Nobel Prize, Perrin founded and became Vice President of the Institut de Biologie Physico-Chimique. Funded by Baron Edmond de Rothschild to the tune of fifty million francs in the absence of state support, Perrin later came to see this event as the much-needed change in fortunes for French science between the wars.

In 1926, de Broglie was nominated for the Nobel Prize, which he won in 1929 for his 'discovery of the wave nature of the electron'.[74] No French physicist had won the Prize between the success of the Curies and Henri Becquerel in 1903 and Jean Perrin's in 1926. Perrin won the award for his research on atoms, and he nominated de Broglie in the same year. The nationalistic edge of these two successes in the space of three years was sharpened by the longstanding hostility between France and Germany and the dire state into which French physics had fallen since the end of its golden age in the first decade of the century. Boosted by national pride over de Broglie's French citizenry, the diffusion of modern physics and its philosophy in France were aided by the 'decontamination' of Relativity and QM of their 'Germanity' through his writings in the *Revue de Métaphysique et de Morale* and elsewhere.

By 1928, when de Broglie was made Professor of Theoretical Physics at the new Institut Poincaré (created by Émile Borel) where he would remain until 1962, his work on QM had won him a prominent position on the world stage and returned to French physics some of its esteem. In 1930, Perrin helped secure through parliament an extra one thousand francs a year as part of the budget of national education, and although he continued to complain that French physics was held back by insufficient state funding, the situation was now changing. As we will see in chapter three, the efforts of Langevin, de Broglie and Perrin began to bear fruit in the mid- to late 1930s under the Popular Front, when French nuclear physics took on global significance as a burgeoning field of research and development.[75]

Reviewing Modern Physics and Philosophy in France

Like everyone else excited by, interested in and anxious to understand the new physics, the Surrealists had to gain access to a very specialised field. Although theoretical physics since Newton has been practised largely through the medium of mathematics, the main problem with the transmission of the ideas arising out of physics in the twentieth century has been the increased level of mathematisation of the discipline, widening the gulf of specialisation between expert and layperson. We heard Heisenberg and others state flatly that events in space-time and at the subatomic level could not be conveyed in either ordinary language or imagery because their effects are not perceived by our senses.

The two most prestigious physics journals of the period were both German. Established in 1799, the conservative *Annalen der Physik* was favoured by right-wing physicists and those still attached to the classical picture of nature, publishing most of the major papers by Einstein and Planck and all of Schrödinger's. By contrast, Helge Kragh associates the *Zeitschrift für Physik*, founded in 1920, with 'liberal and avant-garde views'.[76] More or less ignored by the older generation, the *Zeitschrift* was attractive to younger physicists and others prepared to break with the established language and concepts of physics in the name of quantum discontinuity, such as Heisenberg, Pauli, Born and Jordan. Publication in these journals was affected by the international boycott of German physics in place from 1919 to 1926. With the exception of Langevin, this was nowhere observed with greater fidelity than in a France under postwar reconstruction, unusually resistant to modern physics.

The leading French academic journals carrying dispatches from the front line of *la nouvelle physique* were, in order of ranking, *Comptes Rendus*, which published short papers on every aspect of science culled from weekly *séances* held under the auspices of the Académie des sciences; the *Journal de Physique*, which was the core physics journal in France; and *Annales de Physique*, a smaller format publication echoing the title of the premier German physics journal and originally called the *Annales de Chimie et de Physique*.[77] By 1910, it was obvious that France had been unable to keep pace with other countries in developing an institutional base for scientific research. Her retrograde attitude towards new ideas in physics and the poor state into which her laboratories fell combined with economic instability in the postwar period, leading to the deterioration in the importance of French scientific publications generally. By 1934, *Comptes Rendus* was the highest ranking of the three main French physics journals in the international league at only number seven.[78]

Whatever the size of their specialist audience up to that point, a general interest in physics was stirring, often compromised somewhat by conceptual confusion alongside the obvious mathematical difficulties it brought. These specialist expositions catalogued features of an odd unseen world, where a particle could be found in two places at the same time and the identity of objects was created by the very act of observation: a world over which classical determinism no longer held sway. Revealing phenomena and behaviour beyond, and deeply incongruous with the gross reality to which the senses were subject, it is hardly surprising that most scientists, let alone lay people, were at first hard-pressed to accept or even understand the findings of modern physics.

Outside the specialist journals, those without a background in science could keep in touch with the latest developments in physical science through the translation of recent research into the language of common discourse (as far as that was possible). This task was borne by semi-popular science journals, philosophy reviews and full-length maths-free expositions by philosophers, scientists, and philosophers and historians of science. The pivotal role of Louis de Broglie in the uncovering of quantum mechanical processes played a significant part in the dissemination of quantum physics in France, and he was an important contributor to some of these publications. To give a sense of the tone of popular and semi-popular interpretations of the new physics and its epistemology in France, I will look now at two journals notable in this regard, *Scientia* and the *Revue de Métaphysique et de Morale*.

Carrying articles of a high quality on science and possessing a particularly wide reach, *Scientia* was published monthly in Milan from 1910 (fig. 17). About a quarter of its articles in the 1920s and 1930s took as their subject matter astro-, bio-, or atomic physics. Side by side with a wide variety of research papers in chemistry, biology, mathematics, philosophy, psychiatry, psychoanalysis, sociology, economics, forestry and nutrition, updates for a curious public on new directions in physics were published in several European languages in *Scientia*. However, the journal must have been aimed mainly at a Francophone audience since all its material was translated into French at the back of each number, while its book reviews and contents page were also in French.

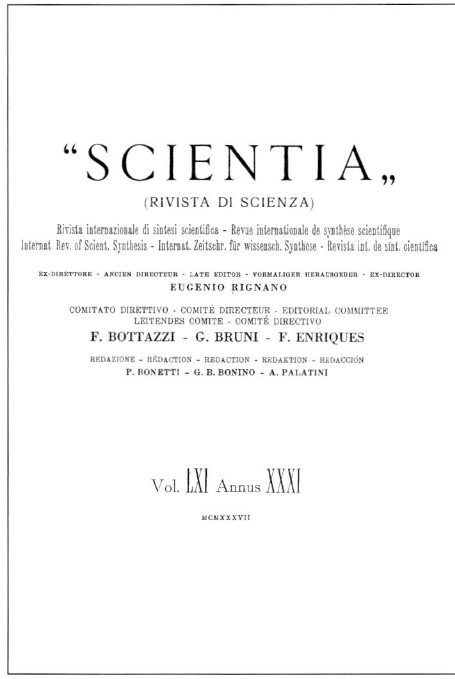
"SCIENTIA"

(RIVISTA DI SCIENZA)

Rivista internazionale di sintesi scientifica - Revue internationale de synthèse scientifique
Internat. Rev. of Scient. Synthesis - Internat. Zeitschr. für wissensch. Synthese - Revista int. de sínt. científica

EX-DIRETTORE - ANCIEN DIRECTEUR - LATE EDITOR - VORMALIGER HERAUSGEBER - EX-DIRECTOR
EUGENIO RIGNANO

COMITATO DIRETTIVO - COMITÉ DIRECTEUR - EDITORIAL COMMITTEE
LEITENDES COMITE - COMITÉ DIRECTIVO
F. BOTTAZZI - G. BRUNI - F. ENRIQUES

REDAZIONE - RÉDACTION - REDACTION - REDAKTION - REDACCIÓN
P. BONETTI - G. B. BONINO - A. PALATINI

Vol. LXI Annus XXXI

MCMXXXVII

17 Cover of *Scientia*.

Of the pioneers of QM, Planck, de Broglie, Pauli, Heisenberg and Schrödinger all submitted popular, explanatory, non-mathematical articles to *Scientia* in the 1930s, guiding readers through the range of findings made in subatomic physics over the preceding years. On another level, articles prompted by recent discoveries in physics gauge its influence on biology and chemistry. Papers on causality and determinism, equally spurred by the profound implications of QM, crop up regularly in *Scientia* authored by philosophers of science such as Léon Brunschvicg.[79] The journal also carried a text by Gaston Bachelard in 1931, a few years before he made an impact on the Surrealist circle.[80]

A conspicuous feature of *Scientia* is the amount of space it allots alongside articles by physicists and philosophers of science to writings on empiricism and logic by thinkers associated with the Vienna Circle such as Rudolph Carnap, Philipp Frank, Otto Neurath, Hans Reichenbach and Moritz Schlick. Until now its unbending positivism has ensured that the Vienna Circle has not been discussed in relation with Surrealism. However, the context of modern physics makes a comparative assessment unavoidable. Operative over a similar pre-war period to that of Surrealism (1924–38), the group was sometimes driven by similar motives to those of the Surrealists, favouring like philosophical concepts and sharing an aversion to (Heideggerian) metaphysics. News of its members' interventions at philosophy conferences and of the group's aspiration towards a single language for all phenomena reached beyond academic circles to infiltrate the Parisian avant-garde. One example of this incursion – unique and astounding for its appearance at this historical juncture – is a 1929 article by Hans Reichenbach in Georges Bataille's review *Documents*, outlining the revision of causality necessitated by modern physics.[81] As I will show, the Vienna Circle's goal of unifi-

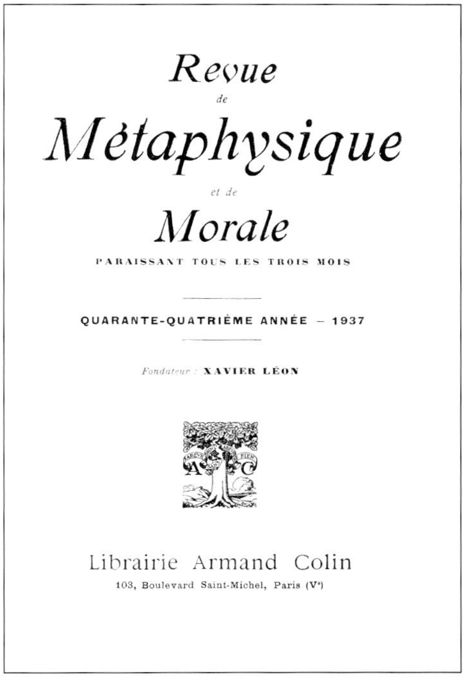

Revue
de
Métaphysique
et de
Morale
PARAISSANT TOUS LES TROIS MOIS

QUARANTE-QUATRIÈME ANNÉE – 1937

Fondateur : XAVIER LÉON

Librairie Armand Colin
103, Boulevard Saint-Michel, Paris (V^e)

18 Cover of *Revue de Métaphysique et de Morale.*

cation was one from which the Surrealists and their friends were not exempt. Its work was known to, and quoted from by Roger Caillois, René Crevel, Max Ernst and Wolfgang Paalen, while most of the scholars associated with the Vienna Circle were called upon by Bachelard in his widely read *New Scientific Spirit* (1934). Attesting to a community of thought in the 1920s and 1930s receptive to a coalition of modern physics, biology and psychoanalysis, I will demonstrate that the link between Parisian Surrealism and the Vienna Circle is a vital one to make.

Reviews of recent books by members of the Vienna Circle take place in *Scientia*, alongside those of publications by all of the physicists and philosophers of science I have named above. The journal also reviews titles by Paul Langevin and Jean Perrin, reader-friendly overviews of the new physics by English astrophysicists and theoretical physicists such as Arthur Eddington and James Jeans, and works by natural philosopher Abel Rey and Émile Meyerson, outlining the ramifications of modern mathematics and physics for philosophy. In the same pages, readers could follow the progress of psychoanalysis in reviews of new books by and on Freud, Jung, Otto Rank, René Allendy and Pierre Janet.[82]

Presenting itself as a 'Revue internationale de synthèse scientifique', the chorus of voices that rises from the pages of *Scientia* is impressive, as is the diversity and authority of its texts, frequently written by leading figures in their respective fields. Aimed at a wide audience, its up-to-date *mélange* of academic articles was meant for the 'educated reader'. Although it published nothing on art and poetry and its general readership contrasted with the more 'selective' audience sought by the avant-garde reviews that the Surrealists controlled and to which they occasionally contributed, *Scientia*'s sober scholarship appealed to radical thinkers whose publications were driven by alternative motives: as a violent alternative to received ideas; as preparation for a forthcoming revolution; as a means towards the development and expression of a doctrine; or as a medium through which to attract 'converts'. By looking now at a French journal close to *Scientia* but more oriented towards philosophy, I can show in greater depth how such publications helped narrow the cultural gap between academic and avant-garde, between science and the humanities, and between modern physics and Surrealism.

One of the journals in which *Scientia* advertised was the *Revue de Métaphysique et de Morale*, which made its first appearance in 1893 (fig. 18). Continuing four times annually up to the present day, with only a brief wartime cessation from 1939 to 1944, the *Revue de Métaphysique* was the most important mainstream French philosophy periodical of the twentieth century. Although its title suggests a combination of philosophy and theology or ethics, throughout the 1920s and 1930s it was dominated by articles on philosophy, physics and the philosophy of science under the guidance of its founder and long-time *Secrétaire de la Rédaction*, the Kantian Xavier Léon. According to one commentator, Léon conceived the *Revue de Métaphysique* as 'a site of convergence for specifically philosophical speculation and scientific and religious critical writing'.[83]

In France between the wars the two academic philosophers most concerned with the implications of Relativity and QM for their discipline were the epistemologist and (self-styled) 'anti-positivist' philosopher of science Émile Meyerson, and the neo-Kantian rationalist Léon Brunschvicg. Both published full-length works on the subject as well as being regular contributors to the *Revue de Métaphysique*; in fact, Brunschvicg was its most frequent contributor in the 1920s and 1930s.[84] Articles by one or the other appear in almost every number, dealing with issues around space, time, probability and causality in philosophy and physics. For example, Meyerson's 'Le relativisme, théorie du réel'[85] discusses favourably recent publications by Langevin and Eddington on Einstein's theory, whilst Brunschvicg's 'Le Temps et la causalité' seeks 'a renewal of the psychology of the intelligence' in the new physics.[86] The quantity of material published in the journal on Meyerson and Brunschvicg as well as by them indicates the substantial impact of modern physics on philosophy, and offers a very clear representation of the debate on epistemology in the period.[87]

Meyerson and Brunschvicg continued to write on the philosophy of modern physics for the *Revue de Métaphysique* into the 1930s, Meyerson up to his death in 1933. Until then Brunschvicg was its most regular and reliable guide, but the most prominent contributor to the *Revue de Métaphysique* through the 1930s would be none other than Louis de Broglie, the French physicist whose 'undulating electron' had cleared a path for the second period of quantum theory. Although the purpose of his first piece for the periodical in 1927 was meant ostensibly to mark the centenary of the physicist Augustin Fresnel, de Broglie used it to sketch the development of quantum theory from Planck's experiments at the turn of the century up to the present day.[88] From that point until the outbreak of the war, de Broglie contributed nine mathematics-free articles to the *Revue de Métaphysique*, which played the most important role of any single non-scientific periodical in communicating Relativity and QM and their broader consequences for philosophy to the French audience. While de Broglie's visibility in the *Revue de Métaphysique* does not entirely eclipse the importance of Meyerson's and Brunschvicg's contributions, the elevation of a physicist to principle contributor of a leading philosophy periodical emphatically confirms the extensive influence of QM beyond the field of physical science.[89]

Although the *Revue de Métaphysique* does not share the internationalism of *Scientia* (all its articles were in the French language and nearly all by Frenchmen), it did give new foreign language books a separate review section. In its review pages and overall content, the range of material covered by the *Revue de Métaphysique* was almost as extensive as that of *Scientia*. In both, as in another important contemporary periodical, the *Revue de Synthèse*, physics, philosophy, psychology and psychoanalysis are allowed close proximity – not just in the placement of contributions from divergent fields side-by-side, but also, and more importantly, in the fusion of different disciplines in individual papers.[90] Although it seems self-evident that research in one area will throw light on problems in another apparently distant one, texts combining psychology, psychoanalysis and modern physics can still surprise. Typical in this regard is the article in the *Revue de Métaphysique* by the Italian bio-psychologist Eugenio Rignano, 'La psychologie dans ses rapports avec la philosophie et avec la science', which discusses the implications of Einstein's Relativity for psychology.[91] Speaking eloquently of the gains being made at the time by subjectivity in the new physics, these bold speculations parallel those of Salvador Dalí, Georges Bataille, Carl Einstein, Henri-Charles Puech and others in the Surrealist orbit, which I address in detail later, suggesting a more complex reception of psychoanalysis (for instance) by Surrealism than previously imagined.

By the time those writers began to use ideas from modern physics, widespread interest in Relativity and QM had made its mark on the publishing industry, as indicated in Michel Souriau's 1934 review of Bachelard's *New Scientific Spirit*, noting the 'alarming quantity of publications on the new physics'.[92] An important factor in this was the series of short, inexpensive, popular science books issued by the publishing house of Hermann under the series heading 'Actualités Scientifiques et Industrielles'. The annual increase in the number of Hermann publications gives an indication of the growing market for popular science expositions: from only nine titles in 1929, its first year, it then decreased to eight in 1930, before rising to ten in 1931, twenty-one in 1932, twenty-six in 1933, and no less than seventy-eight in 1934, the year in which Souriau was writing. Established as the Depression took hold in France, by 1938 the imprint boasted eight hundred publications, nearly all in French, by mainly indigenous but also German and English physicists, chemists and philosophers of science. As one reviewer noticed, the publisher had also 'taken it upon itself to make known in France the work of the Vienna Circle'.[93] Unsurprisingly, Louis de Broglie's influence at Hermann was considerable: he is the author of the first book published in the series (*La Crise récente de l'optique ondulatoire*), while a separate sub-division 'Exposés de Philosophie des Sciences' appeared 'sous la direction de L. de Broglie'.[94]

A discussion of these journals and popular expositions begins my construction of a new context for Surrealism, bringing it closer to modern physics and mainstream philosophy than has been possible before now. Previous scholarship has isolated psychoanalysis as an external, autonomous influence, while the larger picture I have sketched here reveals a polymathic tendency in French intellectual and cultural life. As a fundamentally interdisciplinary 'school', Surrealism came to fruition in a climate of vast exploration, drawing upon and commingling physics, philosophy, psychology, biology, anthropology and sociology to rethink the relationship between external phenomena and inner being, with the twin objective of transforming the world and changing human consciousness. The best place to begin this investigation is with Surrealism's chief spokesperson and theorist André Breton. We have already seen that Breton and Surrealism were 'philosophically' well disposed towards the new physics, but I wish to pinpoint specifically where and how Relativity and QM intersect with and inform Breton's thought and writings.

2

RELATIVITY AND EPISTEMOLOGY
André Breton and Gaston Bachelard

André Breton's writings preceding the *Manifesto of Surrealism* demonstrate clearly enough a receptivity to the new ideas with which physics was challenging the long-accepted view of nature. In the first part of this chapter I discuss those early texts to show the close links between the popular understanding of concepts in Relativity and early Surrealist theory. From there, I look at how the jolt given philosophy in the late 1920s by the monumental claims made by QM were filtered into Surrealist theory in the 1930s, largely through Gaston Bachelard's writings on epistemology. My focus in that second half of the chapter will be Breton's key text 'Crisis of the Object' (1936), written in dialogue with what Bachelard called the 'nouvel esprit scientifique' in his book of the same name.[1] That exchange with Bachelard extended throughout Surrealism in the 1930s and 1940s, and constitutes a significant connection between Surrealism and postwar French philosophy, undisclosed until now.

Paul Valéry and André Breton

As a young poet, Breton was taken under the wing of Paul Valéry, his mentor from 1914 to about 1921. This relationship is an appropriate place to begin a discussion of Breton's sources and uses of modern physics. Sensitive throughout his life to the latest currents in science, it is known that Valéry felt 'atomic, quantum and nuclear physics posed scientists with entirely new problems of theory, experiment, apparatus, interpretation and rethinking of epistemological presuppositions'.[2] He visited laboratories to observe research, fostered friendships with Marie Curie, Jean Perrin, Paul Langevin, Niels Bohr and Louis de Broglie, and developed an expertise that led physicists to value his opinion and invite him to attend their gatherings.[3] In his 1946 eulogy, de Broglie accredited Valéry's attraction to science to its precise ideas and language, ambitious intellectual inquiry and metaphorical potential:

> he could find in the developments of modern physics all that his mind took pleasure in: the clarity of concepts and words, the sudden discovery of vast intellectual horizons, the neat subtlety of theories which, in revealing to us the relativity of

space and time and the uncertainties of the atomic world, have opened to human thought so many strange new perspectives.[4]

De Broglie believed that the 'triumphant conquest of the unknown' in modern science, expressed in 'felicitous metaphors', was an exemplary and principal poetic resource in Valéry's work.[5]

According to Robinson-Valéry, his enthusiasm for modern physics led Valéry to nurture 'a very warm relationship' with Einstein, 'whose Special and General Theories of Relativity powerfully stimulated Valéry's imagination'.[6] References to the Theory of Relativity from as early as 1906, the year following its formulation, are contained in Valéry's *Cahiers*, the notebooks he kept from 1894 until the end of his life.[7] However, his note reading '[w]hat I seek in physics (ex[ample]: p[rinciple] of relativity) is that which tackles the problems of the self', confirms that his interest did not reside solely in Relativity's intended application to the problems of space, time and gravitation, but with a broader definition of 'relativity' that did not exclude the relationship between mind and world.[8] That would be the case, too, for his protégé, Breton, due largely to the importance the latter had already assigned to psychoanalysis.

David Paul indicates the increasing number of references in the *Cahiers* to the geometry of Relativity between 1918 and 1922.[9] These were the years during which interest in the theory peaked, and Breton would have known from Valéry (or the newspapers) about the expedition led by Eddington and Crommelin in the service of Relativity to observe the solar eclipse, visible on 29 May 1919 from the island of Principe off the west coast of Africa and from Sobral in north eastern Brazil. Before setting off for Africa, Eddington described the problem of the eclipse as '*weighing* light'.[10] Appropriately, the fulfilment of this surreal-sounding objective coincided exactly with the writing of *The Magnetic Fields* by Breton and Philippe Soupault in Paris (fig. 19). Teeming with remarkable images assembled from spectacular juxtapositions located in exotic environments, the prose and poetry of *The Magnetic Fields* dispenses with all the usual guideposts of narrative structure and descriptive content. Breton later designated the book as the exemplar of a new, 'pure' mode of expression ('Surrealism'), because the two wrote it using the classic (though controversial) Surrealist technique of 'automatism', a practice engendering a passive state through

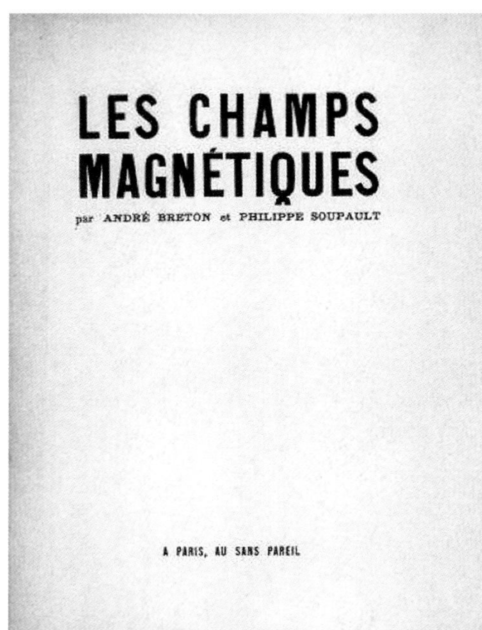

19 André Breton and Philippe Soupault, *Les Champs magnétiques* (1920).

which one is said to hear and record spontaneously the unmediated 'voice' of the unconscious.[11]

Whilst the book as a whole is rightly seen to extend the untamed wildernesses of Lautréamont's revered *Les Chants de Maldoror* (1869), the theme of my own inquiry offers a background against which to propose a new interpretation of certain passages of *The Magnetic Fields*. For although Paris is named towards its conclusion, the dazzling landscapes of the section titled 'Eclipses' suggest nothing less than the intrepid physicists' voyage over sea and land to the forests of Principe and the tropical heat of Sobral:

> The tormented sea lit up these regions; and the instantaneous vegetation and agglomerations of vapours disclosed the heavenly orbs. Celestial activity explored for the first time. The planets furtively drew near and gloomy silences thronged the stars. The hills gather round them the least delays. On the marshes remain only the memories of flights. The need for mathematical absurdities has not been demonstrated.[12]

Far from the urban spaces of modernity, 'Eclipses' is set in a 'north-eastern town' on 'mountains of sand and fossils', where the 'sea changes colour', whilst on 'blood-stained shingly beaches one may hear the tender murmur of the stars'.[13] Naturally, given its title, this section of *The Magnetic Fields* is studded with lunar, solar and stellar imagery, also evoking a sense of anticipation, where 'sentimental clouds made haste. It was the exact and foretold time', and where, 'contacted by shooting stars, the anxious eyes of women have closed for several years. They will no longer see anything but the tapestries of the June sky and the high seas; but there are magnificent sounds of vertical catastrophes and historical events.'[14]

Yet Breton and Soupault do not simply describe a land from afar. They clearly mean to tell of an expedition ('heads lowered, the smart set are leaving on a journey to the centre of the earth. They have been told about buried suns'[15]) to observe these celestial events for experimental purposes: 'The extremest adepts of these theories position themselves on the hill in front of the closing cafés;'[16] '[w]hen one turns one's back on this plain, one perceives enormous conflagrations.'[17] Not only that, but while certain images seem to suggest the photographing of the sun – '[t]he imported chemical riches burned as heavily as incense'[18] – others evoke it as unequivocally as Surrealism can allow: 'The balls of cotton managed to give birth to the sun thrown up like vomit onto the placards. The preceding reference is to chemical peculiarities, to those fine positive precipitates.'[19]

Of course, it could be just a coincidence that 'Eclipses' was written at precisely the moment Eddington and Crommelin registered the eclipse from Africa and Brazil late in May 1919. 'Eclipses' could be just about eclipses. If so, the text and event offer only a propitious twin point of departure for my own exploration of the 'new dawn' heralded by the 1919 eclipse and *The Magnetic Fields*, for they hold much in common. Soon after, Breton described automatism as 'a veritable photography of thought',[20] later referring to its users as 'modest *recording instruments*' in the *Manifesto of Surrealism*.[21] And for him and Soupault – as for Eddington and Crommelin – it must have seemed as though the eclipses they recorded – 'a hole of unknown depth that attracts the gaze of all of us' – marked an interval between the dusk of one worldview and dawn of another.[22]

News spread in the autumn that the Eddington-Crommelin expedition had recorded results squaring with Relativity's predictions, and Einstein's achievement was swiftly acknowledged by large swathes of the scientific community.[23] For complicated reasons connected to the historical context and a personality and biography surrendering themselves to a bewildering variety of 'genius' stereotypes, as well as the 'Revolution in Science' heralded by the *Times* headline of 7 November, Einstein achieved instant international celebrity as a kind of new Copernicus (fig. 7).[24] Meanwhile Relativity became the subject of popular conversation, even as its increasing mathematisation made its finer details more obscure. From that point on, the worldwide Relativity circus entered every locale of cultural discourse and production.

Valéry was quick to respond, translating a sprightly, up-to-date article on the General Theory of Relativity from the 14 November issue of *The Athenaeum*, John Middleton Murry's 'journal of literature, science and the arts', for publication only a fortnight later in the *Nouvelle Revue française*.[25] According to Marguerite Bonnet, Breton's correspondence with Jacques Doucet shows he read Valéry's text with great care.[26] Two features of Breton's work – the prevalence of metaphors drawn from physics and the (hesitant) importation into his early theoretical writings of elements derived from Relativity – were ripened by his closeness to Valéry and the esteem in which he held the older poet. Before dealing with those passages in Breton's early Surrealism that mention Relativity, I will look briefly at his use of scientific metaphors with reference to the writings of Julien Gracq.[27]

Aiming to account for what he calls the 'charged' atmosphere of Surrealism, Gracq shows that a language derived from electromagnetism pervades Breton's writings.[28] He traces the high incidence of such metaphors from *The Magnetic Fields* to *Arcane 17* (1944) with reference to advances made in scientific knowledge in the nineteenth century. Gracq's is a sympathetic study, concerned with discovering an underlying consistency in the direction taken by Surrealism cohering in this metaphorical *leitmotif*: 'it is justifiable to say that a singular thread, a conducting thread par excellence by way of Breton's subconscious traverses Surrealism, from Hegelianism to Freudianism'.[29]

Gracq's introduction of the word *induction* in the scientific context rather than the philosophical one is important for my own argument:

> one of the discoveries of modern physics which could be among those most likely to clarify by analogy the least expressible relations of a 'self' to another could very well be those phenomena of *induction*.[30]

Here, Gracq deploys 'physique moderne' in a less precise and more expansive sense than in my own study, clearly with different aims in mind. But he is onto something in this passage, which, with reference to the concept of induction, I will bring out more fully as my discussion progresses. This term was introduced into science in the early years of the nineteenth century to describe, '[t]he action of inducing or bringing about an electric or magnetic state in a body by the proximity (without actual contact) of an electrified or magnetized body' (*O.E.D.*). Gracq includes Gaston Bachelard in his discussion of Breton's metaphors of induction, yet it is the Bachelard of the 1940s that he calls upon, the author of a 'series of studies on the imagination of matter'.[31] In this book, I am concerned with the Bachelard of the 1930s, who states in *The New Scientific Spirit*, 'true scientific thought is metaphysically inductive.'[32] This quotation caught Breton's eye and is repeated in 'Crisis of the Object', which I examine below in the context of the exchange between Breton and Bachelard in the 1930s. From that perspective it will become clearer how Breton's persistent employment of the imagery of electromagnetism indicated by Gracq also reveals an intellectual kinship with Bachelard. Although Gracq interprets the term in a different way from Bachelard's, outside of what I call 'modern physics', his argument supports the contention that Breton's thought can be related 'd'un bout à l'autre' to Bachelard's.[33]

Focusing specifically on electromagnetism as a product of nineteenth-century physics, it is not surprising that Gracq does not make the link between Breton's poetic theory and Bachelard's reflections on the scientific mind of the twentieth century. However, his omission of Valéry is more perplexing, for among the first physicists that Breton's mentor alludes to in the *Cahiers* are Faraday and Maxwell, who respectively pioneered and brought to completion the classical theory of electromagnetism. Obviously, Valéry's use of metaphors drawn from electricity and magnetism influenced his young protégé Breton, and both classical and modern physics animate the writings of the two poets.[34] Because electromagnetism deals with electricity and magnetism which create fields of attraction and repulsion visible in one way or another, both probably felt it had greater poetic application than the more 'abstract' General Theory of Relativity, which deals with gravitation (where only fields of attraction are created). However, as I move on now to discuss the occurrences of Einstein's theory in Breton's early writings, I will show how they perceive that, too, under the lens of metaphor.

Dada, Breton and the 'Rise and Fall' of Relativity in the 1920s

Einstein's trip to Paris in 1922, prompting widespread, mostly underinformed discussion of physics, temporality, 'relativity', and so on coincided with the earliest texts by Breton that can confidently be assigned to Surrealism. Einstein's legend and theory went before him, and had already begun to help shape the culture of France. Looking back in 1929, an article in the *Revue de Psychologie Concrète* gives a telling account of Parisian intellectual fashions in 1920–21 following the welcome afforded the pioneer of French psychoanalysis, Eugénie Sokolnicka, by writers close to the *NRF*:

> Naturally, from this somewhat modest reception among a small circle, echoes reverberated throughout Parisian society. Soon, there was no longer a salon in the

city, nor a gathering of young intellectuals, nor barely even a café where psychoanalysis was not spoken about; it became the subject of fashionable conversation, as Einstein's Theory of Relativity had been a few months earlier.[35]

Indeed, the *NRF* was so modish that the dramatist Jules Romains could mock-wearily write off Relativity in its pages as the fashion of 'last season' in January 1922, two months *before* the exotic physicist landed in Paris![36] Gaston Gallimard's prestigious journal acted as one site for the Parisian reception of both Relativity and Dada, publishing several of Breton's early Dada texts and steering Dada towards literary respectability, notably in the prescient endorsement of 1920 by its editor Jacques Rivière, 'Reconnaisance à dada'.[37]

The 1916 date for Dada's 'official' beginnings in Zürich, General Relativity and Bohr's quantum atom (1913 really, but near enough for some) has been noticed over the years by historians of art, culture and ideas who have sought epistemological congruity between the concepts that accompanied the new physics and Surrealism's immediate predecessor. Seeking a larger context for Jean Arp's interest in chance, for instance, Harriett Ann Watts claims the physicists' 'obsession with physical indeterminacy echoes a protest of individualism, a declaration of subjectivity' shared by artists, particularly the early twentieth-century avant-gardes.[38] Outside of reminscences by Dadas like Arp and Hans Richter, which themselves offer only blurred comparison between indeterminacy in their early work and that of the physicists, Watts cannot really substantiate the 'echoes' and 'relationships' she sees between Dada art and philosophy and QM.[39]

Her main reference is Lewis S. Feuer, who proposes at length what he calls 'isoemotional' parallels between the Dadas and the physicists of QM in his intellectual history *Einstein and the Generations of Science*, though tangible connections are again noticeable by their absence.[40] Feuer considers Werner Heisenberg a kind of receptacle of youthful post-First World War radicalism, and sources the indeterminacy of HUP to Heisenberg's residence in revolutionary Munich in 1919.[41] Not unreasonably, he infers a cult of indeterminacy, linking Dada's fascination with chance and its wartime lunacy in the safe haven of Zürich with Heisenberg's response to the chaos of the war. But Feuer makes a mess of his 'shared episteme' argument by adding to the mix every other strain of the European avant-garde, from Picasso's rejection of perspective around 1907 to Surrealism circa 1929:

> The efforts of representational art to conform to the familiar patterns of the common sense, visual world were to the Surrealist what models of the classical quantum physicists were to Werner Heisenberg and the new generation; the latter repudiated the efforts to picture the electrons as moving in planetary orbits around the atomic nucleus. André Breton called similarly for the liberation of artists from the limitations of the laws of perspective, gravity, and extension . . .
>
> As Heisenberg was disaffected from the everyday realities of material objects and longed to see them superseded by spiritual mathematical forms, that is, Platonic ideas, so the Surrealist strove to 'burst the bonds of reason of narrow rational-

ism', to permeate the world of reality with that of dreams to engender a 'super-reality.' . . . 'The simplest Surrealist act', wrote Breton, 'would consist of going out into the street, revolver in hand, and firing at random into the crowd as long as one could.' For in generational movements among artists, unlike those among scientists, the underlying emotions of rebellion state themselves in the rawest, crudest, most manifest, and least sublimated way.[42]

Worth quoting at length because he gets Heisenberg and Breton on the same page, Feuer's separate development, shared outlook argument is nevertheless complicated by a more obvious one that should accompany and dilute it, detailing the cultural diffusion of ideas and empirical connectivity between science and the humanities. Upon slightly closer examination, that is to say (and not detaining ourselves with further complications such as histories of chance and acausality in *pre*-modern art and physics), one will discover the Dadas had direct access to modern physics, which at least complicates Feuer's claim of their supposedly *distant*, 'isoemotional' links with the physicists. This is certainly true of the Surrealists who *consciously* manipulated ideas from Relativity and QM, as I show.[43]

Richard Sheppard has laid the foundations for a future study of Dada and modern physics.[44] Finding room for Einstein and Heisenberg in the portentously erudite, encyclopedic context he creates for Dada – ranging, absurdly, from Heraclitus to Humpty Dumpty – Sheppard touches on the reception of Relativity and particle physics by Dadas such as Johannes Baader, Hugo Ball, Hannah Höch, Richard Huelsenbeck and Theo Van Doesburg.[45] Until further notice though, histories linking Dada to Relativity and QM lack empirical purchase, sometimes falling into desperate-sounding manoeuvres to do with Einstein probably frequenting the same Zürich bars as the Dadas.[46]

Weightier ballast is available for historians keen to make such links in the equally speculative efforts of Dada's contemporaries and sympathisers to make connections with physics. In 1921, for instance, Belgian playwright and Dada sympathiser Paul Neuhuys linked Dada in France to the writings of Henri Poincaré and Henri Bergson in a lengthy review of Jean Cocteau and Dada in the avant-garde journal he co-edited, *Ça Ira!*[47] Neuhuys locates Dada in the context of what he calls Poincaré and Bergson's 'transcendent investigations of the human mind', adjacent to the 'relativism' of Euclidean geometry.[48] Speaking of Dada as an 'intuitive philosophy' in the light of Bergson, he concludes that its task is 'to liberate itself from relative concepts of human reason'.[49] Neuhuys cleared the way for Clément Pansaers in the special number of *Ça Ira!* dedicated to Dada later that year, who typically and mistakenly confuses Einstein's Relativity with cultural relativism, endeavouring to chart the development of his own version of Dada partly against the cultural coordinates put in place by Einstein who, he says, 'definitively established his philosophy of Relativism' (*sic*) around the beginning of the war.[50]

Breton and his right-hand man in Surrealism Louis Aragon were only marginally less mechanical in their youthful quest to be absolutely modern. Steering clear of Dada by 1922, they single out Einstein and Poincaré along with Freud in a document composed for their employer at the time, the dress designer and collector Jacques Doucet, that was intended to establish the foundations for an ideal library. Where Kant turned to Newtonian physics to create a new form of reason by breaking loose of Leibnizian rationalism, they tell Doucet, a similar sensibility creates continuity today, with 'Poincaré, Freud, Einstein representative of the successive stages of this idea'.[51]

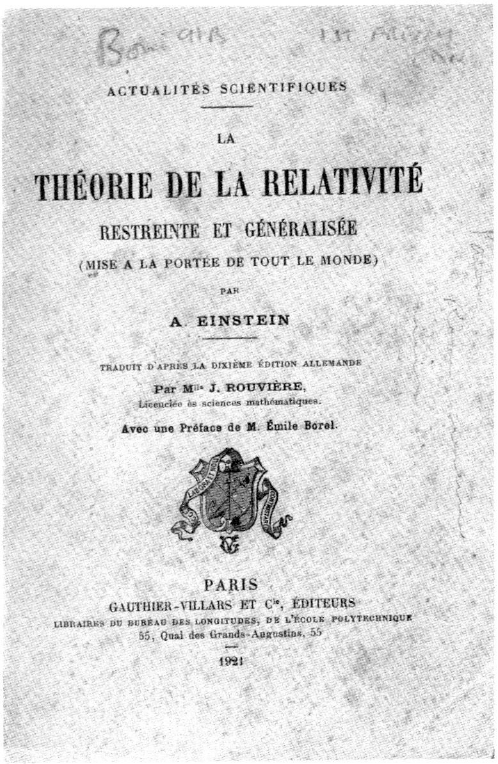
ACTUALITÉS SCIENTIFIQUES

LA

THÉORIE DE LA RELATIVITÉ

RESTREINTE ET GÉNÉRALISÉE

(MISE A LA PORTÉE DE TOUT LE MONDE)

PAR

A. EINSTEIN

TRADUIT D'APRÈS LA DIXIÈME ÉDITION ALLEMANDE

Par Mlle J. ROUVIÈRE,

Licenciée ès sciences mathématiques.

Avec une Préface de M. Émile Borel.

PARIS

GAUTHIER-VILLARS ET Cie, ÉDITEURS

LIBRAIRES DU BUREAU DES LONGITUDES, DE L'ÉCOLE POLYTECHNIQUE

55, Quai des Grands-Augustins, 55

1921

20 Albert Einstein, *La théorie de la Relativité: Restreinte et Généralisée*, trans. J. Rouvière, Paris: Gauthier-Villars et Cie, 1921.

Whatever the jaded view of Relativity from the *NRF* in early 1922, Michel Biezunski observes that in 'the literary chronicles of the time, translations and expositions of Einstein's work as well as Freud's, crowd the review columns'.[52] However maladroitly processed, this information gave a shot in the arm to the most world-weary of French journalists, intellectuals and salon-goers. Breton was no exception, for the vogue for Einstein as well as Valéry's lingering testimony and its stubborn rejection are plainly visible on the surface of his text 'The Disdainful Confession'.[53] A dry run for the Surrealist iconoclasm that would become familiar over the years, this essay was first published in 1923, composed from autobiographical musings written for Doucet over the period Breton's friendship with Valéry began to falter in 1920–21.[54] Accordingly, Valéry is named (alongside Apollinaire) among the former greats replaced by the pantheon of proto-Surrealists – Rimbaud, Lautréamont and Jacques Vaché – who put an end to poetry both as literature and as a career option, favouring instead the street and lived experience.

Here, modern physics can be seen as active in the first stirrings of Surrealism when, with an inexpert deployment of the new language of 'relativity', Breton playfully asserts that age might bring 'a certain displacement of one's centre of gravity'.[55] This is, he claims, a 'completely relative equilibrium' entirely dependant upon the prevalent conditions in which experience is formed.[56] Like Valéry, Breton from the outset uses the voguish term 'relative' in a sense less concerned with the external functioning of nature than with the problems of the *moi*. That is, as a stimulus to posing questions concerning the internal relationships of the 'fragmented self' (posited most urgently by Rimbaud and Freud) and of the fixing of habits by age and experience.

Breton owned the 1921 French translation of Einstein's popular exposition *Relativity: The Special and the General Theory*, which alongside his access to Valéry would have served as a solid introduction to Relativity (fig. 20).[57] However, his account of

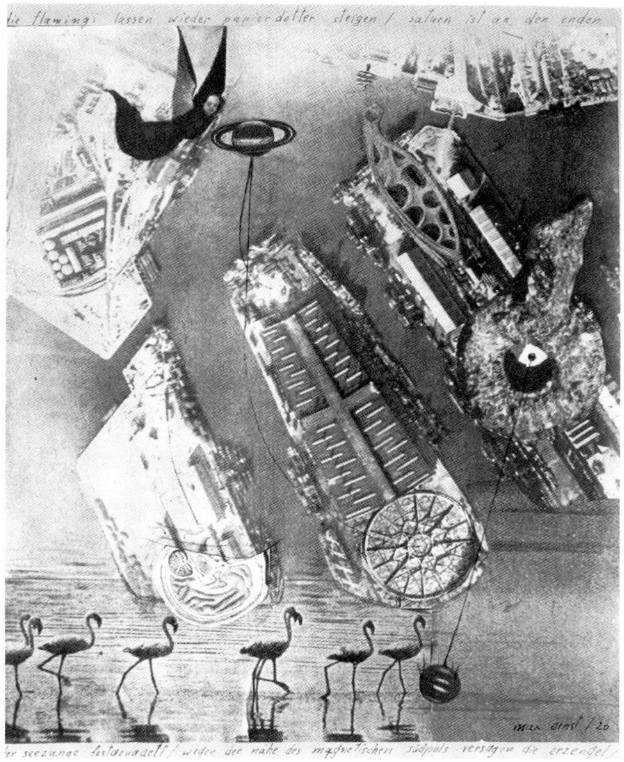

21 Max Ernst, *The Flamingos* (1920). Photographic enlargement of a photomontage with watercolour and ink mounted on paperboard, 29.3 × 23 cm. Private collection, Paris.

the relative simultaneity of events in space-time in 'The Disdainful Confession' is personal and whimsical:

> 'One event can be the cause of another only if they can both be brought about in the same point in space,' Einstein tells us. That is what I have always thought, generally speaking. I deny things as long as I touch the earth; I love at a certain altitude; what would I do even higher? Still, in any one of those states, I have never passed twice through the same point, and saying I touch the earth, at a certain altitude, higher, I am not duped by my images.[58]

Breton is again concerned more with states of mind than with positions in space in this engagingly frivolous appeal to Einstein's theory, begun when he was still participating in Dada activities. If it is true, as Gérard Legrand claims, that in the 1920s, 'Breton was a party to the general misinterpretation by intellectuals of the Relativity of the physicists', then his tone in 'The Disdainful Confession' acts as a warning not to mistake this for an earnest or literal rendering of the new physics of space-time.[59]

This is confirmed in a letter he sent to André Derain around the same time, where Breton writes of his interpretation of Max Ernst's recent work: '[s]ometimes I claim that we owe a pictorial language to him in accordance with the new conception of things advanced by Einstein.'[60] Breton is alluding here to the short text he wrote to accompany Ernst's first exhibition in Paris in May 1921 (fig. 21), in which his handling of Relativity continues to be determined by the spirit of Dada.[61] Voicing his admiration for the imagery of Ernst's collages and paintings, Breton compares their effect on the viewer of *dépaysement* (disorientation) – a category that would become fundamental to Surrealist poetics – to Einstein's description of the 'separate realities' experienced by observers moving relative to each other:

> Belief in an absolute time and space seems about to disappear. Dada does not claim to be modern. It judges futile, as well, the subjection to laws of a given perspective . . . But the marvellous power to bring together, without leaving the field of our experience, two distant realities and from their reconciliation to draw a spark; to put within the grasp of our senses abstract figures endowed with the same intensity, the same relief as the others; and, by depriving us of our frames of reference, to disorient us within our memories, that is what preoccupies Dada for now.[62]

Here, the production of the emotive spark – the central mechanism of what would become Surrealism's theory of the poetic *merveilleux* – is identified at the time of its first brief flaring with the new physics of space-time.[63]

In a vein recalling the ultra-modernism of Marinetti's 1909 Futurist Manifesto, Breton goes on to say in 'Max Ernst' that cinema, modern science and technology have demanded the recognition of new temporalities outside of the lived one, adding, '[w]e eagerly look forward to seeing what this "relative time" we have heard about might be.'[64] Confirming that Breton, like Valéry, believed ideas deriving from the new physics might be applied to ontological problems, he asks (insightfully, given the astounding assertions about to be made by quantum theory) whether the 'fragmentation' of time announces that mankind is preparing 'someday to escape the principle of identity'.[65] Breton saw Ernst as an artist of 'infinite possibilities' whose imagery gave expression to the more expansive reality newly discovered by modern physics, soon to find its full poetic expression in Surrealism.[66]

For the most part, Relativity left Breton's early writings at around the same time Einstein exited Paris, though Breton was not the only one to lose interest in space-time and move on to other business. In October 1924, the very month Breton's *Manifesto of Surrealism* saw the light of day, an anonymous reviewer wrote during the course of comments on Austrian physicist Hans Thirring's *L'idée de la théorie de la Relativité*: 'The ideas developed by M. Thirring are today well known.'[67] As massive an achievement as Einstein's theory was, it could be faithfully chronicled in a single book. Soon put to one side by the bemused or part-comprehending layperson, it could be assimilated quite swiftly along with its implications by the sufficiently unprejudiced expert. Its lack of volume and conceptual diversity in comparison with psychoanalytic theory went hand in hand with a far narrower potential for application to the humanities and culture in general. In spite of Relativity's vast significance for physics and philosophy, diverse cultural relevance was never intended by Einstein, of course, unlike the writings of Freud, which extended into anthropology, religion, literature, art and so on.

For Surrealists like Breton and Aragon, who shared an interest in Relativity, the 1920s ended as they began, with facile comparisons between Einstein and Freud.[68] Aragon's is fairly extensive, though, forming one of the cornerstones of his 1928 *Treatise on Style*, his Breton-pleasing, scatalogical and eloquent poison-pen letter to all things French, literary, cultural, bourgeois, religious, political and military (all things, in fact).[69] Certain passages in the book tell us a great deal about the nature of the poetic attraction (or repulsion) held by Relativity for the Surrealists, though the main point of Aragon's rant is its presentation of Surrealism as a valid means of attaining a marriage of individual style and content. This is brought about in *Treatise on Style* through a fragmented, discordant, sometimes reckless syntax (with typos and misspellings retained in the final text) gridded onto the author's bitter, acerbic perspective on social and cultural attitudes.

Principal among Aragon's complaints about contemporary culture is the one about the appropriation, malformation and subsequent dilution of revolutionary ideas by the mediocre. Hero to the Surrealists and recently claimed by the poet and diplomat Paul Claudel for Catholicism, Arthur Rimbaud's name was mandatory in this context but was soon joined by those of Freud and Einstein:

> These last years have in fact brought us, in their flow of insanities and events, two enormous statues of specialists, which is the kind of statue most likely to be carved up with the names of hoods who want to link their destinies to the kite tails of a more illustrious person at all costs . . . [T]he wily fellow who looks to the shoemaker for the confirmation of his calculations for making eyeglass lenses, uniting the blessed ignorance of the carp to a flair for fashion and the cleverness of camouflage, has no trouble fooling himself before fooling anyone else, and he firmly believes that the rope will hold and everyone will be impressed. I throw these two irons into the fire: Freud, Einstein.[70]

Aragon goes on to decry the vulgarisation of psychoanalysis in popular culture, upbraid contemporary writers for their pretentious borrowings from Freud, and eviscerate critics who claim the writings of their authors predict the discovery of the unconscious. Warming to his theme, he introduces the revered but nonetheless half-digested, cartoon Einstein of the dailies:

> Here is the Naughty Clockmaker, the Dishonest Calendar, the Stammering Genealogist, the Mechanic of Irreconcilable Voyages, the Mohammed of Mathematics, Einstein himself, crowned with couscous and casters. He throws families into disarray, causing those little problems that leave us wondering about Papa's age, and Mama begins to run in order to stay young, and babies no longer suck when they're supposed to: anarchy, anarchy, all is anarchy! Could Einstein be the Antichrist, which would simplify his future dialectical canonification?[71]

Censorious of those who swallow the medium minus the message, Aragon argues that certain metaphors used in the explication of Relativity – 'these swimmers across the Milky Way, these travellers on cannon balls, these ships as long as a day without bread'

– might have a poetic flavour but their kernel of value dissolves with their removal from their original context.[72] Expressing agreement with Wyndham Lewis for the one and only time in his career, Aragon argues that the form of those metaphors was determined 'to propagate new physicomathematical notions which had to get one's attention.'[73] Abandoning personal style, writers short of inspiration decontextualise scientific and poetic metaphors, freeing them of their radical message about the world and man and leaving only the husk of an image. In this way, writes Aragon, '[t]here is no difference here between the fate of Einsteinian imagery and Rimbaldian imagery.'[74]

From Aragon's uninhibited tirade in *Treatise on Style* we begin to get an inkling of why Breton gave up his early dabbling in Relativity. Indeed, it turns out to be Breton's discarded mentor Paul Valéry who Aragon lays into next, given a drubbing for poetic sophisms culled from Relativity. In passing, but not by coincidence, Aragon vents his spleen on one 'M. Fabre', recommending a 'meticulous study' of his book on Einstein, which 'would tell us quite a lot about the misuse of images'.[75] By this, Aragon means the polymath Lucien Fabre, whose 1921 book on Relativity was reliable enough to be prefaced by Einstein himself.[76] Fabre did not exactly endear himself to the Surrealists by accepting the Prix Goncourt in 1923. Along with his closeness to Valéry (who wrote the *avant-propos* to Fabre's 1924 volume of poetry *Connaissance de la déesse* [*Knowledge of the Goddess*]), this gave Aragon further excuse to lump him with the others he despised for their 'disconcerting daily use of the only respectable thinking in the modern world'.[77]

Aragon's hatchet job of Einstein's literary 'abusers' was thorough and contemptuous. Breton washed his hands of Relativity promptly, mentioning Einstein and Freud in the same breath just one more time before the end of the decade in the *Second Manifesto of Surrealism*. Attempting to dilute the doctrinaire aspects of Surrealism for the benefit of a French Communist Party anxious at the thought of Fifth Columnist Surrealists in its midst, Breton argues disingenuously that the epithet 'Surrealists' designated less a statement of principle than a benign title, 'just as the title of "relativists" might have preceded the followers of Einstein or "psychoanalysts" those of Freud'.[78] Although Einstein's great fame was undiminished, the first flush of excitement at Relativity had long tailed off by 1929, and referral to Einstein and Freud as the two great living pioneers of modern thought had stubbornly taken root. We saw that, too, in Breton's early theoretical writings on poetic inspiration, for which the conceptual language of Relativity acted as a significant spur, however malformed by its Dadaistic rendition.[79] But Breton and the Surrealists' enthusiasm for modern physics was fired again around the middle of the 1930s. This was largely due to a single author, Gaston Bachelard, and again had implications for Breton's theory of poetry and art, forming a kind of continuity with these early writings.

Science and Objecthood: Bachelard and Breton

André Breton's 1936 essay 'Crisis of the Object' has always seemed something of a curiosity. In part because of the distaste Breton expresses elsewhere for abstract art,

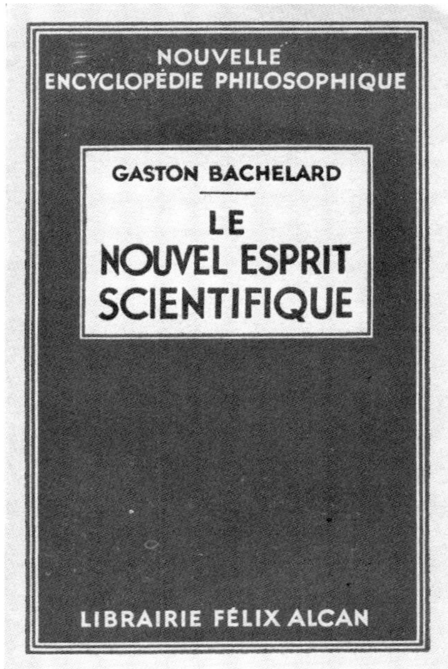

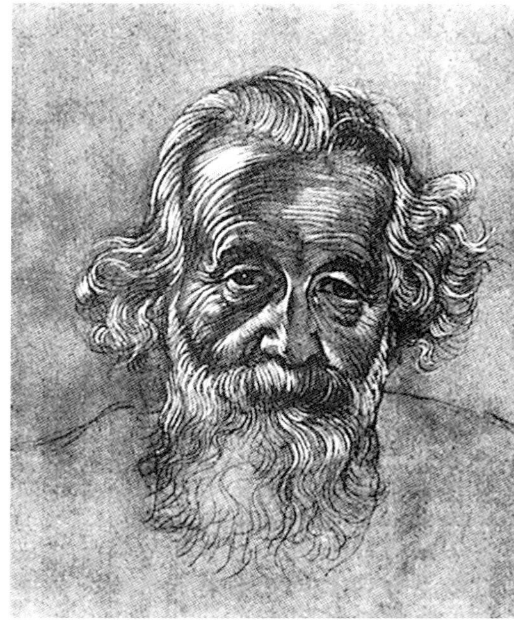

22 (left) Gaston Bachelard, *Le Nouvel Esprit scientifique* (1934).

23 (right) Hans Bellmer, *Gaston Bachelard* (1957). Pencil and gouache on paper, 60 × 40 cm. Present whereabouts unknown.

but mainly because of his attraction to the irrational in all its forms, it seems out of character for him to give over a prominent part of a major statement to a discussion of mathematical demonstration models and the relations between art and science. Although it has drawn its share of critical commentary, analysis of Breton's reflections on geometry, physics and mathematics, which dominate the piece, have been hesitant at best.[80] My argument, of course, will be that the mystery surrounding Breton's most comprehensive statement on the Surrealist object is cleared up by my focus on science and, specifically, mathematised physics and its philosophy, returning the text to its original context.

First I will look at the main inspiration behind 'Crisis of the Object', Gaston Bachelard's 1934 publication, *The New Scientific Spirit* (*Le Nouvel Esprit scientifique*), read by every individual in the Surrealist group and its orbit (fig. 22). One of my main aims is to demonstrate the significance of *The New Scientific Spirit* for Breton and the Surrealists, who developed Bachelard's concept of the 'epistemological break' well before it took on a central function for postwar French philosophy. Rather than waste space with a general introduction to Bachelard's work, which would be more pertinent to the Bachelard scholarship, I will interpret here his connection with the avant-garde, and discuss only those elements of *The New Scientific Spirit* pertaining to 'Crisis of the Object'.

The son of a cobbler, Bachelard worked for the postal service before embarking on an outstanding academic and publishing career (fig. 23). His early training was in physics and chemistry and he taught at the Collège de Bar-sur-Aube from 1919 to 1930. Although his grounding was a scientific one, Bachelard supplemented this knowledge with further studies in philosophy that led to the publication of his doctoral thesis in

1928, completed under the supervision of Léon Brunschvicg and Abel Rey. When *The New Scientific Spirit* appeared, Bachelard had held the Chair of Philosophy at the University of Dijon for four years.

Perhaps because of a background unusual for a successful academic, Bachelard had a maverick streak, and neither his thought nor career developed in the way his institutional position decreed. His sympathies did not lie with the conservatives in science and philosophy, who accepted Relativity only if it could be seen as a continuation of the classical physics of Newton and Descartes. Teresa Castelao lists Meyerson, Brunschvicg, Abel Rey and André Lalande as representatives of the institutionalised Cartesianism dominating the philosophical mainstream in France between the wars.[81] She argues that the resistance towards German thought characteristic of the powerful current of traditionalism in French academic life in the 1920s was shared neither by radical thinkers outside of academia nor by Bachelard:

> a constituency of French avant-garde intellectuals, including scientists and philosophers, greeted the progress of German physical theory with enthusiasm. Bachelard's writings on epistemology belong to this body of thought, undoubtedly on the margins of institutional philosophy.[82]

Although the division between academic and avant-garde was not as clear-cut as Castelao suggests – all of the aforementioned traditionalists held positions on the Comité de patronage of *Recherches philosophiques*, a journal with radical pretensions which I discuss later – Bachelard sympathised with Surrealism and was keen to situate his philosophy of science with the radicals.

In 1934, into the increasingly Hegelianised atmosphere of French intellectual life, Bachelard's *New Scientific Spirit* launched a different kind of binary model to Hegel's dialectic, existing between 'rationalisme' and 'réalisme'.[83] Introducing his study, Bachelard formulates this by asserting that in scientific activity, '[*e*]*xperimentation must give way to argument, and argument must have recourse to experimentation*',[84] where experiment, like the 'pittoresque', belongs to the category of realism, and argument, like the 'compréhensible', pertains to that of rationalism.[85] In contemporary science, he writes, 'the orientation of the epistemological "vector" seems clear. It surely points from the rational to the real.'[86] Since experimentation is the confirmation of reasonable intuition, then the hypotheses dreamt up by the scientist are themselves real.

Bachelard's model for *rationalisme* is non-Euclidean geometry, which crops up again and again in his writings as a kind of ideal manifestation of the powers of the imagination. Non-Euclidean geometries emerged from a centuries-long debate over the validity of Euclid's geometry of two and three dimensions, foundational both for Kantian epistemology and classical physics. Euclid's fifth postulate, which is usually stated '[t]hrough a given point on a plane can be drawn only one straight line parallel to a given line', was long thought unprovable and, therefore, as a temporarily shuttered window onto a possible non-Euclidean geometry. Its repudiation was achieved by three extraordinary mathematicians almost simultaneously, working independently of one another around 1826–32, Carl Friedrich Gauss, Nikolai Lobachevsky and János Bolyai. In non-Euclidean geometries, an infinite number of lines can be drawn through a given point, none of which will intersect a given line. In Lobachevsky's two-dimensional 'hyperbolic' geometry, the sum of the corners of a triangle add up to less than 180 degrees and can be visualised on a 'pseudosphere' or 'surface of constant negative curvature'. A few years later in 1854, Georg Friedrich Bernhard Riemann created another form of non-Euclidean geometry ('elliptical'), which Einstein would call upon to hypothesise the notion of a finite yet unbounded universe. By the end of the nineteenth century, all manner of geometries were available for a great variety of spaces, extending, in turn, the range of surfaces developed to model such spaces to include forms like doughnuts, tunnel interiors and even a space like a venetian blind.[87]

Using the example of non-Euclidean geometry, 'subsuming Euclidean and non-Euclidean geometry alike in an overarching "pan-geometry"',[88] subsequently creating a '[g]eneralized geometry',[89] Bachelard argues in *The New Scientific Spirit* that the particular rationalism with which he is concerned, by which imagination imposes its will on a world previously accessed only empirically by means of the senses, 'can no longer be a closed rationalism'.[90] Equally, the simple macrocosmic systems and groups of facts put in place by Newtonian physics, and even the meaning of established concepts such as 'mass', 'velocity' and 'location', had been analysed and revealed as inherently complex and contradictory in light of the phenomena encountered and solutions offered by subatomic physics.[91] Bachelard declares that modern scientific research must 'amplify what is revealed beyond appearance',[92] 'resolving' philosopher Charles Renouvier's 'second metaphysical dilemma' by arguing that its unknowable subject could be 'materialised' – made into a 'substantialized substantive' – by the constructing epistemology that had enabled modern physics to triumph.[93] Unsurprisingly, given how rarely modern scientific truths agree with those of the past, Bachelard calls on Nietzsche, that most 'unscientific' and unsystematic of philosophers, concerning the assertion of facts that go against the grain of Cartesian 'common sense': 'As Nietzsche says, everything crucial comes into being only "in spite".'[94]

Bachelard claims (and this is the point) that when concepts created by previous systems of knowledge became exhausted, new ones emerge out of the ensuing crisis:

> I shall insist that such new doctrines as non-Euclidean geometries, non-Archimedean measures, non-Newtonian mechanics (associated with the name of Einstein), non-Maxwellian physics (associated with the name of Bohr), and non-commutative (or non-Pythagorean) arithmetic [associated with the name of Heisenberg] are valuable precisely because they confront us with suggestive dilemmas.[95]

This comprehensive revision, or 'philosophy of non-', is what gives Bachelard's thought its revolutionary verve. For if each new scientific generation is driven by the dynamic of the real and rational to create and apply new epistemological models, then the *nouvel esprit scientifique* is animated by an iconoclastic will:

> In order for the bonds of Euclideanism finally to be broken, physicists had to confront a new set of problems with total independence of mind, after liberating themselves from Euclidean presuppositions by engaging in what one might call a kind of psychoanalysis . . . Psychologically, the modern physicist is aware that the rational habits acquired from immediate knowledge and practical activity are crippling impediments of mind that must be overcome in order to regain the unfettered movement of discovery.[96]

Typically and demonstratively disrespectful of convention, Bachelard uses the word 'psychoanalysis' just outside of its Freudian signification, meaning here a dissection, investigation, and/or rejection of constraining routines, beliefs and hypotheses. Bachelard probably read Freud but he is indebted here at the most to the concept of 'repression', and his own use of the term 'psychoanalysis' is really meant to introduce a methodology of the imagination wherein the limitations of rationalism are spontaneously routed. All the same, his words held an obvious appeal for the Surrealists. The same goes for his Mallarméan eulogy to the 'beautiful symbols of mathematics, wherein the possible and the real are joined'.[97]

Breton and the Surrealists – and years later, every French philosopher of note from Louis Althusser to Michel Serres – were particularly taken by Bachelard's new dynamic history of science. In place of the smooth, logical continuity won by an assured, self-aware discipline, Bachelard posits one driven by revolutions, revisions and reconstructions, invited by breaks and aporia opened up by the inadequacy of the reason of the day when confronted by unimpeachable mathematical or experimental evidence. Consequently, Bachelard denies that modern physics grew out of Newtonian physics, observing instead the necessity of the adaptation of previous ideas, or as he puts it, the '*envelopment* of old ideas by new. Intellectual generations', he continues, 'succeed by successive assimilations.'[98] This important insight is elaborated as follows:

> Accordingly, it seems to me that in the interval between the time when one scientific object disappears and the time when a new reality is constituted, there is room for a nonrealist mode of thought, a mode of thought sustained by its own dynamic thrust. Such an interval, critics will say, amounts to no more than an ephemeral moment, which hardly counts alongside periods of firm scientific accomplishment, sound explanation, and solid teaching. Yet it is in these brief moments of discovery that we must learn to grasp the decisive turns in scientific thinking . . . For it is at such times that science must suddenly confront contradictions in the evidence and doubts about its fundamental axioms; it is then that new *a priori* syntheses are proposed, syntheses that, like de Broglie's ingenious theory, create a new reality alongside the old.[99]

The *synthèse géniale* of de Broglie's referred to by Bachelard is, of course, his application of wave-particle duality to all physical phenomena in his groundbreaking doctoral thesis of 1924. Oddly enough (since the opportunity was there), neither Breton nor Bachelard claimed that the near-simultaneous appearance in that year of Breton's *Man-*

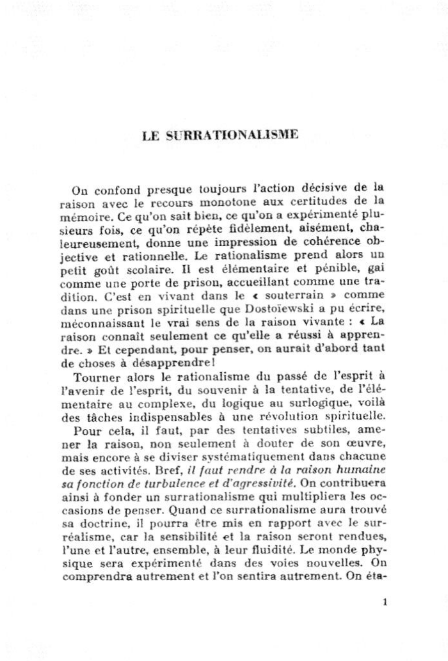

LE SURRATIONALISME

On confond presque toujours l'action décisive de la raison avec le recours monotone aux certitudes de la mémoire. Ce qu'on sait bien, ce qu'on a expérimenté plusieurs fois, ce qu'on répète fidèlement, aisément, chaleureusement, donne une impression de cohérence objective et rationnelle. Le rationalisme prend alors un petit goût scolaire. Il est élémentaire et pénible, gai comme une porte de prison, accueillant comme une tradition. C'est en vivant dans le « souterrain » comme dans une prison spirituelle que Dostoïewski a pu écrire, méconnaissant le vrai sens de la raison vivante : « La raison connaît seulement ce qu'elle a réussi à apprendre. » Et cependant, pour penser, on aurait d'abord tant de choses à désapprendre!

Tourner alors le rationalisme du passé de l'esprit à l'avenir de l'esprit, du souvenir à la tentative, de l'élémentaire au complexe, du logique au surlogique, voilà des tâches indispensables à une révolution spirituelle.

Pour cela, il faut, par des tentatives subtiles, amener la raison, non seulement à douter de son œuvre, mais encore à se diviser systématiquement dans chacune de ses activités. Bref, *il faut rendre à la raison humaine sa fonction de turbulence et d'agressivité.* On contribuera ainsi à fonder un surrationalisme qui multipliera les occasions de penser. Quand ce surrationalisme aura trouvé sa doctrine, il pourra être mis en rapport avec le surréalisme, car la sensibilité et la raison seront rendues, l'une et l'autre, ensemble, à leur fluidité. Le monde physique sera expérimenté dans des voies nouvelles. On comprendra autrement et l'on sentira autrement. On éta-

1

24 Gaston Bachelard, 'Le Surrationalisme' (1936).

ifesto of Surrealism and de Broglie's theory could be explained by recourse to this 'interval' (or *rupture épistémologique* as it would soon be known), separating an old way of thinking about the world from a new one.

However, Breton was undoubtedly nudged into action by Bachelard's complaint that 'we are now at a stage in human intellectual development where we face a real, and useful, crisis in our understanding of the physical world, but there is no corresponding crisis in our philosophical culture.'[100] Taking his cue, Breton went on to mobilise the notion of the breakdown of worn-out structures of knowledge to suggest, in 'Crisis of the Object', the emergence of parallel epistemologies in the histories of art, poetry and science. From this point onwards in fact, Breton's terminology is marked by Bachelard's thought, which subsequently extends throughout Surrealism. Particularly prevalent in Breton's writings in the following years is Bachelard's phrase 'open rationalism' ('*le rationalisme ouvert*'), used towards the end of *The New Scientific Spirit*.[101] This formulated the proposal made in the book's opening pages that rationalism 'should no longer be a closed rationalism', meant to describe a mode of thought unindebted to monolithic systems, favouring instead unexpected relationships and chance configurations.

This expression appears alongside the neologism '*surrationalisme*', introduced by Bachelard into his writings in the brief 1936 tract of that name in the only issue of the 'post-Surrealist' journal *Inquisitions* (fig. 24).[102] Bachelard took the term from an article by the German psychologist Richard Müller-Freienfels carried in *Recherches philosophiques*, an important philosophy journal of the thirties that Bachelard was involved with, which I discuss at length later in relation to Georges Bataille and Arthur Eddington.[103] Coined by Müller-Freienfels to denote a philosophical approach inclusive of both reason and the instincts, 'surrationalisme' was waved before readers of *Recherches philosophiques* again by Michel Souriau before Bachelard steered it towards his own purposes.[104] Although the word does not appear in *The New Scientific Spirit*, it can be viewed as an umbrella term and the three-part text 'Le Surrationalisme' as a kind of précis of Bachelard's thoughts on scientific and poetic inspiration up to that point.

I touch on Bachelard's text again in my chapter on politics as it has some relevance there; here I will outline the elements recapitulating or giving nuance to the arguments of *The New Scientific Spirit* relevant to Breton in 'Crisis of the Object'. Bachelard's article opened the first number of *Inquisitions* in June 1936, the month before 'Crisis of the Object' first appeared in the expensive art periodical *Cahiers d'Art*. Breton's past and present associations with all the members of the editorial board of *Inquisitions* – Louis Aragon, Roger Caillois, Jules Monnerot and Tristan Tzara – assured that he could read 'Le Surrationalisme' in advance of its publication. He probably received a copy from Nicolas Calas, to whom he was close, or one of the other former or peripheral Surrealists who helped make up the fifteen-strong Groupe d'Études pour la Phénoménologie humaine that backed *Inquisitions*.[105]

Bachelard appeals in 'Le Surrationalisme' for 'a spiritual revolution' that will reorient man, freeing him from his role as an inheritor of ideas validated only by habit and repetition, and placing him in a position of anteriority, actively recasting reason in an experimental mode.[106] This much is familiar from *The New Scientific Spirit*, as is Bachelard's attack on 'closed' or, here, 'ossified rationalism', along with his demand that it be critically scrutinised and discarded if found wanting.[107] Again, non-Euclidean geometry acts as the means of its overthrow in the priority that Bachelard gives to the scientific dialectic of Lobachevsky over the 'metaphysical' one of Hegel, both of which appeared around the beginning of the nineteenth century, a coincidence he pointed out in *The New Scientific Spirit*.[108] By contrast with Hegel's dialectic, he claims, Lobachevsky's was linked to a particular idea/event in the history of thought – '[it] dialecticised the idea of the parallel',[109] and in doing so established the 'liberty of reason'.[110] Aware by now of Surrealism, Bachelard writes that, once established, this new *surrationalisme*, 'can be allied with Surrealism'.[111] For now, he laments the misuse of the potentially revolutionary, new non-Euclidean geometries by narrow academic minds, giving nothing less than an invitation to the Surrealists to appropriate this fundamental resource of physical science: 'Unfortunately, no one has made positive, real, Surrealist use of this liberty.'[112] The '*revolution of reason*' promised by *surrationalisme* could be started by making those geometries known to a wider audience outside of academia where their true *subjective* potential had been neglected, he suggests.[113]

Bachelard's declamatory note and cry for the unimpeded exercise of the imagination in 'Le Surrationalisme' are transparently indebted to Surrealism's rhetoric of freedom, yet when he urges his readers to leave the well-trodden paths of thought, he indicates Nietzsche's method as exemplary, as he had in *The New Scientific Spirit*. As an exploratory forerunner testing the very limits of reason, 'Nietzsche recognized at once the backward and methodical nature of sane transmutations: "The most valuable standpoints are always the last to be found: but the most valuable standpoints are the methods."'[114] Finally, and again echoing the terminology of *The New Scientific Spirit*, Bachelard insists that thought must gain a new fluidity free of outworn procedures – only then can 'closed rationalism yield to open rationalism.'[115]

How do these writings enter into Breton's own from 1936, and to what purpose? In the lecture he gave on 1 June 1934 in Brussels, 'What is Surrealism?', Breton speaks of the period 1868–75 as one of considerable importance for both poetry and society: 'It is impossible, looking back upon the past, to perceive an epoch so *poetically* rich, so victorious, so revolutionary, so charged with distant meaning.'[116] Of that period, which also witnessed Rimbaud's brief career, Breton quotes Lautréamont: 'At the time of this writing, new tremors are running through the intellectual atmosphere'.[117] He then goes on to say of these 'tremors', '[t]he problem of their propagation and interpretation remains the same, and as far as we are concerned, remains to be solved.'[118] In the same lecture, Breton holds forth for the first time on what he calls a '*crise fondamentale de l'objet*',[119] later alluding to the object again, in the lecture he gave in Prague in May of the following year, as embodying the primary focus of Surrealist research.[120]

These strands are brought together in 1936 in 'Crisis of the Object', abetted by Bachelard's writings to account for and authorise the emergence of the Surrealist object by embedding it in a larger cultural context. Recalling at once his own Brussels lecture and Bachelard's reflections on the simultaneous appearance of the Lobachevskian and Hegelian dialectics, Breton voices his regret at the lack of a work of comparative history tracing the parallel development:

> during the nineteenth century, of scientific ideas on the one hand and poetic and artistic ideas on the other. There are two extremely important dates in terms of literature that will serve as landmarks: 1830, generally accepted as being the heyday of the Romantic movement, and 1870, the year when both Isidore Ducasse and Arthur Rimbaud began separately to emit the 'new tremors' of sensibility which increased steadily in power and still affect us deeply at the present time.[121]

Breton writes that these two dates have significance in the history of science, too. The first, 1830, saw the discovery of non-Euclidean geometry, which mounted a challenge to the philosophical tradition of Descartes and Kant, and '"opened," so to speak, rationalism.'[122] For Breton, this '"opening" of rationalism' met an earlier '"opening" of realism . . . under the pressure of Romantic ideas, properly so-called.'[123] The second date he gives, 1870, was the year in which mathematicians conceived 'a "generalized geometry" which included Euclidean geometry as part of a comprehensive system'.[124] Breton declares that this was a '*transcended contradiction* which Ducasse and Rimbaud used, in a different field, as a means of achieving a total disruption of sensibility'.[125] Taking his cue from these considerations, Breton concludes: 'modern scientific and artistic thought present us with identical structures'.[126]

Breton's text adheres closely – slavishly even – to Bachelard's. The play of rational and real; the term and concept 'open rationalism'; the example of the discovery of non-Euclidean geometry in 1830 as the outcome of an epistemological crisis; its subsequent integration into a 'generalized geometry' in 1870: each step taken by the intellectual history he gives in 'Crisis of the Object' fits squarely into the footprints left by Bachelard. Is there a comparable example of such a philosophical debt in Breton's writing? Perhaps only that of his perennial loyalty to Hegel. Yet a curiosity of Breton's thought from 1936 is its interest in both Bachelard and Hegel, even as Bachelard offers an outline of scientific progress opposed to Hegel's dialectical synthesis, drawing instead upon the either/or of Niels Bohr's Complementarity.[127] In 'Crisis of the Object', Breton quotes his favourite passage of Hegel's, '"[e]verything real is rational, and everything rational is real"', renewing through these terms Surrealism's repudiation of the Cartesian separation of reason and the imagination, claiming instead that 'today, reason goes so far as to propose the continuous assimilation of the irrational'.[128]

On this, Breton again looks to Bachelard, this time referring specifically to the concept of *surrationalisme* and proposals made in the text of that name:

> That is the sense in which it becomes necessary for *Surrealism* to be accompanied by a *surrationalism* which will act simultaneously as a stimulant and a restraining influence. The recent introduction by Gaston Bachelard into the vocabulary of science of the word *surrationalism*, with the aim of defining a whole mode of thought, lends additional immediacy to the word 'Surrealism', whose acceptation had hitherto remained strictly confined to the world of art. Once again, each of these terms confirms the other, a fact which provides ample proof of the unity and depth of feeling animating all human speculation in our times, whether by poet, painter, or scientist.[129]

Of this new approach, Breton writes of a 'thought process breaking ['pensée en rupture'] with the thinking habits of the past millennium'.[130] In doing so, he anticipates Bachelard's own later use of the phrase '*rupture épistémologique*' to identify a more developed version of the notion.

The powerful impact on postwar French philosophy of Bachelard's seminal idea of a *revolutionary* as opposed to *evolutionary* structure of history, particularly as expounded in his *Formation of the Scientific Mind* (1938), is well known. As with the Surrealists, the receptivity of Georges Canguilhem, Louis Althusser and Michel Foucault was, in part, a question of proximity: Canguilhem admired Bachelard and succeeded him in the Chair of Philosophy at the Sorbonne, and Foucault was a student of both Canguilhem and Althusser.

In place of the reigning developmental model of history in the fields of biology and medicine in which a series of linked discoveries light a path of progress to the present, Canguilhem offers an account of false starts, failures, corrections, revisions and re-evaluations: a plural history of contingencies and expedients.[131] Althusser openly appropriated the concept from Bachelard, using the idea of a 'coupure épistémologique' '*systematically*' with the aim of detaching Marx's later 'scientific' work from the Hegel-inspired writings of his youth, designating the works of 1845 'œuvres de la coupure'.[132] Foucault advances an outline of these developments in the superb introduction to his *Archaeology of Knowledge*.[133] For him, the lesson drawn from 'the *epistemological acts and thresholds* described by Bachelard' is that 'the problem is no longer one of traditions, of tracing a line, but one of division, of limits; it is no longer one of lasting foundations, but one of transformations that serve as new foundations, the rebuilding of foundations.'[134] A lengthier analysis of the extensive use of this concept in French philosophy is out of place here.[135] My intentions are rather to locate the idea in Bachelard's philosophy of QM, fill the gap left by Foucault's historiography by affirming its prior and widespread appropriation in Surrealism, and then give an account of its uses by the Surrealists, led by Breton in 'Crisis of the Object'.

When Breton avows that one of the characteristics of the new scientific and poetic mind is that it is 'no longer reductive but infinitely inductive and extensible', he has in mind the important phrase of *The New Scientific Spirit* I noted earlier: 'true scientific thought is metaphysically inductive.'[136] I discussed Julien Gracq's perspective on Bachelard, showing the consistency of metaphors of induction throughout Breton's work. By locating and privileging in this context the fundamental Surrealist mechanism of induction – of 'magnetic fields' and the *Manifesto*'s 'spark . . . between the two conductors' – Breton hinges Surrealism's poetic theory on Bachelard's theory of the new scientific mind without any contradiction with previous poetic theory and practice in Surrealism.[137]

How does Breton translate these reflections into a 'crisis of the object'? 'This way of thinking', he declares:

> is distinguished above all by the fact that it is dominated by an unprecedented *desire to objectify*. It must be clearly understood, to start with, that both mathematical 'objects' and poetic 'objects' commend themselves to those who created them for entirely other reasons than their plastic qualities, and even if they should happen to satisfy certain aesthetic standards it would still be wrong to try to judge them from this standpoint.[138]

As we saw, with the advent of QM Bachelard identifies in *The New Scientific Spirit* an 'outward' movement taking place in modern scientific thought, from the subjective to the objective, from the mathematical to the experimental, with mathematics increasingly playing the leading role: 'the orientation of the epistemological "vector" seems clear. It surely points from the rational to the real'.[139] Breton sees the Surrealist object as evidence that the same process is finding favour within contemporary art and poetry.

The language of Surrealism emphasised *objectivation*, *matérialisation* and the *concret* during this time, in tune with the broader materialist discourse of French philosophy and politics of the 1930s. The transformation that Surrealist theory underwent by way of this rhetoric is matched in the shift in its art from the distance and visuality of painting to the tactility and intrusion into one's physical space characteristic of the object. This was undoubtedly the outcome of Surrealism's traversal of the passage from idealism to dialectical materialism; its shift from what Breton called in 1934 'a purely *intuitive* epoch' (1919–25) to 'a *reasoning* epoch' (1925–).[140] As he admitted later, during the first of these two periods Surrealism was convinced of '*the primacy of matter over thought*'.[141] It is not surprising, then, that the call first made by Breton in 1924 for 'some of the objects that one approaches only in dreams'[142] could be realised only in the changed atmosphere of the second period – as 'pure desire in concrete form', and as 'nothing less than the objectification of the very act of dreaming'.[143] Dalí's vivid landscapes, of course, were instrumental in the ripening of Surrealist materialism (ironic, given his notorious indifference to political matters), and consistently with this he became one of the most prolific object makers, receiving a special mention for his category of 'objects functioning symbolically' in 'Crisis of the Object'.

The break with the habits of the past rehearsed by the rhetoric of Surrealism and *surrationalisme*, and their refusal of fixity and systematisation entailing a perspective on the world based not on the singularity of identity but on the dynamic flux of relationships, is exemplified for Breton in a particular attitude held towards the object. He regarded the object as the embodiment or signifier of a process, which with Bachelard (and Gracq) he calls *inductive*. Breton believed that the vigour of Surrealism resided in a restless search for new 'charges' wrought from novel juxtapositions. In a Bachelardian passage in 'Crisis of the Object', he declares: 'any intellectual formula which considers itself definitive does so at its own peril, since it thus automatically invites the contradiction with which experimentation is always waiting to confront it. The vital task is to pursue experiments unceasingly'.[144]

This avid 'experimentation' in the quest to expand knowledge – and in doing so extend the domain of what can be called 'real' – by a kind of theoretical 'play', scornful of what had been believed proven in the past, is sanctioned by Bachelard who insists in *The New Scientific Spirit* that *rationalisme ouvert* is:

> to experience genuine surprise at the implications of theoretical speculation. Juvet puts it quite well: 'The surprise created by a new idea or association of ideas is surely the most important element in the progress of the physical sciences, for it is astonishment that excites logic, which is always rather cold, and that forces scientists to make new connections. But the ultimate cause of progress, the reason for our surprise itself, has to be sought in the fields of force that new associations of ideas set up in our minds, fields whose strength measures the good fortune of the scientists lucky enough to bring those ideas together.'[145]

Bachelard refers here to the Swiss physicist Gustave Juvet who had quite a high profile in Francophone scientific scholarship during the period.[146] Without mentioning Juvet or acknowledging Bachelard, Breton leaps on this quotation in 'Crisis of the Object' as confirmation of the contemporary relevance of the Surrealist project: 'Poets and artists join forces with scientists at the heart of these "fields of force" created in the imagination by the reconciliation of two different images.'[147] Comparison between poetic and scientific inspiration is far from rare in science, of course, but the sheer precision with which Juvet's words aligned with Pierre Reverdy's remarks as quoted in the *Manifesto of Surrealism* must have set bells ringing inside the mind of Breton, establishing the relevance of Surrealism to the post-quantum scientific and poetic consciousness through their unambiguous validation of Surrealism's poetic *merveilleux*, a lynchpin of its theory since Breton's early text on Max Ernst.

Arguing that this ability shared by poets, artists and scientists allows them special access to a reality usually concealed, Breton also seizes the opportunity to link the current Surrealist project with the foundational practice of automatism. Applying Juvet and Bachelard's writings on scientific insight to the object, Breton generalises their application to Surrealism by bringing together Bachelard's dialectic of 'real' and 'rational' with Freud's language of 'manifest' and 'latent':

> This ability to reconcile the two images allows such people to transcend the generally limiting factor of the object's manifest existence. With this new focus, on

the contrary, the same object, however complete it may seem, refers to an infinite series of *latent possibilities* which are not peculiar to it and therefore entail its transformation . . . In the words of Gaston Bachelard, 'What is the nature of belief in reality, what is reality as a concept, what is the primordial metaphysical function of the real? Essentially it is the conviction that an entity is more than the sum of its immediate data, or, to put it in plainer terms, it is the conviction that *one will discover more in the reality concealed within the entity than in the immediate data surrounding it*.'[148]

Before their powers of induction enabling scientists, poets and artists to conceive spontaneously new, unexpected, hidden relationships, the ostensible sensorial information by which meaning is attached to objects is stripped away and an underlying, indeterminate *au-delà* imagined into life. In modern experimental physics, scientists found what their approach determined they find, while psychoanalysis attached and detached meaning to objects according to the vagaries of biography.[149]

Bachelard's tracing of science's complexification of the traditional concepts and simple objects bequeathed it by classical physics is closely met, then, in Breton's affirmation of Surrealism's '*total revolution of the object*', achieved when fixed meaning normally ascribed to an object is foregone and a '*change of role*' is allowed.[150] In support of this, Breton appropriates another second-hand idea from *The New Scientific Spirit* (unacknowledged): Bachelard's reworking of the French empiricist Charles Renouvier – making it sound in the process as though the notion of the 'substantialised substantive' is Renouvier's own.[151] Renouvier influenced William James but is virtually forgotten today, and was probably not widely read in the mid-1930s even in France (he died in 1903). Bachelard would have been attracted to his rejection of absolutes and the accent he placed on indeterminism and chance. If Breton had got round to reading Renouvier he would probably have been drawn to those aspects of his thought too, but he definitely received his writings only through this fragment in Bachelard. Just as he never held Juvet's book in his hands, there is no chance he ever seriously read Renouvier.

Given Breton's boundless ambition for Surrealism, his extensive uses of Bachelard's philosophy of science in 'Crisis of the Object' are easy enough to interpret. In Bachelard, he discovered a learned figure in sympathy with some of the aims of Surrealism, whose academic status promised to strengthen its standing and its claim to historical authority in circles beyond those of poetry and art. Bachelard's intellectual negotiation with Surrealism is harder to characterise, though one is clearly in the air in the polemical swagger of 'Le Surrationalisme'. He probably saw in Surrealism a mode of address and a set of theoretical precepts unencumbered by the drier language of academia, useful to the conscription of geometry and physics into his *révolution spirituelle*.

New Adventures in Topology: Surrealist Art and Mathematical Models

Notwithstanding its borrowings, 'Crisis of the Object' is the key statement in the debate on the object as it took place in Surrealism in the 1930s. Written for the exhibition of

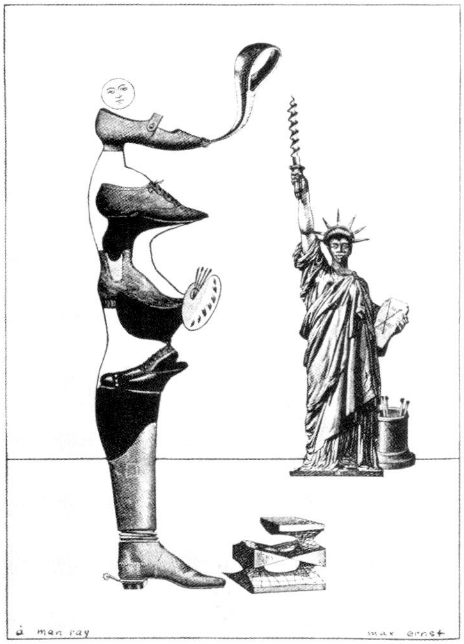

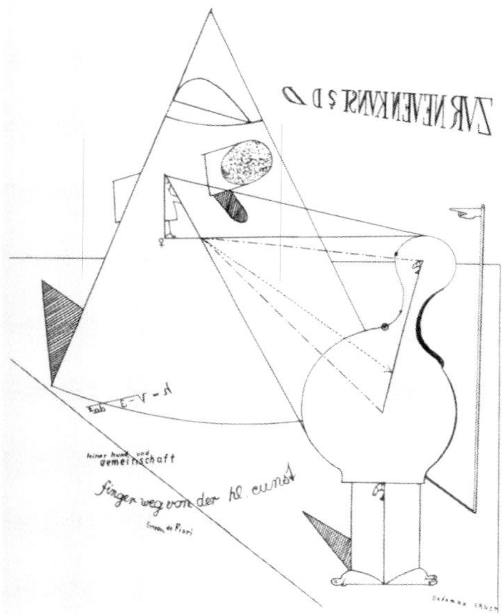

25 (left) Max Ernst, *Portrait of Man Ray* (1935). Collage, 17 × 11.5 cm. Private collection, New York.

26 (right) Max Ernst, from *Fiat Modes: pereat ars* (1919–20). Lithograph, 44.5 × 33 cm.

Surrealist objects held at the galerie Charles Ratton in May 1936, it was published in the special 'object' number of *Cahiers d'Art* in the same month, which represents the main extension of that debate.[152] Its concern is with the role of the object in Surrealism rather than the Surrealist object, a focus given by the exhibit and audience for which it was intended. Consequently, like the Ratton show it gives prominence to mathematical models and natural forms as well as objects fabricated by the Surrealists themselves.

Breton became aware of mathematical models as he was growing convinced by his reading of Bachelard that the epistemology of modern, mathematicised science held a mirror up to Surrealism. Although accounts conflict, Surrealist contact with mathematical models is usually dated to the moment Max Ernst discovered them at the Institut Poincaré in Paris in 1934, incorporating illustrations of them into his collages from that year – notably in the lower part of his *Portrait of Man Ray* of 1935 (fig. 25). Ernst almost certainly knew such models earlier, since the largest collection is in Göttingen, the first home of QM, and an exhibition of models took place in his hometown Cologne in 1919.[153] This is possibly the source of the algebraic inscriptions and playful geometry of certain of the de Chiricoian series of lithographs of that year, *Fiat Modes* (fig. 26). Ernst brought the Institut Poincaré models to the attention of the other Surrealists, primarily Man Ray who photographed them between 1934 and 1936, although this dating, like the narrative of their Surrealist interpretation, requires further study. In spite of the still-veiled back-story, Man Ray's photographs are now well-known, so most of the fragmentary writings on Surrealism and mathematical models exist in the literature on him, even though there is little available English language scholarship on mathematical models *per se*.[154]

The first models were made in Germany, and one writer states that their construction 'went hand in hand with work at the furthest frontiers of research in the area of algebraic surfaces'.[155] Their purpose was to give in three dimensions the diversity of surface made available by geometry through the tracing out of straight lines in space, giving students a concrete, visual alternative to the algebraic surface. Often constructed with movable parts and string to enhance the variety of surfaces they can present ('deformation' is the technical term), the greatest value of the physical representation of geometric objects is to be had in the fields of design, engineering and architecture. Predating abstract sculpture, the earliest models were produced in large quantities from plaster casts made from special modelling clay in the 1860s. However, Man Ray also found rare and unique examples in wood and wire in the Institut Poincaré, made by forgotten mathematicians and their students, essentially trying their hand at sculpture.[156] Their value as pedagogical tools is confirmed by the great increase in their demand in the 1870s, yet a dual purpose – as both teaching tools and curiosities – is already suggested by an exhibition of models as early as 1876 at the Science Museum in London.[157] They were manufactured in numbered series and sold through the publishing house of Ludwig Brill (brother of the mathematician Alexander von Brill) from 1877, which was taken over by Martin Schilling in 1899. The Schilling catalogue of 1911 'contained 40 series with almost 400 models and devices' and as the main source of reference was indispensable to Ernst.[158] Each model is labelled with its algebraic description and a serial number, though Man Ray hides these details from view in his photographs with the aim of freeing them from their scientific origins, even taping over the descriptive label in one instance.

The great period of model making ended during the First World War, and the death in 1925 of the mathematician Felix Klein, who was their main advocate in the classroom, largely put paid to their pedagogical usage. In 1932, Schilling wrote to the Mathematics Institute at the University of Göttingen (home of a collection of about one thousand) with the news that no models had been made for several years. By the time Ernst and Man Ray discovered the six-hundred-strong collection in the model room of the Institut Poincaré in 1934, they had taken on a similar reduced status to that reserved in Surrealism for Native American Katchina dolls. Created with the purpose of making visible the invisible, they had outlived their practical use and migrated to the display shelves as curiosities, more at home in the flea market than the classroom. Man Ray's own version of his encounter with the models is typically idealised. Simultaneously emphasising the *Wunderkammer* angle and the presumed old-fogeyness of science and mathematics, he recounts their discovery 'in dusty cases at the Poincaré Institute'.[159]

The number of photographs Man Ray actually took is still unknown,[160] but twelve are reproduced with Christian Zervos's 'Mathématiques et art abstrait' and Breton's 'Crisis of the Object'.[161] The two texts appear consecutively in the 'object' number of *Cahiers d'Art* in 1936, two models with Zervos's text and the other ten between it and

Breton's. Inspired by the 'incontestably beautiful' mathematical models, Zervos takes them as a point of departure for his assessment of the comparative merits of mathematics and art in a discussion of aesthetics.[162] Predictably, he positions them as polar opposites, arguing that the first is concerned with the intellect while the second springs from eroticism and emotion. At first, this seems quite contrary to the contemporary views expressed by Bachelard and Breton (via Juvet) seeking a commonality in the shared 'emotive spark' of insight. This prologue sets up his subsequent critique of the coldness and lack of lyricism in abstract art, aside from Kandinsky's, of course, which was a regular fixture in *Cahiers d'Art* under Zervos's editorial guidance.

But his essay also reveals Zervos's awareness of the debate on modern physics that was taking hold within Surrealism and the avant-garde, for he makes favourable remarks on what he sees as the 'poetic' nature of QM in contrast with classical physics. The second, he states,

> proceeds by calculation and reasoning, it submits itself to formal rules, to a group of deductive operations. In the science of our day, which holds some points in common with Greek science before the Platonic era, a greater poetry is revealed. Excessive precision is suppressed. The abstract formulae of contemporary science are enriched by anxiety and by the human senses, for the passion underlying our life is incorporated as much into abstract thought as into the aesthetic domain.[163]

Zervos's understanding of QM seems unusually knowledgable in this passage. His mini-account of the loss of absolute precision in measurement in modern physics, and allusion to an element of subjectivity that 'enriched' its mathematical formulation, is obviously meant to invoke the insurmountable problems in observation and calculation that dogged QM, incorporated into its theory as Heisenberg's Uncertainty Principle. But the comparison Zervos makes with Presocratic philosophy – rooting modern developments in physics to the earliest known natural philosophy and science of his homeland Greece – is less common yet equally valid, and I will return to that theme later.

Before I examine Man Ray's presentation of mathematical models, asking how they achieve the separate identity 'simply through a *change of role*' demanded of the object by Breton, I want to discuss the models' original meanings – their manifest content – since Breton was far from indifferent to what was represented in their literal, surface register.[164] We saw that the 'freeing' of rationalism through the discovery of non-Euclidean geometries around 1830, and the formulation of a 'generalised geometry' in 1870 were matched in poetry, according to Breton, by the high point of Romanticism and the emergence of Ducasse and Rimbaud. Taken together, these two events marked a break with previous modes of sensibility and for that reason mathematical models were, for him, souvenirs of an epistemological revolution. Briony Fer's claim that 'these objects were conceptually, as it were, turned on their heads' by the Surrealists, then, recognises only part of the charm they held for Surrealism, and does not acknowledge the connection between 'Crisis of the Object' and Man Ray's photographs.[165] Fer's reading emphasises only the latent aspect of mathematical models, whereas the poetry/science parallels in Breton's text tell us that the mathematical model was the sole instance of an object in the Surrealist theory of the object that was valued for what it manifested – that is, for its original (pedagogical) purpose. The Surrealists admired mathematical models for the same non-Cartesian reasons as Bachelard, who writes in 'Le Surrationalisme' of the mental health that resides in 'those numerous geometries left in darkness by official and pragmatic teachings.'[166] That is to say non-Euclidean geometries emblematised the rejection of the boundaries of 'common sense' Euclidean geometry in pursuit of an expanded notion of the real.

More characteristic of Surrealism is the other half of the equation: the steps taken by Man Ray and Breton to uncover the latent content of the models. Man Ray first separated them from each other and removed them from their display cases at the Institut Poincaré. In *Cahiers d'Art* they appear isolated, growing more singular, exotic, enigmatic and suggestive outside of their scientific and academic contexts. Further 'disruptions' are caused by the flattening effects of lighting, accentuating the Arp-ish biomorphism of the 'Kummer surface' (fig. 27), and by bleaching others into Brancusi-like whiteness (many of these models are coloured, in fact). Deep recesses are carved by dark shadows, breaking up surfaces, while close-ups, varied angles and cropping intervene further between viewer and model, denying the models their scientific 'voice' (fig. 30).

A veteran of both photography and painting by this time, Man Ray mobilises a battery of effects, drawing and commenting ironically on both. His approach to lighting, for instance, seems more appropriate to fashion photography at times, where light can dissolve or accentuate texture, while the models are framed almost as portraits in certain cases. Behind some, horizon lines are suggested, and with others, horizontal surfaces receding into depth recall parodically the conventions of landscape painting. In the first photograph in the series as it appeared in *Cahiers d'Art*, for instance, Man Ray seems to look down through a kind of distorted perspective from the height of Surrealism's racy, urban modernism towards the last great master of landscape painting and classical pictorial composition, slyly parodying Cézanne's remark about handling nature in terms of the cylinder, sphere and cone (fig. 28). In a final disorienting intervention the photographs are given iconic status, slickly mounted and grandly presented one to each large *Cahiers d'Art* page, through which their scale and any relative sense of size are lost.

Mathematical models would be readymades if they were chosen for their lack of aesthetic satisfaction, when in fact a crucial role in their selection was played by visual fascination. They are more akin to 'found objects', which 'speak' through the subjectivity of the observer. In line with the speculation that animates the *trouvaille*, Breton transforms the models in 'Crisis of the Object' by his addition of 'poetic' titles to indicate hidden depths behind their visible, prosaic aspect. In doing so, he compromises the original 'identity' of each, retained through anodyne mathematical inscriptions beneath the photographs in *Cahiers d'Art*.[167] Obviously posed as a 'primitive' artefact the mask-like, Picasso-esque object with wire attachments captioned 'Surface reglée', for example, is titled by Breton *Ainsi parlait . . .* (*Thus Spoke . . .*), inviting the viewer to seek out the model's veiled Nietzschean connotations (fig. 29). The three-dimen-

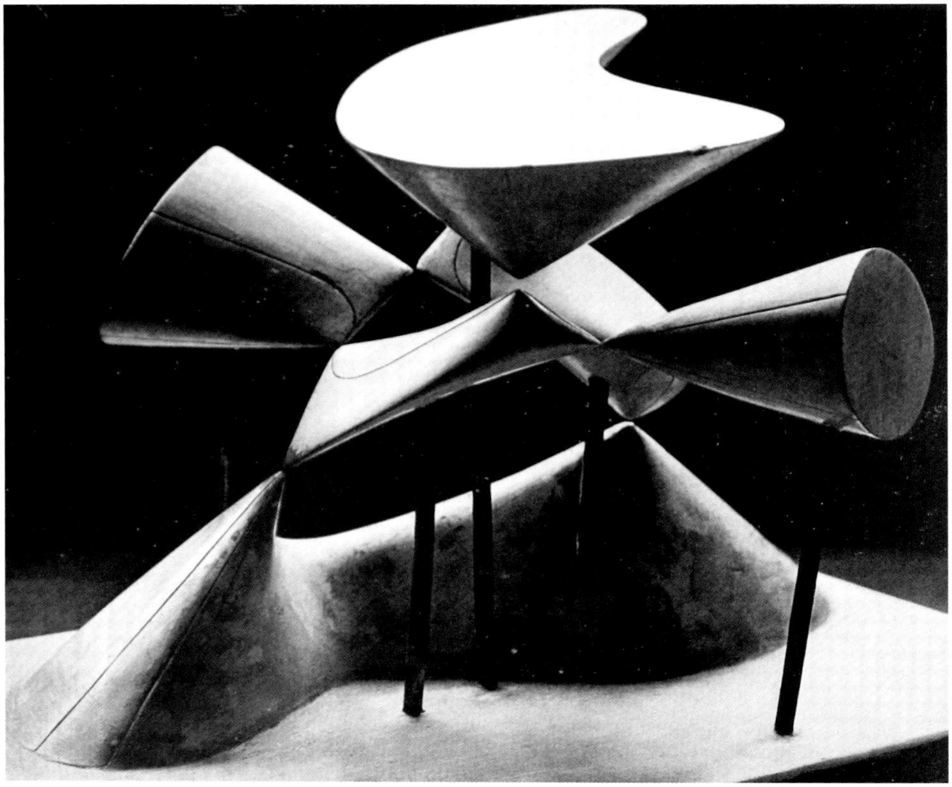

27 (top left)
Man Ray, *Surface de Kummer à seize points double, dont huit réels* (1934–36). Photograph, 29.5 × 24 cm. Man Ray Trust.

28 (bottom left)
Man Ray, *Perspective d'un cube, d'un sphere, d'un cone, et d'un cylindre* (1934–36). Photograph, 29.5 × 24 cm. Man Ray Trust.

29 (bottom right)
Man Ray, *Surface reglée* (1934–36). Photograph, 29.5 × 24 cm. Man Ray Trust.

30 (facing page)
Man Ray, *Surface à courbure constant negative d'ennepar, dérivée de la pseudosphère* (1934–36). Photograph, 29.5 × 24 cm. Man Ray Trust.

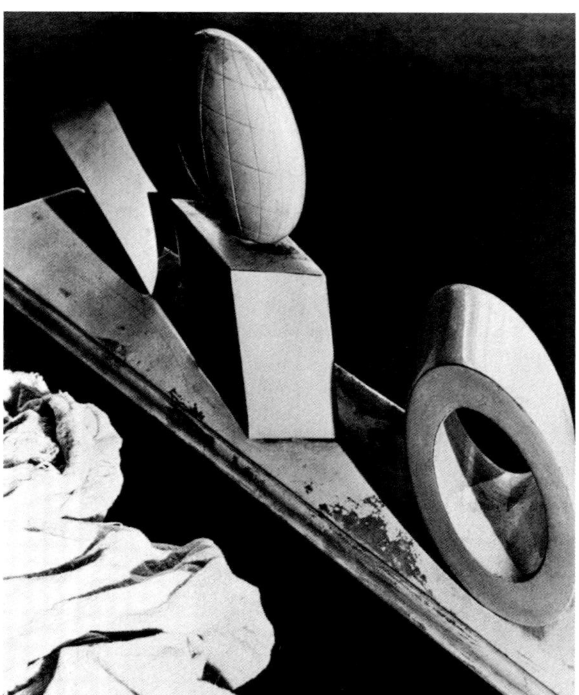

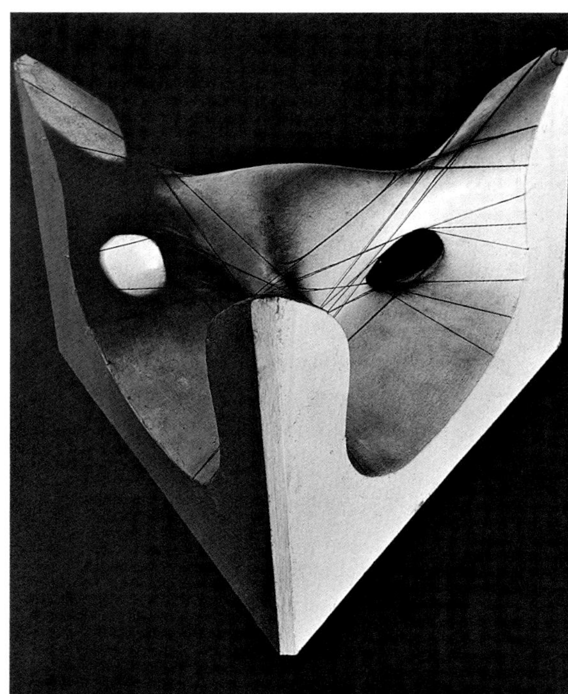

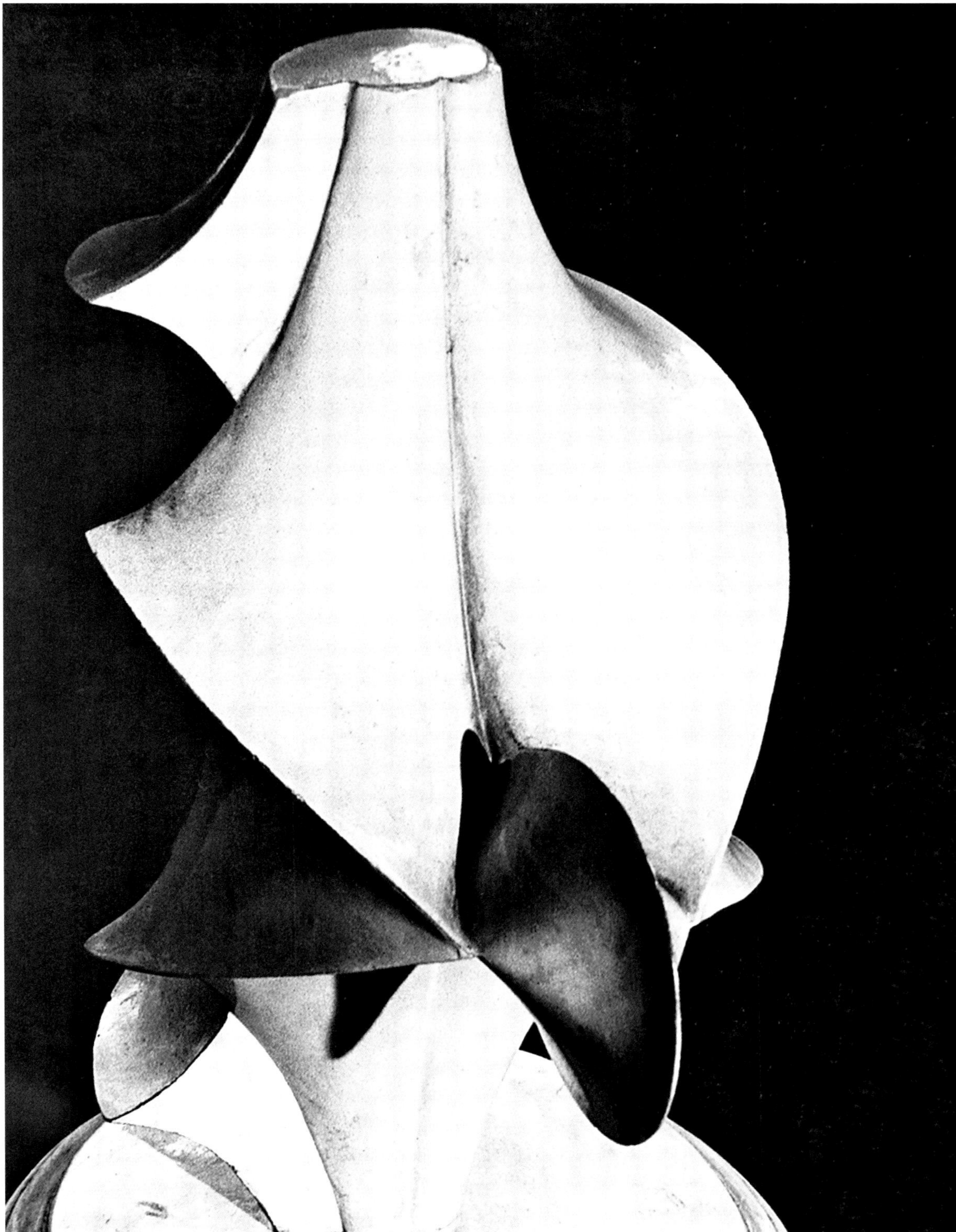

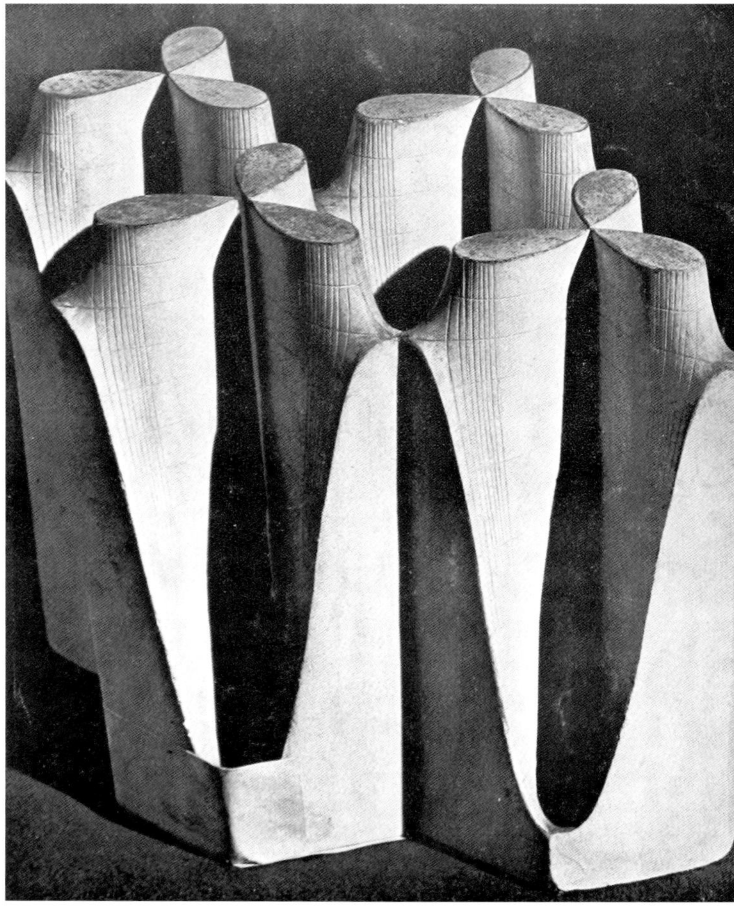

31 Man Ray, *Allure de la fonction elliptique* p^1 *(u) pour* $G2 = 0$ *et* $G3 = 4$ (1934–36). Photograph, 29.5 × 24 cm. Man Ray Trust.

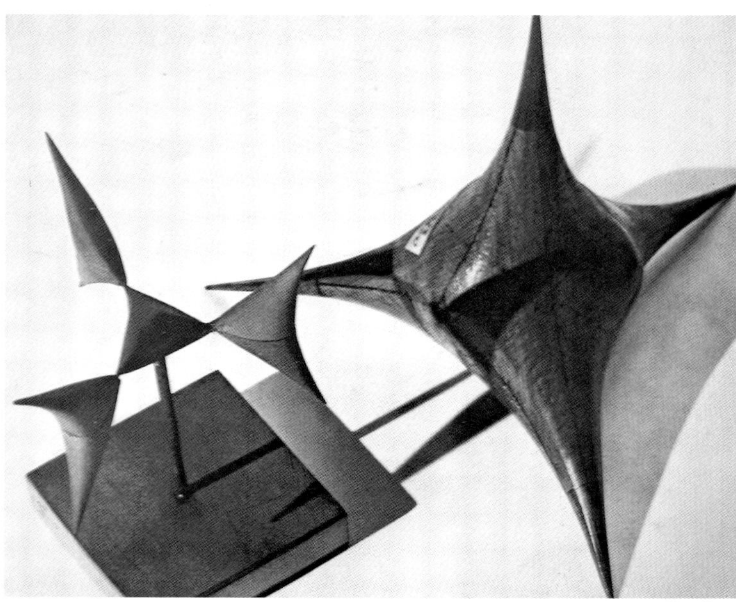

32 Man Ray, *Surfaces algébriques du IVe degré* (1934–36). Photograph, 29.5 × 24 cm. Man Ray Trust.

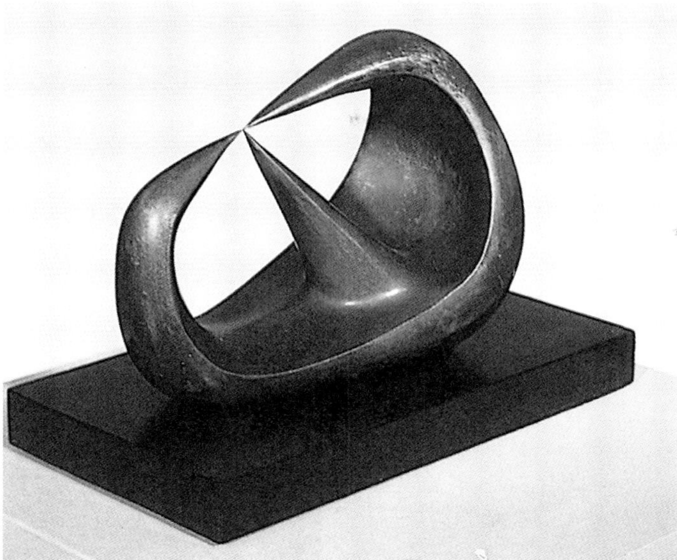

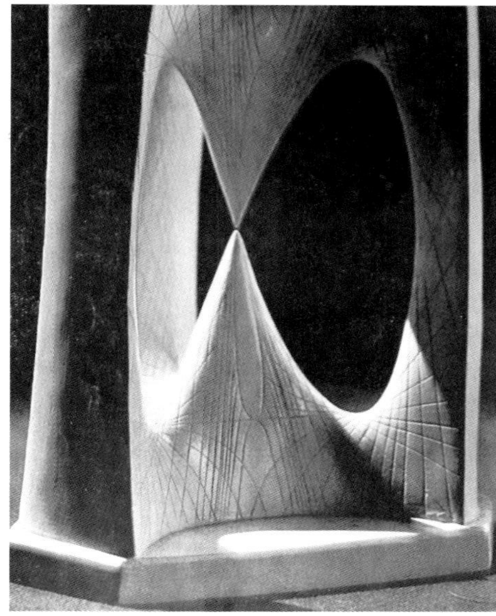

33 (above, left) Henry Moore, *Three Points* (1939–40). Bronze, 14 × 19 × 9.5 cm. Tate Gallery, London.

34 (above, right) Man Ray, *Divers types de points coniques* (1934–36). Photograph, 29.5 × 24 cm. Man Ray Trust.

sional graph accidentally shaped as a cultish gathering, labelled 'Allure de la fonction elliptique p[1] (u) pour G2 = 0 et G3 = 4' (this label is taped over in Man Ray's photograph), becomes *Les Pénitents roses* (*The Rose Penitents*) (fig. 31).[168] Elsewhere, the thorny, toylike 'Surfaces algébriques du IVe degré' gives Breton the wonderful *La Bague de Rosier* (*The Ring of the Rosebush*) (fig. 32). Surrealism's double naming of mathematical models in *Cahiers d'Art* brings out their manifest and latent levels, creating an oscillating play between identities.[169]

Apart from their manifest embodiment of a rejection of classical geometry and space, constituting their 'ideological' attraction to modernist and avant-garde artists, the evocative forms and ambiguous latencies of mathematical models were sufficiently enticing to create a trend among both sculptors and painters from the mid-1930s. Man Ray's photographs acted as a significant spur to artists working within or close to Surrealism, and I close this chapter with a brief survey and interpretation of this minor trend underlying the voguishness of the object, yet to be analysed at length.

Henry Moore acknowledged that some of his sculpture of the late 1930s took its cue from the models he saw at the Science Museum in London in 1938, though given the dates of these it is likely that he set off to look at them after viewing Man Ray's photographs in *Cahiers d'Art*, which make their presence felt mainly in Moore's use of string.[170] Less obvious is the formal similarity between his *Three Points* of 1939–40 (fig. 33) – where three tapering forms just fail to touch at their extremities – and the models in *Cahiers d'Art* captioned 'Surface de Kummer a seize points double, dont huit réels' and 'Divers types de points coniques' (fig. 34). Gordon Onslow Ford even believed Moore's reclining figures illustrated the morphological congruence of the mathematician's and sculptor's practices.[171]

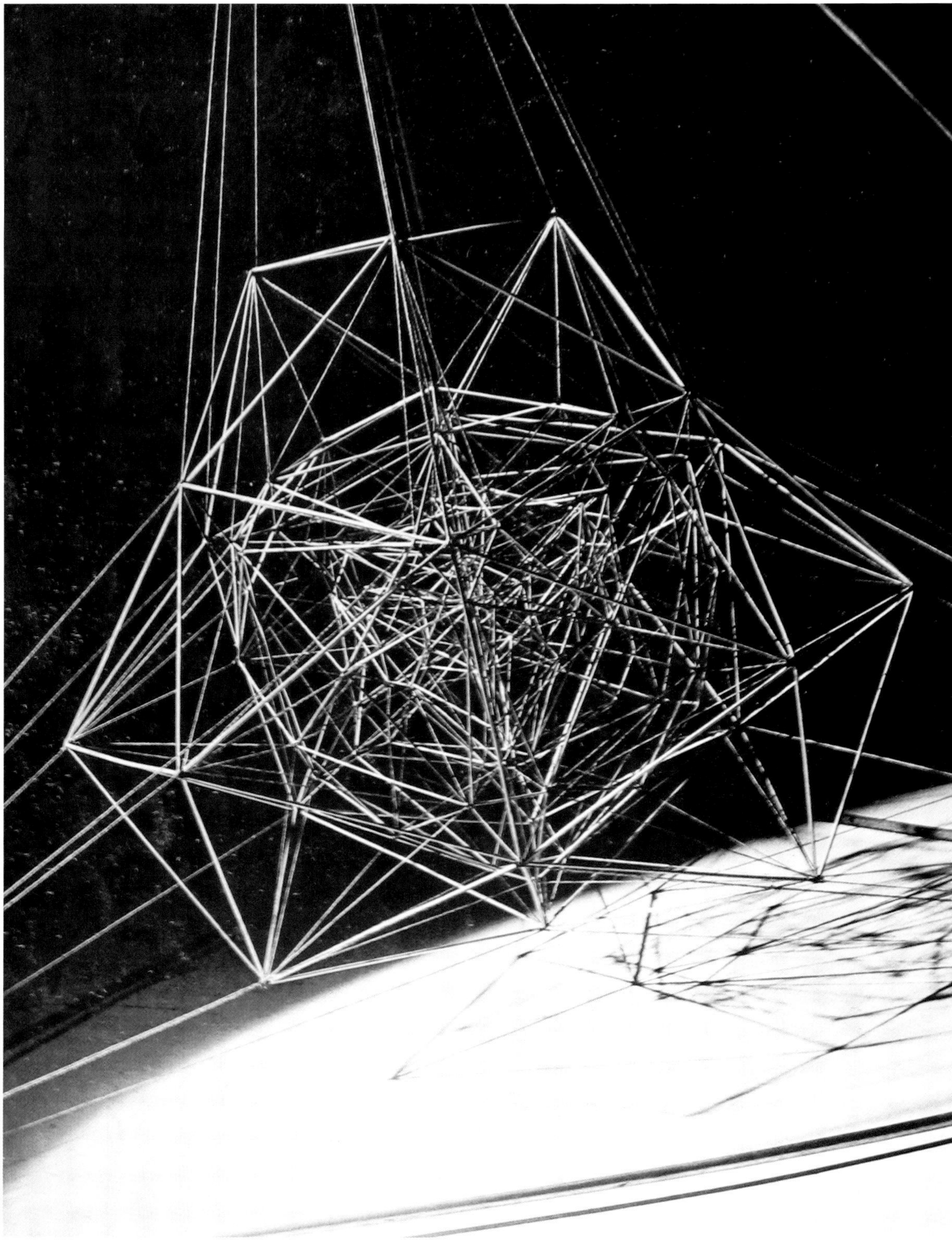

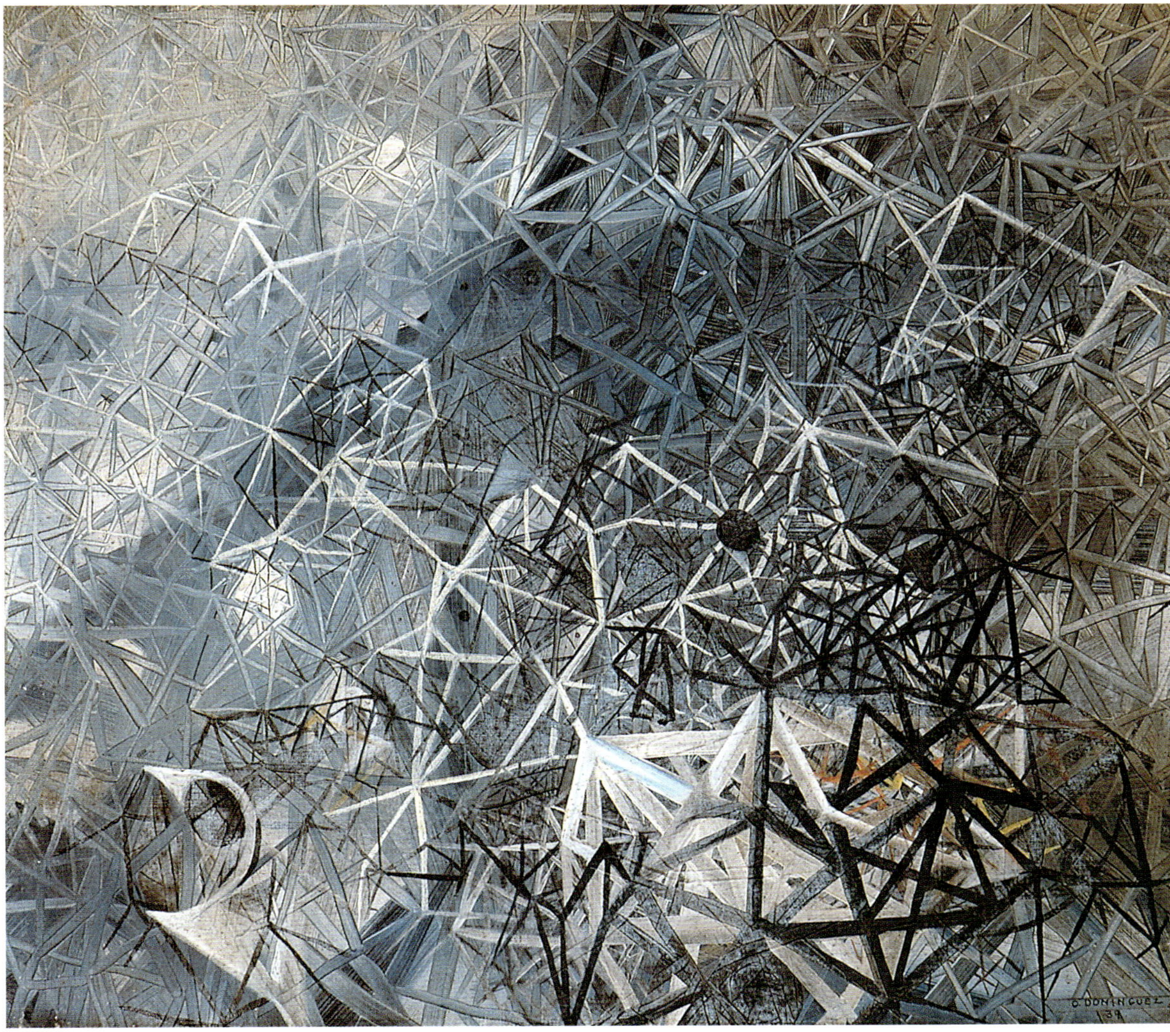

35 (facing page) Man Ray, *Polyèdres*, (1934–36). Photograph, 29.5 × 24 cm. Man Ray Trust.

36 (above) Oscar Domínguez, *Landscape of Arteries* (1939). Oil on canvas, 46 × 54.5 cm. Private collection, Paris.

The iconography of Oscar Domínguez's paintings of the late 1930s and early 1940s also announces his appropriation of the forms of mathematical models. Marcel Jean suggests that the photograph of a polyhedron in Man Ray's series (fig. 35) could have led to the tangled architecture of paintings such as the 1939 *Paisaje de Redes* (*Landscape of Arteries*) (fig. 36), which bridges the landscape and urban space as well as the inner human and outer world.[172] The slightly later series of reclining nudes, particularly the pseudogeometrical *Mujer* (*Woman*) of 1941 (fig. 37), more obviously demonstrates Domínguez's conscription of the morphology of the models – particularly 'Divers types de points coniques' – into a satirical attack on the reclining nude of (Euclidean) classicism.[173]

This model held a particular fascination for Surrealist artists, also making its way slightly altered into Max Ernst's 1948 painting updating Bellini, *The Feast of the Gods*

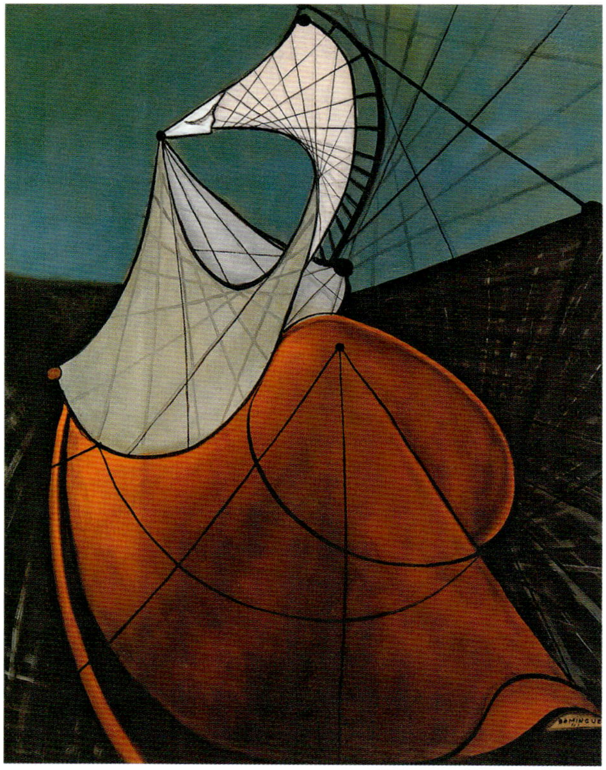

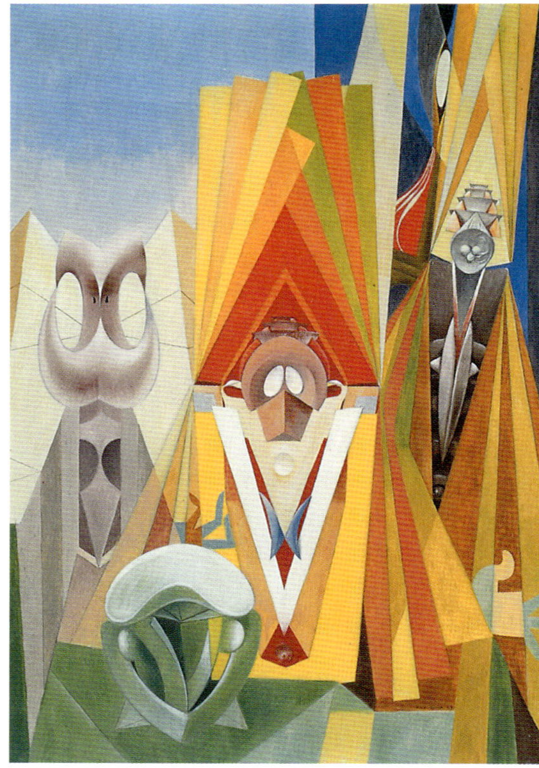

(fig. 38). The heads of all the figures in this painting seem composed in one way or another with generalised reference to mathematical models, whilst the object in the foreground is a revision and amalgamation of elements derived from Man Ray's photographs of the 'Kummer Surface' and the wire and plaster 'Ruled Surface'. Although Ernst was apparently the first Surrealist to admire the models, using them in his collages, he resisted their influence in his painting until 1947 when their forms start to invade his work. The first time he uses them according to the chronology of Werner Spies *Oeuvre-Katalog* is in the painting *Universaphrodite* of that year in which the ovaries of the Aphrodite figure are derived from Man Ray's photograph of a divided torus, captioned in *Cahiers d'Art* as 'Un Plan bitangent a un tore le coupe suivant deux cercles (théorème de Villarceau)' (fig. 39).[174] The eyes in the frottage *Mask* (1947), which reappear in Ernst's 1947 painting named after J. Bell Pettigrew's book *Design in Nature* (fig. 40), are based on the same model, which crops up repeatedly in Ernst's work of that year and the next.[175]

37 (above, left) Oscar Domínguez, *Woman* (1941). Oil on canvas, 65 × 50 cm. Private collection, Madrid.

38 (above, right) Max Ernst, *The Feast of the Gods* (1948). Oil on canvas, 153 × 107 cm. Museum Moderner Kunst, Vienna.

More inventive for its complex yet seamless combination of mathematical forms with other collage elements is Ernst's series of *Paramyths* collages of 1949. Here, he is more pointedly concerned with satirising the notion of the transcendence of the classical ideal, mixing classical figures with evidence of more recent geometric and mathematical innovation. One of the *Paramyths* collages, dating in fact from 1938 (fig. 41), features a wire model similar to one photographed by Man Ray. A nymph who is part

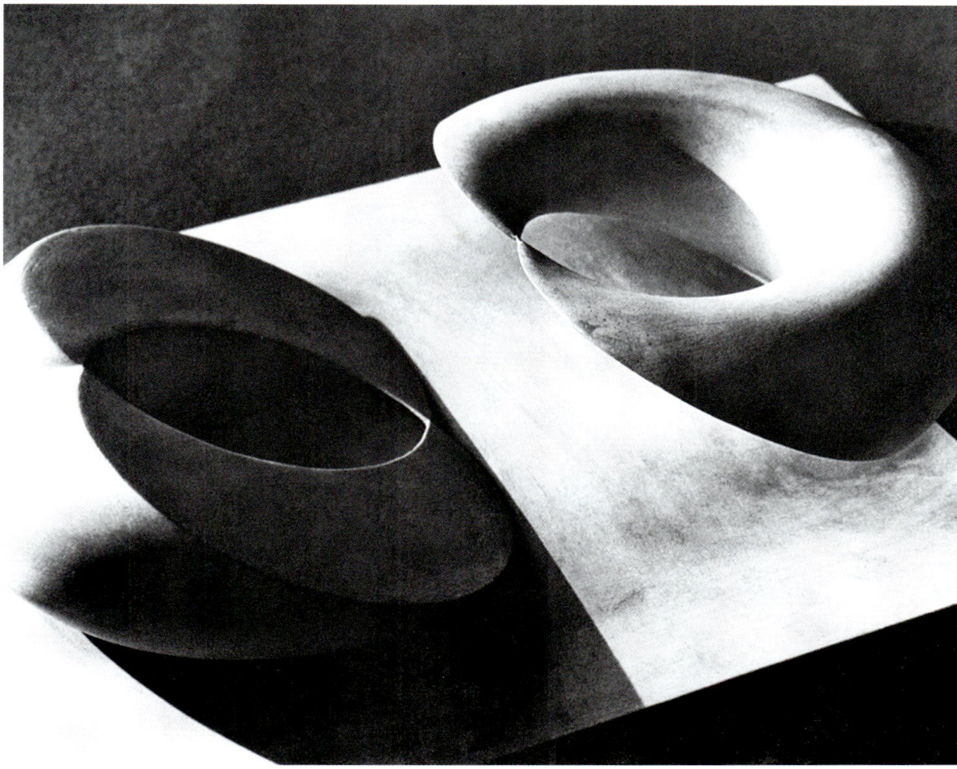

39 Man Ray, *Un Plan bitangent à un tore le coupe suivant deux cercles (théorème de Villarceau)*' (1934–6). Photograph, 29.5 × 24 cm. Man Ray Trust.

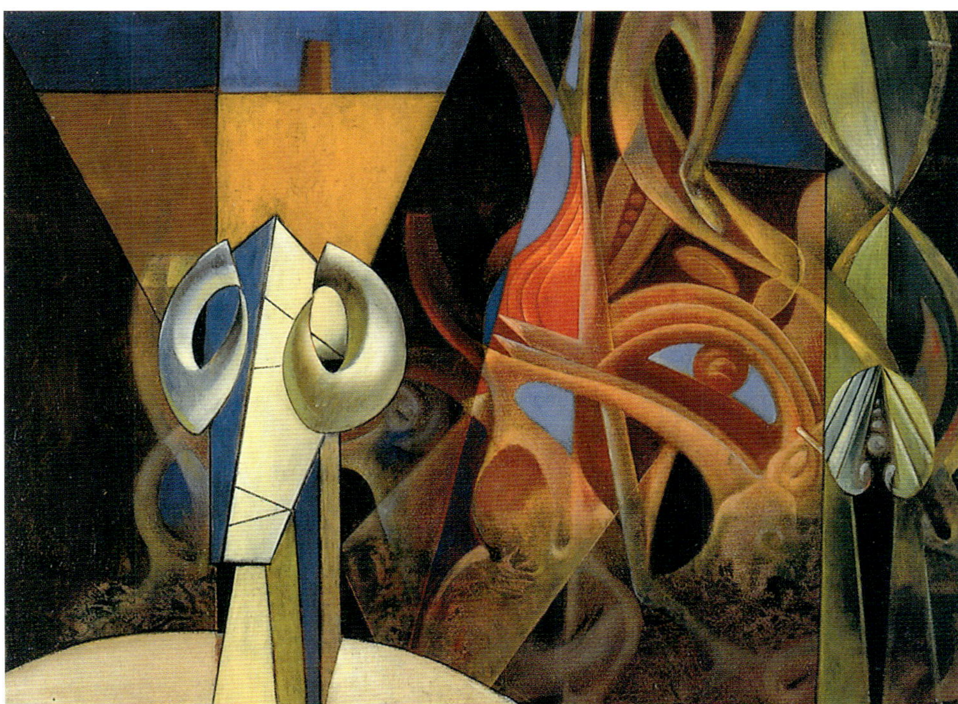

40 Max Ernst, *Design in Nature* (1947). Oil on canvas, 50.8 × 66.7 cm. The Menil Collection, Houston.

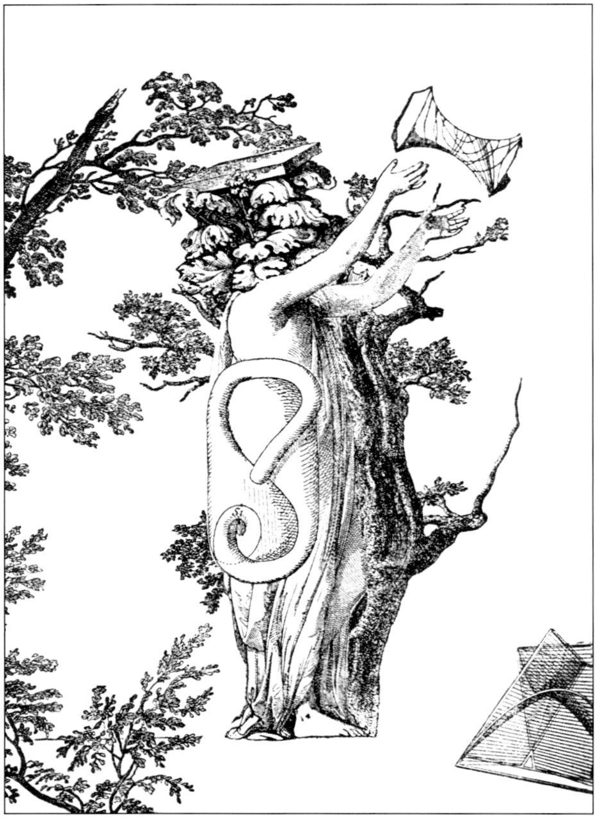

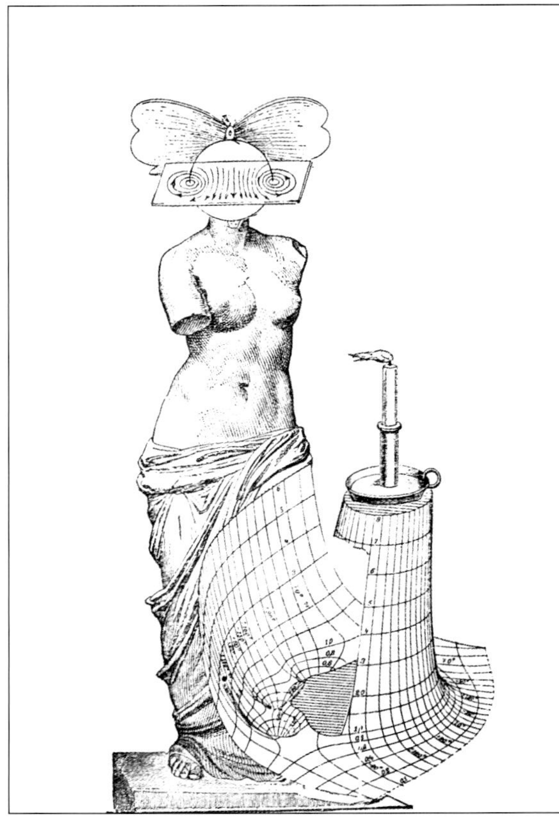

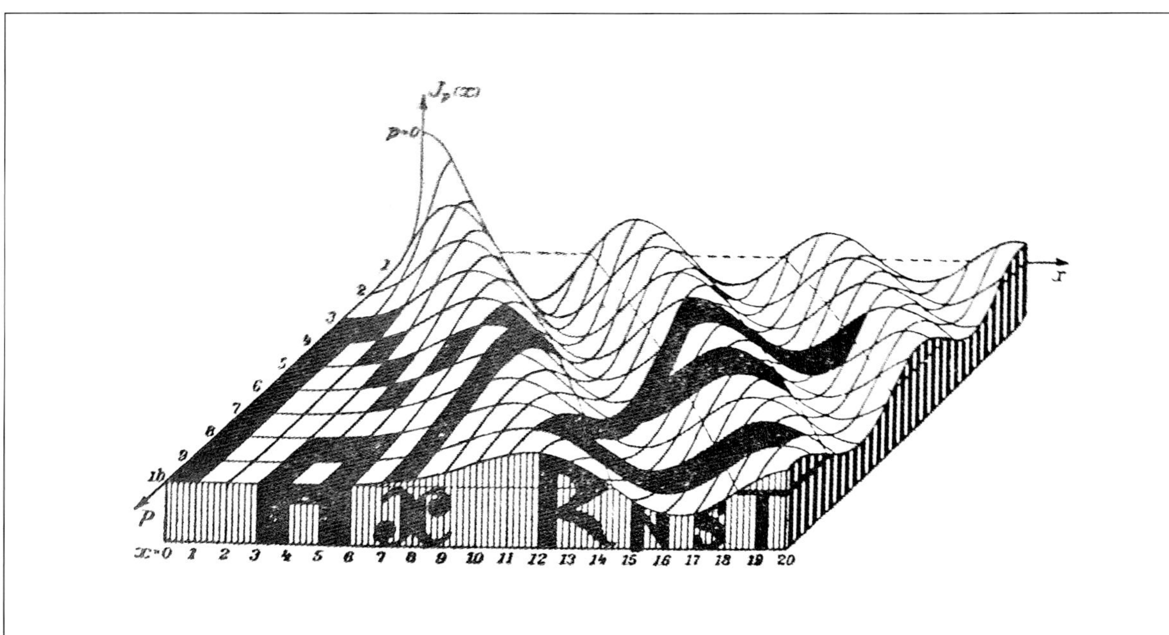
$J_p(x)$
p=0
x
p
x=0 1 2 3 4 5 6 7 8 9 10 11 12 13 14 15 16 17 18 19 20

41 (top left)
Max Ernst, illustration for Comte de Lautréamont, *Œuvres complètes*, Paris, 1938. Collage. Dimensions and present whereabouts unknown.

42 (top right)
Max Ernst, *Once Upon a Time there was a Mouse in Milo* (1948). Collage, 20 × 14 cm. P. H. Taucher, Helsinki.

43 (bottom)
Max Ernst, *Ausstellungssignet* (1948). Ink on paper, 5.6 × 10.5 cm. Present whereabouts unknown.

Klein bottle receives the 'good news' of the discovery of non-Euclidean geometry by means of a pseudosphere deposited from above. In another, titled *Once Upon a Time there was a Mouse in Milo* (fig. 42), the 'ideal form' of the *Venus de Milo* is set against the undulating surface of a three-dimensional graph. At the end of 1948, Ernst incorporated his name into a slightly different kind of graph in the drawing *Ausstellungssignet* for the cover of the catalogue for his exhibition at the M. Knoedler & Co. Gallery in New York, joining his name and recent work with the 'beautiful symbols of mathematics' eulogised by Bachelard (fig. 43).

Man Ray was the original owner of *Universaphrodite*, the first work in which Ernst used elements of the mathematical models. Ernst's return to the models in his paintings of the late 1940s probably has something to do with their appearance at the same time in Man Ray's 1948 series of undemanding paintings, *Shakespearian Equations*. For each of these seventeen canvases, Man Ray took one or more of the models and re-presented them with few alterations, each accompanied by the title of a play by Shakespeare. For example, his painting of 'Surface reglée' was titled *The Taming of the Shrew*, while the polyhedran in oils was named *The Merchant of Venice*. Man Ray claimed this procedure was as arbitrary as Breton's titling exercise in 'Crisis of the Object'. But it is obvious that the title he gave his painting of one model, *The Merry Wives of Windsor* (fig. 44), and the name given the same model by Breton, *Les Pénitents roses*, were both guided by a descriptive rationale.

Man Ray showed the *Shakespearian Equations* paintings for the first time in 1948, an event fresh in Marcel Duchamp's mind when he wrote a brief passage the following year on Man Ray for Katherine Dreier's Société Anonyme catalogue. For there, Duchamp concentrated on Man Ray's photographs of mathematical models and the *Shakespearian Equations* that followed, describing the lacklustre paintings with the generosity only a friend could muster as 'bringing their inherent beauty in contact with organic themes'.[176] Given Duchamp's long-held interest in science, mathematics and the geometry of higher dimensions, the models would have had special resonance for him and it has even been suggested that the polyhedran and similar models inspired Duchamp's *Mile of Twine* installed for the New York exhibition *First Papers of Surrealism* in 1942.[177]

Duchamp's interests were shared by the young Matta, whose work formally and conceptually reveals his closeness to Duchamp from the late 1930s. Before becoming an artist, Matta gained some knowledge of geometry, mathematics and science from his architectural training, part of which took place with Le Corbusier. Max Ernst played on these enthusiasms of Matta's in 1948 during the period Ernst himself was introducing mathematical models into his work, producing a bio- and geomorphic portrait of Matta, *Mattamathics*. Matta recalled that after viewing the reproductions of Man Ray's photographs in *Cahiers d'Art* in November 1936, 'I moved on to another family of irregular, unbroken, geodesic forms.'[178] Less beholden to the 'ideological' ramifications of non-Euclidean geometry, Matta does not conscript their forms into his work literally but exploits their topology in search of ways of depicting the new spaces of

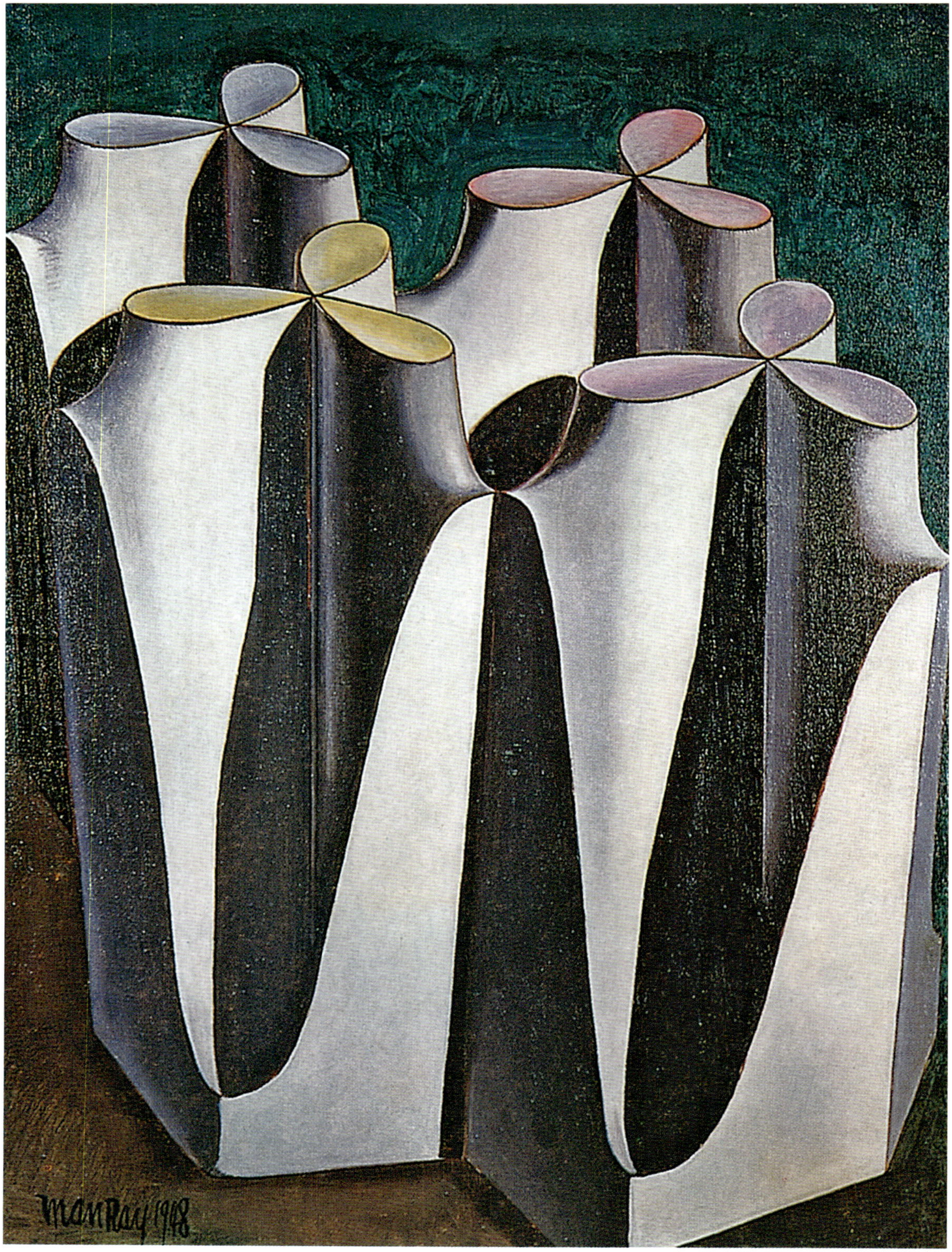
Man Ray 1948

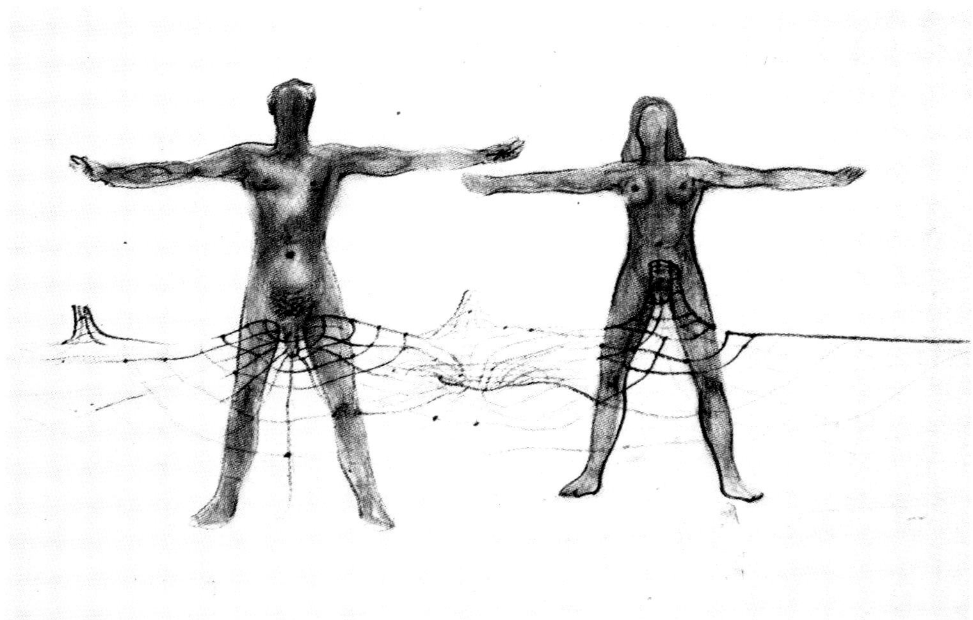

44 (facing page) Man Ray, *The Merry Wives of Windsor* (1948). Oil on canvas, 61 × 45.7 cm. Galleria Il Fauno, Turin.

45 (above) Matta, illustration from unpublished manuscript 'Redefinitions Towards a New Dimension for Painting,' 1939-40. Pastel and pencil on paper, 34 × 28.5 cm. The Nesuhi Ertegun and Daniel Filipacchi Collections.

physics in the old medium of painting. When they are manifested more literally in his canvases they prompt hallucinatory landscapes and act, unexpectedly, as the inspiration for erotic narratives, as in the Tanguyesque landscapes of 1938 and in the 1939–40 illustration from 'Redefinitions Towards a New Dimension for Painting' (fig. 45). In the second, Matta uses the same model Ernst used for *Once Upon a Time there was a Mouse in Milo*, drawing an analogy between the eros of psychoanalysis and the electromagnetic attractions and repulsions of mathematical science. I examine at length Matta's combination of psychoanalysis and physics later.

Soon after viewing Man Ray's photographs in *Cahiers d'Art*, Matta met the English artist Gordon Onslow Ford and the two began research along similar lines. As Onslow Ford remembered it, 'we had been influenced by an exhibition of mathematical objects that we had seen at the Trocadero', continuing, 'my knowledge was based on science, mathematics and seamanship.'[179] Onslow Ford refers here to the exhibition 'Art and Science' held under the auspices of the 1937 Exposition Internationale des arts et des techniques at Jean Perrin's newly established Palais de la Découverte. He said that in the year they met, 'Matta and I were fascinated by the mathematical objects (formulae expressed in 3-dimensional forms) that opened the mind to other levels of space-time . . .'[180] Like Matta, he was captivated by mathematical models and although Sawin has written, '"Divers types de points coniques" was the model that influenced Onslow Ford the most', the chequered, self-intersecting 'landscape' section of *The Luminous Land* of 1943 demonstrates he also shared an interest in the graph-like 'models for function theory' admired by Matta and Ernst.[181]

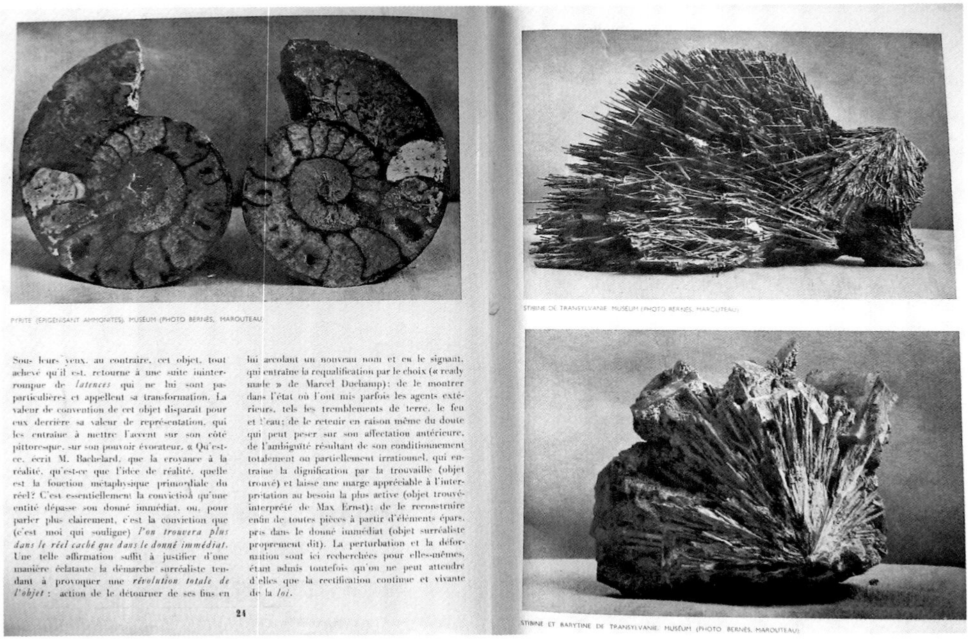

PYRITE (EPIGÉNISANT AMMONITES). MUSÉUM (PHOTO BERNÈS, MAROUTEAU)

Sous leurs yeux, au contraire, cet objet, tout achevé qu'il est, retourne à une suite ininterrompue de *latences* qui ne lui sont pas particulières et appellent sa transformation. La valeur de convention de cet objet disparaît pour eux derrière sa valeur de représentation, qui les entraîne à mettre l'accent sur son côté pittoresque, sur son pouvoir évocateur. « Qu'est-ce, écrit M. Bachelard, que la croyance à la réalité, qu'est-ce que l'idée de réalité, quelle est la fonction métaphysique primordiale du réel? C'est essentiellement la conviction qu'une entité dépasse son donné immédiat, ou, pour parler plus clairement, c'est la conviction que (c'est moi qui souligne) *l'on trouvera plus dans le réel caché que dans le donné immédiat.* Une telle affirmation suffit à justifier d'une manière éclatante la démarche surréaliste tendant à provoquer une *révolution totale de l'objet* : action de le détourner de ses fins en lui accolant un nouveau nom et en le signant, qui entraîne la requalification par le choix (« ready made » de Marcel Duchamp); de le montrer dans l'état où l'ont mis parfois les agents extérieurs, tels les tremblements de terre, le feu et l'eau; de le retenir en raison même du doute qui peut peser sur son affectation antérieure, de l'ambiguïté résultant de son conditionnement totalement ou partiellement irrationnel, qui entraîne la dignification par la trouvaille (objet trouvé) et laisse une marge appréciable à l'interprétation au besoin la plus active (objet trouvé interprété de Max Ernst); de le reconstruire enfin de toutes pièces à partir d'éléments épars, pris dans le donné immédiat (objet surréaliste proprement dit). La perturbation et la déformation sont ici recherchées pour elles-mêmes, étant admis toutefois qu'on ne peut attendre d'elles que la rectification continue et vivante de la *loi*.

24

STIBINE DE TRANSYLVANIE. MUSÉUM (PHOTO BERNÈS, MAROUTEAU)

STIBINE ET BARYTINE DE TRANSYLVANIE. MUSÉUM (PHOTO BERNÈS, MAROUTEAU)

46 André Breton, 'Crise de l'Objet,' *Cahiers d'Art*, Year 11, May 1936, pp. 24–25.

Written during the years in which the form of 'concretion' known as 'biomorphism' oversaw much of Surrealist painting and sculpture, both the content of 'Crisis of the Object' and its presentation in *Cahiers d'Art* owe much to studies of morphology (fig. 46). Artists like Ernst favoured books such as J. Bell Pettigrew's three-volume *Design in Nature* (1908) and D'Arcy Thompson's *On Growth and Form* (1917/1942), as fuel for collage as much as anything else. In particular, 'Crisis of the Object' seems modelled on Édouard Monod-Herzen's *Principes de Morphologie générale*, one of the most celebrated books in Surrealism in the 1930s.[182] A major source of information and imagery for the Surrealists, both Dalí and Matta mention it at some point and use it as a source of illustrations, whilst Wolfgang Paalen owned a copy of the book.[183]

In his *Foyers d'incendie*, Greek Surrealist Nicolas Calas proclaims *Principes de Morphologie générale* one of the most important scientific events of the time, pointing out Monod-Herzen's inspiration in the research of the Indian physicist and plant physiologist Jagadis Chunder Bose (on whom a brief text by Monod-Herzen appeared in the first number of the marginally Surrealist French review *Le Grand Jeu* in 1928).[184] Under the sub-heading 'Le Nouvel Esprit esthétique', Calas explicitly links Surrealist art with the new physics as expounded by Bachelard, before embarking on a lengthy pseudo-scientific discussion (with diagrams) of Monod-Herzen's book. Monod-Herzen is not mentioned in *The New Scientific Spirit*, of course, but Calas maintains his *Principes de Morphologie générale* sets out 'the laws where the new aesthetic spirit is forged'.[185] The exposition of these laws prefaces Calas's bravura attempt at a new synthesis of the physical and the mental, the non-Cartesian *rapprochement* that drove the Surrealist

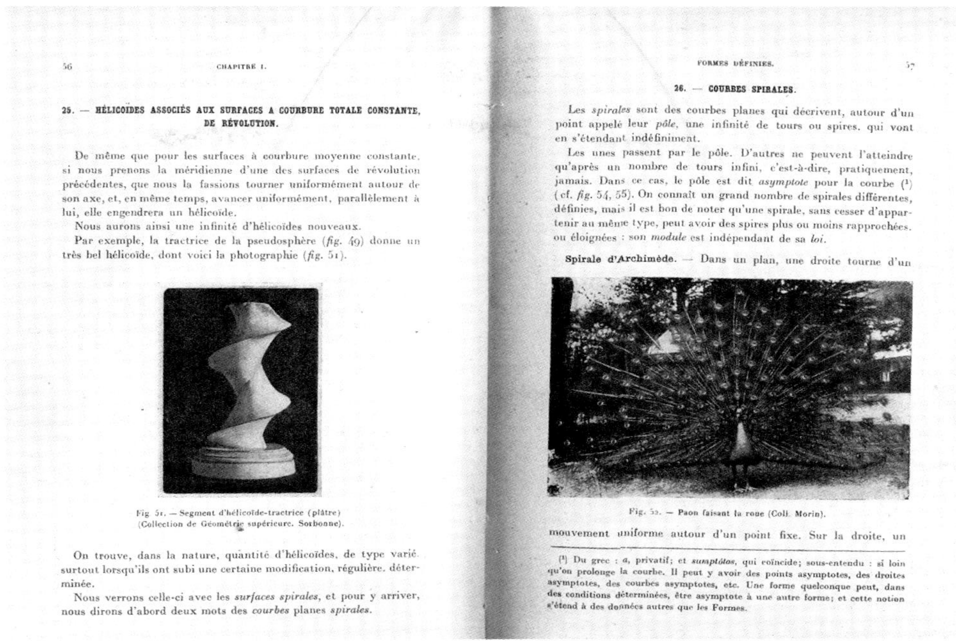

56 CHAPITRE I.

25. — HÉLICOÏDES ASSOCIÉS AUX SURFACES A COURBURE TOTALE CONSTANTE, DE RÉVOLUTION.

De même que pour les surfaces à courbure moyenne constante, si nous prenons la méridienne d'une des surfaces de révolution précédentes, que nous la fassions tourner uniformément autour de son axe, et, en même temps, avancer uniformément, parallèlement à lui, elle engendrera un hélicoïde.

Nous aurons ainsi une infinité d'hélicoïdes nouveaux.

Par exemple, la tractrice de la pseudosphère (*fig.* 49) donne un très bel hélicoïde, dont voici la photographie (*fig.* 51).

Fig. 51. — Segment d'hélicoïde-tractrice (plâtre) (Collection de Géométrie supérieure. Sorbonne).

On trouve, dans la nature, quantité d'hélicoïdes, de type varié, surtout lorsqu'ils ont subi une certaine modification, régulière, déterminée.

Nous verrons celle-ci avec les *surfaces spirales*, et pour y arriver, nous dirons d'abord deux mots des *courbes* planes *spirales*.

FORMES DÉFINIES. 57

26. — COURBES SPIRALES.

Les *spirales* sont des courbes planes qui décrivent, autour d'un point appelé leur *pôle*, une infinité de tours ou spires, qui vont en s'étendant indéfiniment.

Les unes passent par le pôle. D'autres ne peuvent l'atteindre qu'après un nombre de tours infini, c'est-à-dire, pratiquement, jamais. Dans ce cas, le pôle est dit *asymptote* pour la courbe (1) (cf. *fig.* 54, 55). On connaît un grand nombre de spirales différentes, définies, mais il est bon de noter qu'une spirale, sans cesser d'appartenir au même type, peut avoir des spires plus ou moins rapprochées, ou éloignées : son *module* est indépendant de sa *loi*.

Spirale d'Archimède. — Dans un plan, une droite tourne d'un

Fig. 52. — Paon faisant la roue (Coll. Morin).

mouvement uniforme autour d'un point fixe. Sur la droite, un

(1) Du grec : *a*, privatif; et *sumptôtos*, qui coïncide; sous-entendu : si loin qu'on prolonge la courbe. Il peut y avoir des points asymptotes, des droites asymptotes, des courbes asymptotes, etc. Une forme quelconque peut, dans des conditions déterminées, être asymptote à une autre forme; et cette notion s'étend à des données autres que les Formes.

47 Two-page spread from Édouard Monod-Herzen, *Principes de Morphologie générale*, Tome I 1927, pp. 56–57.

enterprise and one that by the 1930s could not ignore the new physics. The importance of morphology, he argues, lies in:

> a synthetic view of research pursued in three different directions: in the field of physical experience by the brilliant research of Bose, in the field of psychic experience by the no less brilliant research of Freud into the symbolism of dreams, and in the field of scientific speculation where morphology has assumed a new position thanks to the physico-mathematical conception we owe to the research of Bachelard. The new scientific spirit is triangular.[186]

Calas goes on to discuss Dalí's paintings and theoretical writings in this context. Dalí's preoccupation with morphology in the 1930s made him some kind of pioneer of the 'new aesthetic spirit' for Calas, who maintains that art can find ever greater inspiration by drawing on the same resources that had reanimated the scientific mind. Cannily joining this epistemological efflorescence and apparent mainspring for the regeneration of Surrealism in the late 1930s to the origins of the movement in the *nouvel esprit* proclaimed by Apollinaire in 1917 in which Surrealism was forged, Calas writes: 'It is high time that we approached aesthetics as we approach art, with a new spirit.'[187]

Containing photographs of mathematical models alongside rock and plant formations, *Principes de Morphologie générale* was passed around the Surrealist group in the 1930s, presenting a blueprint for the encounter between non-Euclidean geometry and natural morphology. Similarly, in *Cahiers d'Art*, 'Crisis of the Object' was interspersed with fantastic mineral formations meant to compare and contrast with Man Ray's photos of mathematical models. This context was supposed to demonstrate the fundamental Surrealist concern with a beauty that emerges and is spontaneously recognised outside of conventional conscious paths of apprehension. In this way, Breton dovetails Surrealist aesthetics with the 'natural' processes of Freud's id (as opposed to the encultured, civilising force of the ego) and the algebra of modern science, asserting the relationship between the *merveilleux* content of Surrealist objects, the extravagant forms produced by nature and the 'accidental' beauty of mathematical models – all supposedly contrived outside of aesthetic preconceptions. When he composed 'Crisis of the Object' in 1936, it must have seemed to Breton as though the revolutionary discoveries of the 'new scientific mind' in the fields of physics, mathematics and geometry sanctioned the direction taken by Surrealism, as it became increasingly clear that the movement had been failed by politics.

3

EPISTEMOLOGY AND POLITICS

Gaston Bachelard and Surrealism

Gérard Legrand claims that André Breton's allusions to Bachelard's *surrationalisme* were 'a little distant', but that hardly does justice either to Breton's borrowings from the text of that name or the larger scope of Surrealism's appropriation of Bachelard's philosophy of science.[1] That extends far beyond what has been imagined till now in the literature, and I have hardly begun to assess the way it conjoins physics, politics and philosophy in the movement, which is the purpose of what follows.

For much of the period I cover, physics was in a state of seemingly perpetual decline in France by comparison with the rest of Europe and the US. From the early 1930s, increased government intervention transformed French physics and by the end of the decade France was once again a world leader thanks largely to the efforts of the Popular Front coalition government. Naturally this had a fairly direct impact on the way in which politically inclined writers and artists like the Surrealists interfaced with science, and that is what I will discuss here with reference partly to some of the recent English language scholarship, which has paid overdue, in-depth attention to all aspects of Surrealism's politics.[2] But I also want to enlarge upon a discussion that I have already touched on, concerning the conceptual language of 'objectification' or 'concretion', which came about through the linking of politics to philosophy by physics, in Surrealism and beyond.

The interwar history of Surrealism's politics can be sketched briefly. Although Surrealism originated in a sense of outrage at the established order, and its theory accumulated around a notion of the overthrow of damaging, outmoded attitudes in the years following the First World War, the Surrealists did not welcome immediately either the Bolshevik Revolution in Russia or Communist Party politics at home. Louis Aragon's dismissive characterisation of the Russian Revolution, given as late as December 1924, as a 'vague ministerial crisis', was soon replaced by serious group discussions about Communism, leading to the Surrealists' protest in July 1925 against French colonialism in Morocco.[3] The greater warmth afforded Communism by the Surrealists continued that year in Breton's review of Trotsky's book on Lenin and by the alliance forged with the independent socialist cultural review *Clarté*.[4] The coalition began well but was never meant as a merger, and obstructions to collective action quickly caused it to break down. However, the partial surrender of independence by alliance members through publication in *L'Humanité*, the cultural journal of the Parti Communiste français, soon led to the total adherence to the PCF by Surrealists Aragon, Breton, Paul Éluard, Benjamin Péret and Pierre Unik in January 1927. From about that moment, stung by accusations of idealism from Georges Bataille and others, Surrealism also sought a means of salvaging a form of materialism from Freudian psychoanalysis that would complement its greater activity on the political level.

The Surrealists turned out to be a handful for the Communist Party, which neither trusted nor understood the Surrealist revolution. Bourgeois by birth, aristocratic by intellectual tendency and cliquish by inclination, the Surrealists were inevitably left in a quandary over how to serve the proletariat and operate within the confines of the Party. Its demand that they turn their hands to journalism or literature was met less with reluctance than incomprehension or profound repugnance. Although Breton and his friends claimed to join the Party as individuals rather than Surrealists, Breton's Communism always came second to his Surrealism, with a big gap in between. The ill fit between the two led to countless reproaches for Breton and Surrealism in the period leading up to the end of the 1920s, ultimately resulting in a stand-off with the Surrealists faithfully expressing their commitment to the Communist cause while retaining a detached bearing outside of the PCF. Surrealist subversion up to 1933 has a wild appearance – violent, scatological and bizarre – quite different in constitution from the systematic class protest under the steerage of Marx and Engels recommended by a PCF frequently bewildered by Salvador Dalí's demented ramblings and the Surrealists' taste for the Marquis de Sade. By the end of the 1920s, Surrealism's revolutionary rhetoric was despised by Communists such as former *Clarté* editor Jean Bernier, who viewed the *Second Manifesto* as a piece of theoretical posturing and took his one-time ally Breton to task for self-importantly proclaiming Surrealism's current position 'as if he were Einstein or Freud'.[5]

Aragon soon showed his colours. All the reports he sent to Breton in late 1930 during his trip to the Soviet Union to attend the Congress of Revolutionary Writers in Kharkov in the Ukraine insist on his unbending fidelity to the Surrealist cause, in spite of growing tension in his friendship with Breton. But the politically naïve and eager-to-please Aragon fell ever deeper under the influence of Communism at the expense of his Surrealism while in the Soviet Union. Just before his departure for France in December, he got himself in a muddle under pressure from the Communists, and together with fellow French delegate and Surrealist Georges Sadoul signed a document not only relinquishing whatever gains had been made for Surrealism at the Congress but also criticising Freudian thought and the *Second Manifesto*.[6]

A fearsome tongue-lashing from Breton back in Paris brought the confused Aragon back into Surrealist line temporarily, but his course was now set. It was confirmed in his crude propaganda poem 'Front Rouge', written in Moscow, published in France, and seized by the authorities in 1931 for its seditious content, including incitement to murder. Defended by the Surrealists on the grounds of freedom of expression yet derided for its zero Surrealist content, 'Front Rouge' also gave Breton the chance to unbottle some pent-up frustrations about the Party's cultural policy. Since *L'Humanité* gave as good as it got, the ties between the two camps now became transparently thin. The 'Aragon Affair' ended with Aragon's repudiation of Surrealism in March 1932 and 'defection' to the PCF, along with Maxime Alexandre, Luis Buñuel and Sadoul, bringing to a close a traumatic period in Surrealism and a deeply damaging episode in its relationship with organised Communism.

Surrealism's involvement in politics was transformed by extraordinary social and political unrest in France in the 1930s, without equal since the 1870s. From 1931, France had been hit harder by the world recession. As a succession of Radical governments struggled to deal with mass unemployment, increasing poverty and general economic decline, a significant portion of the population of France called for a similar style of authoritarian government they had seen accede to power in Germany in 1933. An extremely volatile situation was exacerbated by corruption, misjudgement and inept gestures at the highest levels of government, and by reports in the right-wing press of political conspiracy. A demonstration on 6 February 1934 organised by several extreme right wing and fascist groups took place in Place de la Concorde, turning into an attempted *coup d'état* as 60,000 protesters set off for the Chamber of Deputies, accompanied by several thousand equally disenchanted Communists. The police intervened and a riot ensued, leaving around fifteen dead and two thousand injured (though figures on this vary a little). Although it failed, the violence unleashed by the attempt to bring down the government led many, including the Surrealists, to make decisions or affect compromises that bring home the factional complexity of political life in France in the 1930s. The Neo-Socialist Maurice Déat, for instance, called for a revolution even if it opened the way for fascism in France, while Breton protested against the fascist riots in the tract of 10 February *Appel à la lutte* (*Call to Battle*), calling for a general strike. Written at the behest of a group of radicals, Breton was joined by ninety signatories with many of whom he held little sympathy. A further demonstration on 9 February by the Communists participated in by the Surrealists led to more terrible violence.

These circumstances would lead to the creation of Léon Blum's Popular Front, which sought to unite groups on the left in the face of the rising tide of fascism at home and abroad. In March, several of the Surrealists joined the newly formed Comité de vigilance des intellectuels antifascistes (CVIA).[7] A kind of forerunner to the Popular Front, this alliance had a broad political base and was chaired by the physicist Paul Langevin, who was a *compagnon de route* of the Communists at the time, and by the Socialist Paul Rivet, who was then director of the Trocadéro and a contributor to *Scientia* and *Minotaure*, having been on the original Comité de rédaction of Bataille's *Documents*. In July, the Soviet Union ordered the PCF into an alliance with the Socialists, and these conciliatory moves temporarily calmed the Surrealists' turbulent relations with the Party, though not for long. Alliances among avant-garde groups, too, were breaking down the determined insularity of the Surrealist group, stubbornly maintained by its members since its formation. This process is most evident in the Surrealists' involvement with *Minotaure* from 1933 onwards, the first journal to which they consistently contributed that was not exclusively Surrealist. The same tendency was furthered by the changing political climate, bringing about, for instance, the coalition between several of the Surrealists and Bataille and his friends in the founding of the anti-fascist group Contre-Attaque in 1935.[8]

In May 1935, Breton travelled to Tenerife to help organise an international exhibition of Surrealism. In an interview with the cultural journal *Indice*, he was asked: 'What is the attitude of Surrealism towards the most important theses of dialectical materialism and of contemporary psychology?'[9] This gave him the chance to hold forth on his current thinking on politics, psychoanalysis and physics. Without mentioning a political party, Breton confirmed the Surrealists' commitment to class struggle and loyalty to dialectical materialism, specifically renewing their allegiance to Hegel. On the subject of 'modern psychology', he reasserted the Freudian credentials of Surrealism while, significantly, distancing it from 'the greater part of Freud's philosophy', condemned as '*métaphysique*'.[10] From a rejection of metaphysics to a new appeal to physics proved a short step, for Breton continued as follows:

> Let us remember that three great scientific discoveries (evolution, the cell, and the transformation of energy) permitted the edification of the dialectical materialist system of nature, and correlatively aided the understanding of the general laws of the development of society. I regard the consideration of current scientific developments to be more worthwhile than those of the psychological movement, which always lags behind the former. Of all sciences, modern physics seems particularly to require our attention.[11]

If physics was starting to look more attuned to Surrealism than the 'psychological movement', then in all probability science had also started to seem to Breton a lot less troublesome than politics as a means towards theorising the 'transformation of the world'.

By then the Surrealists were becoming increasingly divided over their role in the era of the Popular Front and largely disenchanted over their dealings with the PCF. In the second half of 1934, a mood of gloom and impending calamity spread through Surrealism's politics in spite of the temporary group solidarity over Aragon's misdemeanours and the shared understanding of the requirement for decisive political activity. Internal disaffection over how best to carry out the second of these led to a major secession from Surrealism, studied in detail by Harris, who writes:

> the separation of Tzara, Crevel, Monnerot, Caillois, Char, and Giacometti from the rest of the group, in the months preceding the International Congress of Writers for the Defence of Culture in June 1935, was effected not only for political reasons but on theoretical grounds as well. The decision taken by this fraction of the Surrealist group to work with the Popular Front, as they chose to do in 1934–5, was premised not only on the imperatives of the antifascist struggle, in which they felt

they had to take part, but also on a belief that Surrealism was slipping back into art and literature.[12]

Breton wished to speak at the Communist-sponsored International Congress of Writers for the Defence of Culture, but ruined his chances a week before it began with his notorious (and still remarkable) physical assault on the Russian journalist Ilya Ehrenburg, who had made the mistake of libelling the Surrealists a year earlier. Unfortunately for Breton, Ehrenburg was on the organising committee for the Congress, and now arranged the censure of Surrealism, blocking Breton's inclusion in the programme in spite of the efforts of the tormented René Crevel, who was also on the committee (as was Aragon). Crevel's disastrous suicide a few days later came about partly from the ill will that marked relations between Surrealism and the PCF. That relationship was now brought definitively to a close by Breton's justification of Surrealism's position, rubbishing of the Congress, and attack on the social and cultural policy of the Soviet Union in the no-turning-back text 'On the Time the Surrealists were Right' of August 1935.[13]

After years of discord, Breton undoubtedly felt some relief alongside the obvious dismay at the severing of ties with the PCF. The Tenerife interview already showed him anticipating trouble and scanning the horizon for new pastures for Surrealism. He was not alone in this, as Harris notes when he records the attitude of those who seceded from Surrealism in search of an enterprise '*more* scientific in its rational investigation of affectivity', in order, ironically, to bring them closer to the PCF.[14] The linking of politics and science under Bachelard's mediation was well under way close to Surrealism, in fact. We can leave Breton for a while, guessing about Surrealism's political future and wondering about the potential of modern physics, and look now at the political and social context for that transaction and its scope within the surrealised avant-garde.

French Physics in the 1930s: Jean Perrin and the Popular Front

Since Jean Perrin won the Nobel Prize in 1926, and increasingly since 1930 due in part to de Broglie's Nobel success in 1929, science in France benefited from a greater interest shown at government level in funding experimental research. From the 1920s, Perrin had conducted an alarmist campaign on behalf of French science, warning that France was in danger of sinking to the level of a third-rate scientific power (with Germany, as ever, the comparative measure and threat). Scientists turned out to be extremely effective political lobbyists with many friends in the Chamber of Deputies, and by 1930 a strong foundation existed for the state to organise and fund research. The pressure Perrin and others placed on government led to an extraordinary revival in the fortunes of science generally and physics particularly, to the extent that historian Harry W. Paul writes that the 'growth of scientific organisations, both major and minor, accelerated in the 1930s, a period of creative reordering unequalled in the history of French science since the 1870s and 1880s'.[15]

Due to his high standing among French scientists and active engagement in left-wing politics, Perrin became an important figure when his friend Léon Blum led the Popular

Front government to electoral victory in 1936 (fig. 48). The Popular Front was committed to greater funding for research, confirmed by the establishment of an Undersecretary of State for Scientific Research in Blum's cabinet. This role was filled initially by the politically inexperienced Irène Joliot-Curie, and then by Perrin himself in Blum's second ministry of 1938. According to Mary Jo Nye, Perrin was very effective in his role, securing remarkable increases in state funding for scientific research.[16] While to an extent it was the outcome of French fear of a technologically and militaristically resurgent Germany, Perrin's achievement is impressive, nevertheless, for being made in the years before massive state funding of physics became the norm in the nuclear age. The interaction of politicians and scientists paid enormous dividends for scientific teaching and research in France in the 1930s, with physics at its head.

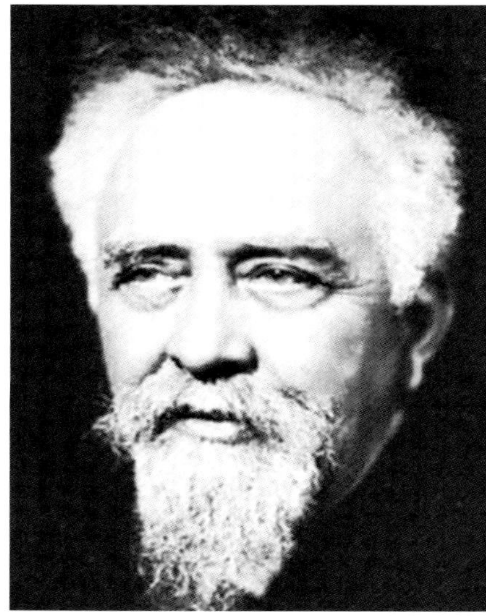

48 Jean Perrin in the 1920s.

During this period, too, the city of Paris was gearing up for the massive Exposition Internationale des arts et des techniques of 1937. Designed to give notice of France's complete recovery from the trauma of the First World War, the Exposition placed France at centre stage in matters of art, science and technology. Perrin presided over a major congress in the year of the Exposition, on biology, chemistry and physics. Attended by twelve hundred scientists (including eleven recipients of the Nobel Prize) from sixty-five countries, it must have seemed as though France had made up all the ground lost since the International Congress of Physics met in Paris in 1900.[17] Keen to emphasise close links between art and science, the Exposition included a vast exhibition entitled 'L'art et la science', the first of several given over to the relationships between art and science held at Perrin's creation, the Palais de la Découverte. Although Surrealist artists were not represented at the exhibition, Éluard, Caillois and Breton all participated in Exposition events and Gordon Onslow Ford and Matta saw the display of mathematical models held there.[18]

A reformist rather than a revolutionary, Perrin was a member of neither the PCF nor the Socialist Party (SFIO), yet his leftist sympathies were deeply felt and widely broadcasted. They combine with a key component of his political lobbying: an attitude towards science as the purest expression of progressive thinking, which Paul calls a 'synthesis of Enlightenment scientism, Littréan aesthetics, revolutionary camaraderie, all tinged by the Marxian hope of liberation from *Entfremdung*'.[19] If such a view was bound to attract the interest of Communists, then his shades-of-Bachelard characterisation of scientists as 'voyants' and his description of the 'physico-chemical *esprit*' as 'developed from a profound taste for Mystery to which risk gives even a charm, from passionate desire for Adventure, and from that love of Poetry and Beauty without which there is no artistic creator' would have found an enthusiastic audience in Surrealism.[20]

Close to Perrin, Paul Langevin was much more the outspoken Marxist and, like Perrin, unwilling to dissociate the radicalism of the new physics from political and social radicalism. Although Marxism has always sought authority through identification with science, the amazing conquests of physics since 1900, culminating in the 1920s, account for the increased urgency of attempts to conscript Relativity and QM into Marxism by writers in 1930s France. In the face of the scepticism of most physicists, Jacques Solomon (son-in-law of Langevin and a physicist himself) made a typical attempt to square modern physics with Marxist materialism in a conference review in *Cahiers du Bolchevisme* in 1938.[21] Arguing that its overturning of established ideas made physics 'une science révolutionnaire', Solomon obediently follows the wayward path trodden by Lenin in *Materialism and Empirio-Criticism* (1909), along which the 'agnostique' Ernst Mach came in for a sarcastic drubbing as a solipsistic, reactionary idealist for his insistence on the role of sensation in the description of space, time and the physical world.[22] Threatened by a short-lived Bolshevik vogue for Mach, in his book Lenin had clumsily attempted to flatten all philosophical, scientific and mathematical stirrings of idealism, subjectivism and relativism from Berkeley through Kant to Poincaré, to confirm the status of a world empirical, objective and absolute beyond sense perception, in the name of dialectical materialism.[23] Solomon lumps Mach and the Vienna Circle together, aligns Lenin's censure of Mach historically with the moment at which 'Langevin and Perrin demonstrated brilliantly *the reality of atoms*', and cack-handedly advances the revolutionary credentials of modern physics by comparing it with alchemy.[24] Reporting Langevin's affirmation of the reality of an external world and upholding of the materialist conception as one of the conference highlights, Solomon complains about the delegates' ignorance, on the whole, of the Great Minds of dialectical materialism, and insists on the essential link between Marxism and modern physics:[25]

> They must be shown that the dialectical materialism of Marx, Engels, Lenin, and Stalin represents the modern form of the scientific conception of the world. It is linked to science in movement. Its contents are adapting, modifying with the progress and the transformations of science.[26]

Attempting to yoke 'scientific' socialism to modern physics, Solomon's essay expresses commonplace views within French Communism in the era of the Popular Front in the argumentative and theoretical bases it touches.[27] I now look at how this discourse plays itself out close to and within Surrealism.

Inquisitions *and the Groupe d'Études pour la Phénoménologie humaine*

In October 1935, a few months after Breton's statement to *Indice* about politics, psychoanalysis and physics, Tristan Tzara, former Dada leader and now freshly disaffected Surrealist, organised a discussion circle, the Groupe d'Études pour la Phénoménologie humaine. The idea was that the group would meet fortnightly to hear and discuss a paper on politics, philosophy or science, and that these would be collected as essays to form the theoretical component of a journal, *Inquisitions* (fig. 49). Of the fifteen members of the Groupe, eight had been or were still just about Surrealists, or were sympathetic to Surrealism. Like *Inquisitions* itself, the collective sprang directly from the complex political situation of the time and was close to the Popular Front. For these reasons, an examination of the organisation, background and content of the Groupe d'Études pour la Phénoménologie humaine and *Inquisitions* forms an appropriate case study for this discussion.

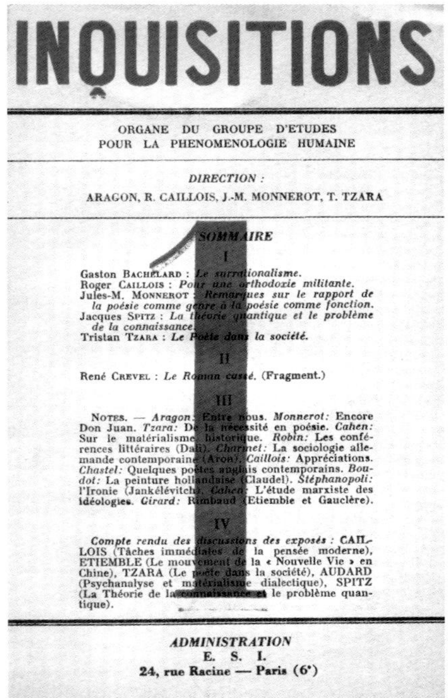

INQUISITIONS

ORGANE DU GROUPE D'ETUDES POUR LA PHENOMENOLOGIE HUMAINE

DIRECTION :

ARAGON, R. CAILLOIS, J.-M. MONNEROT, T. TZARA

1

SOMMAIRE

I

Gaston Bachelard : *Le surrationalisme.*
Roger Caillois : *Pour une orthodoxie militante.*
Jules-M. Monnerot : *Remarques sur le rapport de la poésie comme genre à la poésie comme fonction.*
Jacques Spitz : *La théorie quantique et le problème de la connaissance.*
Tristan Tzara : *Le Poète dans la société.*

II

René Crevel : *Le Roman cassé.* (Fragment.)

III

Notes. — *Aragon:* Entre nous. *Monnerot:* Encore Don Juan. *Tzara:* De la nécessité en poésie. *Cahen:* Sur le matérialisme historique. *Robin:* Les conférences littéraires (Dali). *Charmet:* La sociologie allemande contemporaine (Aron). *Caillois:* Appréciations. *Chastel:* Quelques poètes anglais contemporains. *Boudot:* La peinture hollandaise (Claudel). *Stéphanopoli:* l'Ironie (Jankélévitch). *Cahen:* L'étude marxiste des idéologies. *Girard:* Rimbaud (Etiemble et Gauclère).

IV

Compte rendu des discussions des exposés : CAILLOIS (Tâches immédiates de la pensée moderne), ETIEMBLE (Le mouvement de la « Nouvelle Vie » en Chine), TZARA (Le poète dans la société), AUDARD (Psychanalyse et matérialisme dialectique), SPITZ (La Théorie de la connaissance et le problème quantique).

ADMINISTRATION
E. S. I.
24, rue Racine — Paris (6e)

49 Cover of *Inquisitions*, no. 1, June 1936.

The motley band of former Surrealists on the editorial board presiding over the one and only number of *Inquisitions* in June 1936 is a further indication of more mobile alignments among writers and intellectuals, reflecting the new politics of the Left. It consisted of Louis Aragon, now a leading member of the PCF, whose break with Breton and the Surrealists in 1932 was bitter and irrevocable; Roger Caillois, who departed Surrealism for non-political reasons in December 1934; Jules Monnerot, who was a relatively uncommitted leftist holding residual sympathies with Surrealism, which would emerged more fully later;[28] and Tzara himself, who ended his involvement with Surrealism in a letter to *Cahiers du Sud* in 1935, giving his allegiance to the PCF.[29] Aragon's praise for *Vigilance*, the bulletin of the CVIA (to which the Surrealists continued to give their support), between the covers of *Inquisitions* says much for the curious political atmosphere guiding loyalties during this brief period.[30]

Issuing from the convergence of the politically committed avant-garde and modern science, *Inquisitions* was deeply concerned with the function of the new physics in society and its philosophical potential for politics. Among the five points advertising the intentions of the Groupe d'Études pour la Phénoménologie humaine are two that reveal the importance it deemed to reside in modern science:

> – to find the current of theoretical and affective ideas best equipped to tackle the problems posed by the human sciences
>
> – not to be limited to the literary aspect of this problematic, and to open up a perspective on the intellectual level inclusive of all the sciences placed at the service of man, following the example of the Popular Front on the political level . . .[31]

When *Inquisitions* actually appeared, certain of the aims set out in its *bulletin de souscription* confirmed those of the Groupe. Probably penned by Roger Callois, these go so far as to promise special attention to modern science, and to distance *Inquisitions* from the world of art in which its contributors felt Surrealism had become mired by its involvement with the expensive review *Minotaure*:

> The review 'INQUISITIONS' intends to assemble and compare the findings of those modern scientific or para-scientific studies which, in several of the fields in which they have been taken up, have proved well-suited to facilitate a crystallisation likely to meet some of the principal questions presently in play . . .
>
> Modern science, in the majority of its fields, offers infinitely more deserving and promising foci of interest than the contributions of artistic preoccupations, for example. A task more urgent than the co-ordination of this material could hardly be conceived, it being understood that the principle of co-ordination can only be the directly human aspect of this research, incompatible with retrograde tendencies of all kinds.[32]

Overseen by a politically active editorial board and published by Éditions sociales internationales (ESI), the official publishing house of the Communist Party, *Inquisitions* was driven towards a political and pedagogical goal: to bring the new science to the masses, as confirmed in the mission statements I quote here.[33]

A closer look at *Inquisitions* reveals how it attempted to carry through this objective. The periodical is an art- and poetry-free zone and contains barely any literature (even though it was concerned with the theory of poetry). Of the five theoretical articles it carried in its sole, inaugural number, three are of a scientific nature to a greater or lesser extent. Jacques Spitz read a paper to the Groupe on 4 February 1936, which appeared in the review under the title 'La Théorie Quantique et le problème de la connaissance' ('Quantum Theory and the Problem of Knowledge').[34] Forgotten today in the anglophone world and barely remembered in France, Spitz was educated as a physicist before becoming a novelist in the late 1920s, working in a genre that can be placed somewhere between Surrealism and science fiction.[35] He contributed at least twice to the *Nouvelle Revue française*, with a review of Léon-Pierre Quint's *Le comte de Lautréamont et Dieu* in 1930, and a piece on QM in 1938.[36]

Spitz's article in *Inquisitions* offers an uncomplicated insight into the ramifications of the new physics for philosophy, dwelling at length on de Broglie's wave/particle duality and Heisenberg's Uncertainty Principle, which, as I have described, shows 'the impossibility of considering phenomena as ruled by a strict determinism.'[37] Repeating a fairly common line of reasoning during the period, Spitz avers that its application to the domain of philosophy sheds light on difficulties met by that discipline in its analysis of consciousness. Turning to Bohr's Complementarity, he notes that just as the fullest description of the behaviour of matter can be given only by recourse to *both* the particle *and* the wave models, so an interpretation of consciousness is impossible if it is 'purely physiological (purely objective), or . . . purely psychic (purely subjective)'.[38] Spitz's is not the most sophisticated application of the new physics to philosophy, but it acts as further evidence that a reworking of philosophical dualism under the guidance of QM penetrated Surrealist circles by the mid-1930s. Questioning the Cartesian–Kantian separation of subject and object, it offered an alternative duality to the Hegelian dialectic resolved through synthesis, as in the unification of opposites demanded by Breton in the *Second Manifesto of Surrealism*.

The discussion following Spitz's paper when it was read to the Groupe was a lively one, and Roger Caillois's response is the main source of illumination, showing his familiarity with the new physics and the possibility of its application to philosophy.[39] Caillois's own offering was read to the Groupe on 8 January 1935, titled 'For a Militant Orthodoxy: The Immediate Tasks of Modern Thought' ('Pour une orthodoxie militante: les tâches immédiates de la pensée moderne'), and concerned with the binding of the social collective through science now that myth had lost its social function.[40] Against a background of French economic decline and social unrest, Caillois sketches a general intellectual malaise, leaving not even a once-vigourous Surrealism untouched. By this time, Caillois had become a minor thorn in Breton's side, critical of the 1930s Surrealism of *Minotaure*, which he thought burdened with 'semi-aesthetic activities' and a lack of experimental sobriety.[41] Caillois's paper sustains his demand for greater clarity in intellectual discourse, arguing that the Rimbaudian retreat from reason could no longer be tolerated: '[i]t is high time we broke,' he says, 'with this form of intellectual hedonism.'[42] This requirement is extended in a remarkable footnote claiming that scientific rationalism now held the place once occupied by poetry, using a parallel along similar lines to Aragon's complaint in *Treatise on Style* about the comparable fate undergone by Einsteinian and Rimbaudian imagery:

> In any case, it is significant that the boldest and best-led attempts to destroy the *forms of sensitivity* and the projections of perception came not from poetry but from science, especially from Relativity physics. It has thus provided the most important contribution toward the enactment of the *long, immense, and reasoned derangement of all the senses* called for by Rimbaud.[43]

For Jean Wahl, this striking passage emblematised the Nietzschean energy and aggression of *Inquisitions*.[44] Here the meaning of Caillois's 'orthodoxie militante', advancing the notion of a new methodology underpinning revolt that is scientific, rational and systematic, is crystallised.

On the face of it, this programme seems at odds with that of Surrealism. Yet Caillois's favourable citation in his essay of the 'exemplary' figures Baudelaire, Rimbaud and Lautréamont confirms that his move away from Surrealism in search of scientific exactitude did not enjoin his rejection of the thought of the figures that he and the Surrealists admired and hoped to emulate.[45] Rather, he seeks a new perspective from which their achievements would not simply be surveyed intact, but would be seen in a new way by a modern audience subject to the new reality brought by the architects of modern physics. Caillois writes of the necessity of matching intellectual intuition with experience, citing with approval the book by Philipp Frank of the Vienna Circle, *Théorie de la connaissance et physique moderne*, for its rejection of Cartesian 'common sense'.[46] During a time of disorder, Caillois insists in familiar terms on the requirement of philosophy to undertake the same procedure of 'generalisation' as geometry and physics, 'whereby Riemann's geometry absorbed Euclidean geometry and Relativity physics that of Newton, taking them as particular cases of a more comprehensive synthesis'.[47]

As Caillois indicates in a footnote, the term *généralisation* is taken from Bachelard's *New Scientific Spirit*, also exploited at exactly that moment by Breton in his discussion of the object. I have already covered one aspect of Breton's reading of Bachelard, and I will return to it later in the political context. During this period Caillois was equally under the influence of Bachelard's writings, particularly *The New Scientific Spirit*. More than this, a whole generation of thinkers relied heavily on Bachelard's work, making it the single most important factor in the communication of *la physique nouvelle* to the Surrealist milieu.

Gaston Bachelard and Surrealism

Bachelard's status as a university lecturer did not prevent him forming friendships with radical thinkers outside of academia, and his editorial position on the journal *Recherches philosophiques* from 1934 furthered such communication. In fact, it was at this time that he first came into contact with Caillois, presumably once the young writer had read *The New Scientific Spirit*. That conjunction led to Caillois's appearance in *Recherches philosophiques* and Bachelard's contribution to *Inquisitions*. According to one commentator, it was through Caillois 'in September 1934 that Bachelard truly came to know poets and the Surrealist movement'.[48] Bachelard's correspondence with Caillois and Tzara in 1935 reveals that the two former Surrealists sent their recently published work to Bachelard, and that he responded favourably (he seems also to have corresponded with René Char).[49] Unstinting in his praise of Surrealism, Bachelard was obviously unaware of the rancour likely between individuals in, out, or on the way out of the group, at a sensitive moment in its history when Surrealism could have been fitted with a turnstile. Mutually beneficial exchange was the order of the day between Bachelard and the larger Surrealist milieu in the 1930s. At *Inquisitions*, this came about when Caillois wrote to him in December 1935 requesting a contribution to his and Tzara's new review, to which Bachelard immediately assented. The piece was close to completion by the first week of the following month.

Alongside the contributions to *Inquisitions* by Spitz and Caillois drawing on modern physics and addressing its philosophical and social value was Bachelard's densely argued tract on the theory of knowledge, 'Le Surrationalisme',[50] in which Bachelard asserted that the agency of habit was representative of a malevolent power, to be confronted and defeated so that it could be replaced by anticipation. Human reason, he maintained, should discard its passivity in favour of an active function, which would help to realise a '*surrationalisme*' acting with Surrealism so that 'sensibility and reason will . . . be restored to their fluidity'.[51] Bachelard's text opens the periodical, signifying at once the eminent standing of its author and its manifesto-like importance.[52] Yet in allowing its publication at the front of a supposedly post-Surrealist journal, the editors scored something of an own goal. Bachelard's genial defence of Surrealism's insurgency of the imagination and conjoining of it with all that was most radical in science must have made Tzara and the others squirm, as much as it would have been music to Breton's ears.

The content and tone of 'Le Surrationalisme' were in part determined by Bachelard's apoliticism and recent discovery of Surrealism, which together bring elements of his essay into conflict, too, with Caillois. Bachelard expresses in 'Le Surrationalisme' his preference for the poetic *aperçu* over the drawn out accumulation of facts. At the same time, he employs a bravura tone and rhetoric of freedom familiar from Surrealism. Taken together, message and delivery heighten the drama of his text, and where he italicises his declaration '[*i*]*f, in any experience, one does not risk one's reason, that experience is not worth while attempting*', the voice is less that of an academic than a Baudelairean poet.[53] Here as elsewhere, Bachelard makes much of his impatience with officialdom, convention and regulation. Ironically, where Caillois's interpretation of *The New Scientific Spirit* – sustaining his demand for intellectual sobriety – suggests a passage out of avant-garde and into academia, Bachelard's demand that reason be placed at risk indicates the opposite route.

I noted earlier the importance of that widely read volume for 'Crisis of the Object', to the point of Breton's casual plagiarism of its key points and sources. Aside from Breton, among those in the Surrealist group or on its periphery who acknowledged *The New Scientific Spirit* were Caillois, Nicolas Calas, Michel Carrouges, René Crevel, Jules Monnerot, Denis de Rougemont, Pierre Mabille, Jean Paulhan, Raymond Queneau and Zdenko Reich. From the evidence of their writings, it is clear that Tzara and Dalí read it, and Aragon, Bataille and Char were undoubtedly familiar with Bachelard's reflections on epistemology. A copy of *The New Scientific Spirit* was in Wolfgang Paalen's library, and even that inveterate bibliophobe Marcel Duchamp read passages from it and offered a typically nonchalant critique of some of its assumptions.[54] Its extraordinary currency has barely been noticed in the scholarship on Surrealism until now, yet Bachelard's *New Scientific Spirit* was a pivot for debates among the Surrealists and their friends (and enemies) on science, epistemology, politics, the object and art, as well as a bridge between Surrealism and postwar French philosophy.

The foundations of Bachelard's epistemology are laid in *The New Scientific Spirit*. In my chapter on Breton I gave an account of its contents relevant to his borrowings. Here, to link the book to its philosophical and political context, I wish to introduce one of its main themes – which can be termed 'objectivation', 'concretion' or 'realisation' – by recourse to the particular meaning the author gives the terms *rationalisme* and *réalisme*.[55] Where the first describes the subjective sphere of intuition and interpretation, the second indicates the objective domain of experiment and experience. One cannot exist without the other; both are essential for a conglomerative philosophical account of 'reality' that does justice to mind and nature. Bachelard's assertion at the outset of *The New Scientific Spirit* that 'the application of scientific thought seems to me to tend essentially toward reality', together with his declaration of his intention to 'demonstrate what might be called the realization of the rational or, more generally, the realization of mathematics', describe the 'outward' process of scientific thought in the first quarter of the century.[56] The movement from the rational to the real, or from the subjective to the objective, encompasses the period spanning the 1905 publication of Einstein's Special Theory of Relativity and the 1927 formulation of QM.

Although Bachelard describes a process of scientific enquiry that took place over the preceding thirty years, he expresses it in a language peculiarly of the moment, for the notion of a 'concretion' of subjective 'material' was widespread, existing before and after the publication of *The New Scientific Spirit*. Evident in Breton's 'Crisis of the Object' ('The whole pathos of modern intellectual life resides in this unremitting quest for objectification'[57]), and in Paul Éluard's 1935 phrase 'physique de la poésie', it surfaces in Dalí's expression of 1930, 'true realizations of solidified desires', and in his writings of that decade describing a 'thickening' of space.[58] It is equally evident in the overwhelming tactility of Dalí's pictorial universe of the 1930s, with its carnivorous creatures and soft, viscous structures, as opposed to the airy 1920s automatism of André Masson. Outside of Surrealism, the same tendency towards *matérialisation* is articulated in the 'physicalism' of the Vienna Circle and enshrined in book titles such as Albert Spaïer's *La Pensée concrète*,[59] Jean Wahl's *Vers le concret*[60] and Armand Petitjean's *Imagination et réalisation*.[61]

A closer look at Petitjean's book reveals another perspective on Surrealism, Bachelard and modern physics, in the context of politics (fig. 50). Although *Imagination et réalisation* was never reprinted and is now virtually unobtainable and largely forgotten, it offers a kaleidoscopically fragmented panorama on many of the authors, themes and concepts I look at here.[62] Demonstrating a broad understanding of the modern physics and philosophy of Bohr, Eddington, Einstein, Heisenberg, Langevin, Meyerson, Whitehead and so on, *Imagination et réalisation* refers to several of the Surrealists with whom Petitjean had contact. Born in 1913 and active by 1935 in a cultural enviroment saturated with Surrealism, Petitjean was a young 'non-conformist' and contributor to *Minotaure*. An Anglophile and Germanophile, he showed an inclination 'to the modernism of James Joyce and German *Lebensphilosophie*' according to Claudine Frank-Elster, submitting his writings to the literary journal *Mesures* in the late 1930s alongside an assortment of surrealised avant-gardists, among them Bataille, Breton,

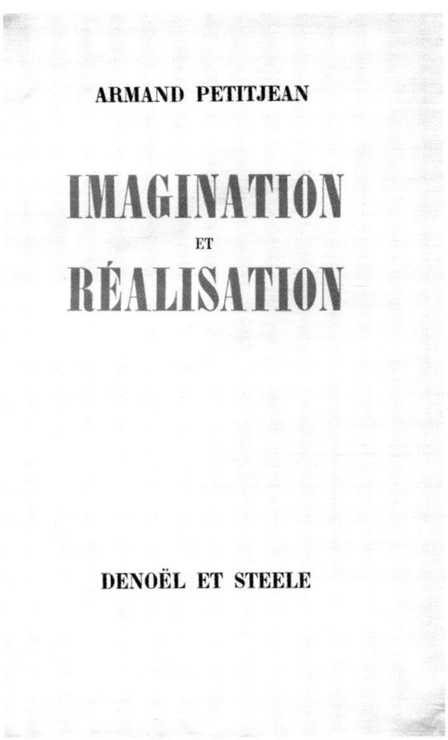

50 Armand Petitjean, *Imagination et réalisation* (1936).

Caillois, Daumal, Éluard, Leiris, Georges Limbour and Raymond Queneau, all of whom published in *Mesures* between 1935 and the war.[63] In one way, *Imagination et realisation* holds up a kind of contemporary mirror to the story I am telling, because Petitjean's ideas are driven largely by Surrealism and modern physics, deserving more attention than I can allow them here.

Petitjean's celebration of the creative, active imagination in *Imagination et réalisation* is largely the outcome of his contact with Bachelard and the Surrealists. Breton responded by quoting from the book in his *Anthologie de l'humour noir*, which is a little odd given Petitjean's criticism of Surrealism and the avant-garde generally.[64] According to Petitjean, the virtue of the imagination does not lie in its capacity to reproduce familiar forms, but in its ability to anticipate new ones. His observation that language touches 'on the very physics of thought',[65] and his aim, expounded in *Mesures*, to elucidate this 'physique de la pensée' alongside a 'psychique de la nature' by means of 'the intermediary of the psycho-physiology of the eye . . . which might allow one to perceive in the mind the great laws governing at once the world and the mind', aim to break with the distinction between the inner and outer, to discover a language that might conflate psychology and physics.[66] To this end his work drew upon Einstein's space-time, the writings of the physicist Gustave Juvet, and Bachelard's *New Scientific Spirit*.[67]

Frank-Elster argues that Petitjean's work as a whole (four books between 1936 and 1946) manifests 'a rather distinctive conjunction, within the French context, of modernism and fascism'.[68] However, this is not really evident in *Imagination et réalisation*, which is more the work of a part-processed, residual leftist revolutionary. The book bears the impress of Surrealism, while its rhetoric is one of stubborn resistance to the movement's cultural primacy and its style, tone and content are shaped in this conflict. Although Petitjean expressed his support for the Popular Front in 1936, and an implicit politics informs his column 'Battle of the Books', which appeared from late 1937 to late 1938 in *Vendredi* (a journal sympathetic to the Popular Front), political sympathies are entirely overridden in *Imagination et réalisation* by its intellectual debt to Bachelard.[69] In this light, Bachelard's very favourable review of the book in the *NRF*[70] and his warm appreciation of Petitjean's examination of 'biological poetry' in his *Lautréamont* amount to a kind of nepotism.[71]

Published in 1939, the year after his *Formation of the Scientific Mind*, and a year before his *Philosophy of No*, Bachelard's interpretation of the great hero of the Surrealists in his *Lautréamont* is deeply affected by the scientific and philosophical context I am outlining. His study of *Les Chants de Maldoror* confirms the parallels discovered in his earlier writings of the decade, between the poet's *métier* and the physicist's, applying to a work of literature the imagination of a philosopher of science.

Bachelard argues that in Lautréamont's universe, 'energy is an aesthetic'.[72] This can be taken as the *leitmotif* of his analysis, for it is the 'muscular lyricism',[73] the 'images of propulsion'[74] and Lautréamont's 'hierarchy of velocities' upon which the scientist's imagination alights.[75] He sees Ducassean poetics not simply as a receptacle of sensations received from the world to be offered benignly to the reader, but as a force drawing an empathetic, physical response. Quoting from the poetry of Breton and Éluard, Bachelard argues that the work of the 'good Surrealists' also engages the reader on a visceral level (although he uses the Bataillean word 'virilité' to stress its force of activation).[76] The multiple references to aggressive activity in *Maldoror* – from the realistic gripping and tearing motions of claws and beaks to the clinging action of the suckers of octopi – perform a similar duty of forcing a reaction from the reader. By enumerating references to energy, velocity and volume in *Maldoror*, Bachelard seeks to demonstrate how the imagination can intrude upon the real. In tune with the materialist discourse of the 1930s, he avows that the realism, violence and immediacy of the book's imagery exert a physical impact.

Another side of *Lautréamont* leaves the sphere of the tactile-physical, to offer sometimes strangely incongruous analogies between the domains of poetry and physics. These passages prefigure Bachelard's subsequent path, where, in the second half of his career, he took a far greater interest in poetry than in the epistemology of physics. Echoing Nietzsche's notion of the 'untimely' philosopher, Bachelard argues that revolutionary imaginations such as those of Lautréamont and Einstein are out of step with their time. Destined to be understood initially only by a few, he compares their followers:

> At the beginning of the era of Relativity, Painlevé says of the fifty mathematicians gathered around Einstein, 'Look, you can *see them understanding*,' in order to demonstrate the soundness of the new theories to those who can't understand them. The same must be said to those who rage about the liberties of the Surrealists: 'Look, you can *see* them understand Lautréamont.'[77]

Complementing his comparison of the reception of radicalism in physics and in poetry, Bachelard illustrates the structure of poetics in terms derived from physics and geometry, claiming *Les Chants de Maldoror* is of a rare order of literary works that cannot be understood in the same language or through the same means of exegesis as are commonly used. Like non-Euclidean geometry, he writes, it obeys a unique set of laws in its formation of an alternative universe.

In his conclusion, Bachelard compares the activity of transformation carried out by the imagination on its forms with the work of the physicist whose quest to understand nature cannot attain to objectivity because it demands an active intervention, altering what it seeks to examine. To this Heisenbergian interpretation of the work of the poetic imagination, and equally anchored in his scientific background, Bachelard adds the assertion that for Lautréamont's metamorphoses to be truly seminal his example would have to engender a '*non-lautréamontisme*.'[78] Succeeding Lautréamont, this new epistemology of poetics would perform the same service of *generalisation* (as discussed in *The New Scientific Spirit* and advocated by Caillois in his *Inquisitions* tract) as non-Euclidean geometry had for Euclidean geometry. This would be neither an act of opposition nor of replacement, but one 'of awakening certain dialectics on the most fruitful levels of Ducassean principles'.[79] Appearing in the year that war was declared, Bachelard's book ended not on a note of despair, but with a Surrealist restatement of the transformative powers of the imagination.[80]

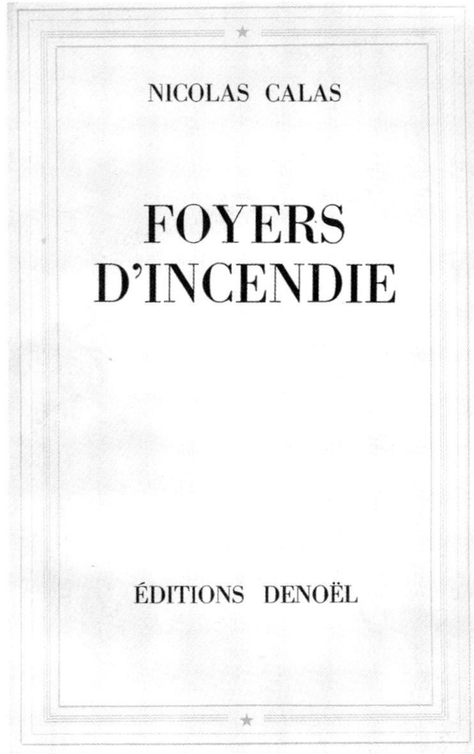

51 Nicolas Calas, *Foyers d'incendie* (1938).

Nicolas Calas turned largely towards Bachelard and *The New Scientific Spirit* when he sketched his version of the prevailing current of contemporary thought in his 1938 book, *Foyers d'incendie* (*Foci of Fire*) (fig. 51).[81] A member of the Surrealist group after his stint with the Groupe d'Études pour la Phénoménologie humaine, Calas gives the materialist argument expounded by Surrealism and a large number of France's intellectuals a literary and artistic inflection. He argues that the 'subjective' attitude, represented in literature by the novels of Joyce and Proust and in art by Impressionism and abstraction, is superseded by the 'will to objectification expressed in art by Surrealism and in philosophy by surrationalism'.[82] Like the Surrealists led by Breton, Mabille and Matta, and probably with an eye on Nietzsche and Bataille's acephalic outpourings, Calas was concerned with the creation of a 'new man', a monistic Promethean being uniting the powers of body and mind.

Notwithstanding its broad modernist scope and literary erudition, *Foyers d'incendie* hits all the right buttons from Trotsky to Roussel to gain Surrealist approval. Published two years after Petitjean's equally ambitious super-survey *Imagination et réalisation*, Calas's book sounds like a conscious attempt from within Surrealism to give a rival account of the intellectual landscape. With Bachelard's *nouvel esprit* up for grabs and assimilable by a number of competing factions, Calas takes careful aim at Surrealism's rivals for Bachelard's legacy. Those who contrived the notion of 'l'art impur', for instance (such as Caillois, though he is not mentioned), were untouched 'by the grace of the "new scientific spirit"'.[83] Calas advocates Bachelard's thought over the return to favour in some quarters of the philosophy of Descartes and the continuing relevance of Bergson, the second perceived as the philosophic support for the 'subjectivist aesthetic' of modernism, which can only 'sunder us further from the concrete'.[84]

Showy, omnivorous erudition replaces analysis at times in *Foyers d'incendie*. Calas's skimmy arguments and superficial conclusions – often those steering Surrealism and science onto each other's path – frequently drop to a level of extraordinary silliness.

Even the genial Bachelard would have balked at Calas's contrived and (unintentionally) surreal analogy 'a painting by Tanguy is tragic, a scientific experience is tragic, Newton's apple is tragic' (in as much, apparently, as the desire unleashed by the impact of each is 'tragic').[85] Yet Calas's *Foyers d'incendie* was as widely read in its day as Petitjean's *Imagination et réalisation*, even though the mini 'forest fire' it punned on in its title and created among its readership could not outlive its social, political and intellectual context, and was extinguished without trace in the early 1940s.[86]

For the first half of the 1930s a small cell existed in Surrealism comprising Salvador Dalí, the novelist René Crevel and the poet Paul Éluard, all of whom shared in the language of 'concretion' we have seen emerge here. Dalí is less relevant than the others to the political context, and I look at him in my final chapter. Here, I relate those writings by Crevel and Éluard that draw on modern physics and Bachelard's writings to political and philosophical concerns.

It is obvious that Crevel shared the same preoccupations as a broad band of contemporary French intellectuals from the terminology of his 1933 essay in *Le Surréalisme au Service de la Révolution* (*SASDLR*), 'Notes en vue d'une psycho-dialectique' ('Notes Toward a Psycho-Dialectic').[87] Applauding the 'very concrete desires' uncovered by Freud's research, Crevel hails the same '[r]eturn from the abstract to the concrete' celebrated later by Breton in 'Crisis of the Object'.[88] However, Crevel prioritises the political potential of 'concretion', writing: '[t]hanks to the dialectic the materialists have recaptured contact with life', allowing each 'the possibility to act on his universe'.[89] Crevel discusses in the same article the recently translated Einstein–Freud correspondence of the previous year, *Why War?* Like Einstein and Freud themselves, Crevel was not interested at this stage in the likely encounter between physics and psychoanalysis, but places their letters in a socio-political context, pessimistically pondering their contents in the course of a defence of revolutionary action and dialectical materialism. Crevel seems curious already about physics, indicating that analytic methods had brought about 'the immense progress of natural science'.[90]

For him, too, the appearance in 1934 of Bachelard's *New Scientific Spirit* was a major event, acting as the motor for his 1935 'Au Carrefour de l'amour, la poésie, la science et la révolution' ('At the Crossroads of Love, Poetry, Science, and Revolution').[91] First published in April 1935, it was probably conceived as a review of Éluard's recent volume of poetry *La Rose publique* of December 1934.[92] As indicated by its title, Crevel's text is concerned with combining different currents prevalent in French thought in the mid-1930s. Crevel writes that this synthesis can take place only through the exclusion of religion, which he attacks at the opening of the piece. His view that the 'new reason' demanded by modern physics holds out the chance of a final repudiation of Christianity was expressed by several of the Surrealists, including Crevel's friend Dalí.

By this time Crevel was ready to posit a relationship between psychoanalysis and physics. In his typically baroque style, he proclaims that just as contemporary psychology 'discovers, explores, conquers the mountains of moving desires, collects the unavowed geysers, the coral of the unconscious', so, the 'flood of invisible but no less real energy' uncovered by physics promises to 'penetrate physico-chemical mysteries to know, dominate, use the powers'.[93] In attempting this analogy, Crevel voices another version of Surrealism's disposition to give modern expression to the relation of inner and outer, microcosm and macrocosm, through psychoanalysis and physics. In both its theoretical writings (Mabille) and painting (Dalí and Matta), this connection is made repeatedly in Surrealism from the 1930s.

More explicitly than Calas, Crevel wanted to link modern physics to philosophical and political materialism within a framework privileging Surrealism as a pioneering epistemology. Deriving social significance from the philosophy of science by comparing the 'primacy of matter' of dialectical materialism with the scientists' restless reconstitution of the physical world, he (mis)quotes a Bachelardian passage of Jean Perrin's concerned with the brief life span of every concept in physics once removed from its experimental context when it 'ends by losing its utility, even its significance'.[94] One of the leading figures of the CVIA, Perrin's quest to popularise and politicise science had found a large audience in France from government level down, and an enthusiastic advocate in the Communist Crevel, who was one of the seven Surrealists who signed the CVIA's manifesto in March 1934. Crevel's cry, 'Scientific socialism. Socialism *by* and *for* science. Science *by* and *for* socialism', was a response to Perrin's quest to spread knowledge of science to the masses, as well as a declaration of his adherence to the 'scientific socialism' of Marxism.[95]

But Crevel goes further afield in his attempt at synthesis, aiming to link dialectical materialism and physics through the Logical Positivism of the Vienna Circle. At first, it might seem paradoxical that Crevel should aim at a Surrealist synthesis of physics and politics with recourse to positivism, which was supposedly antithetical to Surrealism, as I demonstrated in my opening chapter with reference to Breton's *Manifesto of Surrealism*. Yet the 1930s materialism of Surrealism found an unexpected partner in Rudolph Carnap's dissection of metaphysics in books such as *La Science et la Métaphysique devant l'analyse logique*, admired by Crevel and other Surrealists for its analysis of language, and its criticism of the despised metaphysics of Heidegger.[96] I discuss this unexpected confluence of Surrealism and Logical Positivism at greater length in chapter five.

Crevel presses his case with reference to Bachelard's *rupture épistémologique*, greeting with enthusiasm the passing of Newton's mechanistic universe and the introduction of a measure of indeterminism into physics. He appraises Breton's *hasard objectif* as an almost scientific notion, or a kind of 'lyrical science', sustaining 'cold', 'objective' science with the warm pulse of human subjectivity, making Breton's theory a 'calculation of sentimental probabilities, algebra of nuances'.[97] Like Bachelard and Breton, Crevel suggests that poets, artists and scientists view the world in similar ways, by means of comparable forms of intuition:

> How and why to continue opposing science and poetry, since poets and scientists, as much by untrammelled flashes of insight as by carefully thought-through

logical and experimental deductions, illustrate, affirm, and prove the universal law of reciprocity, its action on the life of the body and soul and the necessity of its triumph.[98]

Inspired by Bachelard, Crevel attempts to join synchronically the dots of Surrealist poetry, Communist Party politics and modern physics, rather than seeking a diachronic, ancestral lineage for each:

> modern science has corrected the views of Newtonian astronomy which, according to Gaston Bachelard (*The New Scientific Spirit*), gave its rigour to the doctrine of Kantian categories and its absolutes to *a priori* forms of space and time.
>
> Éluard's rose is as different from Condillac's rose as twentieth century dialectical materialism is from eighteenth century mechanical materialism.[99]

Fashioning a contemporary episteme from the poetry, politics and science of modernity as manifested in Surrealism, Marxism and Bachelard's new epistemology, Crevel cuts them loose from Condillac's epistemology and Newton's mechanics, mainstays of Enlightenment reason.[100]

Burdened by an unwieldy, ornate prose style and revealing little commerce with scientific precision or even academic sobriety, Crevel gets in a mess over poetic and analytic modes of expression. His laborious recital of references and jumble of divergent ideas are the work of an author torn by mixed loyalties at the time of writing. Crevel had taken Dalí's side when the artist was threatened with expulsion from the Surrealist circle in February 1934. By September, in the wake of civil disobedience on the streets of Paris and the ascendancy of the left to power, Crevel was extremely apprehensive about Surrealism's seemingly imminent elevation to canonical status, and fed up with its arty leanings (its plush journal *Minotaure*, its international art exhibitions, its literary diversions, its posh patrons). Even as his essay shows an undiminished regard for Breton's writings, Crevel was requesting in letters to Surrealism's leader that he point the movement back in the direction of experimental and political activity. In addition, Crevel was a contributor to the cultural review *Commune*, edited by Aragon and sponsored by the PCF, during a period in which Surrealism and Communism were at each other's throats. Now he was required to square his continuing devotion to Surrealist theory with a newfound enthusiasm for Logical Positivism in the guise of Carnap's writings. 'Au carrefour de l'amour, la poésie, la science et la révolution' is one of Crevel's last essays, and his sad death a few weeks later left its conflicts unresolved.[101]

Dated 'Davos, février–mars 1935', Crevel's text was written at Davos platz in the Rhaetian Alps in eastern Switzerland, the location of one of that country's most important medical centres, the Clavadel Sanatorium (a clinic still stands there today). Crevel had suffered from tuberculosis since 1926 and journeyed to Davos periodically to gain relief from the illness. Paul Éluard also suffered from tuberculosis, and a letter to his ex-wife Gala (whom he first met at the clinic) reveals he was there at the same time.[102] Inspired by Éluard's collection of poems, Crevel's essay was probably checked over by the poet as it was being written (the manuscript bears alterations in Éluard's handwriting, in fact).

Can we assume that Éluard, too, read Bachelard? The title and 1935 date of his *Minotaure* text 'Physique de la poésie' ('Physics of Poetry') strongly suggest he took a cue from Crevel.[103] Whatever the case, the subject matter of that article – ostensibly book illustration, but really the power of poetry to inspire artists to produce an original vision from its imagery – together with Éluard's injunction 'Let language take concrete form!', reiterates the anti-idealistic sentiments expressed by Éluard's contemporaries, though here the demand is made of language.[104] Insisting that literal representation by poet or artist is anachronistic, Éluard demands that depiction transform the subjective vision into an object: 'Only with closed eyes can one see what one wants.'[105]

Although Éluard's text can be related to the materialist discourse in the philosophy, politics, theory, poetry and art of the period, its title does not give convincing evidence of an interest in physics or its epistemology. But a direct link between Éluard and Bachelard exists. While Bachelard admired Éluard's poetry more than any other and quoted his verse in his publications from the 1930s onwards, Éluard was aware of the 'new reason' Bachelard promulgated in 'Le Surrationalisme'. Accordingly, when Éluard published extracts from a forthcoming work in *Minotaure*, he included several passages from Bachelard's influential text, side-by-side with others by Surrealism's distinguished precursors and current practitioners.[106]

In the year of Crevel's 'Au carrefour de l'amour, la poésie, la science et la révolution' and Éluard's 'Physique de la poésie', Tristan Tzara approached the question of language's practicality and adaptability in its relationship to thought in his *Grains et issues*.[107] This work turned out to be an extended farewell letter to Surrealism, expressing views on language and thought in some ways close to Dalí's, which nevertheless rejected automatism and were thus guaranteed to annoy Breton.[108] Where Crevel's text rhapsodises about the 'new partnership' of science and politics, and Éluard sees language as the experimental base upon which subjectivity could become 'concretised', Tzara's brief discussion goes a little further out on a limb, arguing that the inadequacy of an antiquated language to deal with the 'new reality' was one of the factors impeding the enactment of the proletarian revolution.

In *Grains et issues*, Tzara seizes on the expediential concepts of modern physics as evidence of the limitations met by thought when it seeks to accumulate experimental facts and communicate knowledge that outpace language. Tzara writes that 'langage de nature rationaliste' was popularised in the eighteenth century expressing the *physique mécaniste* of that era of the French Revolution, and was now surpassed by the advancing current of scientific thought, 'the ideological demands of which, pushed to the fore, prepared that same revolution'.[109] The problem with language, he continues, is that it has failed to develop since that time in the wake of scientific advances. Therefore, as dramatised by the seemingly paradoxical results gained by Relativity and

QM, 'the recent notions of bodies exceeding three dimensions, of space-time, matter-radiation and indeterminism' were misleadingly couched in the old language.[110]

His editors assert that Tzara 'had a longstanding interest in scientific progress, particularly in the field of physics',[111] and the former mathematics student Tzara reveals more than a lingering interest in science here. He demonstrates some understanding of the problems faced by modern physics in trying to explain phenomena that seemed to defy common sense in the limited language, imagery and concepts of classical physics. Furthermore, his comments suggest more than a passing acquaintance with Bachelard's remarks in *The New Scientific Spirit* on the failure of the language of philosophy to keep pace with the new concepts devised by science in the latter's vain pursuit of a definitive, panlogical account of the building blocks of matter (that is, the determination of their nature as either particles or waves). Increasingly disaffected with 'literary' Surrealism, by 1935 Tzara's thoughts had turned towards a fuller participation in revolutionary politics, and he apparently came to believe that only a new poetic language that could keep pace with and even explain science could be the language of the coming revolution.

As Elmer Theodore Peterson notes in his oddly titled *Tristan Tzara: Dada and Surrational Theorist*, Tzara's notion of the 'experimental dream' was contrived during this period and commended by Bachelard in 'Le Surrationalisme', showing that he 'wished to give his writings the trappings of serious investigation in this domain'.[112] Tzara was the chief animator of *Inquisitions*, and the experimental orientation of that journal, together with the changing focus of Tzara's writings in the era of the Popular Front, is discussed at length by Steven Harris.[113] However, Peterson's view that Tzara was 'enchanted with the scientific' is not borne out by the latter's non-scientific contributions to *Inquisitions*, which are rather unexpected considering that the journal represents arguably the high water mark of interest in science by writers of the larger Surrealist milieu.[114] At the time, it seems, Tzara was more concerned with taking on what he saw as Surrealism's quasi-mystical, non-materialistic preoccupation with categories such as *hasard objectif*, which grated with his interpretation of scientific socialism even as the appeasing Crevel sought a fit between the two, using Bachelard's *nouvel esprit scientifique* as a conduit connecting them.

Although the impact of *The New Scientific Spirit* on the Surrealists has not been fully assessed by historians before now, it was recognised at the time. In 1937, with Roger Caillois in his sights, Zdenko Reich spoke of 'a new philosophical current emerging in France', which had developed since 'G. Bachelard extended this principle into a kind of psychology of the modern scientist, foundation of the new scientific spirit.'[115] Writing in a Communist periodical, and an ex-Surrealist himself by then, Reich did not spare Roger Caillois's post-Surrealist, non-Communist attempts to lay the foundations for a new philosophy using new structures of thought derived from Bachelard within a framework thought 'bourgeois' by Reich.

Breton: Modern Physics and Politics

Although Breton himself did not write another essay so indebted to Bachelard as 'Crisis of the Object', in the years following its publication he returned hypnotically to its concepts and conclusions, soon turning them to more markedly political considerations, like his former comrades. To the language of 'open' and 'closed' rationalism, he often added the passage from *The New Scientific Spirit* detailing how the crisis in knowledge, or 'epistemological rupture', resulted not in revised but completely new ways of thinking: 'non-Euclidean geometries, non-Archimedean measures, non-Newtonian mechanics (associated with the name of Einstein), non-Maxwellian physics (associated with the name of Bohr), and non-commutative (or non-Pythagorean) arithmetic'.[116] Bachelard's insight became a regular fixture in Surrealist discourse, stretching well into the postwar period.

In 'Nonnational Boundaries of Surrealism' (1937), Breton seeks to justify the extravagance of Surrealist exhibitions (specifically the 1936 International Surrealist Exhibition in London) in the context of revolutionary ferment in Spain and France.[117] Recapitulating the argument given in 'Crisis of the Object', he writes of his attempt there to match a *rationalisme ouvert* that 'defines the present position of scientists (as a result of the development of non-Euclidean and subsequently of a generalized geometry, of a non-Newtonian mechanics, of non-Maxwellian physics, etc)' with a *réalisme ouvert* or Surrealism.[118] Breton spoke more pointedly the following year of the relation between the epistemological break and politics in one of the talks he gave in Mexico during his 1938 lecture tour.[119] Emphasising the importance of an alliance between Marxism and science, Breton seeks an alternative to scientific socialism in a genuine coalition of revolutionary action and physics, damning the outworn habits ingrained in Marxist theory. He told his audience that in order to remain vital the urgent task of the Marxist dialectic was to 'update itself in relation with the achievements of science, which it has failed to do practically for more than fifty years.'[120] Repeating the remarks he made in his 1935 *Indice* interview, he adds once again that non-Euclidean geometry, non-Newtonian mechanics and non-Maxwellian physics had 'greatly modified over the last half-century the philosophical position of scientists', and it was time for politics to catch up.[121] He also reaffirms the parallel he saw between the mode of 'discovery' of the Surrealists and of those who formulated modern physics, asserting it was a '*sign of the times*' that 'poetic *Surrealism* is accompanied by a scientific *surrationalism*, which is its double and measure'.[122]

After the war, Breton and the Surrealists continued to repeat the now-familiar 'non-' list, formulated as a 'philosophie du non' by Bachelard himself at the end of the decade.[123] The tellingly titled 1947 tract *Rupture inaugurale*, written to mark the International Exhibition of Surrealism in Paris that year, included the same list from *The New Scientific Spirit*, expanded to define once again the non-traditional pool of ideas the Surrealists believed was the counterpart to their project.[124] Henri Pastoureau claimed the text was primarily his work,[125] intended as a clarification of Surrealism's political position by placing its main concerns – myth, *humour noir* and *amour fou* –

at the cutting edge of a new epistemology shared by 'une géométrie non-euclidienne, une physique non-maxwellienne, une biologie non-pasteurienne, une mécanique non-newtonienne'.[126] As Surrealism clarified its irreversible departure from the Communist Party of 'M. Sartre', it continued to align its quest to create the total human being with the progress of modern science by means of a rhetoric derived from Bachelard.

By the time of *Rupture inaugurale*, Tzara had gained pre-eminence over Breton as the premier poet of the atomic age in the postwar vilification of Surrealism and recuperation of Dada in France, at least in the eyes of Lettrism leader Isidore Isou. Tzara's tired-of-Surrealism lecture at the Sorbonne in April 1947 'Le Surréalisme et l'Après-Guerre' – which Breton hijacked, gaining a new batch of recruits for Surrealism as a result – makes a (tenuous) link between Dada and modern physics, probably the prompt for some of those speculations by historians of Dada I referred to in my previous chapter:

> In this, on a level which had nothing systematic about it, we found ourselves in agreement with those men of science who, around the same period, with meticulous rigour, took up again the most run-of-the-mill experiments of physics, discovering their deficiencies and constructing from there that monumental edifice which is, today, modern physics.[127]

Tzara's recollection of Dada's aim of total renewal subsequently provided the excuse for some Breton-bashing by Isou. Already fascinated by a long-deceased and consequently benign Dada, and hoping to draw upon its poetic resources to challenge a still-alive though ailing Surrealism, Isou sided with Tzara against Breton a few months after Tzara's attack.[128] Privileging the Dada-Lettrist phonetic poem over Surrealist automatism, Isou took his cue from Tzara, arguing that poetry should be judged against the measure of its '*quantum de creation*', and avowing that Lettrist poetry is '*atomic poetry and represents our century*'.[129]

Pierre Mabille: Modern Physics, Politics and Alchemy

In my opinion, Breton's negotiation with Bachelard's writings in the mid-1930s was facilitated in part through discussions with Pierre Mabille, one of those within the Surrealist group most sympathetic to science. Behind Mabille's theoretical development of Bachelard's epistemology towards political ends lies his close friendship with, and deep respect for, both Breton and Bachelard. A doctor of medicine, Mabille joined the Surrealists and met Bachelard in 1934, the year of *The New Scientific Spirit*. Bachelard felt closer to Mabille than to any of the other Surrealists and often went to meet him at one of the favourite cafés of the Surrealists, the Deux Magots. On receiving a copy of Mabille's book *Egrégores*, Bachelard commented on the similarities he found between their respective outlooks.[130] Mabille's library contained almost all of Bachelard's writings and he even attended several of his courses at the Collège de France, nurturing a reverence for Bachelard that Surrealists usually reserved for Breton.[131]

In his review of Bachelard's *Formation of the Scientific Mind* (1938), written in 1938 or early 1939, Mabille approves his notion of the profound change undergone by modern scientific thought.[132] Offering a similar argument to Breton's 'Crisis of the Object', Mabille, too, argues that this new structure of thought extends outwards from science into the contemporary humanities. Regarding it as evidence of a radical switch from one shared set of knowledges to another, he writes: 'a revolution occurred in artistic sensibility and its modes of expression'.[133] But Mabille goes on to broaden Bachelard's hypothesis further than Breton dared in 'Crisis of the Object', declaring that the socio-political revolution that began in the middle of the nineteenth century had entered its decisive phase in the preceding thirty years. This conception of the violent culmination of a social revolution as evidence of an 'epistemological rupture' – only faintly delineated by Mabille but unmistakable – compares with the reflections of ex-Surrealists like Crevel and Tzara in 1935. The context of Mabille's review effectively demanded that interpretation, for it appeared in the second number of *Clé*, the journal set up by Breton as the mouthpiece of the Fédération internationale de l'art révolutionnaire indépendant, and the last gasp of Surrealism's revolutionary enthusiasm.

From his comments, there emerges a sense of where Mabille's thought differs from others who drew later on the notion of the *rupture épistémologique*. When compared with the interpretation of Bachelard's future commentators, Mabille's seems wedded to humanism. This is undoubtedly due to Mabille's involvement in the Surrealist group, leading him to an interpretation relatively loyal to Bachelard's original, broadly humanist conception. Where French thinkers of the postwar era who remained aloof from existentialism, from Barthes and Foucault onward, regarded humankind as the effect of more or less indifferent agencies, throughout its history, Surrealism optimistically offered a central role to humanity as potentially the wilful creator of its own destiny. This is the way Mabille read Bachelard's *Formation of the Scientific Mind*, extrapolating from the book parallel and 'revolutionary' changes in science, art and society and offering them as evidence that the human race was faced with the 'necessity for a total mutation of its being'.[134]

By the time of the Second World War, Surrealism had become deeply disappointed at the inability of leftist politics to reach a revolutionary solution in the social sphere, increasingly seeking solace in magic, alchemy and the occult as a means to 'transform the world' and 'change man'. At first, we see a transitional blend of revolutionary politics, humanism, magic and modern physics in Mabille's thinking:

> The real tradition of humanity is not one of acceptance but of revolt. It sets humans in opposition to God, it pitches them against the sky, from which they must steal fire to serve their own purposes. The path goes from Prometheus to the hermetics, from them to modern physicists. The latter use the most precise methods and possess the most powerful instruments. But they are nonetheless the legitimate heirs to the tradition of the marvellous.[135]

Mabille's writings had long seemed indebted to the humanism of the late Middle Ages and the Renaissance. And like Matta's paintings, at times they updated the centuries-old identification of the microcosm with the macrocosm – of the human body with the world or universe – by proposing a new, modern symmetry between the inner and outer, mapped by modern psychology and physics.

During the war, Mabille's interpretation of modern science became more tightly bound to his fascination with alchemy and hermeticism, as noted by Laville: 'the modern alchemist, Mabille will write in 1945 . . . shuts himself up in his laboratory with the most advanced instruments of scientific research.'[136] A commonplace response to modern physics from Surrealism (and the humanities generally), this is the kind of fanciful-sounding comparison with alchemy that infuriates mathematicians, materialist philosophers, logical positivists and many physicists. Yet we saw earlier that the physicist Jacques Solomon made a similar assertion, while Louis de Broglie casually made the same observation a few years earlier (in popular mode, admittedly), writing that the changes undergone by atoms in unstable radioactive substances – realising 'the transmutation of elements dreamed of by the alchemists of the Middle Ages' – had been copied artificially in Ernest Rutherford's laboratory.[137]

Mabille's was not an empty analogy, then, and Laville speaks for him and the centrality of science to his thought as follows:

> Modern science remains the first point of reference, placed at the service of a spirit of synthesis originating in the Renaissance, which is restricted to inspire the choice of a path. The thread of Tradition orients the mind, which often discovers, rationally, that non-Euclidean geometries or quantum theory are not strangers to pre-positivist thought. If the mind calculates according to the rules of modern science and reason, the thread of Tradition functions in the entirely different mode of homology . . . Homology, which is also called analogy, is at once a form of scientific discovery and an essential poetic resource . . . its ultimate conclusion is an original humanist conception, a form of the art-science synthesis, which Pierre Mabille considered necessary for the equilibrium of a civilisation.[138]

The same idea of privileging homology/analogy was shared by Breton and Bachelard (and Gustave Juvet), manifested in theoretical terms as induction. According to Laville, Mabille's Surrealist humanism and optimism led him to believe, in the face of global warfare, that humanity stood on the threshold of a post-Cartesian age:

> Mabille affirms that a new conception of the world is about to dawn . . . It is time, he tells us, to participate in a new representation which must be heralded by the artist and scientist . . . monism is the order of the day: soul/matter – body/mind are the dualities of a discarded era. The 'new scientific spirit' is its moral code, to give rise to the Golden Age is its aim.[139]

Mabille persisted in this view during the war, even as it became obvious that Surrealism was unravelling. In a prefiguration to some of our contemporary writings on physics and religion, Mabille even maintained modern science was a confirmation of Hindu cosmology.[140]

Although things changed later, the thought of Mabille and Breton is closely aligned up to the end of the 1930s over Bachelard's epistemology and modern physics. This is evident when Breton attempts to breathe life into a jaded Surrealist art in 1939 by turning to Relativity and the fourth dimension in 'The Most Recent Tendencies in Surrealist Painting'.[141] If that sounds unconvincing at first, then Breton's notion, that a 'transition to the fourth dimension is undertaken on the psychic rather than the physical plane' in Victor Brauner's paintings, seems utterly preposterous.[142]

However, Mabille's single most famous text, on Brauner's loss of an eye the previous year, was to hand, giving a 'scientific' interpretation of the artist's accident, which the Surrealists believed was predicted in several of his canvases of the preceding years.[143] Against the background of Breton's own study of precognition or *hasard objectif* in his book *Mad Love*, Mabille conjectures that it might be possible to free oneself from the restrictions imposed by the five senses, and 'rise beyond visual Euclidean dimensions'.[144] Calling on philosophy and psychology to create a new epistemology in the wake of physics, what follows is reminiscent of Bachelard's claim in *The New Scientific Spirit* that '[s]cience, in effect, creates philosophy'[145] as well as Breton's *Indice* contention of 1935 that psychology lags behind physics:

> These ways of thinking are already used in the abstract sciences but timidity and fearful inertia prevent us from placing them at the service of our knowledge of the everyday life of man. It is clear that psychology, philosophy, morality stagnate by comparison with the conquests of science. The reactionary will to keep the properties of man in check within the framework of an outdated Cartesianism, moreover, is one of the reasons behind the contemporary crisis.[146]

Published in the same number of *Minotaure* as Breton's article, Mabille's essay was obviously read by Breton and discussed with him. For his part, Breton was probably grateful for the scientific underpinning given his *hasard objectif* by Mabille's efforts, and happy to fold the idea of Brauner's psychological departure from Euclidean space in 'L'oeil du peintre' back into 'Most Recent Tendencies', as a psychological displacement to the fourth dimension. Physics and its philosophy still had a lively existence in Surrealism at the end of the decade, then, but politics was on the wane and about to be thoroughly marginalised.

J. W. Dunne and Surrealism

This discussion has moved far from politics, but so had Surrealism, preferring to link modern physics to mediumism rather than Marxism. For that reason, J. W. Dunne's much-loved book, *An Experiment with Time* (1927), sparked their enthusiasm from about this point on. Dunne's unclassifiable work was another important source for those concerned with the implications for psychology of Einstein's space–time. A pilot and inventor, Dunne had experienced what he thought were precognitive dreams, and reluctant to seek answers in the occult he had turned to established science. In particular, he sought explanation from the point of view of the 'observer' existing in a particular 'frame of reference' as posited by Einstein's Relativity. Half dream description and half physical theory, Dunne's book concludes that we are made up of more than one 'I'. The I that sleeps is *subject* to time, whereas a separate I or 'observer' emerges in sleep, detached from ordinary time. This second I can objectively see the whole of the temporal continuum, past, present and future, as though it were a fourth spatial dimension. Consequently, Dunne concludes, this I can witness future events.

Rising above everyday space-time, Dunne's observer enjoys a view bearing striking resemblance to Brauner's as described by Breton, yet the idea had been in the air for a while, going back as far as H. G. Wells's discussion of the fourth dimension in *The Time Machine* (1895). Einstein's abolition of absolute time and the popularisation and extension (not to mention distortion) of four-dimensional space–time in books such as Dunne's enabled the transplantation of the escape-from-everyday-temporality out of fiction and into the real world of the Surrealists. Although the first 1927 edition of Dunne's book met with popular success, the second in 1929 was virtually ignored. However, by the time it reached its third edition in 1934, the new model of the universe constructed by Relativity and QM, and popularised by Arthur Eddington and others, had taken root. Subsequently, *An Experiment with Time* became a huge success, even drawing a supportive response from Eddington himself, which was reprinted in some future editions of the book.

Well-known to the Surrealists and their friends, Dunne's book was read and part translated in 1932 by Raymond Queneau,[147] mentioned by Armand Petitjean,[148] present in Wolfgang Paalen's library, and later recommended by Marcel Duchamp among the notes collected in the *White Box*.[149] Breton could have read *An Experiment with Time* in 1948 when it was translated into French as *Le Temps et le Rêve*, though he might have known about it earlier (fig. 52).[150] Mention of its 'rêves prémonitoires' in Michel Souriau's review in the final 1937 number of *Recherches philosophiques* probably reached Breton, especially since his own *Mad Love*, which has much in common with Dunne's book, was published that year.[151]

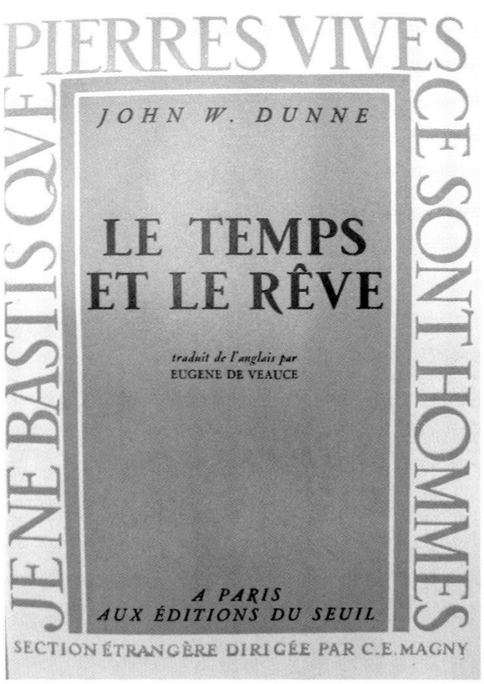

52 J.W. Dunne, *Le Temps et le Rêve* (1948).

Whatever the case, Dunne's stirring description of one of his premonitory dreams and its sensational realisation in the 1902 volcanic eruption on Martinique – 'as dramatic', he wrote, 'as any lover of the marvellous could desire' – could not escape Breton's notice long.[152] Asked in an interview in 1950, '[w]hat are your most immediate projects?' he indicates a desire to 'return to the problem of dreams, on which John W. Dunne's book *An Experiment with Time* seems to have shed new light'.[153] Breton continued his promotion of Dunne's writings well into the 1950s, linking them closely to his own notion of *hasard objectif*. Consistent with his 1939 argument about Brauner accessing some kind of psychic fourth dimension, he explicitly compares Mabille's account of Brauner's accident with Dunne's theory in 1955, averring that coincidence and even the possibility of the existence of a human soul could be investigated fruitfully by means of 'that still unknown *other dimension*, in which John W. Dunne invites us to search for the secret of *serial time*'.[154]

When he expresses his enthusiasm for Dunne's book in 1950, Breton revokes some of his politically-inflected materialism of the 1930s, regretting that, in his exploration of the relations between dreams and waking life in *The Communicating Vessels* of 1932, 'I sacrificed too much to the "materialism" of the hour.'[155] Breton was casting his mind back to Surrealism's swerve from idealism towards the end of the 1920s due to its greater commitment to Marxist materialism, as noted here, but also under immense pressure applied by Georges Bataille. By 1950, the polemic with Bataille lay in the distant past, and Bataille himself was long estranged from Marxist politics (to which he had never fully converted) and as deeply engrossed in mysticism as Breton and the Surrealists were in magic and the occult. Also around that time, though, Bataille published his *Accursed Share* (1949), demonstrating an interest in physics that had continued since at least the time of his youthful sniping with Breton, while at the same time he welcomed Louis de Broglie onto the general editorial board of his journal *Critique*. My next chapter sets out the first account of Bataille's preoccupation with modern physics, given in the context of his fractious yet ambivalent relationship with Surrealism and his controversial attraction to mysticism.

4

ASTROPHYSICS AND MYSTICISM

Georges Bataille and Arthur Eddington

The editor's gloss on Georges Bataille's 1938 meditation on social formations, titled 'Attraction and Repulsion I: Tropisms, Sexuality, Laughter and Tears', in Denis Hollier's collection *The College of Sociology (1937–39)*, makes the astonishing assertion that Arthur Eddington (1882–1944) alluded to Bataille's work, noting the influence upon Bataille's thought of 'Eddington (to cite a physicist, and a physicist who cites Bataille)'.[1] The text 'Attraction and Repulsion' was among those read out by Bataille at the short-lived College of Sociology, whose lectures were given or attended by many soon-to-be-celebrated super-intellectuals, such as Walter Benjamin, Roger Caillois, Pierre Klossowski, Alexandre Kojève, Michel Leiris, Denis de Rougemont and Jean Wahl. Most of those involved with the College of Sociology, including Bataille, had been close to Surrealism, and their respective interpretations of sociology, ethnology, German philosophy, Marxism, psychoanalysis, biology and modern physics were frequently mediated by the intellectual authority enjoyed by Surrealism at the time.

Like Bataille's other essays in the 1930s on the structure of societies and the human mind, 'Attraction and Repulsion' draws upon models developed in psychoanalysis, physics and sociology. Here as elsewhere, Bataille launches into a psychoanalytically inflected hypothesis of a repulsive sacred core to society composed of menstrual blood, bodily putrefaction, sexual organs, tears and laughter. He avows that 'the very dark, repulsive nucleus, around which all turbulence revolves, has created the principle of life out of the category death, springing out of falling.'[2] Transgressive assertions like this are standard fare in Bataille's writings, the atmosphere and general themes of which constellate in patterns utterly at variance with Eddington's. It comes as no surprise and a bit of a relief, then, to discover that the fantastic image presented by the editor's note in Hollier's collection, of the Quaker bachelor Eddington quoting from, say, Bataille's *The Solar Anus*, exists only due to a translator's error.[3]

Beyond this beguiling mistranslation, however, there is evidence of some shared ground between the two, even though their respective treatment of the same material differs sometimes dramatically. I wish here to situate the thought of Bataille and Eddington in relation to the theme of celestial mechanics, giving special attention to the role played by mysticism. The idea is to give a more complex and conciliatory reading of the interaction between science and the humanities than the poles apart view given by

C. P. Snow and his descendants. That transaction forms the main inquiry in what follows, dominated by a discussion of Eddington's *The Rotation of the Galaxy* (1930) and Bataille's 'Celestial Bodies' (1938). To begin with, I sketch in some relevant remarks about Bataille's stake in Surrealist theory and practice, though I keep that to a minimum because an established and expanding industry is already devoted to that subject. More significant are Eddington's availability to the Surrealists and the survey I give of how the Surrealists read him, together with my account of Bataille's introduction to and deployment of the new physics. The lack of literature on the Surrealist encounter with Eddington's writings is to be expected given the English astrophysicist's relative obscurity in the present day. However, it is remarkable that at the time of writing not a single text exists on Bataille's reading of modern physics, even though scholars of Bataille such as Francis Marmande have prodded readers in that direction.[4] The overall theme and structure of my book demanded I narrow this chapter into a Bataille–Eddington case study in the face of my growing understanding of the immense interpretive possibilities offered by Bataille's engagement with modern physics, a subject requiring book-length consideration.

Early Bataille: Surrealism, Sociology and Quantum Mechanics

At its most turbulent, Surrealism expressed a uniquely visceral variety of iconoclasm. Its influence continues to be felt in cultural studies and critical theory today partly because of its delayed reception in the late 1950s and 1960s by writers such as Jacques Derrida, Michel Foucault and Julia Kristeva. In Bataille's work, they saw the supreme extension of Surrealism into a dark, violent, anti-Enlightenment 'un-knowing', yet it was only after Bataille's death in 1962 that this assessment took place, because he had been little known in France and almost entirely unknown abroad since his first texts dating from the 1920s.

Bataille did not receive any formal philosophical training, but his discovery in 1922 of Nietzsche, a writer never fully integrated into the Surrealist pantheon, at the Bibliothèque Nationale where he worked as a librarian from that year, had a formative effect on his thought.[5] He was close enough to Surrealism in the 1920s to translate anonymously some mediaeval 'poèmes incohérents' for publication in *La Révolution surréaliste*,[6] though he did not join the Surrealist group, reporting years later in the third person a 'mutual hostility between himself and André Breton'.[7] By then, through copious reading, Bataille had equipped himself with a set of themes that would appear consistently in his work over the years, among them violence, eroticism, blindness, darkness and obscurity. A year in psychoanalysis in 1926–27 helped him find his voice with his brutal pornographic novel *Histoire de l'oeil* (1928), narrating an obsessive and sadistic sexual relationship culminating in the strangling of a priest whose gouged-out eye is placed first in the heroine Simone's anus then in her vagina. The early, important text *The Solar Anus* dates from that period (1927, published in 1931).

The excesses of Bataille's work alarmed even Breton and the Surrealists, whilst Bataille opposed his 'base materialism' to what he took to be their idealism, still imbued, he argued, with rationalism. In 1929–30, his journal *Documents* rivalled those of Surrealism, as his confrontation with the Surrealists reached a head in Breton's portrayal of Bataille in the *Second Manifesto of Surrealism* (1929–30) as an 'excrement philosopher' and 'staid librarian'.[8] Bataille responded by criticising Breton's priestly language as 'Icarian illumination', arguing, 'the blinding celestial vault, when it becomes a psychological obsession, implies spiritual elevation', but also destruction.[9]

Bataille used *Documents* as a platform to develop ideas that later became central to his work, already revealing a curiosity towards, and limited understanding of, recent research in QM. However, when it surfaced in *Documents* it was due more to Bataille's collaborators on the journal, such as his co-editor Carl Einstein whose knowledge of modern physics was extensive and even formative to his art criticism, as will become apparent. In the second number of *Documents*, Einstein gave an 'ethnological' account of André Masson's paintings indebted to the emerging recognition of indeterminism in physics following the formulation and publication of Heisenberg's Uncertainty Principle two years earlier. In the wake of that extraordinary scientific event, and in the midst of *Documents*' bitter feuding with Breton's Surrealism, Einstein attempts to wrest Masson's art from the nineteenth-century synthetic 'chemistry of the intellect', which Breton had borrowed the previous year from Edgar Allan Poe's poetry theory to identify it with what he (Breton) called 'Masson's science'.[10] Einstein sets about this by concocting an explanation for Masson's work that is a coalition of Planck's discontinuous packets of energy quanta and Heisenberg's indeterminacy, claiming that the artist's 'hallucinatory' images 'create a breach in the order of mechanical processes; they introduce blocks of "acausality" into that reality that had once presented itself absurdly as one'.[11] It is no coincidence that Einstein's essay on Masson in *Documents* is followed immediately by a text I noted in my first chapter, Hans Reichenbach's 'Crise de la causalité'.[12] One of many written on the subject during the period, Reichenbach's article complements Einstein's by explaining how the new indeterminism introduced by physics 'signifies a radical rupture with the causal representation made of the world by classical physics'.[13]

Further evidence of an absorption of the new physics into Bataille's *Documents* (a process entirely absent from the contemporary house journal of the Surrealists, *La Révolution surréaliste*) is the Nietzschean writer Arnaud Dandieu's review of the essay by Émile Meyerson, 'Le Physicien et le Primitif', which had appeared in a recent number of *Revue philosophique*, a journal Bataille knew well (and to which Dandieu contributed).[14] Dandieu's review is a symptom of *Documents*' part-Freudian, part-ethnological investigation into what it assumed were the 'primitive' elements underlying the veneer of civilised behaviour and thought.[15] Like Carl Einstein, and typically for writers of the time learning about QM, Dandieu calls attention to 'the recent works of M. L. de Broglie' and 'quantum theory, which reintroduces discontinuity and the irrational into science'.[16]

Remembered today, if at all, for co-authoring with Robert Aron in 1931 the ripe anti-American rant *Le Cancer américain* (par for the course for French intellectuals in the 1930s), Dandieu was one of the main figures polarising debate on the new physics in philosophy circles.[17] This is evident in the contribution to *Recherches philosophiques* by Claude Chevalley and Alexandre Marc.[18] Focusing chiefly on Dandieu's furious condemnation of economic rationalism, it assesses the contemporary rejection of reason in French thought, giving Surrealism its only (passing) mention in *Recherches philosophiques*.[19] As Chevalley and Marc point out, Meyerson was an important figure for Dandieu, who praised Meyerson's philosophy of science at the expense of Bergsonism (part of a wide-ranging philosophical debate on rationalism yet to be assessed either by Surrealism's or Bataille's commentators), as in his 1930 text 'Espace' in Bataille's *Documents*.[20]

Bataille was busily familiarising himself with the same conceptual language, and although the themes and tone of his writings show little sympathy with those of Meyerson, he began a piece on that philosopher during this period that is the earliest clear evidence of his interest in the philosophy of physics and mathematics. Spurred on presumably by conversations with Dandieu to whom Bataille was close (both were then employed by the Bibliothèque Nationale and they would organise together a small tricentennial Spinoza exhibition that ran from December 1932 to January 1933), he characterises Meyerson's work as 'apparently entirely dependent on science' in the context of the 'new union' between science and philosophy, 'in which philosophy plays the role of enthusiastic servant.'[21] However, Bataille's text peters out and remained unfinished and unpublished in his lifetime, probably because he still lacked a sufficiently rounded understanding of the new epistemology emerging from physics to engage fully with Meyerson's work and also because his own thought had not yet found its full intellectual purchase.[22]

This arrived over the next year, derived (it is claimed) primarily from sociology and anthropology, enabling Bataille to forge the thematics of 'base materialism' into a set of major statements that took *dépense* (expenditure) and general economy as their focus, reaching their classic formulation in the 1930s. Inspired initially and chiefly by his 1931 readings of Émile Durkheim's *The Elementary Forms of Religious Life* (1912) and Marcel Mauss's 'Essay on the Gift' (1925) – especially Mauss's account of spectacular episodes of destruction in tribal societies of the American north west – Bataille posits the idea of an immoderate energy inherent in all human beings, social groupings and planetary bodies demanding an obligatory expenditure and waste.[23] This exceeds the restricted economy that underpins and contains a Western thought and materialism circumscribed by the notion of equivalence and exchange, and even the 'regrettable' overproduction and disposal of consumable goods in the developed world. Instead, it expresses the *necessity* for a violent, destructive consumption that is irreducible to any system of particularity, determinism or totality. Such a necessity is understandable only in terms of a general economy:

> Economic activity, considered as a whole, is conceived in terms of particular operations with limited ends. The mind generalizes by composing the aggregate of these operations. Economic science merely generalizes the isolated situation; it restricts its object to operations carried out with a view to a limited end, that of economic

man. It does not take into consideration a play of energy that no particular end limits: the play of *living matter in general*, involved in the movement of light of which it is the result. On the surface of the globe, for *living matter in general*, energy is always in excess; the question is always posed in terms of extravagance.[24]

From the early 1930s onwards, ideas of wasteful expenditure and heterogeneity motivated and nuanced all of Bataille's thought and activity.

Without going into much detail, one of Bataille's best commentators, Arkady Plotnitsky, writes: 'Bataille's work on general economy was significantly influenced by modern science and, specifically, quantum physics, often directly invoked by Bataille himself.'[25] Of Bataille's writings that, he claims, contain 'important direct references' to quantum physics, Plotnitsky lists 'Materialism' (1929), 'The Critique of the Foundations of the Hegelian Dialectic' (1932), *Inner Experience* (1943), *On Nietzsche* (1945) and *Guilty* (1961). In other words, he claims that Bataille's writings from the earliest to the latest are conversant with QM.[26]

Observance of Bataille and the Surrealists' concern with modern physics can reopen debates long thought settled. For instance, comparison of 'Materialism', Bataille's 1929 critique of the artificiality of scientific practice, with Breton's May 1935 interview in Tenerife with *Indice* shows them unwittingly reaching agreement over the respective value of psychoanalysis and modern physics.[27] Where Breton seems to reject that part of Freudian thought not lending itself to the active process of 'concretion', Bataille argues that materialism should be defined by recourse to psychoanalysis rather than physics. Materialism, he writes, must be based on:

> psychological or social facts, instead of on artificially isolated physical phenomena. Thus it is from Freud, among others – rather than from long-dead physicists, whose ideas today have no meaning – that a representation of matter must be taken.[28]

Bataille does not clarify what he means by a 'psychological fact', and it is significant that his own brand of materialistic thought would soon favour the greater empiricism of sociology and modern physics (the second favoured over the Newtonianism of 'long deceased physicists'). Through the red mist of the *Second Manifesto*, Bataille's 'Materialism' exemplifies for Breton 'an obnoxious return to old anti-dialectical materialism, which this time is trying to force its way gratuitously through Freud.'[29] Yet under closer inspection, there is no disagreement. Where Bataille favours the methodology of psychoanalysis (as would Breton), repudiating classical physics for its reliance on a suspect methodology (as, again, Breton would reject its 'Newtonian–Cartesian' foundation), Breton dismisses Freud's 'philosophy' rather than his symbolism (the 'concrete' symptom or, in Bataille's words, 'representation of matter'), favouring the latter for its susceptibility to materialisation. A 1930s context for their dispute that takes account of modern physics as well as psychoanalysis shows that Bataille and Breton only *seem* to make statements that contradict one another.

Science, Alexandre Kojève and Recherches philosophiques

Bataille's involvement in Marxism could have brought him closer to the Surrealists, but his expression of those views through contributions to the journal *La Critique Sociale* (1931–34) was frequently made at the Surrealists' expense, serving rather to increase his distance from them and consolidate his relationship with Raymond Queneau, who wrote with him 'The Critique of the Foundations of the Hegelian Dialectic'.[30] Queneau had been a member of Breton's circle in 1928–29, and as an able mathematician was among the best equipped of those associated with the Surrealists to comprehend the technical aspects of the new physics. His interest in the emerging literature on the subject comes across strongly in his assiduously kept reading lists, which record him in the late 1920s and early 1930s reading several works by Meyerson, including *La Déduction relativiste* and *Réel et déterminisme dans la physique quantique* (read by Bataille in 1935), Eugène Bloch's *L'Ancienne et la nouvelle théorie des quanta*, Jean Perrin's *Les Atomes* and, in January 1935, Gaston Bachelard's inevitable *New Scientific Spirit*.[31] However, there is little evidence that Queneau wove his up-to-date knowledge of contemporary advances in physics into his fiction. Although the content of his *Dimanche de la Vie* (1951) was inspired in part by his Hegelianism and the meticulous organisation of certain of his novels benefited from his knowledge of mathematics, his 'vast scientific culture' and 'literary culture' were kept separate for the most part, whereas Bataille, Breton, Caillois, Calas, Dalí, Mabille, Matta and Paalen all sought to inscribe the new scientific spirit into their art or theory.[32] Although he wrote on modern physics and mathematics for *La Critique sociale*, Queneau expressed no desire to transport concepts from Relativity or QM into a literary theory or any other, as Bataille attempted in his writings.[33]

Ironically, Queneau is probably best known these days to most Anglophone scholars for compiling, transcribing and editing someone else's work – the legendary series of Paris seminars led by Alexandre Kojève, which Queneau attended with Bataille. The seminar held at the École pratique des hautes études was begun by Alexandre Koyré in 1926, but when circumstances led Koyré to discontinue his own involvement he suggested that Kojève, his friend, fellow Russian and regular at the seminar, take over. Kojève's disquisition on *The Phenomenology of Spirit* was thus initiated, spanning the same years as the Surrealist-run journal *Minotaure*, 1933–39. During that period, Kojève introduced a new Hegel to a whole generation of intellectuals in France, many of whom frequented his seminar, among them Raymond Aron, Bataille, Pierre Klossowski, Jacques Lacan, Maurice Merleau-Ponty and Queneau. According to Gérard Legrand, Breton's curiosity drew him there between 1935 and 1938, though there is no evidence for this and it is contested by other commentators.[34]

Also at that time, Bataille was deeply involved with the anti-Fascist group he created in 1935, Contre-Attaque, which included several Surrealists in its ranks and confirmed his *rapprochement* with Breton after several years of dedicated bickering.[35] Bataille's new closeness to a Surrealist milieu enthusiastically getting to grips with Bachelard would have helped him gain some access to recent developments in physics, largely through what the French call the *vulgarisation*, of which Arthur Eddington was a master. But Bataille was closer to another circle more actively associated with research and study in physics, hinted at in an autobiographical statement he made whilst at work on his *Sur Nietzsche* during the war: '[f]rom '33 (I think) to '39 I followed Alexandre Kojève's course devoted to the critical analysis of *The Phenomenology of Spirit* . . . During the same period, through immense reading, I stayed abreast of scientific activity.'[36] Does a connection exist between the two halves of this statement: Bataille's attendance at Kojève's Hegel seminar and his ravenous reading in science?[37]

Kojève's interpretation of *The Phenomenology of Spirit* nurtured a small Hegel cult, now a key moment in the history of twentieth-century French philosophy. Breton's interview with *Indice* confirms that his renewed interest in physics began in the period when the Surrealists were discovering Bachelard, and while Breton himself was attending (perhaps) Kojève's seminar. Although Bachelard offers an alternative to the Hegelian dialectic indebted to QM and was no enthusiast of Hegel, there is a rare argument made for Hegel as a precursor to QM in Kojève's interpretation of Hegel's elaboration of the methods of the philosopher and scientist in the course of a seminar in the 1934–35 academic year, which I look at in a moment. Bataille was certainly present and, according to Legrand, Breton might have attended.[38]

Kojève's discussion of the philosopher and scientist could only have been carried out by someone with a familiarity of recent developments in quantum physics. Hardly ever mentioned in the literature on Kojève, that expertise was gained between 1928 and 1931 when he worked towards the completion of his doctorate in philosophy at the Sorbonne and the École pratique des hautes études, studying mathematics and modern physics privately.[39] Kojève's first full-length work, *Zum Problem einer diskreten Welt* (*On the Problem of a Discontinuous World*) (1931), offered to the ÉPHÉ as his doctoral thesis (and refused), was concerned with the new physics of the quantum, and in the same year he gave a series of lectures on QM to the Société de philosophie scientifique. His 1933 doctoral thesis, presumably based on these lectures, was titled *L'Idée du déterminisme dans la physique classique et dans la physique moderne*, completed under the guidance of natural philosopher Abel Rey (who had also been Bachelard's supervisor) at the Sorbonne. The text would not be published until 1990.[40] Kojève's biographer argues that these two early works on physics, along with the 1931 *Athéisme*, 'form at the time the firm outline of a philosophical system that will occupy Kojève all his life'.[41]

This knowledge lies at the basis of his interpretation of Hegel's dialectic in the sixth to ninth lectures in the 1934–35 seminar. Kojève claims that Hegel's criticism of the scientist is aimed at those who seek an '*aperçu*' of an object secured by a distant, autonomous subject.[42] In place of this, Hegel posited in the *Phenomenology* '"the simple-or-undivided Whole (*Ganze*) which has an overview of itself (*sich übersehende*)"'.[43] Kojève says the first approach is that of pre-Hegelian 'pseudo-philosophy' and 'pseudo-science (Newtonian science), which reflects *on* the Real while placing itself *outside* of the Real'.[44] This was rejected by Hegel in favour of the second – a passive, naïve meditation on the object of study: '[t]he "method" of the Hegelian Scientist,' Kojève insists, 'consists in having no method.'[45]

Given Kojève's profound understanding of modern mathematicised physics, it is not surprising that his preliminary remarks on Hegel's scientist are followed by an interpretation strongly influenced by the philosophy of quantum theory. Accidentally echoing the Surrealists' call to 'transformer le monde',[46] a task which until then had been, for them, the province of revolutionary politics, Kojève introduces the scientist-observer, 'a man who lives within Nature and is indissolubly bound to it, but is also opposed to it and wants to transform it', adding, 'science is born from the desire to transform the World in relation to Man'.[47] 'Scientific experience', he goes on to say, 'perturbs the Object because of the active intervention of the Subject, who applies to the Object a *method* of investigation that is his own and to which nothing in the Object itself corresponds.'[48] According to Kojève's Hegel, the true result of scientific research, then, is not a set of objective facts but an interaction between the observer and the phenomena under investigation, and science never attains 'this autonomous real, this "thing in itself" of Kant–Newton'.[49] The concern of 'la science vulgaire' is 'not with the concrete real, but with an *abstraction*' – a 'disturbed', because objectified, version of the 'Real' object.[50] The same goes for 'vulgar' psychology, nosology and psychology, which attempt equally artificially to isolate the object of study.

Having spoken up to the brink of Heisenberg's Uncertainty Principle, Kojève now plunges in, referring explicitly to the discoveries of QM in support of his interpretation of Hegel's scientist:

> This interpretation, on which Hegel insisted very much, is currently admitted by science itself. In quantum physics it is expressed in mathematical form by Heisenberg's relations of uncertainty. These relations show on the one hand that the experience of physics is never perfect, because it cannot achieve a description of the 'physical real' that is both complete and adequate (precise). On the other hand, the famous principle of 'complementary notions' follows from it, formulated by Bohr: that of the wave and the particle, for example.[51]

Kojève goes on to say that while the findings of physics are established as contradictory when placed in common sense language, in their algebraic form there is neither uncertainty nor contradiction. Therefore, although quantum physics admits that its findings are correct while being 'contradictory' when placed in written and spoken language, it is still guilty of 'vulgar science' because only 'le Discours philosophique' can give access to Truth: 'it alone is related to the *concrete* Real – that is, to the totality of the reality of Being.'[52] Because Bachelard believed that philosophy should alter its language to accommodate the new concepts demanded of physics by the explanatory force of the mathematics, he would have been in complete disagreement with Kojève (and Hegel) on this point.

Kojève's comparison between Hegel's phenomenological rejection of the subject–object distinction and the conclusions reached by quantum theory in the mid-to late 1920s is instructive for the history of epistemology and twentieth-century French philosophy. It suggests that Hegel's return from a century of near exclusion from French thought to initiate phenomenology's long challenge to the dominance of Cartesianism, and the widespread enthusiasm for quantum theory, both had their source in the same challenge being laid to the hitherto invincible epistemology of the Enlightenment.[53]

Evidence for this is forthcoming from another source, adding to our understanding of Bataille's scientific context. As I showed, this challenge to the Cartesian bias of mainstream philosophy was made *sotto voce* at the *Revue de Métaphysique et de Morale*, a publication Bataille knew well.[54] And several contributors to that journal held places on the Comité de patronage of a more radical philosophy periodical largely forgotten today, *Recherches philosophiques*, among them Meyerson and Léon Brunschvicg, neo-Kantian André Lalande, philosopher-sociologist Lucien Lévy-Bruhl and the aforementioned Abel Rey (fig. 53). The diverse specialties represented by this group set out an editorial position responsive to subject matter of wide scope.

But a glance at its editorial board – made up of Henri-Charles Puech, Albert Spaïer and Alexandre Koyré – brings us closer to Bataille and Surrealism and shows why the tendency towards eclecticism of the *Revue de Métaphysique et de Morale* was not only consolidated in *Recherches philosophiques* but slanted towards a convergence of the physical and human sciences, German philosophy (notably phenomenology) and psychology. Commended by Raymond Aron because it 'instilled in France, and in the French language, many of the German teachings',[55] *Recherches philosophiques* was co-founded by Koyré and Spaïer and supported by Kojève, Bachelard and Wahl (all of them German speakers) specifically as an outlet for German phenomenology, then receiving little space in either of the two main philosophy reviews, the *Revue de Métaphysique et de Morale* and *Revue philosophique*.[56] Discussion of *Recherches philosophiques* aids us in tracking Bataille's philosophico-scientific parameters in the 1930s, because it was apparently in the vicinity of the Koyré/Kojève seminar on Hegel that the idea for a new forum for contemporary thought in France was realised. Crucial to the subject of Surrealism and modern physics as a whole, *Recherches philosophiques* has received barely any scholarly attention. Here I can only sketch an account of it relevant to Bataille's interest in modern physics, beginning with its editorial board.

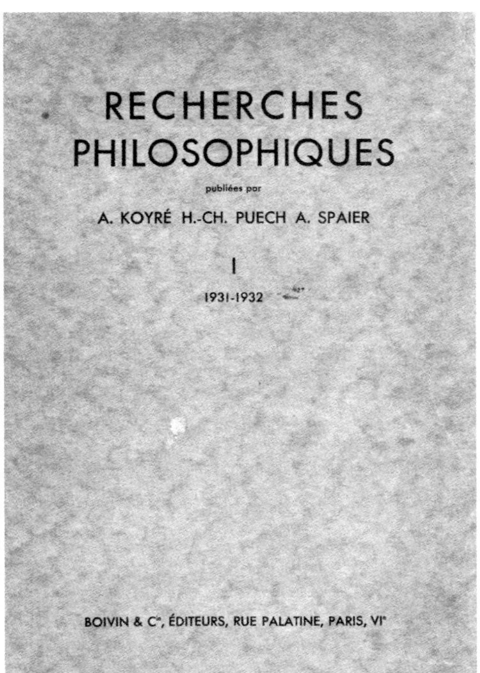

53 Cover of *Recherches philosophiques*.

Puech was an orientalist and historian of comparative religion, folklore and ancient philosophy who became editor of the *Revue de l'histoire des religions*.[57] Of the same generation as the Surrealists, Puech had contacts with the avant-garde from the 1920s and contributed to Bataille's *Documents*, though his exchanges with Bataille have not yet been assessed by historians.[58] During the run of *Recherches philosophiques*, he published a notable article in *Minotaure*, 'Signification et représentation', which demonstrates his concern with the discursive and conceptual relationship between physics, psychoanalysis and phenomenology.[59] Puech shared his interest in psychoanalysis with his colleague on the editorial board of *Recherches philosophiques*, the philosopher Albert Spaïer, who was the main animator of the publication. Spaïer was an admirer of Freud and saw to it that psychology and psychoanalysis received attention in the main body of the periodical and in its book review pages.[60]

The third editor of *Recherches philosophiques* was the philosopher and historian of science Alexandre Koyré, whom I have already introduced. Like Puech, Koyré possessed a commanding knowledge of both physics and psychology and had established relations with the avant-garde prior to the founding of the journal. A Russian emigré who arrived in France in 1919, it was largely down to Koyré that the journal was read by avant-garde writers in the Surrealist orbit and received their contributions. His early work explored religious ideas and the philosophy of Descartes, while his 1929 doctorate in philosophy examined the thought of Jakob Boehme.[61] However, Koyré will be remembered as a historian who viewed science not in isolation but as a product of its age, and as a set of facts comprehensible only in the broader context of the thought of its age.[62] Koyré's contact with the avant-garde probably began some way into his teaching of Hegel at the École pratique des hautes études, perhaps when Kojève began to attend in 1932 or in the wake of Bataille's initially sceptical arrival at the seminar around the end of the 1920s. It would have been during his tenure at the ÉPHÉ that Koyré met Puech, who was its directeur d'études from 1929, and presumably it was there, too, that Bataille secured Puech's cooperation over articles for *Documents*.

Like *Documents*, *Recherches philosophiques* is the site of an interaction between the academic and avant-garde. Yet as we saw, Bataille had not taken a serious interest in the natural sciences at the time of his editorship of the more ethnologically inclined *Documents*, and consequently modern physics and its philosophical interpretation make only fleeting appearances there. By contrast, *Recherches philosophiques* emerged a little later under the guidance of an editorial board aware of the new physics, when knowledge of Relativity and QM were beginning to bed down among non-scientific French-speaking writers. Although it appeared only once a year, running for six issues of 450–600 pages until 1937, during its short life it gave exposure to pioneering articles on physics, philosophy, psychoanalysis and sociology, while offering others endeavouring to combine two or more of these disciplines in often demanding, speculative, ambitious ways.

While *Recherches philosophiques* never secured the collaboration of a prominent quantum physicist like Louis de Broglie, several individuals with first-hand knowledge of the groundbreaking discoveries of the preceding quarter of a century in physics give reliable accounts of its main developments. Bachelard contributed three articles on physics and philosophy, and reviewed relevant new German, French, Italian or English physics books in every issue.[63] A text by Émile Meyerson, 'La notion de l'identique',[64] appeared posthumously, and Meyerson is himself the subject of a piece by R. Johan, showing some continuity with the philosophically more conservative *Revue de Métaphysique et de Morale*.[65] Other articles in *Recherches philosophiques* concerning themselves with recent developments in physics include Spaïer's 'Pensée et étendue' in the first number;[66] Maurice Janet's account of closure in mathematics and science in his 'La finalité en mathématiques et en physique' in the second issue;[67] and immediately following Janet, Hermann Jordan's 'L'indeterminisme vital et le dynamisme des structures causales', on indeterminism, a subject attracting attention in every scholarly field since the formulation in 1927 of Heisenberg's Uncertainty Principle.[68] The same second number contains Michel Souriau's discussion of the different meanings invested in the term 'matter' by contemporary philosophers and scientists. Relying principally on Arthur Eddington's *œuvre* throughout, Souriau's article 'La Matière et le concret' even seems titled after the inspired contrast made by Eddington in his introduction to *The Nature of the Physical World* between the everyday solid objects we encounter and science's 'thinly scattered' tables, chairs and pens made of active atoms.[69]

Albert Spaïer died in 1934 and his obituary appeared in the third issue, by which time Kojève's Hegel seminar had brought Jacques Lacan onto the scene. Spaïer's interest in psychoanalysis, together with the contacts Lacan made at the Hegel seminar, ensured his 1932 doctoral thesis was reviewed in that number of *Recherches philosophiques*, though it was making the rounds elsewhere and was well-known to Surrealists like Dalí and Crevel.[70] More often than not, though, the journal was attracted to writers who observed philosophical trends dealing with new ideas of space, time and matter emerging from physics. Like Souriau's work, Eugène Minkowski's contributions to *Recherches philosophiques*, beginning with his 'Le problème du temps en psychopathologie', examine the subjective experience of temporality whilst revealing the thematic investment made by the journal in the debate on the relationship between the subjective experience and objective description of matter and time.[71] The same concern is highlighted in the metaphysician Raymond Ruyer's 'Sur quelques arguments nouveaux contre le réalisme', appearing in the fourth number and driven by a clear understanding of contemporary physics, observing that 'by necessity, the idea of exteriority has two very different senses depending on whether one views it from the perspective of the psychologist or physicist'.[72] The fifth, penultimate number of *Recherches philosophiques* is entirely dedicated to the subject of time with Ruyer, Minkowski and Jordan pursuing their reflections on the question, while the essay by the philosopher of science and future teacher of Albert Camus, René Poirier, 'Temps spirituels et temps matériel', also demonstrates its author's interest in the theme of 'subjective' and 'objective' time.[73]

The journal underwent a slight reorientation after Spaïer's death in 1934, when Puech and Koyré were joined on the editorial board by Bachelard and Jean Wahl (who as Léon Brunschvicg's nephew was unusually well-connected). They were undoubtedly responsible for the inclusion from the fourth number onwards of articles by participants in the soon-to-be-founded College of Sociology, initially Roger Caillois[74] and Pierre Klossowski,[75] whilst another with close links with the Surrealists, René Daumal, also contributed.[76] The fifth issue carried a second article each by Klossowski and Caillois[77] along with Koyré's review of the ever-popular Eddington, comparing his discussion of group theory in *New Pathways in Science* with those of Hermann Weil (*sic*) and 'P. Juvet' in *La structure des théories physiques modernes* (*sic*), averring 'we have never encountered such an amusing, original and non-mathematical exposition of the subject as Sir A. Eddington's'.[78]

This 1935–6 number of *Recherches philosophiques* saw Bataille's only contribution. Given the context I have developed here, it is not surprising that his text draws upon recent research in modern physics, nor that Bataille updated himself on the primary literature on quantum theory during its preparation by reading Hans Reichenbach, Niels Bohr and Paul Dirac.[79] In 'The Labyrinth', he builds further on ideas first explored in his key text published in *La Critique sociale* in 1933, 'The Notion of Expenditure', contesting the idea of a unitary ontology and arguing that Being, absent from the human but projected into the social, is no more to be found in the atom than in the increasingly unknowable electron (in the era of QM), from which it nevertheless seems to take its structure.[80] Echoing arguments made in *Documents* by contributors Puech and Carl Einstein on the contingent, expedient, illusory, manufactured structure of both nature and self, Bataille challenges the notion of communication between 'beings' on the basis that each individual 'state' of the person is an ephemeral effect of language, just as 'the atom composes its unity from variable electrons'.[81]

Bataille's line of inquiry was inspired by Paul Langevin's *La Notion de corpuscules et d'atomes*, to which he refers in a footnote (fig. 54).[82] This publication gave him the material to link a series of analogies on the macrocosmic level, based on the role of the indeterminate electron in substantialising the 'being' of the atom: '[a] particular being not only acts as an element of a shapeless [*informe*] and structureless whole . . . but also as a peripheral element orbiting around a nucleus where being hardens'; 'In the case where many cities abdicate their central function in favour of a single city, an empire forms around a capital where sovereignty and the gods are concentrated.'[83] Published as a slim volume

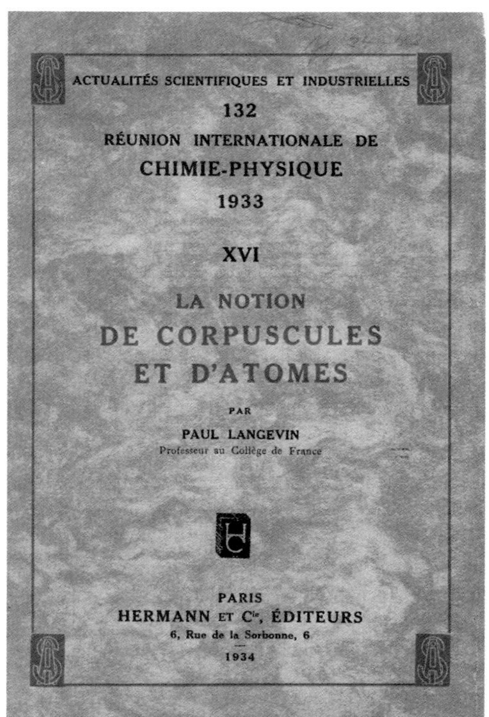
ACTUALITÉS SCIENTIFIQUES ET INDUSTRIELLES
132
RÉUNION INTERNATIONALE DE
CHIMIE-PHYSIQUE
1933
XVI
LA NOTION
DE CORPUSCULES
ET D'ATOMES
PAR
PAUL LANGEVIN
Professeur au Collège de France
PARIS
HERMANN ET Cie, ÉDITEURS
6, Rue de la Sorbonne, 6
1934

54 Paul Langevin, *La Notion de corpuscles et d'atomes* (1934).

by Hermann in 1934, Langevin's book had a high profile at the time, described as a 'tour de force' by Koyré in his review in the previous number of *Recherches philosophiques*[84] and also reviewed in *Scientia*.[85] Concerned in part with demonstrating how the atomic theory of QM demanded an intersection of physics and chemistry, Langevin's tract was ideal reading matter for the beginner seeking to get to grips with the latest theory of the electron. The general reader could easily have comprehended the sheer novelty of the new scientific spirit heralded by Langevin, with recourse to Eddington:

> Eddington, who is a great scientist, expresses it like this in his book titled *The Nature of the Physical World*: 'It will perhaps be said that the conclusion to be drawn from these arguments from modern science, is that religion first became possible for a reasonable scientific man about the year 1927 . . . If our expectation should prove well founded that 1927 has seen the final overthrow of strict causality by Heisenberg, Bohr, Born and others, the year will certainly rank as one of the greatest epochs in the development of scientific philosophy.'[86]

Like a damp stained wall, quantum theory can conjure up just about any view of the world if stared at long enough, and Eddington would not be the last to find confirmation of his religion in its paradoxical design.

On the following page, Langevin gives a brief account of this new (Heisenbergian) indeterminacy alongside an abbreviated version of Bohr's Complementarity. It is this new non-concept of the electron – indeterminate until observed (ontologised) as wave or particle – that shapes Bataille's argument on being and insufficiency in 'The Labyrinth', where he argues that humans are driven to act in order to fill the void of life with a sense of their own being. In his quest to overcome his feeling of insufficiency, the individual produces an unusable surplus, which breaks with the circumscribed logic of production and expenditure. In this way, the concepts produced by modern physics play a key role in Bataille's theorisation of general economy in the 1930s.[87]

Although this appearance in the fifth number of *Recherches philosophiques* was Bataille's one and only outing in its pages, his closeness to the journal is continued in the next 1936–7 number, where a materialist critique taking account of Meyerson's work appears by Bataille's recently deceased friend Arnaud Dandieu, co-written with Denis de Rougemont, who was concurrently active in the newly founded College of Sociology.[88] However, by this time, new trends in French philosophy ensured a lesser role for physics in *Recherches philosophiques*. While Wahl's thoughts on Heidegger and Kierkegaard had been aired as early as the second number, and Heidegger himself published in the first (reportedly on Bataille's suggestion to the editors), Jean-Paul Sartre's rise to prominence now saw his brand of existentialism begin to make its impression on intellectual life in Paris.[89] Beginning a series of major statements (some of which were carried in periodicals like the *Revue de Métaphysique et de Morale*), in this final number of *Recherches philosophiques*, Sartre takes his leave of Husserlian phenomenology and sets the tone for the existentialism of *Being and Nothingness* with his 'La transcendance de l'Ego: Esquisse d'une description phénoménologique', while his book *L'Imagination* is acclaimed in lengthy articles and book reviews here and elsewhere.[90]

As in the *Revue de Métaphysique et de Morale*, new publications by all of the main physicists involved in the formulation of Relativity and QM come in for scrutiny in the book review pages of *Recherches philosophiques*, where the dominance of physics is evident. Koyré himself reviews books on philosophy and physics, as does one A. Kojevnikoff (Kojève) who rounds out his assessment of Eddington's *Expanding Universe*, Jeans' *New Background of Science* and Hermann Weyl's *Open World*, saying 'the revival of interest in philosophy among modern physicists is not due merely to chance, but is here a natural consequence of the very character of these theories'.[91] Although (because?) Bachelard was the chief reviewer, all of his books receive favourable attention, including, of course, *The New Scientific Spirit*, described by Souriau (now a fellow member on the periodical's Comité de rédaction) as 'useful for the mass of riches it holds in a small volume, for the clarity and exactitude of its conclusions'.[92] As we saw, that opinion was soon to be shared by a larger audience of intellectuals, writers and artists.

The diverse subject matter of *Recherches philosophiques* (loosely based on the successful formula developed by Xavier Léon at the *Revue de Métaphysique et de Morale*) made it a prime site for the surveyal of modern physics by an audience beyond the immediate world of the philosophy of science, where the profound importance of its discoveries had been quickly understood. The roles played by Meyerson, Puech, Koyré and Kojève in making the new physics known to avant-garde writers were crucial, as were those of popularisers such as Bachelard, James Jeans and Eddington.

Surrealism and Arthur Eddington

Widely reviewed in French science and philosophy periodicals, Eddington's books naturally provided a useful guide to contemporary science for the Surrealists and Bataille, as did his reader-friendly articles in journals like *Scientia*, where he published four times between 1910 and 1923 (fig. 55).[93] Eddington was regarded by French philosophers of science as a reliable and lucid expositor, and his and Langevin's recent publications on Einstein's theory are discussed seriously and at length in Meyerson's article in the *Revue de Métaphysique et de Morale*, 'Le relativisme, théorie du réel'.[94] Eddington's high profile among French writers and artists during the period I am discussing is further confirmed in the *Nouvelle Revue française* column, 'La Pataphysique du mois'. Written by René Daumal, who published in *Recherches philosophiques* and who had been close to Surrealism, it kept the *NRF* audience abreast (sometimes ironically) of the work of Eddington, Werner Heisenberg, Erwin Schrödinger and other luminaries between June 1934 and February 1935.[95]

In a lengthy 1935 assessment of the French translation of *The Expanding Universe*, a reviewer hits on one of the chief characteristics of Eddington's work, describing it as

55 Arthur Eddington, 1931.

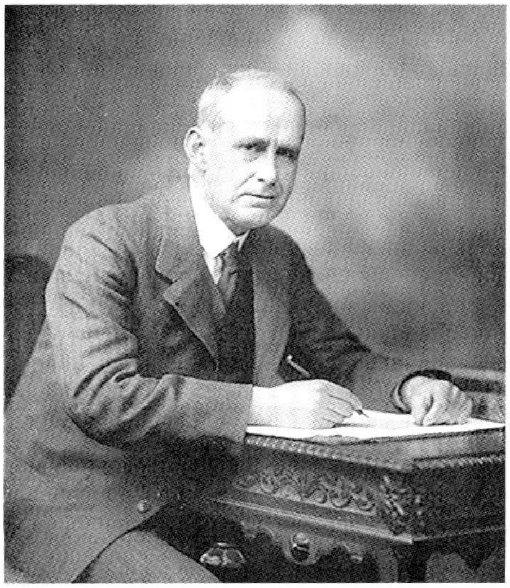

'très amusant'.[96] Among the best-known popularisations of modern physics between the wars, Eddington's books held a peculiar appeal for the Surrealists for the way they intersperse, elucidate and 'soften' scientific observation with quotations from Shakespeare, Lewis Carroll, Jonathan Swift and others, commending his writing to audiences more exercised in the norms of literature than those of science. Eddington's great love of Carroll's *Alice* stories made him a kind of incipient Surrealist, and Chandrasekhar's comments on Eddington's word-games are particularly suggestive in this connection:

> I recall Eddington telling me that at school one of the games he played was to make up English sentences which were grammatically correct but which made no sense, in the manner of Lewis Carroll. An example he gave me was
>
> *To stand by the hedge and sound like a turnip.*
>
> In later life, Eddington was wont to introduce such sentences in his more serious writings to make a point. Thus in his Swarthmore Lecture, *Science and the Unseen World*, we find:
>
> *Human personalities are not measurable by symbols any more than you can extract the square root of a sonnet.*[97]

References in *Space Time and Gravitation* (1920), *Matter in Interstellar Space* (1929), *The Nature of the Physical World* (1930), *The Expanding Universe* (1933), and *New Pathways in Science* (1935) to Shakespeare, Carroll and Swift (all listed by Breton at one time or another as British precursors to Surrealism) confirmed Eddington's relevance to Surrealists like Salvador Dalí, who exploited Relativity's conceptual oddities in his paintings and writings from the beginning of the 1930s.

A classic example of a Surrealist reading of Relativity aided by Eddington's imagery is the one by Oscar Domínguez and Ernesto Sábato. In his 'Most Recent Tendencies in Surrealist Painting' of 1939, Breton discusses the 'lithochronic surfaces' of Domínguez, such as his 1938 *Paysage cosmique* (fig. 56).[98] These experimental paintings aimed to capture something of the nature of an event in space–time as understood after Einstein, and began to appear in 1938 in the form of 'cosmic landscapes'. They were the outcome of Domínguez's meeting with the young Argentine radical Sábato, who was serving a fellowship at the Curie Laboratory in Paris by day while associating with the Surrealists at night.[99] Sábato's articles on the work of Max Planck, George Gamow and Eddington appeared in the Argentine journal *SUR* in 1945, and his expertise is evident in the text 'The Petrification of Time', written with Domínguez in 1938.[100]

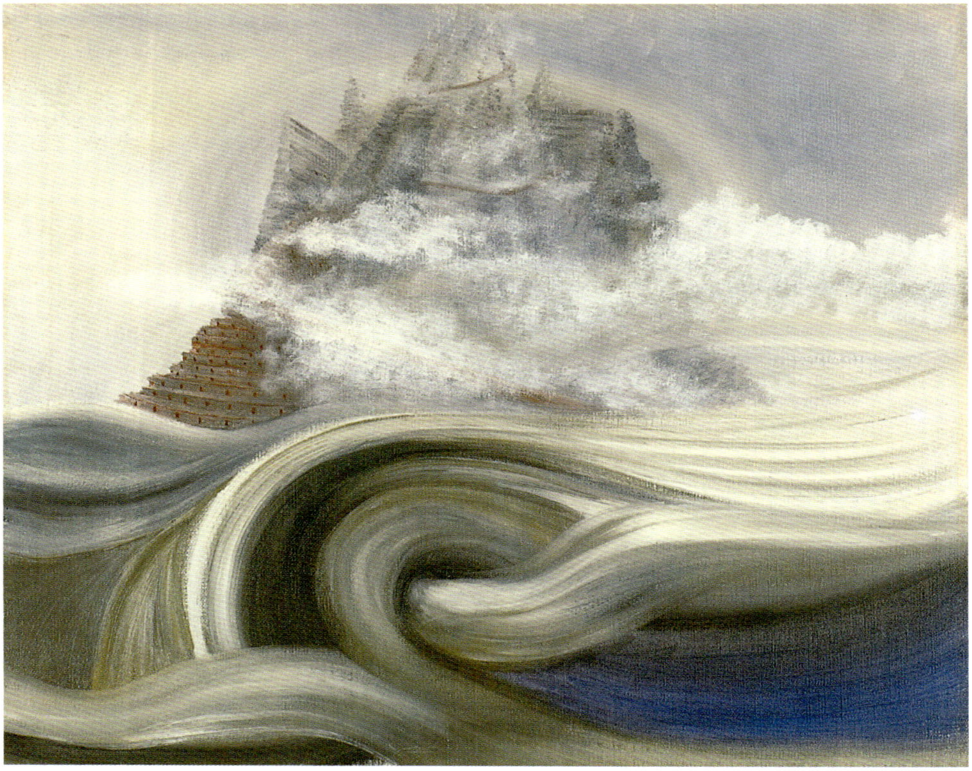

56 Oscar Domínguez, *Paysage cosmique* (1938). Oil on canvas, 38 × 46 cm. Private collection, Paris.

The first half of this brief article is concerned entirely with the mathematical structure and intuitive nature of 'four-dimensional' space-time. Sábato argues that Einstein's conception of space-time had introduced the notion of a four-dimensional continuum that could be represented mathematically, but also acted as a spur to imaginative speculation. Referring to Eddington's picturesque and 'surreal' (not to mention Dalínian) notion in *Space Time and Gravitation* of a man with one eye moving at 200 000 kilometres per second in relation to the other eye being able to attain a vision of the fourth dimension, Sábato concedes that present-day mortals would have to make do with artistic conjecture.[101] Appearing to draw upon Edwin A. Abbott's *Flatland*, Sábato goes on to say that it is possible to imagine the emergence of four-dimensional surfaces in our three-dimensional universe by analogy with the emergence of three-dimensional forms into a flat world inhabited by two-dimensional beings. This, too, was probably taken by Sábato from Eddington, who cites *Flatland* in *Space Time and Gravitation* as '[p]robably the best known exposition of the fourth dimension'.[102] Although the word 'petrification' in the title of their article suggests the concretion I noted in other quarters, the series of 'lithochronic' surfaces that came out of the musings of Sábato and Domínguez were far from Dalínian 'hyper-realism', having more in comon with the Bergsonian 'smeared' temporality of Futurism (fig. 57).

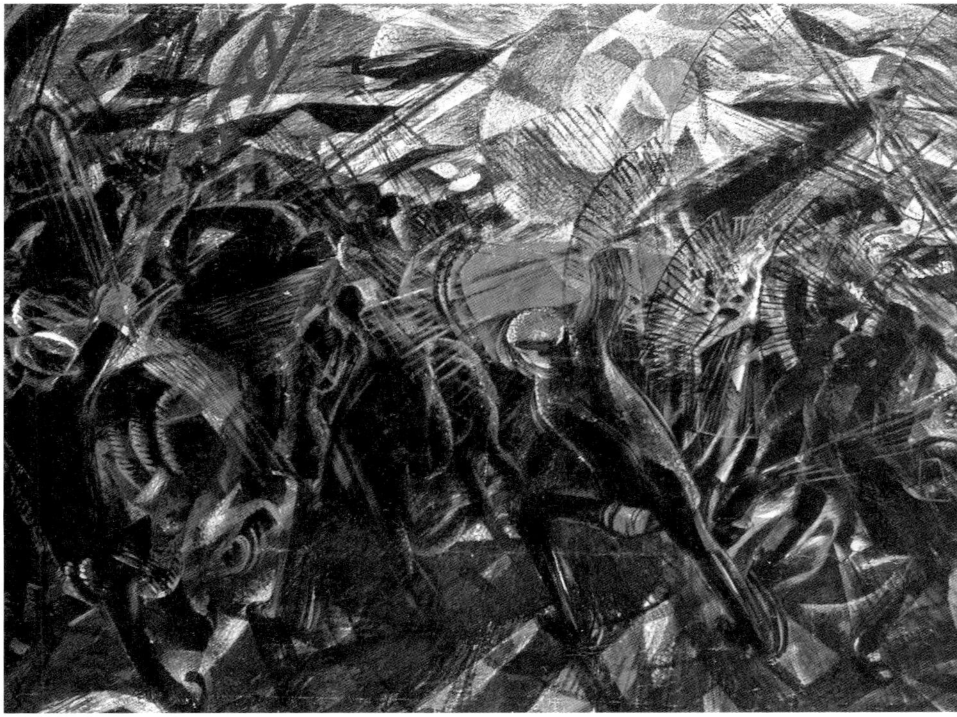

57 Carlo Carrà, *Funeral of the Anarchist Galli* (1910–11). Oil on canvas, 185 × 260 cm. The Museum of Modern Art, New York. Acquired through the Lillie P. Bliss Bequest.

Eddington's coalition of 'nonsense' literature and bizarre physics, admired by Surrealists like Domínguez, is deployed by Bataille in concert with a mass of other strands of popular and semi-popular science culled from his voluminous reading. Future understanding of the development of his economic and sociologic theory must acknowledge his exploration of atomic, astro-, bio- and geophysics. This is particularly true for his College of Sociology-era writings of the late 1930s, such as his talk of 20 November 1937, co-authored with Roger Caillois, 'Sacred Sociology and the Relationships between "Society," "Organism" and "Being"', which seeks to reconceptualise the idea of 'society' by sketching a kind of hierarchy of what Bataille called the 'être composé', ranging from the atom to modern Western society.[103] His concept of the 'compound being' leads Bataille to reject simplistic analogies based on the 'organism', regarded by him as too unitary and inflexible a definition. For this reason, he is resistant to scientific approaches wedded to concepts ('mass', 'group', 'totality', 'body'), ill-equipped to adapt to complex patterns of behaviour, singling out and quoting at length the *Scientia* article of that year by the astronomer Émile Belot.[104] For Bataille, Belot forged a rigorous yet narrowly conceived analogy between the evolution of stars and that of organisms on the human scale. Belot himself draws upon the writings of James Jeans, who Bataille had recently read in translation in a publication to which Eddington among others also contributed, *Discussion sur l'évolution de l'univers*.[105]

Alongside Jeans and Eddington, Michael Whitworth places Alfred North Whitehead at the top of his league table of the English-speaking audience's favourite popular science writers in the early 1930s, and judging from the reviewing habits of the periodicals I have looked at, something of this can be extended to the French context and specifically to Bataille's writings.[106] Combining a sophisticated understanding of the new physics and philosophy with a reader-friendly style, Whitehead's fame was secured with *Science and the Modern World* (1926), and his books were regularly reviewed in French philosophy journals, including the *Revue de Métaphysique et de Morale*. A task of his major work of 1929, *Process and Reality*, is to reach a definition of 'society' through the classification of many different kinds of society, arguing, 'our present epoch is dominated by a society of electromagnetic occasions.'[107] Whitehead's discussion in the same book of the competition between 'special societies' and the hierarchy of societies that he perceives as present in the world – 'individual electrons, protons, individual molecules, societies of molecules such as inorganic bodies, living cells and societies of cells such as vegetable and animal bodies' – compares with the model employed by Bataille in 'The Labyrinth' and throughout the period.[108] Bataille read English and certainly knew of Whitehead's writings, anyway, through Jean Wahl's 1931–2 *Recherches philosophiques* text 'Vers le concret' and from his reading of Wahl's 1932 book of the same name, assessing comparatively the philosophies of William James, Whitehead and Gabriel Marcel.[109]

Rehearsing events in the astral realm alongside terrestrial affairs, whilst observing behaviour at the molecular and atomic levels, Bataille's essay 'Sacred Sociology and the Relationships between "Society," "Organism," and "Being"' brings to mind passages in Eddington's popularisation, *Stars and Atoms* (1927), which insists that 'the road to a knowledge of the stars leads through the atom; and important knowledge of the atom has been reached through the stars.'[110] Bataille probably knew the book, as its French translation of 1930 was reviewed in the *Revue de Métaphysique et de Morale* in 1931, where the anonymous reviewer declares the study of the stars had been advanced by new discoveries in atomic research, 'which cement the macroscopic with the microscopic'.[111] In another celebrated book of the period, *The Expanding Universe*, Eddington summarises the latest research on the movement of the stars, laying emphasis on behavioural patterns organised around attraction and repulsion, concepts Bataille thematises in his College of Sociology talk 'Attraction and Repulsion I: Tropisms, Sexuality, Laughter and Tears'.[112] By that time, Bataille's thoughts had turned fully towards celestial mechanics, general economy and the 'society of the stars', and in the 1938 article 'Celestial Bodies' he draws directly on Eddington's writings to argue his case.

Eddington and Bataille: Icarian Illumination and Elimination

Given as the Halley Lecture in May 1930, Eddington's *Rotation of the Galaxy* contains little of the playfulness that animates *The Expanding Universe* and *Space Time and Gravitation*, yet this is the key source footnoted by Bataille in 'Celestial Bodies'.

In the book, Eddington gives a brief history of the debate on the motion of the galaxy, before cautiously outlining the limitations placed upon the reliability of its observation by noting the risk taken 'in inferring the nature of the complete system from the small sample within reach'.[113] Once the velocity of the earth is accounted for, he argues, the movement of the stars takes on a calculable pattern: 'they are by no means those of an unorganised crowd'.[114] Eddington takes the asymmetrical design of clusters of stars moving at great speeds as evidence of rotation, calculating the timescale for a full turn of the wheel of the galaxy at 250 million years. Such calculations assumed an approximately steady state for the stellar system:

> It may perhaps be thought that too much of a fetish has been made of the 'steady state' in the various mathematical theories of the galaxy, but that is a misconception of their aim. Although formally the mathematician may seem to be designing a model stellar system which will last forever, that is only his way of tackling the design of a model which will last long enough to fulfil the obvious requirements of the problem.[115]

This contingency of the steady state (of which Bataille took careful note) is echoed in Eddington's closing remarks, where he writes: '[e]very engineering instinct we possess protests that such disk-like arrangements of matter are precarious, and their stability is not to be trusted.'[116] He concludes with a wistful comparison that anthropomorphises our galaxy as 'a young man in a hurry'.[117]

Although Bataille does not take issue explicitly with Eddington in 'Celestial Bodies', my earlier discussion of general economy demonstrates that Bataille's thought departs entirely from conventional philosophic and scientific notions of the stellar system (or anything else) as a particular body subject to measurement – as a fetish, unity or form. In 1929, the year before the appearance of *The Rotation of the Galaxy*, Bataille published a short statement in *Documents*. Its title, 'Informe' (usually translated as 'Formless'), introduces an adjective specifying debasement beyond form, concept or knowledge:

> What it designates has no rights in any sense and gets itself squashed everywhere, like a spider or an earthworm. In fact, for academic men to be happy, the universe would have to take shape. All of philosophy has no other goal: it is a matter of giving a frock coat to what is, a mathematical frock coat. On the other hand, affirming that the universe resembles nothing and is only *formless* amounts to saying that the universe is something like a spider or spit.[118]

Bataille's attack here upon what Martin Jay has called 'the fetish of form, which was so dependent on visual distance',[119] is a manifestation of his mistrust of the visual. It led him to link his own 'base' materialism to 'the Gnostic principle of darkness that was opposed to the Hellenic worship of clarity and light', a view entirely at odds with the mathematical rationalism of Eddington.[120] Similarly, his 'Rotten Sun', published in *Documents* in 1930 at around the time Eddington gave the Halley Lecture, offers an alternative to the rationalist, heliocentric interpretation of the sun. Against its 'poetic meaning of mathematical serenity and spiritual elevation' in the 'mind of one whose weak eyes compel him to emasculate it', Bataille's essay poses the sun as actually *viewed* by an individual, implying animal and human sacrifice, blindness, madness, horror, 'mental ejaculation, foam on the lips, and an epileptic crisis.'[121] Jay notes: '[t]he only light cast by the potlatch ceremonies he found so fascinating was produced by the flames consuming the wealth destroyed.'[122] And the negation of the sun, or its transformation into a 'blind spot', is a significant trope in Bataille's general economy.

While Eddington writes in *The Rotation of the Galaxy* of the generation of temporary mathematical models acting as rungs on a ladder towards a more complete understanding of the stellar system, Bataille regards such endeavours as determined and shaped by an always absent 'unknown'; they are 'futile but pointed labours' as Jonathan Strauss has put it.[123] Strauss argues further that, for Bataille, science, like philosophy, systematises a homogeneous, restricted economy, in the sense that its 'geometrical regularity' (the demands of its symbolic codes, languages or 'mathematical frock coat') allows no space for arbitrary or irrational phenomena, driven as it is by ingrained habits of assimilation and integration. Panlogism, monism, causality and recourse to origins in the Big Bang theory or the 'mind of god' are symptomatic of what Bataille called restricted economy: humankind's will to circumscribe through signification, which is also 'the concealment of that for which it longs.'[124] By contrast, a general economy accepts the existence of an *other* as its formative object of representation. This 'other' remains irredeemably beyond recuperation by the intellect and its inventions, but its inclusion within the general economy can 'account for' heterogeneous elements in the universe, society or human (though they remain unaccountable). Strauss continues:

> It is precisely what science cannot see, what it obscures, that point to which it has blinded itself but which in exchange allows it to see everything else, that is the meaning of science . . . the heterogeneous can never be represented, is always obscured, as if in a cloud of dust, by that which would designate it, is always mistaken by a writing which remains within its own terrestrial limitations and must inevitably fall short.[125]

Bataille believed that heterogeneity – the irrecuperable excess beyond human limits that cannot find form in language, knowledge or thought – could be experienced through, and therefore 'represented' as, 'horror, violence, hatred, sobs, crime, disgust, laughter and human love'.[126]

Evidently, in choosing these limit conditions of human experience, Bataille's aim was to strip away elevated 'dignity' and holistic 'integrity', not just from scientific discovery, but from creative endeavour more generally. In Eddington's thinking, such creativity finds expression in a relatively benign language, greatly uncongenial to Bataille's idiom, as 'a striving born with our consciousness or an Inner Light proceeding from a greater power than ours'.[127] As I noted earlier, whilst the wrangling between Bataille and Breton was at its most intense in the late 1920s, Bataille objected to Breton's 'elevated' language, pouring scorn on its 'Icarian naiveté'.[128] Bataille implicates Surrealism in 'the call to order of a heaven that today demands the most imbecilic elevation of any man's spirit', opposing the imperialist eagle with Marx's image of the 'old mole', a symbol of proletarian revolution in his *Eighteenth Brumaire of Louis Bonaparte* (1852).[129] Curiously, Eddington's own affirmative image of Icarus in his 1920 British Association Address depends on a misinterpretation of the myth, recasting Daedalus's son as an inventor:

> Cautious Daedalus will apply his theories where he feels most confident they will safely go; but by his excess of caution their hidden weaknesses cannot be brought to light. Icarus will strain his theories to the breaking-point till the weak joints gape. For a spectacular stunt? Perhaps partly; he is often very human. But if he is not yet destined to reach the Sun and solve for all time the riddle of its constitution, yet he may hope to learn from his journey some hints to build a better machine.[130]

This passage is fairly typical of Eddington's routine rhetoric of humility as to the efficacy of science's 'explanation' of the universe, symptomatic of his Quakerism and the value he placed in mystical experience. But it also conveys his 'optimistic pessimism', presenting science as a progressive endeavour moving one step back to take two forward. As with the provisional but necessary 'steady state' model of the galaxy defended by Eddington in *The Rotation of the Galaxy*, failure is figured here as an adjunct to success, as simply delaying the inevitable attainment of a goal ('he is not *yet* destined to reach the Sun'), and therefore as a useful and necessary factor in the economy of scientific progress. While his remarks appear at first to carry some of the fatalism that can be ascribed to Bataille's view of scientific activity (the 'futile but pointed labours' of Strauss's phrase), this elevation of failure to a virtue, as useful to knowledge, is quite antithetical to its stubbornly negative sense of wasteful expenditure (of time, money, energy and so on) without profit in Bataille's resolutely defeatist scheme.

Equally divergent from Bataille, of course, is Eddington's Christian–Quaker belief in a divine light that illuminates and dignifies humanity. Associated with the light of the sun and stars, this flares up in Eddington's 'Icarus' passage in his metaphor of the theoretical weakness that will be 'brought to light'. Eddington often attuned the 'divine light of truth' to the progress of humankind in its 'natural' thirst for knowledge, writing in *Science and the Unseen World*, '[i]n science as in religion the truth shines ahead as a beacon showing us the path.'[131] By contrast, Bataille's characterisation of the sun in *The Solar Anus* is framed in a far from heroic or elevated way, describing 'the filthy parody of the torrid and blinding sun',[132] a black excremental hole like the 'dark, repulsive nucleus' of 'Attraction and Repulsion', which he posits at the core of every body on every scale. No wonder Bataille favoured the idea of the 'old-mole' revolution, which 'hollows out chambers in a decomposed soil repugnant to the delicate nose of the utopians'.[133] But if the traditional conflation of the stars with the heavens (enshrined in the words 'céleste' and 'celestial', which retained a residual religious significance in the French and English languages) give a stability to Eddington's metaphors, how do religious connotations, together with Eddington's rendering of the movement of the stars, translate in Bataille's text 'Celestial Bodies'?

Bataille's essay was published in the expensive French art review *Verve*, accompanied by three ink drawings and two lithographs by André Masson (fig. 58) and two more lithographs by Kandinsky.[134] The remarkable drawing by Masson titled *The Labyrinth* (1938) dominates the double page spread in *Verve* (which was published in both English and French). The labyrinth idea often acted in Surrealism as a metaphor for the unconscious, but here it might almost be an illustration for Bataille's *Solar Anus* while aligning in a simple image individual identity, in the form of a human fingerprint, with the celestial bodies and rotating galaxy of Bataille and Eddington. Bataille's inquiry into loss and the wasteful outlay of energy, first mapped in 'The Notion of Expenditure', is extended onto the cosmic scale in 'Celestial Bodies', and is quite at variance with Eddington's account of the behaviour of stars.[135] In *The Rotation of the Galaxy*, Eddington claims that Saturn's rings 'have a rough resemblance to the disk-like form attributed to our galaxy'.[136] This analogy is accepted by Bataille at the opening of his article,[137] but then inverted to assert for the solar system, as he had for the universe in 'Informe', a disfigured indeterminacy, evading unitary representation:

> If the sun were driven by a steady movement of gravitation around a centre, the universe to which Earth belongs could, on the whole, be considered analogous to the closed system of our planetary revolutions (only the order of magnitude would have changed). Whereas the movement described by the sun and its planets about the galactic centre opens, as it were, into and through the sky. When, indeed, we do perceive this movement within infinity – driving universes like our own – it in no way resembles the stable and geometric rings of Saturn; it seems hurled into space like a whirling explosion.[138]

The behaviour of the galaxy suggests to Bataille 'the swirling suns of our light shows', and, like the mesmeric, luminous, terminatory feats of the Potlatch ceremony, Aztec sacrifice and myth of Icarus, 'a dazzling explosion of fireworks which vanished after a few seconds into the night'.[139]

Conscripting *The Rotation of the Galaxy* and modern science in general into an attack upon anthropocentrism, Bataille notes that the immensity of the universe they describe allots barely any significance to human beings. 'According to Eddington', he writes in 'Celestial Bodies', 'the precariousness of a universe like ours is reflected in the way it looks.'[140] But Bataille goes on to argue that this *looking*, this assumed distance from – and observation and particularisation of – an external object, is a disguised form of that same anthropocentrism. Science, he maintains, has artificially set

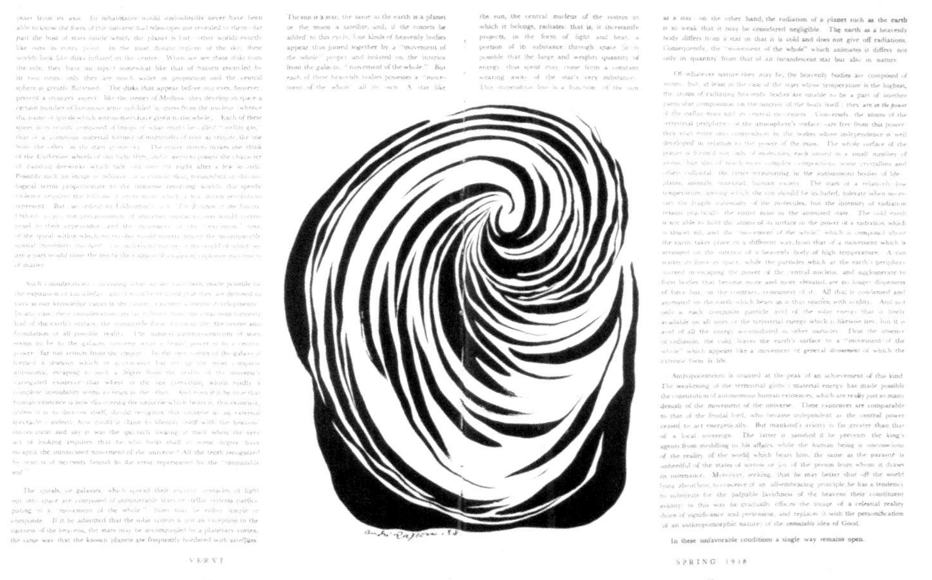

58 Georges Bataille, 'Corps célestes', *Verve*, vol 10, no. 2, March–June 1938, 98–99.

the earth apart as a still point in the galaxy, when really it belongs within its 'mad spin':

> A domain has been marked off directly with the galaxy's whirlwind, such that it appears constituted as not subject to the reality of the universe, as entirely autonomous, to the point of perfect immobility set within the mad spin of it all. And even if human existence is really in the process of now discovering the universe that sustains it, this existence must acknowledge the universe as a spectacle external to it or deny itself. How, indeed, can it claim to identify with the rapture of the heavens, acknowledging itself as a spectacle viewing itself, when the fact of looking presupposes that the viewing subject has somehow escaped from the rapturous movement of the universe? All that we recognise as truth is necessarily linked to the error represented by the 'stationary earth'.[141]

This is the (dark) core of Bataille's argument, setting out starkly where his attitude towards science, truth and knowledge differs from Eddington's. For Bataille, the subject of thought cannot be separated off from the object of its inquiry, because he believes thought (*pensée*) and expenditure (*dépense*) mutually exclusive. Insofar as thought must have its object and aim, wasteful expenditure ceases; and in as much as intellectual energy is expended *to bring science its results*, there is no meaningless forfeit. Where the rationalism of Eddington's closed system allows the fallen Icarus (or his replacement) to rise again by virtue of the knowledge gained from his fall, the logic of Bataille's general economy of outlay without return demands Icarus's pointless sacrifice.

Varieties of Mysticism

If, as Bataille's commentators argue, the 'truth of the universe is expenditure, which is not graspable, never fully thinkable', how could human beings gain access to this 'truth', and through what 'knowledge' if not through thought?[142] This brings us to the religious, mystical side of Bataille's writing, expressed in 'Celestial Bodies' in a language and imagery of intoxication, festival, dance, avidity and ecstasy. Here Bataille marks the transgressive passage he believes human beings must take from avaricious greed to infinite excess; or as he expressed it: '[t]he true gift of the self, ecstasy – which is not only a link between the sexes – marks, nonetheless, the limit of greed, the chance of escape from cold movement, and of rediscovering the joy of sun and spiral.'[143] Susan Rubin Suleiman writes that in the transgressive experience 'purposeful action, or work, becomes subordinated to free play', which she calls the 'excessive, quasi-mystical state we associate with religious ecstasy and generally with the realm of the sacred'.[144] Bataille felt that only through loss, a loss adjacent to death, could human beings know the black light of divinity, 'regain the free movement of the universe . . . dance and swirl in the full rapture of those great swarms of stars'.[145]

Bataille's extraordinary vision of a festive stellar dance in 'Celestial Bodies' is closely linked to his activities and reading at the time of writing. Following the collapse of Contre-Attaque in 1936, with deep divisions on the Left reopening and another war seemingly inevitable, Bataille formed a secret society that he described later as 'non-political, religious, [a]nti-Christian, Nietzschean'.[146] It published four numbers of a journal, *Acéphale*, while the College of Sociology manifested its other public face. The group met in secrecy, sometimes to carry out sacred rites at a privileged place, until Bataille's crazy suggestion (apparently serious) that the group perform a human sacrifice in Paris unsurprisingly alienated several of its members and led to the breakup of the group at the beginning of the war. Appropriately, the figure of Acephalus, who in losing his head is freed of his reason, was used as the emblem for Bataille's secret society (illustrated by Masson), which was active at the time he wrote 'Celestial Bodies' in 1938. That group's 'religious, anti-Christian, Nietzschean' orientation is evident in Bataille's spectacle of the heavens as a festive, Dionysian, orgiastic disbursement of profligate energy.

Bataille borrowed Nietzsche's *Gay Science* (1882) from the Bibliothèque Nationale in May 1936, and he quotes from the book in 'Nietzsche and the Fascists' of early 1937 and again in 'The Obelisk', a text exactly contemporary with 'Celestial Bodies', published in April 1938.[147] The 'gay science' of Nietzsche's title has been taken to suggest 'light feet', 'dancing' and 'laughter', all of which bring us close to the spirit of Bataille's text.[148] So, too, does Nietzsche's attitude towards religion. For although his endlessly quoted pronouncement 'God is dead' first appears in that book, Nietzsche's rejection of Christianity was not a denial of divinity.[149] Rather, Nietzsche aimed to combat what he called '[a]fter-effects of the most ancient religiosity', such as causality, ignorance, dependency, negativity, suicide, the idea of sin and limiting epistemological structures.[150] The section in *The Gay Science* titled 'Delight in Blindness', in which the wanderer says 'I love my ignorance of the future', is meant to indicate Nietzsche's anti-Christian indeterminism,[151] as is the suggestion of 'Preludes of science', where science is placed in a lineage of interpretation that includes magic and alchemy.[152] There is, then, a distinctly Nietzschean inflection to Bataille's rejection of goals and his wish to 'escape from those traps which goals represent'.[153] Bataille adopted, he explains, 'an attitude which follows in the wake of religious experience', but is 'freed from the quest for means', finding its fulfilment in experience alone.[154]

'Celestial Bodies' can and should be read as a piece of religious writing, as should much of Bataille's *œuvre*. In fact, from the moment of his maturity, Bataille revealed a religious yearning. He converted to Catholicism and lived a pious life from 1914 until 1920 when he lost his faith, and his first published text dating from around 1918 is a six-page eulogy to the bombed Notre-Dame de Rheims, never mentioned by Bataille and only found after his death.[155] His discovery of Nietzsche two years after his rejection of Christianity began a lifelong fascination with *l'expérience intérieure*, and Sartre's gibe at the 1943 book of that name – Bataille's longest text to date – as 'un nouveau mystique'[156] would not have been taken by Bataille in the critical spirit in which it was intended.[157] Bataille's ruminations in *Inner Experience* on ecstasy, the void and a God of extravagant, unlimited excess – one who 'is already in the Nothingness of His insatiability' and 'knows nothing of the extent of His thirst' – would have found a sceptical reader in Eddington, to say the least.[158] Yet Bataille's *Recherches philosophiques* essay, 'The Labyrinth', is folded into that book, confirming the close association in his thinking between mysticism and science.[159] Eddington alludes insistently to the same connection, and it is through mysticism that a closer relationship between the divergent temperaments and interests of Eddington and Bataille may be discovered.[160]

As a Nietzschean, anti-Christian and non-technical rendering of the conduct of the stars, it is hard to imagine 'Celestial Bodies' offering anything citable to Eddington. However, Eddington did not hesitate to use the term 'mysticism', and is even introduced by his biographer as 'one of the greatest astronomers of his day and a profound mystic in his approach to all knowledge'.[161] In his contribution to the 1930 BBC symposium on 'Science and Religion', Eddington stresses human will and purposiveness, arguing that while 'our bodies are pieces of stellar matter' in 'an inorganic world of chance', the 'spark from the divine spirit' that inhabits all humans enables us to understand our 'responsibility for truth.'[162] Contrary to Bataille's 'acephalic man', who loses his head to reach truth through experience, Eddington regards God-given reason as fundamental in the search for truth.[163] While he accepts the quantum-mechanical 'incompleteness' of physics, Eddington's view was obviously wedded to a notion of origins, ends and progress, for he writes of the physicist who 'has carried his work to greater perfection than formerly' and who 'advertises for someone to complete it'.[164] But here, Eddington also accepts that there is no question of physics giving 'the whole truth about experience',[165] as he did in the 'Science and Mysticism' chapter of *The Nature of the Physical World*. Adapting Niels Bohr's Principle of Complementarity (aired for the first time in autumn 1927) to a scientific and poetic description of the movement of waves on the sea, he concludes in the latter, 'it is impossible to allow that the one kind of appreciation can adequately fill the place of the other'.[166]

Eddington's and Bataille's mysticisms draw nearest to each other when we consider Eddington's handling of the distinction between 'symbolic knowledge' and 'intimate knowledge' in the 'Science and Mysticism' passage. On the level of symbolic knowledge, Eddington locates dogmatic theology and 'geloeology' (the science of laughter), whilst their counterparts on the level of intimate knowledge are mystical experience and laughter itself.[167] Eddington finds laughter, 'a not unfair analogy for our mystical feeling for Nature, and I would venture even to apply it to our mystical experience of God'.[168] Bataille, by contrast, sees laughter as no mere analogy for religious experience, but as an expression of what is unknowable, maintaining that he 'could recapture all the impulses of religious experience, mingling them with the experience of laughter without feeling that religious experience to be in any way impoverished'.[169] The essential 'un-knowing' of Bataille's indefinitely unfinished 'theory' of *dépense* signifies a different indeterminism from that of Eddington, which continues to demand the 'thought of the unity of the whole which is essential to a complete theory', even after the formulation of quantum theory.[170] Yet each of them believed that a mystical experience could lead to some *other*, beyond the 'veil of symbols'[171] or 'mathematical frock coat' designed by thought, and that such experience could find its representation in laughter.

Eddington's discussion of 'Science and Mysticism' ends playfully with the much-quoted passage that gives a physicist's view of the 'difficulty' of entering a room, detailing the speed of each body involved, their insubstantial nature, considerations of entropy and so on.[172] Walter Benjamin, who was close to the College of Sociology, viewed the perilous journey of Eddington's physicist as emblematic of the experience of the modern individual confronted by the overwhelming forces of Western industrialism. Quoting at length from this page in his 1938 article on Kafka, he declares: '[w]hen I refer to the modern big-city dweller, I am speaking also of the contemporary of today's physicists', rounding out his assessment with a remarkable appraisal of Eddington's accidental 'modernism': '[i]n all of literature I know no passage which has the Kafka stamp to the same extent.'[173] Like Bataille and the Surrealists, Benjamin learned about developments in modern physics in part from Eddington's writings, though he had recourse to *Recherches philosophiques* too, even reviewing the fourth 1934–5 number, with essays by Daumal, Caillois and Klossowski, and reviews of Bachelard's *New Scientific Spirit* and Langevin's *Notion de corpuscules et d'atomes*.[174]

Just as Benjamin saw a parallel between Kafka's world and a modern age gripped by forces of science and technology 'preparing to do away with the inhabitants of this planet on a considerable scale', the Surrealists took a dim view of scientific accomplishment with the onset of the Cold War.[175] Typical is an article by Robert Benayoun in the fourth number of the Surrealist journal *Médium* in 1955.[176] In his introduction to the writings of the American author Charles Fort, he argued that when scientific statements begin to sound more like fiction than fiction itself, and 'when the great mathematician Eddington finds clearly laid out in Lewis Carroll's *Jabberwocky* "the essential unknownness of the fundamental entities of physics," it is high time we read Charles Fort.'[177] Fort's near-Surrealist anthology collected scientifically unexplainable phenomena, and his criticism of established physics could be explicit. Among the Jarryesque extracts from his book carried by *Médium* was one comparing Einstein's Theory of Relativity ('le sublime dans l'inutilité') to a 1931 experiment with phosphorus at the University of California, which had produced luminous pigs.[178] The inclusion of Fort's writings in their journal crystallises the Surrealists' view of physics in the postwar period, demonstrating that the literary attractions of the scientific *vulgarisation*, which had initially drawn Bataille and others to the genre, were now sufficient grounds for its scornful dismissal, along with the rejection of physics as a discipline. I look more at this later.[179]

Bataille did not join the Surrealists in their anti-nuclear protest in the era of the Cold War. After *Inner Experience*, he published further book-length works while editing the review *Critique* from 1946 until his death in 1962, when an assessment began of his writings by thinkers of the calibre of Maurice Blanchot, Roland Barthes, Michel Foucault and Jacques Derrida, all of whom aired their early writings in *Critique*. At *Critique*, Bataille continued to work with physics in his writings, which lost none of their religious content, culminating in *The Accursed Share* of 1949, where he gave his fullest treatment to the theme of the profitless expenditure of energy.[180] Of fundamental importance to his research for this book was his friendship from 1935 with the physicist Georges Ambrosino, who Francis Maramande credits alongside Queneau and Roger Caillois as one of the key figures in helping Bataille stay abreast of the progress of modern physics.[181] Ambrosino was employed by de Gaulle's Commissariat à l'énergie atomique, and had been involved with both *Acéphale* and the College of Sociology before writing articles on atomic and nuclear physics for *Critique*.[182] Appearing in the inaugural number, the first of these was his laudatory review of *De la Mécanique ondulatoire à la théorie du Noyau* (1943 and 1945) by Louis de Broglie, who with his brother Maurice joined Koyré and around twenty others on the extended general editorial board of Bataille's journal in 1951.[183]

In a famous footnote to the preface of *The Accursed Share*, Bataille underlines his continued interest in modern physics in a debt of gratitude, declaring that his book 'is also in large part the work of Ambrosino', and expressing the hope that they will soon resume together their study 'of the movement of energy on the surface of the globe'.[184] Theirs was a genuine exchange, contesting C. P. Snow's pessimistic vision of two cultures that 'can't talk to each other'.[185] Indeed, through comparison with Eddington, Bataille presents a much more complex relation of rejection, interpretation, negotiation and agreement between the arts and science than Snow ever conceived, which I introduced at the beginning of this chapter through a most Bataillean device – a translator's error in Hollier's *College of Sociology* selection.

Incidentally, in the same section of the French version of Hollier's book, another error – a misprint this time – ensures that Eddington's *The Expanding Universe* is rendered as *The Expending Universe*! It is as though the constellation of Bataille and Eddington, seemingly so improbable, is demanded by the general economic logic of mistranslations and typographical errors, which resists the closed rationalism of textual production.[186]

5

QUANTUM MECHANICS AND PARTICLE PHYSICS

Matta, Wolfgang Paalen, Max Ernst

My heavy investment in historical research is in line with one of the main aims of my study: to show that the polemics of those arguing for the incompatibility of the 'two cultures' must inevitably place theoretical considerations before historical ones.[1] By this, I mean that categorical assertions claiming to figure the antagonism that exists between science and the humanities always come up against historical evidence that paints a more conciliatory picture. Because of the general critical tendency – flying in the face of the increased interdisciplinarity of our own day – to keep the two separate, marginalised debates of the past that can shed light on this issue await their place in history. I have already pointed to several broad-ranging exchanges within the avant-garde and academia in 1930s France on the new challenges laid to traditional forms of rationalism by modern physics, psychology, philosophy and art. Concordant with this, and far from demonstrating a simple misunderstanding or dismissal of science, I showed that a complex set of responses was forthcoming from the Parisian Surrealist group.

In what follows I use QM and particle physics as guides to deal with another set of exchanges within and around Surrealism. Beginning in the mid-1930s, I outline some art-historical responses on the borders of Surrealism that take account of QM and psychoanalysis simultaneously. From there, I try out another case study, comparing the art and writings of two Surrealist artists, the Chilean Roberto Matta and the Viennese Wolfgang Paalen, ultimately focusing on Paalen alone. Both artists were drawn towards abstraction from the mid-1930s in the wake of exchanges on Relativity and QM, and I argue that these are crucial to an understanding of their work and its development. As will become clear, with the disintegration of the Surrealist group at the beginning of the Second World War, Paalen and Matta sought to decode the profound discontinuity of life itself at its most elemental level. I conclude with the discussion of Surrealism and Logical Positivism that I promised earlier with reference to debates of the mid-1930s on QM and materialism, in which I suggest a new context for Max Ernst's work of the period.

Art Theory, Modern Physics and Psychoanalysis in France, 1934–1935

I mentioned earlier that from the moment of Max Planck's discovery of the discontinuous emission of energy as radiation by electrons at the turn of the century, quanta gained a controversial reception in the physics community over the succeeding quarter of a century, culminating in the full formulation of quantum theory in 1927. That the same period saw the modification by psychoanalysis of the human individual – as though viewed in a mirror facing a disintegrating world – into an entity rudimentarily composed of a similarly fragmented interior, did not escape the attention of French thinkers of the era with avant-garde connections.

In 1934, the year of Gaston Bachelard's *New Scientific Spirit*, Georges Bataille's co-editor on *Documents*, the art critic Carl Einstein, published his *Georges Braque* in French.[2] For twenty years Einstein had been a prescient and well-informed writer on tribal artifacts, also playing an important role in introducing Cubism to a larger German and French public. In *Georges Braque* he draws upon his extensive knowledge of modern physics and the philosophy of science with the idea of mapping an epistemological field that would accommodate quantum theory, psychoanalysis and Cubism. Einstein achieves this by means of a language strikingly similar to Bachelard's – of breaks, ruptures, breaches, fragmentations and discontinuities – which we would now associate with the rhetoric of poststructuralism and postmodernism. Elucidating the shattered motifs of Braque's painting (fig. 59), Einstein argues that modern physics shows the physical objects we encounter 'are only a kind of resistance, a barrier in the stream of a functional world'.[3] He claims that the Cubists' break-up of forms in their paintings corresponds to 'a rent between man and the conventional conception of the world, a reconstruction of the structure of sight'.[4] Within 'this rupture of continuity', Einstein believes can be found a new reality, 'breaking the chain of unitary causality'.[5]

Einstein goes on to argue in *Georges Braque* that the tendency towards acausality and heterogeneity that he perceives as a dominant feature of modern art, together with the language of fragmentation and dispersal to which it gave rise, have their source in a new epistemology emerging from physics. Consequently he can present Cubism as evidence of a paradigm shift in a revolutionary rather than evolutionary history of art:

> If one wanted to arrive at a new understanding of and a new form for man and the universe, it was important above all to break the banal image of reality, to break, thus, oneself, with conventional history. This enterprise was risked by several poets and men of science, as it was by two Cubists whose boldness amounted to a historical event of the highest importance.[6]

> It is at once painful and humiliating to see historians of art consider the facts of the history of art as isolated phenomena; a treatment making of art a monstrous world, adored like a divine miracle. It should be incorporated once and for all into a historical context; recall for example that Riemann and Lobachevsky had already, much earlier, expanded the notion of space beyond the limits of unilat-

59 Georges Braque, *Violin and Pitcher* (1910). Oil on canvas, 117 × 73 cm. Öffentliche Kunstsammlung Basel, Kunstmuseum, gift of Raoul La Roche, 1952.

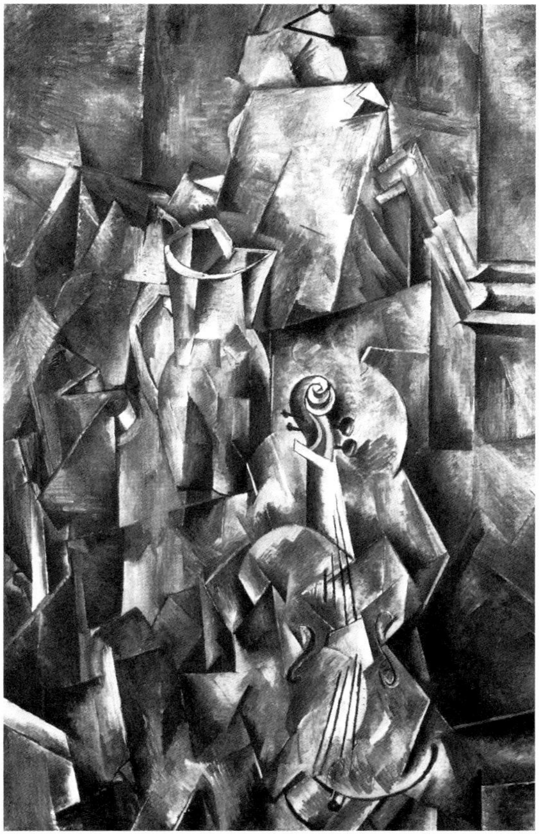

eral description, that the fixing of the soul with static 'substances' and perspective no longer has any currency, and that grammar, no longer suitable to psychic trends and social structures with its declensions and conjugations, is from now on deposed from its reign. Quantum theory tore apart the unitary continuity of the universe, and ultimately cast grave doubt on causality, the heart and soul of all former science. The conscious I is today no more than a simple façade, ready to founder on the spontaneous act . . . So-called Cubism is a perfectly normal phenomenon in a general conception of things undergoing radical transformation.[7]

For my own study, Einstein's combination of the discourses of modern physics and psychoanalysis to account for the direction of art is particularly striking. Taking the form of a farewell to humanism, this coalition is evident in several places in *Georges Braque*, where Einstein compares the 'broken surfaces' of Cubism with the discontinuity of QM, alongside his assertion, first, that 'the end of a "stable" and circumscribed subject is equivalent to an elimination of the anthropocentric standpoint'[8] and, secondly, that 'the belief in a hermetic I, of its definitive figuration, must be abandoned'.[9] Although he does not mention psychoanalysis by name here, it is pretty obvious that Einstein means to draw a parallel between the complexity that physics had uncovered in the structural properties of the natural world, and the manifold agencies governing the formation of the individual ego, described by Freud.[10]

Carl Einstein was not alone in believing that the move towards abstraction in art was steered by the same emergent forces as those driving advances in physics and psychoanalysis. Later in 1934, the historian of comparative religion and ancient philosophy Henri-Charles Puech, whom we bumped into in my last chapter as an editor of *Recherches philosophiques*, published an article titled 'Signification et répresentation' in the Surrealist-dominated review *Minotaure*.[11] Like Einstein, Puech expands on the problems of artistic representation in a world in which both object and subject seemed less comprehensible (in the classical senses of 'totality' and 'wholeness') with each passing day. By its retreat from the naturalism that traditionally marked painting, Puech believed modern art had stolen a march on metaphysics and literature whose research still tended towards a mode of visual representation indebted to the everyday physical world of Newton.

In those fields, he argues, it is maintained that human and world, subject and object, can best be understood through the descriptive representation of one by the other, emphasising their separation. According to Puech, man's quest for meaning had left him:

> seeking to situate himself in a unity which *comprises* him and which he *could comprehend*, and not managing to visualize himself to himself to the extent to which he visualizes the universe. In any case, there was no divorce between reality and man, but a correspondence between a microcosm and a macrocosm, or identity between Being and Representation, community of objective world and subject, within something that showed itself in the end capable of being represented.[12]

However, this mode of humanistic representation had started to come undone, Puech maintains, with the new conceptualisation of space and time recently initiated by modern science and philosophy. By its assertion that each event in space-time is experienced in a different way by each observer witnessing it, Einstein's Theory of Relativity abolishes forever the notion of the absolute conditioning of phenomena. Because the cognition of every act is now subject to an infinite number of variables, modern physics puts paid to the validity of a static contemplation, 'according to an unchanging background, which, blocking sight and forming a screen, would allow representation'.[13]

To this, quantum theory adds another layer of uncertainty as to the trustworthiness of sense perception:

> Second, much worse failure, and close to collapse: the most recent developments in physics seem not to allow representation to rediscover 'invariants' *this side* of apparent reality in the form of *content*. Solids dissolve and are reconstituted as fluids, matter as radiation, particles as 'wave packets.' The exact delineation of reality gives way to a vaguer consideration of unities more or less arbitrarily defined: strict determinism vanishes before a calculation of probabilities trying hard to come to terms with chance. In future, in place of points of reference – now abolished – that the atom as the 'final thing' supplied to representation, there is nothing more – to still cling to an illusory scrap of an image – than the slow and indeterminate swelling of waves of space, or of something that could still be like pure space, as the ultimate expression of reality.[14]

With its overthrow of the notion of the simultaneity of events occurring within a single, temporal 'structure', and its affirmation of the 'formlessness' of visible phenomena, modern physics narrows the field of naturalistic representational possibilities.

Puech goes on to say that a similar case arises from propositions put forth by psychoanalysis, where complex latent forces compose the biography of each individual, 'singular events, accidents, associations that one drags along behind oneself'.[15] As a consequence, he maintains, '[t]he individual is a *history*, who is present within him and who is *only for himself*.'[16]

To these blows delivered by physics and psychoanalysis Puech notes a final one recently added to epistemologies of 'totalisation' by contemporary philosophy's radical examination of the subject-centred worldview, where:

> realist or 'existentialist' philosophies substitute for visualization based on the opposition of an object-spectacle to a subject-spectator, the idea – if it is still an idea – of a reciprocal *presentation* of reality and man, of a tested *presence*, unamenable, however, to any objective projection detached from whoever tests it.[17]

Attempts in science and philosophy to reach the 'essential' beneath appearances are carried out 'by extremely complicated means, to express it through more and more abstract technical symbols'.[18] In other words, Puech declares:

> it seems that there has been a rupture between modern man and scientific man, between the visualization that man makes from day to day of himself and the universe which surrounds him, practical and familiar, and, on the other hand, the reality which he is able to imagine, at the very most, as *something*, and of which science gives him as nothing more than an abstract splendour or just a means of technical manipulation.[19]

It is here that Puech believes contemporary art was able to take on its singular role. Abandoning all reference to external co-ordinates, it could forsake representation for presentation: a 'manifestation or system of *significative* forces'.[20] The image in painting should no longer be compared to 'a more or less idealised copy, but to a magic object or a mental *symbol*',[21] at once '*realism* and symbol'.[22] The new task of painting, he avows, is not to represent or suggest 'in the banal sense of "symbolism"', but 'to *impose* and to act'.[23] Puech writes that perhaps these new discoveries and this new art will lead to 'une révolution totale' in our vision of the world, 'which, moreover, can no longer be only a vision'.[24] The resemblance of Puech's remarks to Carl Einstein's of the previous year betrays their intellectual sources – as well as being, like Bachelard, an editor of *Recherches philosophiques*, Puech contributed to Einstein and Bataille's *Documents*.[25]

Making Waves and Particles: Wolfgang Paalen and Matta

Around the same time Carl Einstein and Puech were pondering the nature of connectivity between art, physics, psychoanalysis and philosophy, parallels between the scientific and artistic communities were being sought in Christian Zervos's introduction to a large-scale *enquête* carried by *Cahiers d'Art* in 1935. Issuing from what Zervos

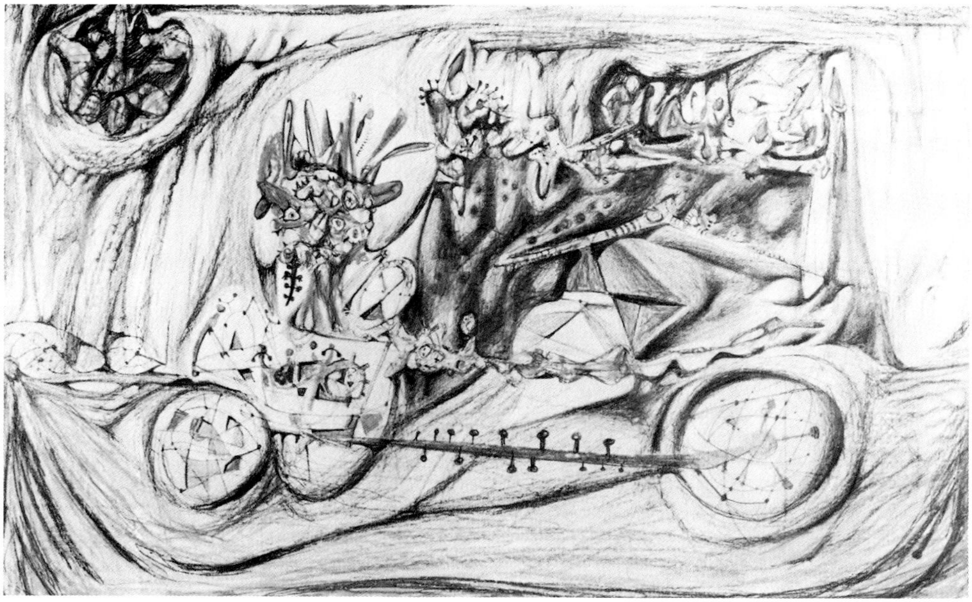

60 Matta, untitled drawing (1938). Crayon and pencil on paper, 32 × 49.5 cm. Private collection, New York.

viewed as a generally felt requirement among artists for a 'revision of plastic values', the inquiry asks how artists are acting upon it, and what kind of demands they believed it was making on younger artists already faced with important decisions in a period of political turmoil.[26] One of the questions Zervos asks was prompted by the then-recent reception by philosophy of new discoveries in physics: '[c]ould it be time to integrate form into content, following the example of philosophy which has already begun to broaden the concept of the world by integrating little by little scientific experience into thought?'[27] The response of Wolfgang Paalen, who would soon turn to Surrealism, anticipates his subsequent science-inflected imagery, stating that the work of art must lie outside historical determinism and the old science: 'that is to say, to force the audience to a contemplation which connects it, like it or not, to the great common current, rising higher than all our known history and leading it to a concept of the world inaccessible with the crude tool of scientific determinism'.[28]

Two closely related themes emerge here that would carry great importance in Paalen's work over the following years. One is the idea that our understanding of the universe is restricted by the mechanistic view handed down by pre-twentieth century classical physics. This is crucial to Paalen's thought, and he goes on to develop it in his art with the idea of showing that the apparently diverse products of a culture are circumscribed by what is perceived possible in a given epoch. Paalen's second, double-edged theme states that art must detach itself from simplistic categories of realism that are historically determined, and also from its submission to utilitarianism, which he regards as the legacy of the old physics. In its place, he asserts later, art must abandon predetermined *description* and learn from modern physics and its philosophy that both nature and culture are the manifestation of a shifting, behavioural flux, which must be the subject of art.

Paalen also speaks in Zervos's *enquête* of the need for common theoretical ground in artistic practice, acting, as it had in science, as a basis for action. However, he is pessimistic on this point, believing a 'spiritual community which can lay claim to universality exists only in the field of science',[29] while conceding that not all avant-garde artists would agree on a single programme. This search for a 'communauté spirituelle' would lead him into the arms of the Surrealists in the following year. With his remarks of 1935, Paalen's ambitions for his own practice start to come into focus, but his immersion in Surrealism over the next four years did not allow him to generate an imagery enabling his depiction of these ahistorical forces. That would only arise at the end of the decade with his break from the group.

Usually said to be the last of the great Surrealist artists, Matta arrived in Paris from his native Chile in 1935, the year of Zervos's *enquête* in *Cahiers d'Art*. Joining the Surrealists in 1937, he engaged with modern physics in a less theoretical and more intuitive way than Paalen would over the next few years. His 'psychological morphology' series constituted his first contribution to Surrealism, already revealing his desire to develop an iconography that draws on the imagery of physics and psychoanalysis simultaneously, to express the link between the inner and outer 'worlds'. In an untitled drawing of 1938, for instance (fig. 60), he seeks a vital formal yet abstract language indebted to the automatism of the Freudian unconscious, while suggesting the inseparability of the interior drama from the forces governing the natural world by his inclusion of the familiar image of an electron circling the nucleus of an atom towards the lower right of the drawing. During the war, he went on to pursue an imagery meant to suggest the inseparability of the subject and object, in search of a new non-anthropomorphic, non-realist language that would express the inner and outer 'splintering', also observed by Carl Einstein and Puech. But Einstein saw Cubism as the parallel outcome of the epistemological crisis that gave rise to modern physics and psychoanalysis in the first years of the century, throwing into doubt the notion of an enduring reality constructed from permanent features. Matta's art, by contrast – later described by Nicolas Calas as a 'transparent Cubism' – was kind of reflection on this 'crisis', made two decades after it took place.[30]

Matta joined the Surrealist group shortly after a grave period in its relationship with the Communist Party. As we know, the Surrealists' dual ambition to act upon the inner human and the outer world was summarised in the slogans 'changer la vie' and 'transformer le monde'.[31] If it was still believed in 1937 when Matta joined them that the first of these goals could be achieved by recourse to psychoanalysis, it is equally certain that by then Breton no longer harboured any illusions that the second could be secured by the organised politics of the day. It is no coincidence, then, that the Surrealists became increasingly interested in physics as a means towards comprehending the world, and of subsequently transforming it, between 1934 and 1936, exactly the moment at which their political ardour was on the wane.

Surrealism's identification of its own thought with that of the Middle Ages helped bring about this tactical switch. The Surrealists' attraction to magic and alchemy, to courtly romance and Arthurian myth, was mediated by Romanticism and Symbolism, and bypassed the rationalism of the Renaissance and Enlightenment that they despised. Particularly relevant to the Surrealists' theory, poetry and painting is the enthusiastic welcome they offered the medieval belief in the relationship between man as microcosm and the world as macrocosm. Early evidence of this theme within the Surrealist milieu is given in Michel Leiris's 1929 *Documents* text, 'Notes sur deux figures microcosmiques des XIV et XV siècles' (fig. 61), whilst imagery indebted to the idea soon found its way into the art of both André Masson and Matta.[32] However, whereas Masson's work took on an anecdotal or 'literary' style, immersed in the same medieval alchemical and hermetic sources as Breton, Matta transformed the concept by creating a new imagery for contemporary man, the world and the universe mediated by psychoanalysis and modern physics.[33]

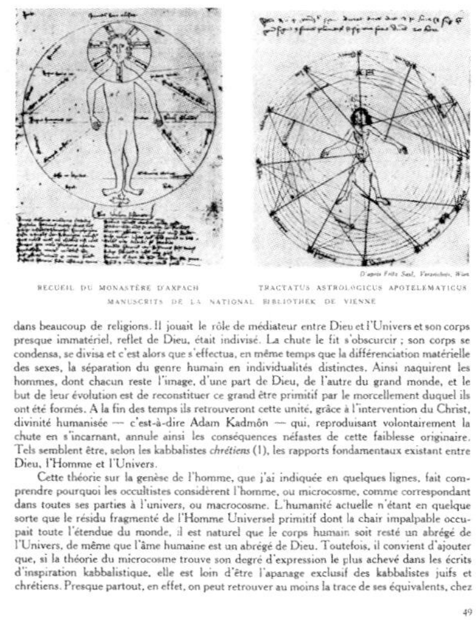

D'après Fritz Saxl, Verzeichnis, Wien

RECUEIL DU MONASTÈRE D'AXPACH — TRACTATUS ASTROLOGICUS APOTELEMATICUS

MANUSCRITS DE LA NATIONAL BIBLIOTHEK DE VIENNE

dans beaucoup de religions. Il jouait le rôle de médiateur entre Dieu et l'Univers et son corps presque immatériel, reflet de Dieu, était indivisé. La chute le fit s'obscurcir ; son corps se condensa, se divisa et c'est alors que s'effectua, en même temps que la différenciation matérielle des sexes, la séparation du genre humain en individualités distinctes. Ainsi naquirent les hommes, dont chacun reste l'image, d'une part de Dieu, de l'autre du grand monde, et le but de leur évolution est de reconstituer ce grand être primitif par le morcellement duquel ils ont été formés. A la fin des temps ils retrouveront cette unité, grâce à l'intervention du Christ, divinité humanisée — c'est-à-dire Adam Kadmôn — qui, reproduisant volontairement la chute en s'incarnant, annule ainsi les conséquences néfastes de cette faiblesse originaire. Tels semblent être, selon les kabbalistes *chrétiens* (1), les rapports fondamentaux existant entre Dieu, l'Homme et l'Univers.

Cette théorie sur la genèse de l'homme, que j'ai indiquée en quelques lignes, fait comprendre pourquoi les occultistes considèrent l'homme, ou microcosme, comme correspondant dans toutes ses parties à l'univers, ou macrocosme. L'humanité actuelle n'étant en quelque sorte que le résidu fragmenté de l'Homme Universel primitif dont la chair impalpable occupait toute l'étendue du monde, il est naturel que le corps humain soit resté un abrégé de l'Univers, de même que l'âme humaine est un abrégé de Dieu. Toutefois, il convient d'ajouter que, si la théorie du microcosme trouve son degré d'expression le plus achevé dans les écrits d'inspiration kabbalistique, elle est loin d'être l'apanage exclusif des kabbalistes juifs et chrétiens. Presque partout, en effet, on peut retrouver au moins la trace de ses équivalents, chez

49

61 Michel Leiris, 'Notes sur deux figures microcosmiques des XIV et XV siècles', *Documents*, no. 1, April 1929, 48–52.

We saw earlier this 'updating' of the ancient relation between man and the world by recourse to psychoanalysis and modern physics in the writings of René Crevel and Pierre Mabille, and in this chapter by Carl Einstein and Henri-Charles Puech. Calas noticed Matta's faithfulness to, and 'scientific' modernisation of, the Surrealist quest, writing that in his art, 'the correlations of the need to change one's life with the wish to transform the world is best expressed in a spatio-temporal reality that is four-dimensional'.[34] Rather than relating the human limbs or organs or humours to the four elements or to the constellations, then, Matta generated a new dynamic imagery of energy, transition and growth, that draws on psychoanalysis and physics to link the spheres of the sub-atomic, the human unconscious and the celestial.[35]

Matta, Modern Physics and Psychoanalysis in the 1940s

Matta left Paris for New York on the wise and timely counsel of Marcel Duchamp at the beginning of the war in 1939. He was received by local New York artists as a painter with a reputation and perfect knowledge of Surrealist theory, which soon caused him to dream of leading his own avant-garde group. Not surprisingly, this ultimately strained his relations with the Surrealists, who were predominantly of the preceding generation. Matta's endeavour to combine a new imagery derived from modern physics with one based on the language of psychoanalysis was, in part, the outcome of this love-hate relationship with Surrealism. Robert Motherwell observed that Matta's inter-

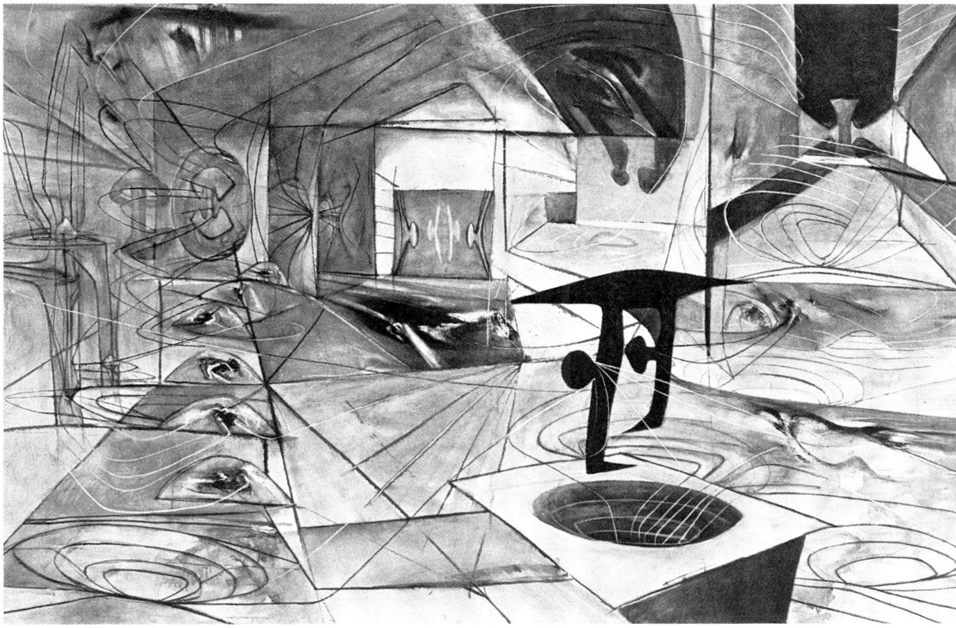

62 Matta, *The Onyx of Electra* (1944). Oil on canvas, 127.2 × 183 cm. The Museum of Modern Art, New York.

est in modern physics was strengthened by his rivalry with the older Surrealists: 'he wanted to show the Surrealists up', Motherwell said, 'as middle-aged grey-haired men who weren't zeroed into contemporary reality'.[36]

In this sense, 1943 was a key year in the development of Matta's art. According to Malbert it was in that year that Matta 'said that Einstein was as important as Freud for the modern artist'.[37] Also in 1943, wrote James Thrall Soby, 'Matta declared his only quarrel with the Surrealists was that they disregarded the phenomena of the *physical* sciences', a claim by Matta (or Soby) that flies in the face of the evidence offered by the 1930s writings of Breton and the Surrealists.[38] Soby continues: '[h]e felt that a new school of painters could evolve from contemporary physics, as an earlier school (Surrealism) had evolved from modern psychology'.[39]

Matta admired Man Ray's photographs of mathematical models in *Cahiers d'Art* in 1936, and even said he responded to them in his paintings of the time.[40] His fascination with the curved non-Euclidean geometry of mathematical models and the spaces they suggest remained with him in New York during the war. Motherwell recounts how he went with Matta to see the collection of models held at Columbia University.[41] His reacquaintance with curved surface geometries in 1943 helps to explain the change he acknowledged his art underwent in that year, 'from a sort of burning fire, mineral lights kind of thing into a space that was described by geodesic lines and waves'.[42] In an article presenting Matta as an artist born out of, as well as bearing 'the uneasy revelations of our scientific age', William Rubin observed the continuing influence of mathematical models on the forms presented in some of his canvases of the war years, such as *The Onyx of Electra* of 1944 (fig. 62).[43]

From a loose interpretation of non-Euclidean geometry and the 'disintegrating' world proposed by particle and atomic physics, Matta gained a new conception of the underlying plurivalence and 'fragility' of the material world (as Eddington and Benjamin saw it). Its lack of an essential describability in terms of separate, isolated forms is translated in his work into a non-anthropomorphic iconography of transparency, fragmentation and insubstantiality, often suggestive of dynamism and energy. Matta's art of the 1940s gives voice not to a Cartesian world of things – of objects and subjects fixed in time and space – but to a single 'being-in-the-world'.[44] For Matta, the new space and its geometries acted as a spur to experiment with forms of pictorial space that do not assume the permanent, solitary, stationary eye of Renaissance perspective, existing at the 'point' of a 'cone' of light.

Expressing his admiration for this aspect of Matta's art in 1944, Breton returns to the same Bachelardian categories – the evocation of a modern *rupture* with the classical *rationalisme fermé* – he had called upon in 'Crisis of the Object':

> he invites us ceaselessly to enter a *new space* which has deliberately broken away [en rupture délibérée] from the old conception of space because the latter is meaningful only to the extent that it is distributive of elementary, closed bodies. It is of little importance that this new way of seeing and showing things has, at one time or another, sought support in the scientific insights of psychological morphology, Gestalt theory, astrophysics, histology and molecular physics. The need to call upon the support of the most modern resources simply expresses the aspiration to extend the field of vision.[45]

This facet of Matta's work was singled out by Duchamp, too, in comments made in 1949 for Katherine Dreier's Société Anonyme catalogue, though Duchamp clarifies its scientific debt somewhat:

> His first and most important contribution to Surrealist painting was the discovery of regions of space hitherto unexplored in the realm of art. Matta followed the physicist in the search for new space which, though depicted on canvas, was not to be mistaken for another three-dimensional illusion.[46]

Always ready to take a swing at 'retinal art', Duchamp concludes his brief remarks with a punning compliment for his friend, evoking the French 'profondeur' to distinguish the 'intuitive' non-Euclidean 'depth' of Matta's canvases from the supposedly flat, purely optical work of his American contemporaries, championed by Clement Greenberg and in the ascendant at the time: 'Matta', writes Duchamp, 'is the most profound painter of his generation.'[47]

Alain Jouffroy maintains that Matta's *œuvre* is marked by his contact with Duchamp, particularly where it denounces 'the limitation of the view to the retinal image or to its memory'.[48] They became reacquainted when Duchamp arrived in New York in 1942, and Matta mentioned later the 'déjeuners hebdomadaires' they shared until 1944.[49] In my opinion, Octavio Paz overdoes it a bit and simplifies the role of Duchamp when he writes of the period: 'in Surrealist painting a radical metamorphosis was in progress,

63 Matta, *La Vertu noire* (1943). Oil on canvas, 76.5 × 182.5 cm. Tate Modern, London.

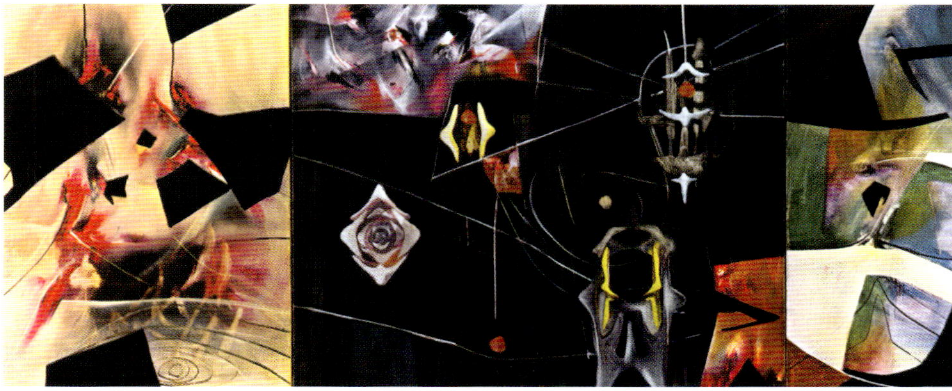

initiated by Duchamp and realized by Matta, a fusion of eroticism, humour, and the new physics'.[50] However, the impact of Duchamp's work and thought on Matta is fairly clear to all. In his later canvases of 1943, Soby sees a new attempt to disrupt the visual architecture of perspective with a switch from the multiple vanishing points of de Chirico to a strategy more indebted to Duchamp's destabilising of the visual.[51] This led to a series of triptychs that give the audience an extremely ambiguous viewing position, such as *La Vertu noire* (fig. 63), painted in 1943 and seen for the first time that year at Matta's third show in New York, held at Pierre Matisse's gallery. The uncertain orientation presented in these canvases (are we looking up, down or across?) frustrates the typical reliance on the visual sense, forcing the viewer back into his/her own body. Reviewing the exhibition, Rosamund Frost evidently draws upon Matta's own remarks to open her column (while losing the zest of his delivery): '[i]n the same way that modern philosophy is based on physics, so the modern artist, if he is a serious artist, is pledged to interpret scientific facts – called phenomena – which are the true realities of life. So thinks and so paints Matta Echaurren.'[52]

In works of 1943 – such as the painting *La Révolte des contraires* (fig. 64) and the drawing *Cataglyph the Great* – Matta attempts to create a kind of non-visual space in a single pictorial language, intended to call forth the energy of both the inner and outer worlds as described by psychoanalysis and physics. An iconography of whirlpools and maze-like geometry evokes at once Freud's technical language of centres cathected with energy linked by paths along which excitation is discharged, and the random scatter of subatomic particles. These pictures recall Matta's 1965 description of the non-Euclidean space he improvised upon, 'which might be compared to the space of temperatures in a room, of meterological space'.[53] Matta spoke similarly of his best-known painting *The Vertigo of Eros* (fig. 65), completed around the same time:

> The reference I was making once again, was to non-Euclidean space, where all the ordinates and co-ordinates are moving in themselves, because the references to the 'wall', shall we say, of the space, are constantly changing. They are not parallel to a Euclidean cube, to which most previous painting referred.[54]

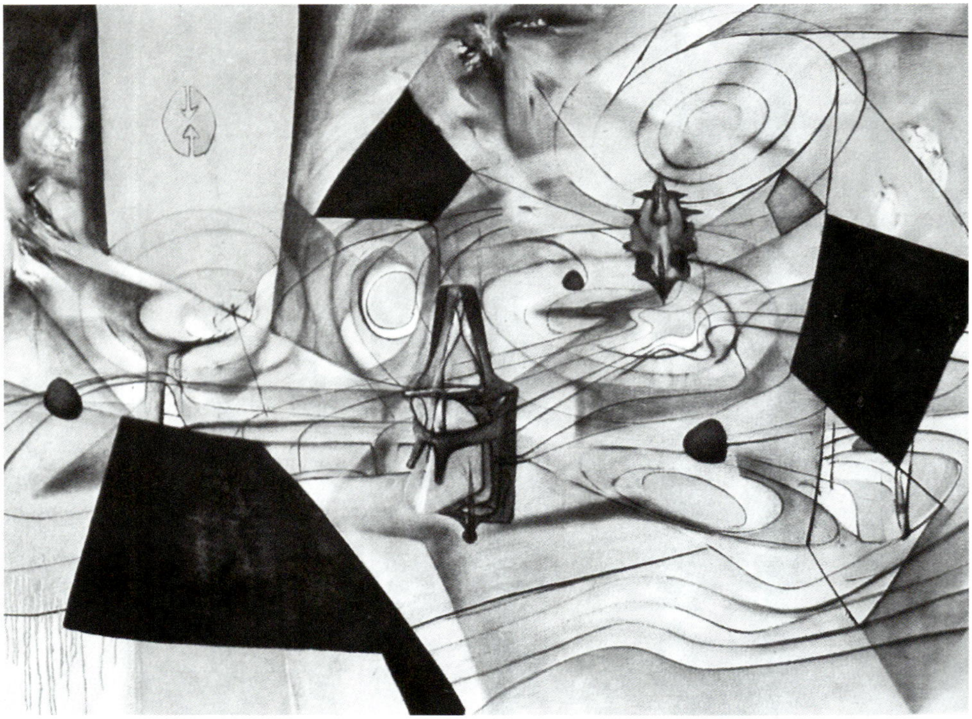

64 Matta, *La Révolte des contraires* (1943–4). Oil on canvas. Private collection.

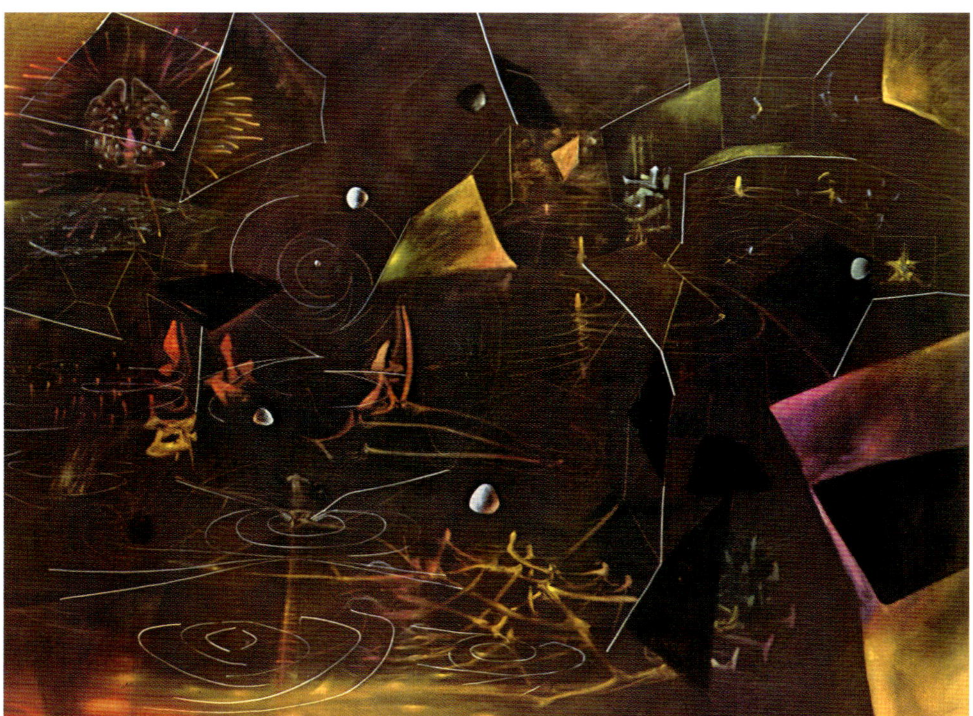

65 Matta, *The Vertigo of Eros* (1944). Oil on canvas, 195.7 × 251.6 cm. The Museum of Modern Art, New York.

While the 'Freudian' title of *The Vertigo of Eros* suggests Matta wanted the work to be thought of as an infinite, chaotic, unmapped 'mindscape', the hovering particle-like objects orbiting circular scratches on the surface of the canvas are also reminiscent of the subatomic 'world', which physics had found to be composed primarily of empty space. The direction taken by Matta's art can be understood only when viewed in the context of the prominent discourses of psychoanalysis and modern physics, also used as guiding points of reference for the decoding of the modern *epistème* in the mid-1930s writings of Carl Einstein and Henri-Charles Puech.[55]

Paalen, Dyn *and Complementarity*

At the same time Matta moved to New York at the beginning of the war, Paalen also left France for America. Swiftly relocating to Mexico with his wife, the artist Alice Rahon, he cut his ties with Surrealism and immersed himself in the philosophy that was issuing from modern physics. Somewhere along the way he produced his first picture unequivocally inspired by his increasing knowledge of modern physics, the gouache *Between Matter and Light* (fig. 66), which breaks with the imagery of his Surrealist years and foreshadows that to come. The reason behind this change of direction is confirmed in Gustav Regler's biography of the artist, largely ghostwritten by Paalen himself:

> Already at the end of 1939, he had painted a series of gouaches in which a new concept of space was announced by strange figures formed by the vibration of spiraloid rays of colour. Since 1939 he has been concerned with the philosophical implications of post-Einsteinian physics 'in order to find out what is right with science and why it works – and what is wrong with art in our time and why it does not work.'[56]

Pierre Mabille adds that Paalen's work was 'entirely transformed' during the period, seeking to plot 'a new space, that of the fourth dimension if I understand correctly.'[57]

On the first point, about the transformation of Paalen's work, Mabille *had* understood, but on the second he had got it all wrong. Paalen's art cannot be reduced to some kind of vague manifestation of the fourth dimension.[58] Initiated here with a dynamic interaction of vaporous, abstract swirls and coils in indeterminate space, Paalen's new style was largely inspired, in fact, by Louis de Broglie's 1937 book *Matière et lumière*, published in English as *Matter and Light* in 1939, the year of Paalen's 'conversion'. Ranging over physics and philosophy, de Broglie's book explains that the physicists' pursuit of the fundamental constituents of nature led them to accept *both* particle and wave theories: 'the interconnection between the corpuscles of Matter and their associated waves is known today under the name of Wave Mechanics'.[59] As we saw in my first chapter, unlike the group comprising Werner Heisenberg, Paul Dirac and Wolfgang Pauli, the quarter of the physics community that included de Broglie and Erwin Schrödinger persisted in its search for a non-mathematical, 'visualisable' description of the microcosm. Although his book deals with forms of radiation invis-

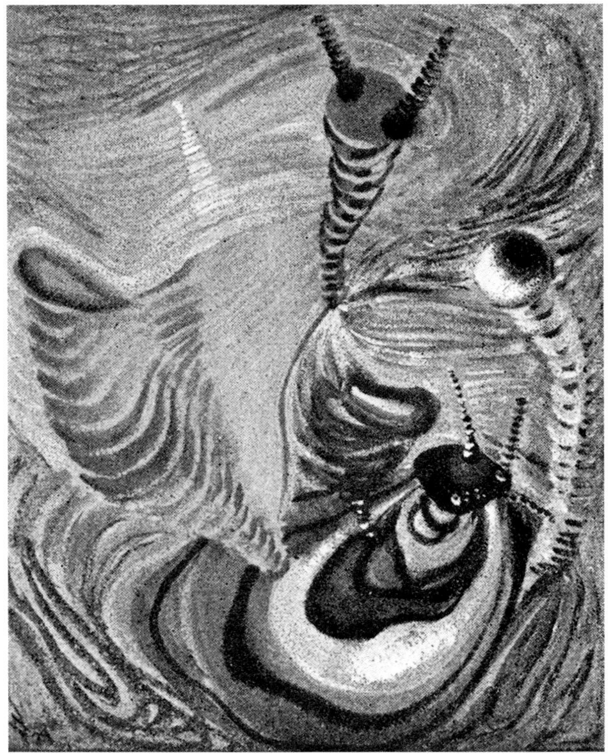

66 Wolfgang Paalen, *Between Matter and Light* (1940). Gouache on sandpaper, 33.7 × 29.5 cm. Fundación Wolfgang y Isabel Paalen, Tepoztlán.

ible to the naked eye, such as infra-red, ultra-violet and cosmic rays, de Broglie insists that their behaviour allows us to intuit for them either/both a wave and particle formation. Having absorbed the lessons of Breton's 'internal model' and the Duchampian 'nonretinal' during his spell with the Surrealists, Paalen took his cue from wave mechanics in works dealing less with descriptive and conventionally realist imagery than with intuitive forms conjured by the artist, suggestive of a fundamental rhythmic energy beyond or behind historically contingent motifs – as he said he would in 1935.

Paalen's work of 1940 continues to explore the spaces inspired by the new physics. With a repeating spiral pattern and circles and spheres appearing to blend into each other like de Broglie's 'interconnection between the corpuscles of Matter and their associated waves', the conté crayon *Discovery of Infra-Space 1* (fig. 67) shows that the subatomic universe 'glimpsed' by the physicist as the primary force behind nature now played the same role in Paalen's art. As well as using *Matter and Light* as the foundation for a new imagery, Paalen quotes from the book in the course of the theoretical texts he published in *Dyn* in the 1940s. Between the covers of that journal and in the paintings of those years, he turned to physics with a vengeance in his elaboration of what he called his 'Philosophy of the Possible', and in his creation of a unique 'cosmic' imagery.

67 Wolfgang Paalen, *Discovery of Infra-Space 1* (1940). Conté crayon, 78.7 × 58.4 cm. The Museum of Modern Art, New York.

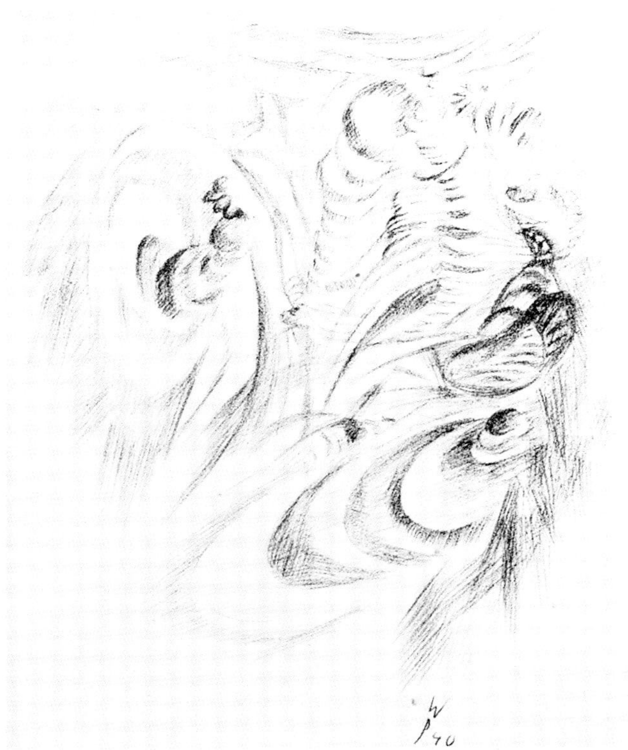

Communicating the new artistic departure taken by his art and writings through the six numbers of *Dyn* (1942–44), Paalen hoped like Matta to create an avant-garde movement to rival Surrealism (fig. 68). Conceived and edited by Paalen and bankrolled by his long-time mistress Eva Sulzer, *Dyn* contains art, art theory, anthropology, poetry, book reviews and a major inquiry on dialectical materialism. An edition of 1,500 of each number was distributed internationally by Paalen's own Quetzal Editions from Coyoacán to César Moro's bookstore in Mexico, to Wittenborn and the Gotham Book Mart in New York, and to Zwemmer in London, from where it reached the continent. *Dyn* was a huge undertaking for Paalen, who churned out paintings and articles in his own name and under the pseudonyms 'Carter Stone', 'Charles Givors', 'Jean Caroux' and 'John Dawson'. These were meant to give the impression of a burgeoning art movement, which, according to Edward Renouf, his assistant editor for the early numbers, Paalen desperately wanted to start.[60] Labelling the journal a 'monologue', José Pierre declares that it was nevertheless, 'this solitude which creates the beauty and greatness of *Dyn*'.[61]

Broadly speaking, the appearance and format of *Dyn* are modelled on *Minotaure* and *Cahiers d'Art*. Written in French, Paalen's texts are translated into English in the first half of the journal then printed in French in a smaller typeface in the second half.

This reversal of the usual hierarchy of text and translation (see, for instance, *Scientia* and *Documents*) shows that Paalen's thoughts were aimed primarily at an English-language New York audience. The one exception is his 'Farewell au Surréalisme', which can be read in the first number of the journal, only in French, and in a larger typeface – all of which speaks towards an audience of French Surrealists, unhappily marooned in New York for the duration of the war.

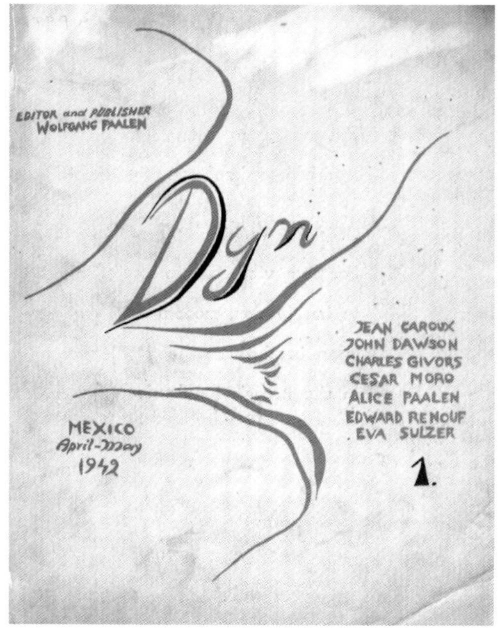

68 Wolfgang Paalen's journal *Dyn*.

Chiefly a vehicle for Paalen's opinions, in the course of its run *Dyn* carries essays by Gustav Regler, César Moro, Henry Miller, Anaïs Nin, Alfonso Caso, Valentine Penrose and Robert Motherwell. It reproduces works of art by Alice Rahon, Henry Moore, Manuel Alvarez Bravo, Brassaï, Marc Chagall, Alexander Calder, Robert Motherwell, Matta, Gordon Onslow Ford, William Baziotes and Jackson Pollock.[62] For many, *Dyn* remained a Surrealist review. Pierre insists 'the preoccupations reflected by *Dyn* are all as Surrealist as those of *VVV*',[63] and looking back without rancour years after ideological disagreements and an on-off friendship with Paalen, Breton wrote that 'la critique vraiment indépendante' would benefit from a perusal of *Dyn*'s theoretical texts.[64]

Close to Paalen during the *Dyn* years, Gordon Onslow Ford confirms that modern physics was Paalen's main inspiration at the time,[65] and essays from the journal collected in 1945 under the title *Form and Sense* emphasise the side of Paalen's theory inspired by physics.[66] Indeed, *Dyn* announces in its title it will showcase an art and philosophy taking science as their point of departure. In his fictional interview with 'Carter Stone' in the sixth number of the review, Paalen explains his chief concerns:

> It seems to me that we have to reach a *potential* concept of reality, based as much on the new directives of physics as on those of art, a concept that I call *dynatic* (from the Greek word *tó dynaton*: the possible). [Footnote: From which the name of this review DYN is derived.] A Philosophy of the Possible which would understand art as a rhythmic equation of the world, indispensable complement of the logical equation that science makes. For only the cooperation of the two will be able to create a new ethic capable of finishing with metaphysical and religious obscurantism. It means *dissociating once and for all imagination and metaphysics*; to understand that imagination creates reality as much as it is created by reality, that the images of art are neither vain reflections, nor blueprints for tools, but blueprints for man himself.[67]

The marriage of art and science through Complementarity; the centrality of rhythm; the rejection of metaphysics; the Bachelardian–Nietzschean–Surrealist primacy given the imagination and the exercise of its will; the elaboration of a 'Philosophy of the Possible': these are the themes treated in Paalen's essays in *Dyn*. Undoubtedly familiar with the reflections of Carl Einstein and Puech – his library contained a complete run of both *Documents* and *Minotaure* – his texts in *Dyn* on modern art and QM echo theirs. However, I can show now that Paalen's surprisingly complete knowledge of the new physics and its philosophy also led to independent insights remarkably close to those made by the physicists responsible for the development of quantum theory.

In an article published in *Dyn* titled 'On the Meaning of Cubism Today', Paalen argues that the development of Cubism is inseparable from the science of its day because both progressed according to what was made available by contemporary mental habits:

> The simultaneous overturning of fundamental notions in painting and in physics gave rise to comparisons that were rather vague but sometimes intuitively right. Already in 1913 Apollinaire spoke of non-Euclidean geometry and of the fourth dimension apropos of Cubism. Broadly speaking, one might say that the analogy between the new painting and the new physics consists in that elements formerly held as cognitive or conceptual a-prioris enter as constitutive factors in the very structure of the edifices of art and of science.[68]

By making the second (broad) assertion about an underlying epistemic structure moulding possibilities in art and science, Paalen avoids any suggestion that Picasso and Braque consciously incorporated ideas emerging from Relativity into their work. He does not argue that Cubism is a 'scientific art', then, preferring the now familiar assertion that 'the great thoughts in the same period often follow parallel ways without their authors being aware of it'.[69]

According to Paalen, Cubism pointed the way for twentieth-century art by its rejection of traditional methods of representing three-dimensional space. However, it retained the subject, leaving it to others such as Kandinsky to complete the historical task it began by 'deliberately abandoning the subject'.[70] For Paalen, this advance beyond the subject in abstract painting like Kandinsky's had a parallel in the 'progress' made by QM beyond Relativity:

> The crisis of causality in science seems to me to correspond with the crisis of the subject in art . . . Quantitative physics [sic], in perceiving that the causal concept becomes inapplicable in the microcosmic domain – and painting, in abandoning the causal development of plastic relations – is the same revolution.[71]

Like Carl Einstein, Paalen sees Cubism as the outcome of the radical new conception of reality that emerged simultaneously across several fields of knowledge. As an ex-Surrealist now, he did not regard Surrealism as its end result but believed that its logical conclusion lay in abstract art – in the evacuation of subject matter by artists from their work once they no longer see the need for a point of departure in nature. Without traducing either Cubism or Relativity (he had just painted Albert Einstein's portrait, in fact), he regards them as overtaken by linked, progressive trends: in a word, Kandinskyian abstraction and QM.

The pioneering roles played by artists, poets and scientists in steering our knowledge of the world by means of the divergent concepts of their respective fields is discussed extensively by Paalen in the key text 'Art and Science'.[72] Introducing his theme, Paalen analyses the Goethe–Newton controversy, which articulates, he believes, 'one of the major conflicts in our civilization'.[73] By means of the decomposition of an incident light through a prism, Newton's optical experiments (made public in 1704 in the *Opticks*) showed supposedly pure white light was complex, in fact, and composed of seven discrete monochromatic colours. Following his own experiments towards the end of that century, Goethe denied this was the case, claiming that the 'manipulation' of nature by the intervention of the tools of the scientist and their reduction to mathematical formulae could not be a substitute for the judgment of the senses. Although the experimental base of Goethe's objection to Newton depended upon a quite different methodology, the violence of his attack in the second, polemical part of his 1810 *Farbenlehre* (*Theory of Colour*) shows the question at issue was complicated by the different ways the poet and scientist perceived and conceived the world.

Rather than taking sides in this long-in-the-tooth dispute, Paalen indicates the shortcomings of the procedures respectively adhered to by poet and scientist. Reproaching Goethe, because in 'refus[ing] to consider light otherwise than as a qualitative whole' he was unable to complete his investigation,[74] he takes Newton to task, too, for his solely '*quantitative tendency*' (which Paalen says 'finds its most noxious form in capitalism'),[75] admitting 'practically nothing but the quantitative interpretation of reality'.[76] Although Newton's experiments have long been accepted and Goethe was wrong on the question of the division of light, Paalen writes that the legacy of the disagreement continues to find an echo in 'the great misunderstanding' between art and science.[77]

He goes on to insist that the poet/artist's sense-led, intuitive, *qualitative* comprehension of nature should be accepted alongside the scientist's 'objective', abstract, *quantitative* investigation, in an overarching theory of knowledge. Consequently, he argues for the institution of a kind of Complementarity between the two fields of art and science in their respective researches, conscripting that term as it was used by Niels Bohr in the formulation of his Principle of Complementarity of 1927. As we know, Bohr took this step because the everyday language suitable to the 'objective' description of macrocosmic classical systems hindered attempts to describe phenomena observed in microcosmic quantum systems, where the means of observation disturbed the object of study. By its anticipation of the discovery of atomic behaviour grounded in the old language of waves and particles, one modern experiment would find that matter behaves in a way suitable to a wave description while another following a different experimental route would order matter into a particle 'composition'. For example, a light beam passing through a double pinhole takes on the characteristics of a wave, whereas the same beam subjected to an instrument containing a collection of recoiling scatterers reacts as though composed of particles.[78] The outcome of this seemingly insurmountable problem was Bohr's affirmation, under the rubric of Complementarity, of the inseparability of the observing subject and the phenomena under examination.

Paalen's deployment of Complementarity in *Dyn* in the context of the art–science debate seems speculative at first but receives support from an authoritative quarter. In May the previous year, Werner Heisenberg gave a lecture in Budapest titled 'The Teachings of Goethe and Newton on Colour in the Light of Modern Physics' ('Die Gothesche und die Newtonsche Farbenlehre im Lichte der modernen Physik').[79] Published that year, Heisenberg's lecture began along similar lines to Paalen's essay, claiming that recent discoveries in nuclear physics shed new light on the old dispute between Goethe and Newton, long since thought solved in Newton's favour. He indicates the strengths and weaknesses of Newton's theory: 'it enables us to construct accurate optical instruments which open up parts of the universe not normally accessible to our senses,' while, 'it is plainly of no assistance to a better appreciation of the world of colour surrounding us'.[80] Yet Heisenberg maintains that Goethe's criticism of experimentation – that it demanded nature be converted 'artificially' into a raw material – means his theory cannot aid in the control of phenomena and its adaptation to practical usage like Newton's, while the intuitive approach laid out in his *Farbenlehre* – systematised 'to serve the artist, particularly the painter'[81] – comes closer than Newton's mathematically fixed colour wavelengths to our everyday experience of colour and light:

> In the section 'The Sensual-Moral Effects of Colour,' [Goethe] deals for example with symmetrical arrangements of colours according to polar relations. He presents an arrangement of the six primary colours in a regular hexagon or circle divided into six equal parts: red, blue-red, blue, green, yellow and orange, in this order. Each colour in this circle lies opposite its complementary, thus red opposite green and blue opposite orange.[82]

Heisenberg duly proposes that a cultural 'complementarity' between the theories of Goethe and Newton already exists, in the sense that 'many painters gathered knowledge and enrichment from Goethe's theory,' while '[w]ithout a mathematical theory of light there would never have been a telescope or a microscope'.[83]

Here, Heisenberg introduced the still-recent quantum theory, brought to completion by Complementarity, avowing: 'our experiments are not nature itself, but a nature changed and transformed by our activity in the course of research'.[84] Seeming closer at first to Goethe's anthropocentric procedure than Newton's technical exploration of light, Heisenberg shows here some sympathy with the German poet's position. Like Paalen, who declares in 'Art and Science' that 'life does not consist of abstractions', Heisenberg speaks (slightly ironically given the mathematical formalism of his matrix mechanics) of the perils of a scientific activity in which mathematical abstraction is allowed to eclipse direct sense experience of the world.[85] However, he adds, it would be foolhardy to suggest that accepting such a position entails a rejection of Newton's discoveries. But how to adjust our way of thinking so that neither of the two apparently irreconcilable theories is discarded?

One way to answer this is to say that, as with the acceptance of the mutually exclusive theories of wave and particle, both are correct answers *within the terms of the inquiry* or within the context of a particular approach. To quote Heisenberg again, both are valid once we take into consideration 'the way in which a question is put and the method of research employed'.[86] The division between them is accepted as a reality – as the axis around which our understanding of the world revolves. Heisenberg argues that minds attuned to the theory and method of QM will agree with Goethe that his theory cannot be reconciled with Newton's, but will argue that both are required under different circumstances if we are to understand nature given the diverse and even contradictory evidence to hand. Heisenberg's argument, then, offers a clear parallel with the claim made by Paalen the following year in 'Art and Science':

> as no man can hold himself satisfied with the affirmation that light is merely a number of vibrations, it is highest time to realize that the poet speaks as truthfully as the scientist when he claims that light belongs to the realm of vision. And to the extent that we believe that through a scientific formula we can know all there is to know about a thing, we are surely more backward than the 18th century. . . . But – and herein lies the great misunderstanding – the poet, in his turn, is wrong if he acknowledges nothing but the emotional quality of things, and if he puts this quality in false opposition to the study of their functioning. He is always right so long as he speaks *as poet*.[87]

Following a remarkably similar route, Heisenberg and Paalen reach very similar conclusions, each extending Bohr's Complementarity to accommodate the quite distinct versions of nature offered by art and science (as did Eddington in *The Nature of the Physical World*).

I do not think Paalen was familiar with Heisenberg's 1941 lecture or its publication in *Geist der Zeit* that year. Information concerning Heisenberg's wartime activities was difficult to come by from Paalen's new home in Mexico, and no physics or philosophy journals were found in his library after his death. As a German-speaking 'degenerate' artist in exile with Jewish family connections, it is barely conceivable that he possessed sufficient contact to stay abreast of developments in the philosophy of science in Germany around the time at which he composed 'Art and Science' in Mexico. Paalen's grasp of implications of QM for the philosophy of science seems all the more impressive for that.

Further evidence of this is given where Paalen traces the quantity/quality (Newton/Goethe) division back to the Greeks for whom, he claims, quality dominated. In the thought of Democritus, Paalen sees an 'anticipation of the atomic theory':

> For in saying that 'sweet is sweet, bitter is bitter, cold is cold, only by convention' Democritus seems to anticipate Galileo for whom such qualitative perceptions are nothing but 'names.' However, for Greece this theory remained a *philosophic* dis-

covery; actually antique science does not measure but classifies qualitative entities. While modern science is born of the experiments of Galileo that begin to establish universal 'laws' based on the measurable.[88]

Galileo's laws could have acted as a counterpoint to the qualitative tendencies of philosophy, he argues, but instead the balance shifted too far in the other direction, bringing about a model of the universe dominated by the quantities of physics, which Paalen finds equally intolerable.

Again, a similar view was expressed by Heisenberg, this time a good few years before Paalen, in his 1932 lecture 'On the History of the Physical Interpretation of Nature' ('Zur Geschichte der physikalischen Naturerklärung'), in which he invoked Democritus and Leucippus in search of philosophical themes in antiquity shared by modern atomic theory.[89] Paalen could have read Heisenberg's paper when it was published in 1935,[89] yet elsewhere, in lectures given six years after Paalen's 'Art and Science' in 1948, both Heisenberg and Erwin Schrödinger use the same Democritus quotation as Paalen – 'sweet is sweet, bitter is bitter, cold is cold, only by convention' – to illustrate the foreshadowing of modern physical theory by the Presocratics.[90] Paalen's duality quantity/quality in 'Art and Science' allowed him to make acute observations of the current philosophical potential of Complementarity and the ancient roots of contemporary physical theory that demonstrate surprising agreement with those of Heisenberg and Schrödinger.[91]

Like the other articles published by Paalen in *Dyn*, the idea behind 'Art and Science' was to clarify the differences between his own theory and those of Surrealism. This mission was given greater urgency by the emergence, also in 1942, of a voice for Surrealism in the New World in the shape of the journal *VVV*, around which Breton hoped the Surrealists would regroup. Paalen's elaboration of the leading role played by artist and scientist in the creation of concepts guiding our knowledge of the world is carried out at the expense of Surrealism, which, he claims in 'Art and Science', 'is on the wrong track when it tries to poetize science, which can only lead to mysticism'.[92] While recognising that there *was* an engagement with science taking place within Surrealism, Paalen believed the incorporation of Relativity and quanta into a poetic view of the world meant the infringement of one field upon another. By contrast, he was pressing for art and science to remain separate without either side yielding to a theoretical dogmatism that insisted on a monopoly of the truth.

In the same third number of *Dyn* of 1942, Paalen extends his criticism of Surrealism's 'misuse' of science to Nicolas Calas in his review of Calas's book *Confound the Wise*, published that year.[93] In a harsh assessment, Paalen accuses his former friend of a 'manifestly irresponsible eloquence' used to 'ape the jargon of science' and opine 'at random in the old mystico-metaphysical cant'.[94] Familiar with the thought of the Vienna Circle, Paalen echoes the concerns of Hans Reichenbach over the poeticisation of concepts emerging from physics, and his words of caution against the unrestrained application of scientific imagery transplanted into the field of poetry. Paalen dramatises his admiration of Vienna School materialism at the expense of Surrealist 'metaphysics' (quite the contrary of Surrealism's own view of its theoretical practice, as I will show) by placing on the page across from his scornful appraisal of Calas's efforts his mostly favourable review of Reichenbach's *From Copernicus to Einstein*.[95] Here, he gives an overview of Reichenbach's writings after noting attempts made by scientists such as Planck, de Broglie, Bohr, Schrödinger, Einstein, Heisenberg and Reichenbach himself, 'to integrate the new perspectives of physics into contemporary thought'.[96]

When Paalen reviews *VVV* in the following number of *Dyn* under the name of 'Charles Givors', he returns to the same themes to develop his commentary on Surrealism, arguing that its weakness lies in its failure to engage systematically with the new epistemology emerging from modern physics.[97] Insisting, 'it is necessary to be abreast with [*sic*] contemporary science and philosophy',[98] Paalen quotes Breton's prediction of a 'spectacular discovery that . . . will take place in the sphere of *physics*', in his 1942 lecture to the French students of Yale University published in *VVV*.[99] But Paalen's aim was not to praise Surrealism for its assimilation of the new physics, but to contrast, in a moment of self-congratulation concealed behind a pseudonym, what he saw as Breton's merely gestural appeal to its teachings with his own complex, philosophically inflected reception of QM:

> The hope [in Surrealism] for a dialectical miracle corresponds to the expectation of 'a spectacular discovery in physics.' As if it were not the characteristic of great discoveries in modern science not to be spectacular; in fact, they become so only through a poorly understood vulgarization which often obscures their real significance. But the great, the formidable discovery is there, before us, only it has nothing spectacular about it, and does not allow itself to be caught in a picturesque slogan. This discovery is the new physics of which Louis de Broglie says that there have been few seismic disturbances in the history of the intellectual world that could be compared to the revolution brought about by it [*sic*]; its influence will be immense; its philosophic consequences are barely beginning to be glimpsed by a few: c.f. 'Art and Science' by W. Paalen (*Dyn* 3).[100]

The whiff of ambition is pretty powerful here, recalling Paalen's 'manifesto' 'Farewell au Surréalisme', which appeared in the first number of *Dyn*, comprising an attack on dialectical materialism, Marxism and Surrealism, and causing Breton, predictably, to blow his top.[101] In the brave new world of quantum theory and Complementarity's either/or, Paalen's message in *Dyn* is that dialectical synthesis will be consigned to the past with all the other 'isms', to be replaced by an art and art theory apprised of the 'conquests and methods of science'.[102]

How does Paalen's enthusiasm for wave mechanics and Complementarity translate in his painting? An aphorism of Paalen's spells out what his art owes to the new representation of nature emerging from quantum mechanics: '[n]ot to paint *after* nature, but to work according to its great rhythms; not to follow its fortuitous aspects, but to grasp

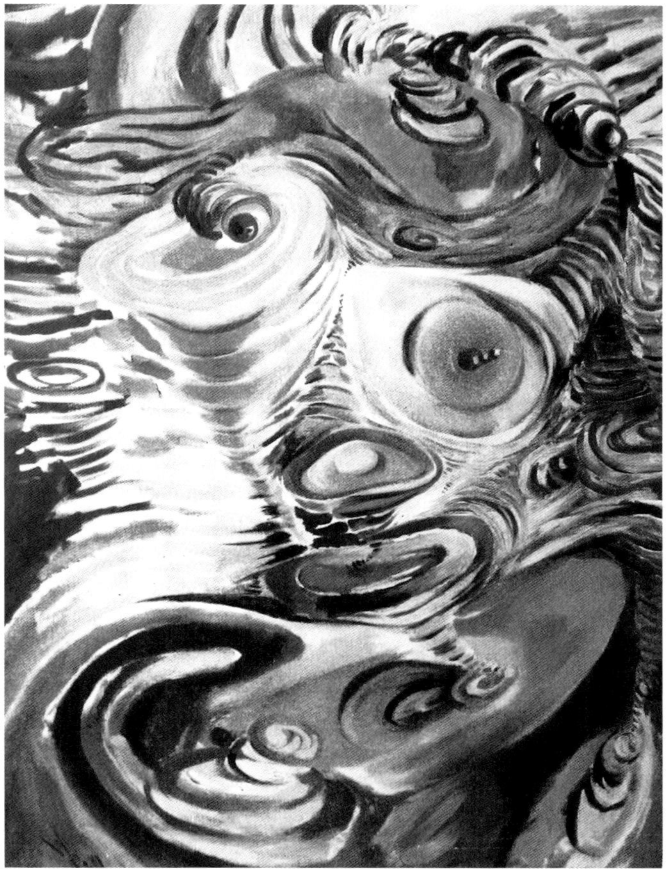

69 Wolfgang Paalen, *Figure pandynamique* (1940). Oil on canvas, 129.5 × 96.5 cm. Fundación Wolfgang y Isabel Paalen, Tepoztlán.

its universal procedures.'[103] The ripples pulsing across the surfaces of Paalen's canvases of the *Dyn* years bear witness to his pursuit of nature understood not in the form it inhabits in landscape painting, but in its behavioural, vibratory sense – the microcosmic flux of reality revealed by the wave mechanics of de Broglie and Schrödinger. Typical is *Figure pandynamique* (fig. 69) of 1940, to which Paalen must have attached some importance as it is the first image, reproduced as a colour plate, to greet the reader in the inaugural number of *Dyn*. Furthermore, it appears opposite the introductory essay 'The New Image', which unveils Paalen's new, physics-based philosophy of art. The forms packed tightly into the pictorial rectangle of *Figure pandynamique* have a watery fluidity expressed in cool colours where the general dominance of greys, blues, dark greens and purples is undiminished by Paalen's moderate use of yellow. Here, the idea of Complementarity is presented (rather literally) by a set of repeating rings sometimes ending at the 'tail' with a sphere representing, presumably, a particle. A second reading of the painting in the context of Complementarity can take a cue from Paul Feyerabend, who uses Wittgenstein's duck-rabbit image viewable as one or the other but not both at once: 'why should it not be possible to say conflicting things about "the same situation"', Feyerabend asks, 'and yet be right?'[104] Although Paalen's painting can be viewed as a set of indeterminate waves, an anthropomorphism is suggested in the title *Figure pandynamique*, sanctioning an alternative but not simultaneous reading of the painting as a ghostly face.

Although he was unusually well informed of developments in modern physics, Paalen's painting rarely gives the impression of radiating effortlessly out of his theoretical speculations like Klee's and Kandinsky's (the artist-theorists he admired). During this period particularly, they seem forced by intellect and unmodulated by feeling. Whatever their quality, though, they do not necessarily place Paalen at odds with the Surrealists. These are not simply formal experiments, and he had not defected to abstraction. Paalen is definitely painting *something* in these works, and throughout the *Dyn* years his objective for his art continued to be the development of a diverse formal and chromatic language that would express nature as rhythm.

While Carl Einstein and Henri-Charles Puech argued that Cubism and abstraction were the product of the same new rationalism that shaped modern physics and psychoanalysis, the art of both Matta and Paalen aimed to extend the Cubist project in the wake of this debate, seeking to escape entirely the ocularcentrism of the Renaissance subject and reject the concrete existence of an external object. Matta's painting avoids the clumsy literalism of Paalen's, ironically *because* of its cavalier (mis)appropriations from other fields, including modern physics. The strength of Matta's work lies in a kind of 'science envy', convinced that the languages of psychoanalysis and physics could be translated into imagery expressing the infinite complexity of both mind and universe. Entertaining a view of science that contrasted sharply with Paalen's sober theory, he nevertheless achieves intuitively the '*participation sympathique*' recommended by Puech in 1934 as a means of artistic discovery. That view is crystallised in Matta's assertion, 'the real is made up of oscillations, rays and waves. To unveil the spirit of a thing, or a being, it is necessary to enter into the family of waves in which it appears. This is a molecular and nuclear conception, where everything is a nucleus of energies. Invisible, thermal, dynamic, magnetic, sonorous revelations.'[105]

Quantum Mechanics: From Ernst Mach to Max Ernst

I have established that Matta's association of the new physics and new psychology was not at all unusual for the time. We saw several Surrealists in the early and mid-1930s have a go at blending the two discourses, and now I will relate these to similar efforts by materialist philosophers and even physicists concerned with quantum interactions. An important precursor and intermediary for this was the nineteenth-century Austrian physicist and philosopher Ernst Mach. Rejecting the absolutes of Newtonian mechanics as much as the *a priori* space and time of Kant, Mach's thought reads as a kind of proto-phenomenology. With the twin objective of unifying science and eliminating metaphysics, Mach repudiated in principle any distinction between propositions based on experience and *a priori* laws, arguing that everything science claims as fixed and objective was subject, in fact, to constant re-examination through experience and open to subsequent revision.[106] While it is not accurate to say that Mach denied the existence of an objective world, it is true that he rejected Kant's concept of the 'thing-in-itself', tying objects closely to sensation. For this reason, he famously denied the material existence of atoms insofar as they lay beyond sense perception and were 'metaphysical'; that is, 'things of thought'. Insofar, that is, as their description was not 'reducible to propositions containing only perception terms as predicates', in the words of Philipp Frank.[107]

I mentioned that Carl Einstein's practice as an art historian drew upon discoveries of twentieth-century physics, for which he had had a passion since his time as a student.[108] Einstein agreed with the philosopher Konrad Fiedler that the same requirements for research, creativity and the invention of a worldview underlie both scientific and artistic work.[109] But Einstein's main philosophical and scientific source was Ernst Mach. When he outlined the origins of his art theory and the path it had taken in a letter to art dealer Daniel-Henry Kahnweiler in June 1923, he aligned his thought closely with Mach's: '[o]n the theoretical level', he concedes, 'it is perhaps Mach to whom I am closest.'[110] Liliane Meffre offers a general summation of how Carl Einstein might have received Mach:

> For Mach, who abolishes all barriers between the physical and psychical, no opposition exists between the internal world of the subject and the external world, fully formed from any two elements, space, time, colours, sounds. The 'thing-in-itself' does not exist, 'it is sensations and their combination which give rise to our concepts.' . . . Mach replaces the concepts of cause and effect with the notion of function which allows one to represent much better the relationships between elements. This notion of function is at the base of Einstein's theory of art.[111]

The new emphasis Einstein believed art theory should place on the interaction of depicted motifs rather than on the identities of objects themselves was discussed by him primarily with reference to Cubism's 'post-perspectivalism'.

How do elements of Mach's thought relate to QM and subjectivity? Because of the emphasis Mach's positivism placed on the limitations and formative aspects of the subject's observation – and even though Mach himself rejected Einstein's Relativity for certain of its non-empirical logical deductions such as the constancy of light velocity – some of those central to the formulation of QM admired his ideas, as indeed had the young Albert Einstein.[112] Wolfgang Pauli was Mach's godson, in fact, and amongst the most sceptical of that generation, though his positivism was not of the same order as Mach's.[113] The Machism of Erwin Schrödinger developed in his early years in Vienna. In a letter of 1940 to Arthur Eddington, he identified himself with the tradition of Machean positivism,[114] and in 1959 he gave a perfect expression of a perception at once Machean and quantum mechanical, insisting: '[i]t is the same elements that go to compose my mind and the world. The subject and object are only one.'[115]

Of this first generation of physicists concerned with quanta, Pascual Jordan took the most from Mach. Later in life he declared, '[Mach] strongly influenced me. Basically, even today I am still a follower of Mach, though of course I am critical on points of detail. But fundamentally I have constructed my physical thinking on Mach's conceptions.'[116] Jordan was one of those closest to Niels Bohr, who is usually regarded as the father figure of QM, spending some months with him in Copenhagen in 1927 after completing his doctorate with Max Born in Göttingen in 1924 (fig. 70). Two years after his 1927 lecture in Como presented Complementarity as a formulation of non-synthetic wave-particle duality, Bohr began to make statements linking psychology and biology to his core quantum mechanical concept.[117] His confirmed Machism meant Jordan was interested in making the same relationships, and he entered into dialogue with Bohr on the matter in 1931.

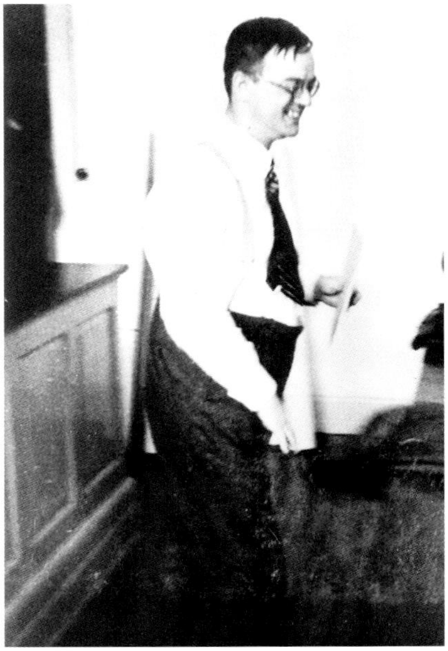

70 Pascual Jordan, Copenhagen, 1936.

Briefly put, Jordan posits the idea that on the Newtonian, observable level, 'the statistical acausality of atomic reactions' averages out to a basic causality for inanimate matter, whereas for animate matter, 'the acausality of specific atomic reactions *amplify* to a macroscopically operative acausality'.[118] For Jordan, then, the free will that characterises humans, animals and plants is a direct echo of the same acausal constituents that form inanimate matter and yet disappear for the latter when raised to the macroscopic level. Jordan believed that his 'amplifier theory' (as it became known) gave QM a general application, extending to biology and psychology. Gentle disagreements with Bohr did not prevent Jordan publishing on the subject in 1932 in Arnold Berliner's journal *Die Naturwissenschaften* (fig. 71). Originating in 1913, at one time or another this journal carried articles by Planck, Einstein, Bohr, Schrödinger, Born, Heisenberg and Pauli, and this is where Surrealist artist Max Ernst came into contact with Jordan's writings on biology, psychology and QM.

Ernst mobilises Jordan's theory in his brief essay titled 'What is Surrealism?' ('Was ist Surrealismus?'), published in 1934 to accompany his exhibition of that name held at the Kunsthaus in Zürich from 11 October to 4 November. Ernst attempts to answer that question by rethinking the role of the artist-observer in the creative process, separating Surrealism from traditional models of philosophical inquiry and attuning its practices and findings to those of QM. Ernst's experimental technique of *frottage* creates suggestive, patterned surfaces from rubbings made with pencil or charcoal over wood, which are then identified and actualised through additional elements applied by Ernst, in the same way a Rorschach ink blot can be visually interpreted initially then worked up so that it conforms to that interpretation (fig. 72). In 'What is Surrealism?' Ernst presents *frottage* images both as the raw material of the unconscious and as pseudo-scientific, parodic illustrations of a natural history book. Alongside other experimental methods devised by Surrealism such as automatism, he commends the technique as

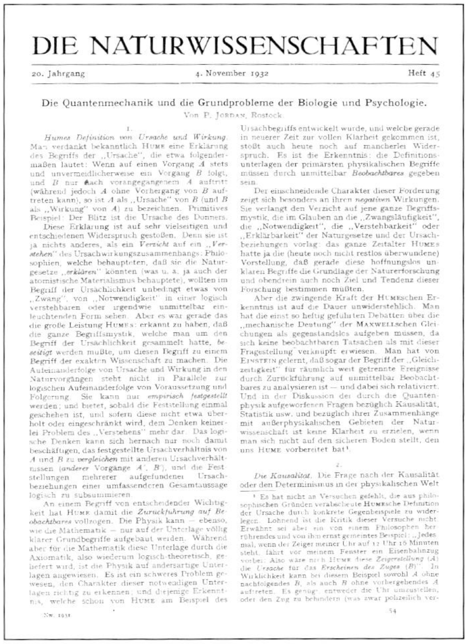

DIE NATURWISSENSCHAFTEN

20. Jahrgang 4. November 1932 Heft 45

Die Quantenmechanik und die Grundprobleme der Biologie und Psychologie.

Von P. Jordan, Rostock.

1.

Humes Definition von Ursache und Wirkung. Man verdankt bekanntlich Hume eine Erklärung des Begriffs der „Ursache", die etwa folgendermaßen lautet: Wenn auf einen Vorgang *A* stets und unvermeidlicherweise ein Vorgang *B* folgt, und *B* nur nach vorangegangenem *A* auftritt (während jedoch *A* ohne Vorhergang von *B* auftreten kann), so ist *A* als „Ursache" von *B* (und *B* als „Wirkung" von *A*) zu bezeichnen. Primitives Beispiel: Der Blitz ist die Ursache des Donners.

Diese Erklärung ist auf sehr vielseitigen und entschiedenen Widerspruch gestoßen. Denn sie ist ja nichts anderes, als ein *Verzicht* auf ein „*Verstehen*" des Ursachwirkungszusammenhangs: Philosophien, welche behaupteten, daß sie die Naturgesetze „*erklären*" könnten (was u. a. ja auch der atomistische Materialismus behauptete), wollten im Begriff der Ursächlichkeit unbedingt etwas von „Zwang", von „Notwendigkeit" in einer logisch verstehbaren oder irgendwie unmittelbar einleuchtenden Form sehen. Aber es war gerade das die große Leistung Humes: erkannt zu haben, daß die ganze Begriffsmystik, welche man um den Begriff der Ursächlichkeit gesammelt hatte, *beseitigt* werden mußte, um diesen Begriff zu einem Begriff der exakten Wissenschaft zu machen. Die Aufeinanderfolge von Ursache und Wirkung in den Naturvorgängen steht nicht in Parallele zur logischen Aufeinanderfolge von Voraussetzung und Folgerung. Sie kann nur *empirisch festgestellt* werden; und bietet, sobald die Feststellung einmal geschehen ist, und sofern diese nicht etwa überholt oder eingeschränkt wird, dem Denken keinerlei Problem des „Verstehens" mehr dar. Das logische Denken kann sich hernach nur noch damit beschäftigen, das festgestellte Ursachverhältnis von *A* und *B* zu *vergleichen* mit anderen Ursachverhältnissen (*anderer* Vorgänge *A'*, *B'*), und die Feststellungen mehrerer aufgefundener Ursachbeziehungen einer umfassenderen Gesamtaussage logisch zu subsummieren.

An einem Begriff von entscheidender Wichtigkeit hat Hume damit die *Zurückführung auf Beobachtbares* vollzogen. Die Physik kann — ebenso, wie die Mathematik — nur auf der Unterlage völlig klarer Grundbegriffe aufgebaut werden. Während aber für die Mathematik diese Unterlage durch die Axiomatik, also wiederum logisch-theoretisch, geliefert wird, ist die Physik auf andersartige Unterlagen angewiesen. Es ist ein schweres Problem gewesen, den Charakter dieser notwendigen Unterlagen richtig zu erkennen; und diejenige Erkenntnis, welche schon von Hume am Beispiel des Ursachbegriffs entwickelt wurde, und welche gerade in neuerer Zeit zur vollen Klarheit gekommen ist, stößt auch heute noch auf mancherlei Widerspruch. Es ist die Erkenntnis: die Definitionsunterlagen der primärsten physikalischen Begriffe müssen durch unmittelbar *Beobachtbares* gegeben sein.

Der einschneidende Charakter dieser Forderung zeigt sich besonders an ihren *negativen* Wirkungen. Sie verlangt den Verzicht auf jene ganze Begriffsmystik, die im Glauben an die „Zwangsläufigkeit", die „Notwendigkeit", die „Verstehbarkeit" oder „Erklärbarkeit" der Naturgesetze und der Ursachbeziehungen vorlag: das ganze Zeitalter Humes hatte ja die (heute noch nicht restlos überwundene) Vorstellung, daß gerade diese hoffnungslos unklaren Begriffe die Grundlage der Naturerforschung und obendrein auch noch Ziel und Tendenz dieser Forschung bestimmen müßten.

Aber die zwingende Kraft der Humeschen Erkenntnis ist auf die Dauer unwiderstehlich. Man hat die einst so heftig geführten Debatten über die „mechanische Deutung" der Maxwellschen Gleichungen als gegenstandslos aufgeben müssen, da sich keine beobachtbaren Tatsachen als mit dieser Fragestellung verknüpft erwiesen. Man hat von Einstein gelernt, daß sogar der Begriff der „Gleichzeitigkeit" für räumlich weit getrennte Ereignisse durch Zurückführung auf unmittelbar Beobachtbares zu analysieren ist — und dabei sich relativiert. Und in der Diskussion der durch die Quantenphysik aufgeworfenen Fragen bezüglich Kausalität, Statistik usw. und bezüglich ihrer Zusammenhänge mit außerphysikalischen Gebieten der Naturwissenschaft ist keine Klarheit zu erzielen, wenn man sich nicht auf den sicheren Boden stellt, den uns Hume vorbereitet hat[1].

2.

Die Kausalität. Die Frage nach der Kausalität oder den Determinismus in der physikalischen Welt

[1] Es hat nicht an Versuchen gefehlt, die aus philosophischen Gründen verabscheute Humesche Definition der Ursache durch konkrete Gegenbeispiele zu widerlegen. Lohnend ist die Kritik dieser Versuche nicht. Erwähnt sei aber ein von einem Philosophen herrührendes und von ihm ernst gemeintes Beispiel: „Jedesmal, wenn der Zeiger meiner Uhr auf 12 Uhr 16 Minuten steht, fährt vor meinem Fenster ein Eisenbahnzug vorbei. Also wäre nach Hume diese *Zeigerstellung* (*A*) die *Ursache* für das *Erscheinen des Zuges* (*B*)". In Wirklichkeit kann bei diesem Beispiel sowohl *A* ohne nachfolgendes *B*, als auch *B* ohne vorhergehendes *A* auftreten. Es genügt, entweder die Uhr umzustellen, oder den Zug zu behindern (was zwar polizeilich ver-

Nw. 1932 54

71 Pascual Jordan, 'Die Quantenmechanik und die Grundprobleme der Biologie und Psychologie,' *Die Naturwissenschaften*, vol. 20, 4 November 1932, 815–21.

evidence of 'an exploration of a new and incomparably wider field of experience in which the borders between the so-called interior and exterior worlds (as defined by classical philosophy) are becoming increasingly blurred and will probably disappear altogether one day'.[119]

Pressing *frottage*'s claims to innovatory relevance, Ernst continues as follows:

> The revolutionary significance of this initially absurd-seeming description of nature is pointed up by the fact that analogous results have been achieved in modern microphysics. After measuring the path of an electron moving without application of external force, and then measuring its location, P. Jordan concluded: 'But this distinction [between exterior and interior world] is robbed of one of its main props by the experimental refutation of the idea that facts are present in the exterior world that have an objective existence independent of the process of observation.'[120]

In other words, like the physicists of QM, the Surrealists were charting a zone that could no more be satisfactorily designated 'objective' than it could 'subjective'. The acausality of both unconscious and subatomic phenomena had forced a breach with the classical descriptive rationale served by those unitary and mutually exclusive categories. Ernst goes on to say of the Surrealist painters:

> they move freely, daringly and naturally on the physically and mentally quite real ('surreal'), if still largely undefined, frontier between interior and exterior world, registering what they see and experience there, and intervening where their revolutionary instincts suggest. In fact, the fundamental opposition between meditation and action (again, according to the view of classical philosophy) disappears along with the fundamental distinction between exterior and interior world, and the universal significance of Surrealism lies precisely in the fact that, in the wake of this discovery, no area of life can remain closed to it.[121]

Ernst speaks from a long-standing knowledge of psychoanalysis and obvious interest in the irrational aspects of nature, which he often contrasts with the mechanised rationalism of modernity, as in the series of *Garden Aeroplane Trap* canvases of the mid-1930s (fig. 73).

Jordan's own linking of biology, psychoanalysis and the philosophy of QM is not very well known, even to historians of science, yet it constituted a significant strand of his thinking as the 1930s progressed. Essentially, he argues that only an understanding of

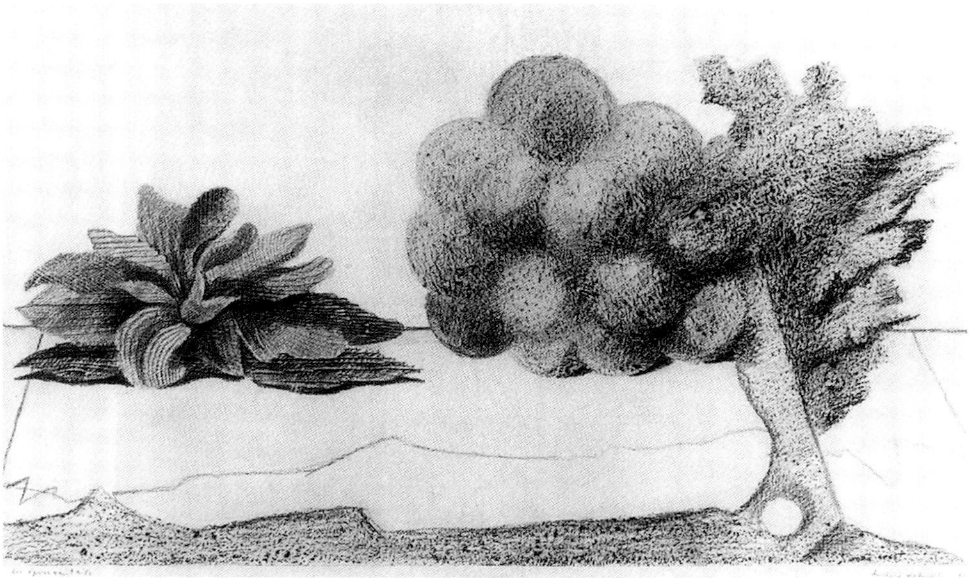

72 Max Ernst, *Scarecrows* (1925). Frottage, pencil and watercolour on paper, 26 × 43 cm. Private collection.

unconscious processes allows the prediction of decisions made by human beings, yet these processes are themselves altered by active inquiry or observation, just as observation in QM interferes with subatomic processes.[122] Jordan's writings within the context of debates on biology, psychoanalysis and observation in QM offer, then, some interpretive leverage for Ernst's nature paintings completed during this period. In the last section of this chapter, I demonstrate the unexpected proximity of such debates to positivism.

Materialism: From Logical Positivism to Surrealism

Jordan's views were debated (and criticised) at length as a 'new vitalism' at a major conference held in Prague in 1934 on QM and biology.[123] This event was arranged as a prelude to an International Congress of Scientific Philosophy, which took place at the Sorbonne in September 1935, promoted by the French government and by several of France's scientific institutions. In June of the following year, Niels Bohr spoke at a related conference held in Copenhagen, giving special attention to the problem of causality, biology and QM.[124] Ernst and the Surrealists had some knowledge of these debates, all of which were organised and steered by the neo-positivist Vienna Circle. Given wide coverage by European philosophy periodicals, the Vienna Circle gained international prominence from these events.[125] Details of the conferences were published in their journal *Erkenntnis*, where Jordan's speculative 1934 essay on biology, psychoanalysis and QM appears, as do the largely negative responses it garnered at the Prague conference in that year.[126]

Jordan's *Erkenntnis* essay opens with an extended appreciation of Hans Reichenbach – who was close to the Vienna Circle, was published regularly in *Die Naturwis-*

73 Max Ernst, *Garden Aeroplane Trap* (1935). Oil on canvas, 54 × 74 cm. Musée national d'art moderne, Centre Georges Pompidou, Paris.

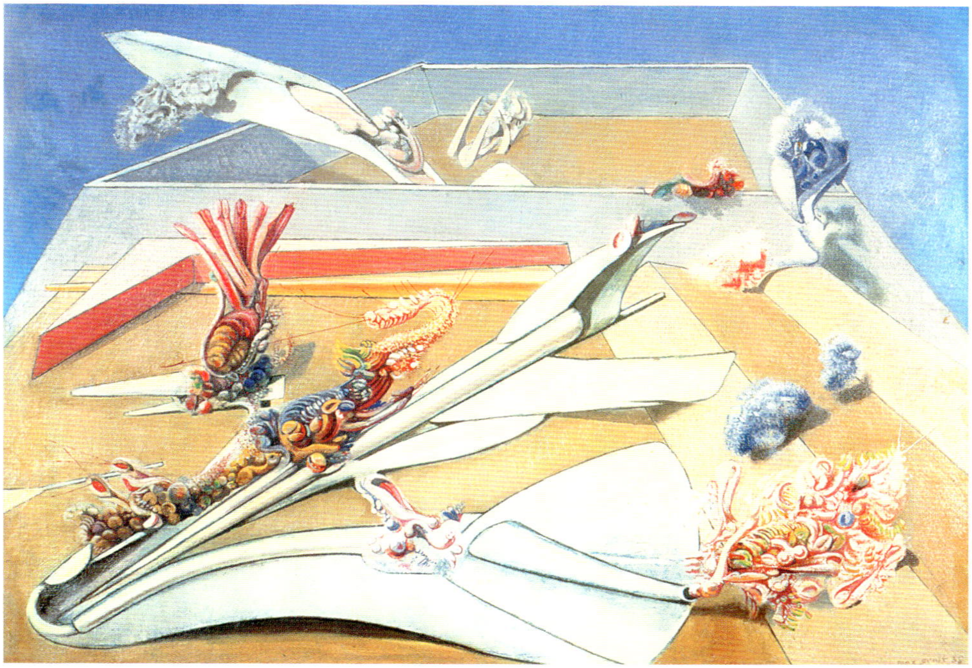

senschaften, and whom we saw pop up unexpectedly in Bataille's semi-Surrealist *Documents* – and Ernst Mach, both of whom, he claims, anticipated the epistemological implications of QM.[127] A similar weave to Mach's of non-Newtonian yet radically positivist themes is central to the thought of the Vienna Circle, which formed around Moritz Schlick in the early 1920s in the philosophy department at the University of Vienna. During the same period, Freud had an academic position in the psychology faculty and Erwin Schrödinger was a student in the physics department. The Viennese connection can and has been used to link Wolfgang Paalen's brusque, post-Surrealist rejection of metaphysics to Mach and the Vienna Circle.[128] Mach held the Chair in Philosophy in Vienna till 1900, and his persistent call for the elimination of all metaphysics from scientific endeavour led Philipp Frank to dub him 'the real master of the Vienna Circle'.[129] In the late 1920s and early 1930s, the group developed a neo-positivist empiricism demanding the same rigour for philosophy as that applied by science. To this end, mathematical logic – the research of Russell, Whitehead and Wittgenstein – was employed in the analysis of language.

Under the aegis of 'physicalism', the Vienna Circle sought the unification of science aimed at by Mach, giving it its alternative title, The Unity of Science group. Its leading intellectual light, Rudolph Carnap, claimed all scientific statements could be reduced to a concrete, physical language: '[t]his thesis,' he writes, '(termed "Physicalism" by [Otto] Neurath) is allied to that of *Materialism*, which respectable philosophers (at least in Germany . . .) usually regard with abhorrence.'[130] Inspired by Mach and expanding his brand of unforgiving, coarse-grained positivism, Carnap claims '[t]here exist fundamentally but one kind of objects, viz. physical events',[131] even going on to declare: '[p]sychology is a branch of physics.'[132]

Now, their negative responses to Jordan's speculations suggest that psychoanalysis in its vulgar form would have been of little interest to the sceptics of the Vienna Circle. Moreover, their sober, empirical programme appears utterly inimical not only to the Surrealists' free play with the symbolism of psychoanalysis specifically, but with Surrealism's taste for the exotic, delirious and irrational in general.[133] The positivism of the Vienna Circle is hardheaded, seemingly incompatible by extension with the very aspirations and principles that motivated and determined Surrealism historically and dictated its theoretical and practical orientation. Indeed, in my first chapter I quoted Breton from the *Manifesto of Surrealism* to this effect (ironically as a means to link Surrealism with the philosophy of modern physics, which was a principal scene of engagement for the Vienna Circle), where he broadcasts his loathing of what he calls 'the realistic attitude, inspired by positivism, from Saint Thomas Aquinas to Anatole France'.[134]

However, we also saw that some years later in 1935, Breton, on behalf of the Surrealists, condemned as '*metaphysics*' what he called 'the greater part of Freud's philosophy', and this is where the project of Surrealism comes to intersect with that of the Vienna Circle.[135] The materialism dominating French thought of the period as well as Surrealist theory and practice in the 1930s, which I have indicated on numerous occasions in the context of the philosophy of physics, bears close comparison with the Vienna Circle's main characteristic, its rejection of metaphysics in favour of a radical 'physicalism'. Attuned to the 'concretising' sentiment of Surrealism, the Vienna Circle's logical positivism held an unexpected allure for Surrealist artists and writers, as paradoxical on first sight as their attraction to modern physics.[136]

At the end of the 1930s in the *Revue de Métaphysique et de Morale*, André Burnier acknowledged the still quite recent triumph of science-based philosophy. Introducing his essay as one that will attempt to throw light on 'the conditions, the scope, and the limits of the problem of knowledge,'[137] he continues:

> Undoubtedly, the history of philosophy offers us a similar experience and each of us has profited from it; but it is today, if not exhausted, then at least surpassed by the incessant progress of philosophy allied with science.[138]

Linking Surrealism to QM and the philosophy of science, then, turns out to be a way of reconsidering the movement in the broader, unexpected context of developments in mainstream French and even European philosophy, beyond the obvious ones of Hegelianism and Marxism. In this way, the thought of individuals such as Émile Meyerson, Léon Brunschvicg, Alexandre Koyré and even A. N. Whitehead and the Vienna Circle receive a rare outing in a study of Surrealism. Indeed, Meyerson's name has cropped up throughout this book, since I noted in my first chapter some grounds for a consideration of Surrealism and Meyerson along the lines of philosophical rationalism, a link that was posited (for the one and only time, as far as I know) as early as the 1920s, by the non-conformist writer André Berge.[139]

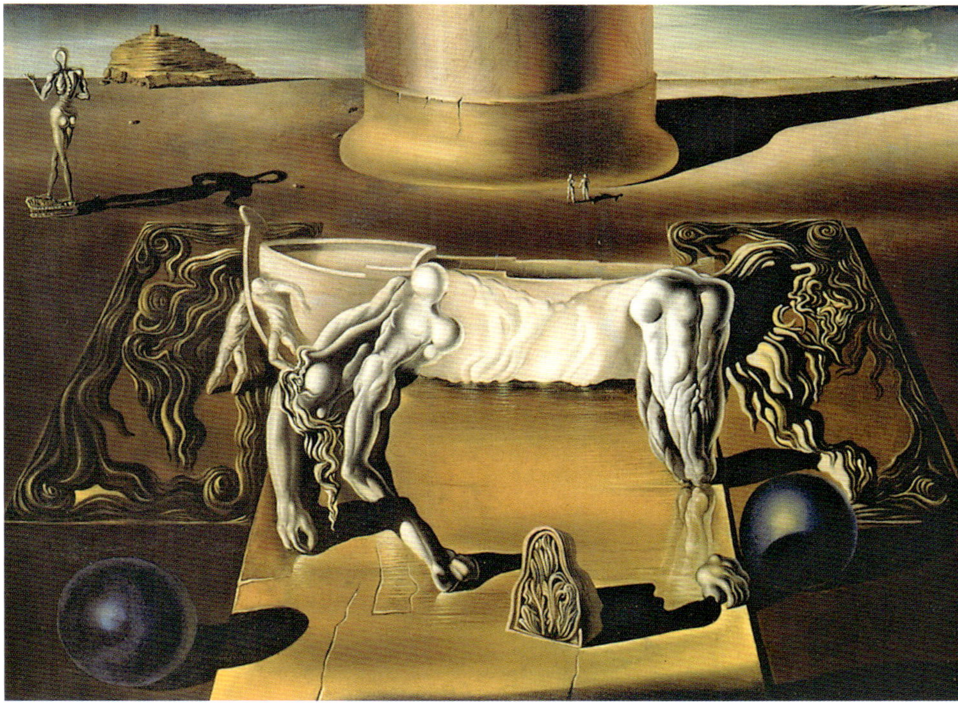

74 Salvador Dalí, *Paranoiac Woman-Horse (Invisible Sleeping Woman, Lion, Horse)* (1930). Oil on canvas, 50.6 × 65.2 cm. Musée national d'art moderne, Centre Georges Pompidou, Paris, Association Bourdon.

I will bring this section and chapter to a close by recording two more attempts comparable with Ernst's to figure Surrealism as a kind of positivism in the light of the new concept of observation inaugurated by QM. One is made by the young French physicist and colleague of Louis de Broglie, Marie-Antoinette Tonnelat, in an essay written on the occasion of de Broglie's sixtieth birthday in 1952.[140] Seeking to characterise 'objective reality' as the outcome of formative peculiarities of observation according to quantum theory, she writes:

> In this way one arrives at this rather paradoxical result: a positivism strictly observed which, by a bad play on words, could be described as *Surrealist*, cannot attain a physically permanent and autonomous reality. In physics as in painting, Surrealism denies the possibility of a description which does not carry *explicitly* the stamp of the observer.[141]

By the time Tonnelat made these remarks in 1952, the word 'surrealism' had taken on a loose currency way beyond anything Breton had expected or desired. In fact, one of the reasons behind Breton's 'history of Surrealism' series of radio interviews, 'Entretions', broadcast from March until June of that year, was to counter popular misconceptions of the movement, though the high visibility they momentarily gave Surrealism in France may well lie behind Tonnelat's analogy.

If she had in mind a particular Surrealist artist it was probably not Ernst, in spite of his own very similar comparison nearly twenty years earlier of Surrealist and quantum mechanical reality. Ernst's comparison was between the physicist's and artist's creation of reality, whereas Tonnelat's comparison is between the physicist and the viewer of a Surrealist painting. There is more chance that Tonnelat was thinking of Salvador Dalí, who by the 1950s had become a kind of Surrealist 'classic' in the collective cultural consciousness. The imagery of Dalí's method of 'paranoia-criticism', meant to lend itself to multiple alternative readings, fits the idea of Complementarity and Tonnelat's 'observer-creator' far better than Ernst's work.

Initiated and developed at the height of his own interest in modern physics, Dalí's paranoia-criticism is an important element in his psychoanalytic reading of Jean-François Millet's painting *The Angelus* (1857–59) in his study *Le Mythe tragique de l'Angelus de Millet* (*c.*1933/1963).[142] Claiming to see the mechanism of paranoia at work in scientific practice, Dalí links it to a kind of fetishism, where the measure of reality ascribed to nature is decided upon through the fixation of the scientist's attention on an exclusive 'window' on phenomena. Like the audience of a multiple-image painting by Dalí, or the observer of the arbitrarily patterned surface of a damp wall, the scientific observer *decides* what is 'objective', he maintains. Dalí then calls on Erwin Schrödinger in support of his assertion that the supposedly objective facts presented by science are preconditioned by a subjective experimental decision:

> How, I ask, not to discern the presence of the paranoiac mechanism in the extraordinarily deterministic phenomenon of 'experimental' choice, which prefaces investigations of the natural sciences? How, I repeat, can such a mechanism not be acting there, when the study of delirious paranoiac particularities lets us witness *an essential change in the objective world, a change that is presented as sudden, and which absorbs, by its instantaneous associative power, all of our attention and our affectivity, which stay irresistibly fixed to a 'certain' number of facts and objects, to the detriment and exclusion of all the rest?* It seems to me that such phenomena, composed also of intense determinations of choice, cannot be in vain; cannot, to a more or less pronounced degree, mark the intervention in this factor, which is, in my opinion, conditional of scientific experimentation, which, to use the exact words of E. Schrödinger consists in '*the momentary disposition of our interest and of its determinant influence in the direction of later work*'.[143]

Just as Ernst sought epistemological validation for *frottage* by associating its hallucinatory yet materialistic practice with the challenge set by QM to the classical demarcation of subject and object (as articulated by Jordan), Dalí claims to perceive the subjective, deterministic, prefigurative mechanism of paranoia in the prescriptive experimental choices made by QM (as explained by Schrödinger).[144] Far from a one-off, this recruitment of modern physics is a regular feature of Dalí's art and theory in the 1930s and beyond.

6

RELATIVITY AND THE FOURTH DIMENSION

Salvador Dalí and Roger Caillois

Although Einstein's theory quickly became a side issue for André Breton in the 1920s, it was reintroduced into Surrealism from 1929 in the form of Salvador Dalí's spectacular *mélange* of Relativity and psychoanalysis. In the following, necessarily selective account of Dalí's deployment of Relativity physics in his art and writings, I will interpret its role in the development of his artistic imagery and methodology, and chronicle its value to him as a metaphor of modernity. When Dalí's work of the 1930s is placed in dialogue with Roger Caillois's science-inflected writings of the same period, it becomes evident that the manner in which Surrealism should assimilate modern physics was momentarily the issue upon which was staked its orientation and claim to serious investigative status. Although fairly recent scholarship has given partial treatment to the exchange between Dalí and Caillois, my claim is that the relationship between, and the nature of, Dalí's *délire* and Caillois's *rigueur* can best be understood within the context offered by modern physics.[1] That discussion is meant to connect with my commentary on the early conscription of the Theory of Relativity into Surrealist theory and criticism by Breton. I argue that Relativity's significance to Surrealism in its formative years created the conditions allowing for its re-emergence in Surrealist art and theory in the 1930s, giving a measure of 'authorisation' and 'validation' to artistic languages like Dalí's, indebted to Relativity's concepts and metaphors.

Eddington, Relativity and Dalí

By now it should be obvious how a movement like Surrealism could sustain an engagement with a supposedly strictly empirical and positivist science such as physics. We have seen on numerous occasions how Surrealism's investment in an expanded notion of the rational and the real found authoritative support in the description of the very strange constitution of time, space and matter offered by Relativity and QM. As early as 1918, accessible books and articles began to appear introducing large audiences to the new aspects of reality ushered in by Relativity, including its insistence on the contraction and extension of bodies in motion, and the idea that time passes at quite different rates in separate 'systems', 'bodies' or 'frames' of reference. Quantum theory did not make its way so readily or swiftly into the popular consciousness, yet reader-friendly accounts of its findings and epistemological ramifications appeared in French from 1926.[2]

Earlier, I described how the Surrealists, like other non-experts, had recourse to popular writings such as those of Arthur Eddington, in which scientific rigour was combined with a more literary taste for the fantastic. Typical is Eddington's book *Space Time and Gravitation* of 1920, which contrasts the gross reality taken in by the senses with the subatomic or 'hidden' one uncovered by modern science. On jaunty form in the following passage, Eddington 'surrealises' behaviour on the macrocosmic scale through its exaggeration in imitation of the microcosmic operation of nature, asking his reader to imagine travelling upwards vertically at 161,000 miles per second and consequently shrinking to half size due to the 'Fitzgerald-Lorentz contraction':

> As you lie in bed, you are, say, 6 feet long. Now stand upright; you are 3 feet. You are incredulous? Well, let us prove it! Take a yard-measure; when turned vertically it must undergo the Fitzgerald contraction, and become only half a yard.[3]

The subject himself is unaware of the changes to his physical properties because his retina contracts by half in the same direction and therefore sees both his shrunken body and the yard-measure as doubled, and therefore as not changed at all. An external viewer must be posited to witness these bizarre events, imagined by Eddington in *Space Time and Gravitation* as follows:

> Suppose that by development in the powers of aviation, a man flies past us at the rate of 161,000 miles a second. We shall suppose that he is in a comfortable travelling conveyance in which he can move about, and act normally and that his length is in the direction of the flight. If we could catch an instantaneous glimpse as he passed, we should see a figure about three feet high, but with the breadth and girth of a normal human being. And the strange thing is that he would be sublimely unconscious of his own undignified appearance. If he looks in a mirror in his conveyance, he sees his usual proportions; this is because of the contraction of his retina, or the distortion by the moving mirror . . . But when he looks down on us, he sees a strange race of men who have apparently gone through some flattening-out process; one man looks barely 10 inches across the shoulders, another standing at right angles is almost 'length and breadth, without thickness.' As they turn about they change appearance like the figures seen in the old-fashioned convex mirrors.[4]

Eddington rounds out a similar 1918 account of the behaviour of moving bodies under the new reign of Relativity as follows: '[w]e thus get a glimpse of what, from our present point of view, must be called the *real* world, strangely different from the world of appearance,' adding, 'you can almost see a living-picture of this real world reflected in a polished door-knob.'[5]

As we saw, this 'real' Einsteinian world set out by Eddington in *Space Time and Gravitation* lies at the foundations of Surrealist theory, in Breton's adoption of the concepts and imagery of Relativity in his pre-*Manifesto* writings of the early 1920s. The first glimpse of Surrealism's spark of the poetic 'marvellous' occurs in the context of Relativity in the essay he wrote for Max Ernst's one man exhibition in Paris in 1921. Synonymous with Ernst's work, the 'real' world ushered in by the new physics became available to Breton and his friends that very year, thanks to the translation into French of both Einstein's *Relativity: The Special and the General Theory* and Eddington's *Space Time and Gravitation*.

Salvador Dalí could read Eddington's *Space Time and Gravitation* in Spanish translation from the following year. Although no theoretical physics was taught in Spain in the early 1920s, Dalí had plenty of access to Relativity as a student in Madrid. The Mathematical Society of Madrid was the centre of discussion of Relativity, which was also debated in the popular science journal *Madrid Científico*. In addition, the Jesuit science periodical *Ibérica* published the writings of Josep (José) M. Plans, who taught in Madrid the only course on Relativity at advanced level in Spain and was the translator of Eddington's *Space Time and Gravitation*. According to Thomas F. Glick, the best summaries of Einstein's lectures in Spain were written by Plans, and published in the Madrid Catholic daily newspaper *El Debate*.[6] Dalí undoubtedly saw Einstein speak, because one of the lectures on the physicist's tour to Barcelona, Madrid and Zaragoza between February and March 1923 took him, on 9 March, to the élite Residencia de Estudiantes, where Dalí had enrolled as a student the previous September.[7] Dalí would have shared in the general incomprehension of the non-specialist audience of the technical details of Relativity, but the great event would have fired his and his young friends' enthusiasm for Einstein and his theory.

Equally important was José Ortega y Gasset's brief introduction to Einstein's talk, situating Relativity in the history of Western thought. Reprinted as 'Mesura a Einstein', this was one of the two most important essays published in Spain during the period on the scientific and philosophic context of Relativity. The other is Ortega's 1922 essay 'El sentido histórico de la teoría de Einstein', published in *El tema del nuestro tiempo* (1923).[8] As well as helping to steer the reception of Relativity, Ortega played a large part in introducing psychoanalysis into Spain. A fluent reader of Freud in German (he gave a simultaneous translation of Einstein's German lecture at the Residencia) with extensive publishing contacts, it was largely down to Ortega that Spanish translations of Freud began to appear as early as 1922.[9] Ortega's *Revista de Occidente* was Spain's leading intellectual review at the time, and was widely read at the Residencia. A major source of European thought and culture, it introduced an audience that included Dalí to Relativity and psychoanalysis ('devoured' by Spaniards according to Ortega in the *Revista de Occidente*), as well as Surrealism from 1924, when it reviewed Breton's *Manifesto of Surrealism*.[10]

Dalí's longstanding knowledge of Surrealism and the imagery of artists like Ernst made his initiation into the group in 1929 fairly straightforward. But his mature style, which dates from that moment, was hardwired into Surrealism anyway, because as he entered that circle he was busily familiarising himself with some of the stranger concepts of Relativity to which Breton had been exposed as he sketched in the background of Surrealist theory ten years earlier. In 'The Sanitary Goat' of 1930, one of the four texts published that year under the title *La Femme visible*, Dalí argues with all the zeal befitting a new recruit that in the current intellectual climate the evidence given by the senses of the outside world could no longer be taken on trust.[11] Complaining that poetic thought is retarded by its 'impressionistic' relationship with external reality (which he associates with Bergsonian 'anti-intellectualism'), he calls upon Einstein's Relativity as a replacement, claiming that certain of its discoveries in the physical sphere can be transferred to the mind:

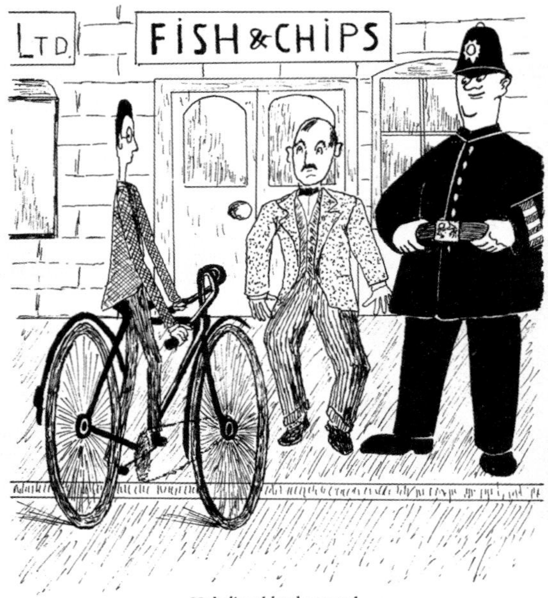

Unbelievably shortened

75 John Hookham, illustration for George Gamow, *Mr. Tompkins in Wonderland* (Cambridge, 1940).

> When Einstein notes the differences between the metre measurement and the metre made of a material which we ourselves must use, in doing so raising the possibility of a physical geometry, he achieves on a very special level of thought interferences the brutal rupture of psychological constants, creative in this case of great symbols and abstract, non-physical simulacra . . . Just as we should count upon the physical dilation of measurements whether common or not, we must count, in parallel or not, on the psychic dilation of ideas; there would no longer be grounds to call upon abstract criteria . . . The new geometry of poetic thought demands a physical revision and accommodation of the order of those to which Einsteinian physics subjects all measurements.[12]

Maintaining that the transformation of physical bodies observed by Relativity could be analogous with the extension and expansion of formerly fixed notions regarding reality, Dalí argues here that a simulacral 'new geometry of poetic thought' would refashion the objects of the imagination.

In this passage, Dalí is describing phenomena that are among the most peculiar features of Relativity physics: the *increase* in the mass of a body with its increase in velocity, and its concomitant *contraction* in length. The second is the Fitzgerald-Lorentz contraction, dictating that one observer moving relative to another at a velocity comparable to the speed of light will discover that the length of his partner has shrunk in the direction of motion.

76 Salvador Dalí, *Female Nude* (1928). Oil, sand, and gravel on panel, 64 × 76 cm. The Salvador Dalí Museum, St. Petersburg, Florida.

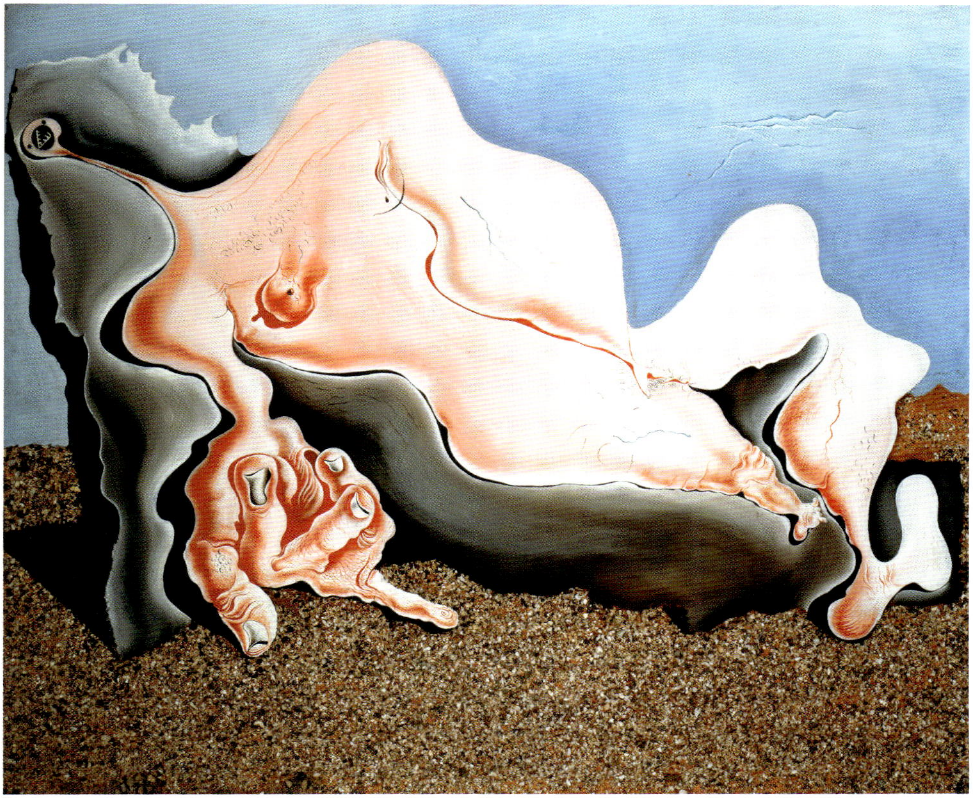

Can such notions garnered from Dalí's writings provide an interpretative key to the imagery of his painting? As we have observed, popular expositions of Relativity like Eddington's revel in their description of a strange new world. Although the changes to physical bodies recounted by the theory are invisible to the naked eye, of course, the temptation to exaggerate them by raising them to the visible human scale is irresistible. Such was the case in popular imagery too, where Relativity was depicted as an alternative terrain or world of its own. An example is the set of illustrations by John Hookham for the physicist George Gamow's popular writings of the 1930s and 1940s, which elucidates the exotic properties of time, space and matter outlined by Einstein's theory (fig. 75). Depicting a world absurdly subjected to the forces that Relativity describes, it amplifies them onto the Newtonian level of everyday experience.[13]

The phenomena described by Relativity – such as the constant variations taking place in the forms of physical bodies, on however small a scale – led Dalí to import and amplify the elastic, contorted world of Relativity into his early Surrealist canvases as soft, viscous, attenuated forms, sometimes suggestive of anamorphosis, as in his *Female Nude* and *Bather* both of 1928 (fig. 76). While his remarks on Relativity in 'The Sanitary Goat' are concerned more with the theory's implications for metaphysics than artistic imagery, the evidence offered by the paintings of the late 1920s reveals Dalí

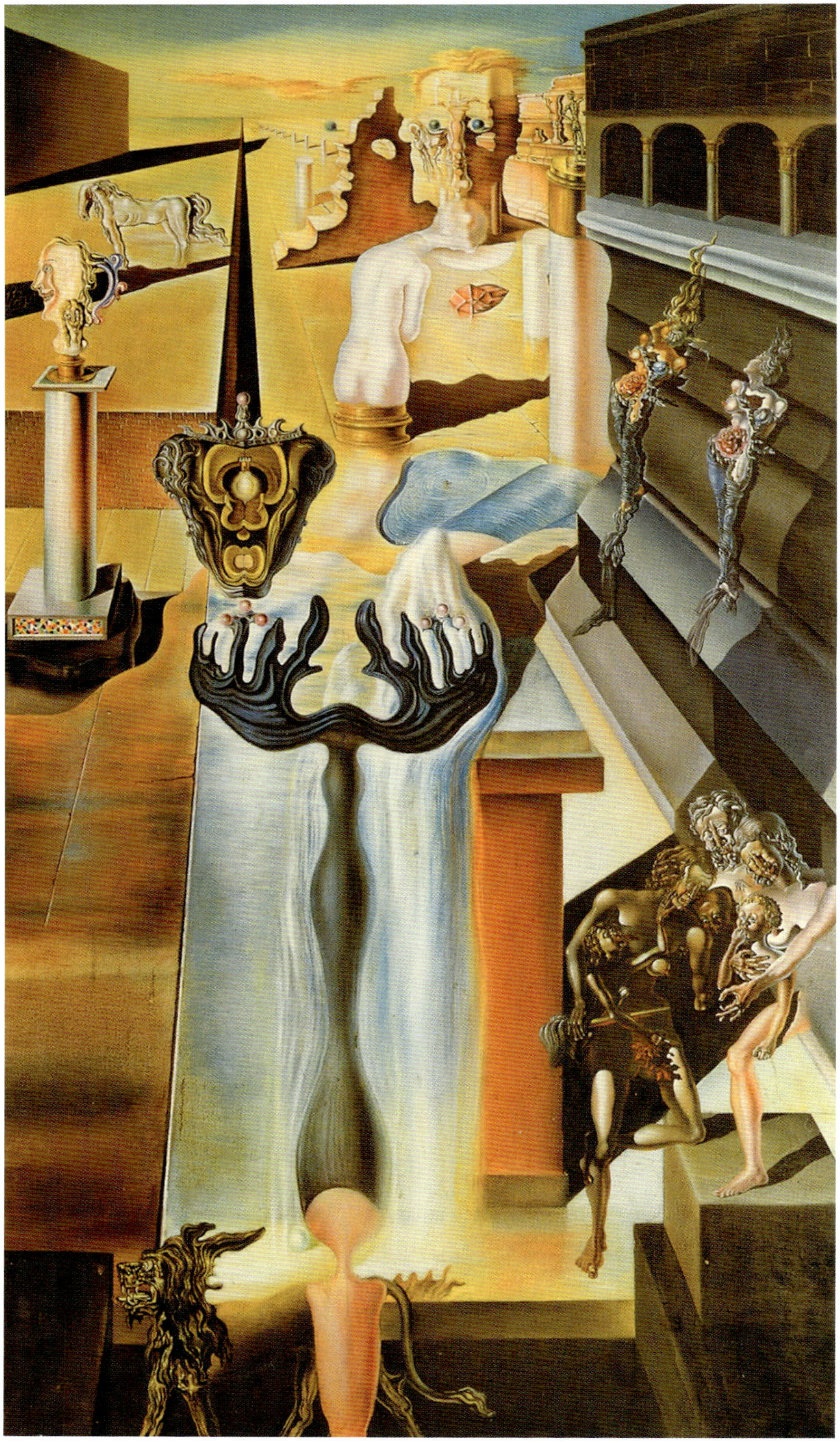

77 Salvador Dalí, *The Invisible Man* (1929–32). Oil on canvas, 140 × 81 cm. Museo Nacional Centro de Arte Reina Sofía, Madrid.

78 Salvador Dalí, *The Great Masturbator* (1929). Oil on canvas, 110 × 150 cm. Museo Nacional Centro de Arte Reina Sofía, Madrid, gift of Dalí to the Spanish state.

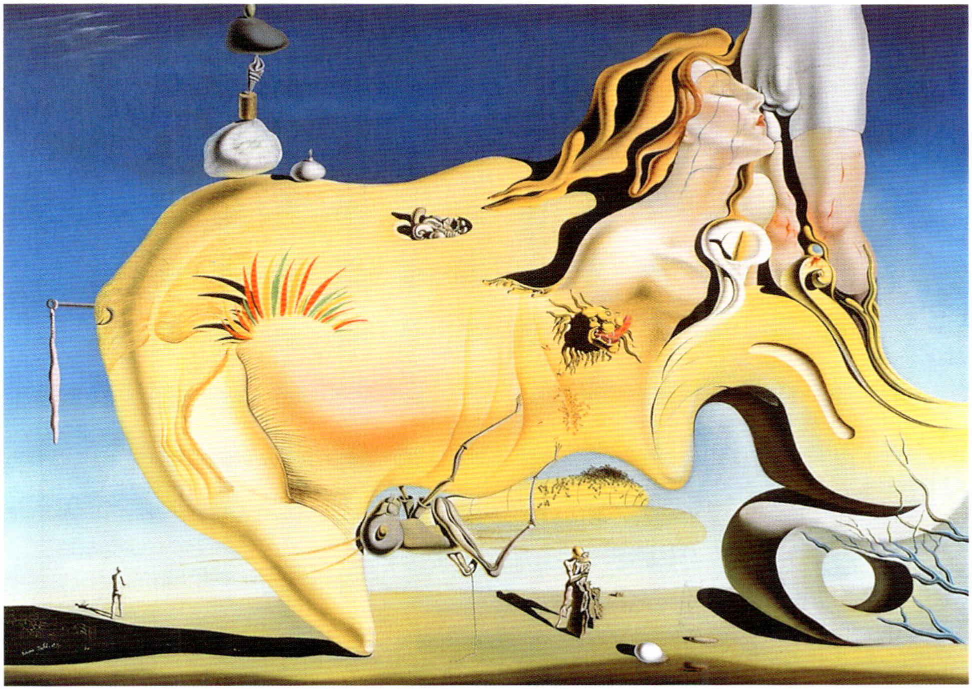

seeking a language in Relativity to depict the 'real' world discovered by Einstein, and introduced to an enchanted public by Eddington.

This is confirmed on closer inspection. Dalí goes on to say in 'The Sanitary Goat' that: 'Physics in this case is of the kind that one can believe in – that one can believe or interpret, even when it forms part of the delirium [dans le délire]: the physics which must create the new geometry of thought will be precisely . . . the delirium [délire] of paranoiac interpretation.'[14] His association of Relativity with paranoia here is highly suggestive. Just as the origin of Breton's theory of the poetic marvellous is found adjacent to Relativity in his 1921 text on Ernst, so, too, the theory and practice of Dalí's principle 'method' of paranoia-criticism – whereby two or more discrete images are combined (with variable success) into a single whole – dates from the time during which he was gaining inspiration from Einstein's theory.[15] Given this, it is possible to hypothesise a close link between paranoia-criticism and Dalí's interest in Relativity, which not only sheds light on the peculiar temporality of Dalí's paintings, but also his concept of four-dimensionality.

My interpretation rests on what I am assuming are a series of pictorial analogies developed by Dalí out of Relativity, a theory in which the term 'frame' or 'system of reference' refers to the unique pocket of four-dimensional space-time inhabited by every person, object and event. With paranoia-criticism's multiple imagery, Dalí could focus illusionistically on a point seized by the three spatial co-ordinates, through which events flood, creating a pictorial analogy of Relativity's four-dimensional architecture of space and time. In this way, he blended past and future events in a single place (frequently

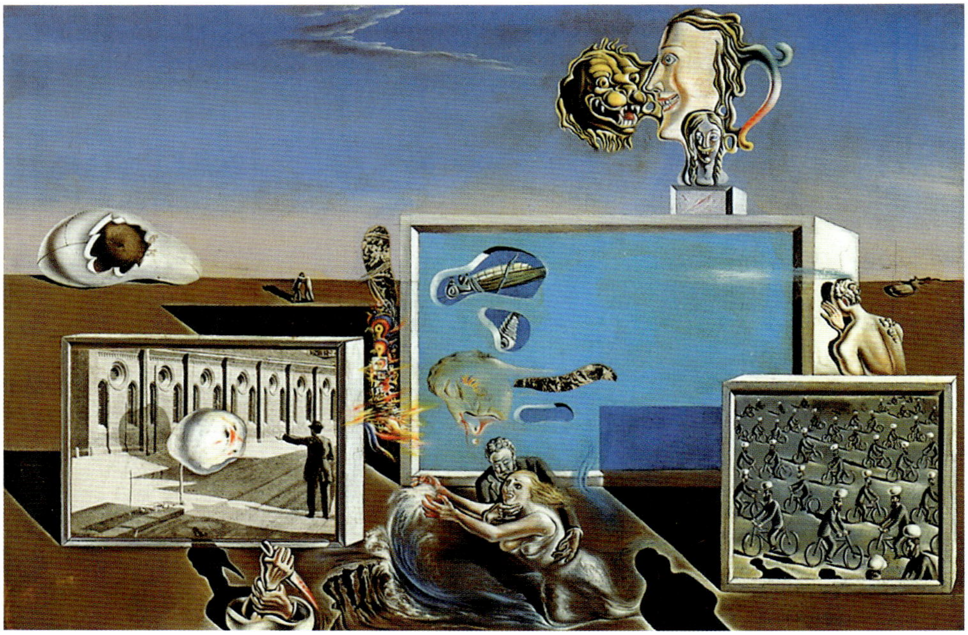

79 Salvador Dalí, *Illumined Pleasures* (1929). Oil and collage on board, 23.5 × 34.9 cm. The Museum of Modern Art, New York (The Sidney and Harriet Janis Collection, 1967).

the beach) to memorialise sites of nostalgia and speculation. I am proposing that this Einsteinian conceit assisted in the constitution of the multiple images of paranoia-criticism, from *The Invisible Man* of 1929–33 (fig. 77) to *Apparition of Face and Fruit Dish on a Beach* (1938), as well as Dalí's hybrid forms, as seen in his *Portrait of Paul Éluard* (1929) and *Great Masturbator* (fig. 78).

Dalí mined Relativity's hallucinatory potential in another way from 1929, using multiple frames within the same image, as in *The First Days of Spring* and *Illumined Pleasures* (fig. 79), to compartmentalise separate realities. Dalí's frames within frames are usually discussed within the context of Freud's 'atemporal' unconscious, yet once his paintings are viewed against the background of his broader interests of the period as expressed in his writings, they emerge as very 'literal' renderings of the metaphorically termed 'multiple frames of reference' posited by Relativity physics, and mentioned by Breton in his 1921 text on Ernst.[16]

One author has it that, in its detail, Relativity can 'acquire a certain "Sibyline strangeness" and surrealistic features.'[17] From the commencement of his mature work, Dalí was undoubtedly seduced by these traits.[18] However, I believe that, underlying Dalí's attraction to the metamorphic and juxtapositional possibilities that Relativity offered his art is the potential it creates for bringing a different kind of temporality into painting to the Bergsonian *durée*.[19] Like Ernst's, the more 'concrete', 'sculpted', spatialised time of Dalí's painting emerged from a conceptual language related to Einstein's space-time and consciously opposed to the 'smeared', 'blurred' temporality of the *durée*, associated then as now with 'painterly' avant-garde languages like those of Impressionism, Cubism and Futurism.[20] In his Surrealist period, Dalí's fourth dimension was always time, never the fourth dimension of *n*-dimensional geometry or its mystical and popular accomplices. And only the context of Relativity can explain his obscure remark a few years later in the pages of *Minotaure*, where he insisted upon Relativity's fourth, temporal dimension as a veritable prospect onto the surreal, calling it the 'delirious [délirante] and Surrealist dimension par excellence'.[21]

This typically Dalinian 'insight' appeared in *Minotaure* late in 1934 in the curious text 'Aerodynamic Apparitions of "Beings-Objects"'. A classic of Dalí's 1930s cartoon (anti-)intellectualism (apparently he was capable of making even Breton laugh), this can be read as a lampoon of the interdisciplinarity of periodicals such as the *Revue de Métaphysique et de Morale* and *Recherches pilosophiques*. In a preamble to a discussion that will ultimately centre on the experience of space and movement in the age of Relativity, Dalí insists luridly on the pleasure to be had from squeezing the pus from blackheads in one's nose. The inspiration for this was probably the passage in Freud's landmark paper of 1915, 'The Unconscious', in which a patient who has become withdrawn and depressed due to the poor condition of his skin expresses his pleasure at the 'spurting' of squeezed blackheads.[22] However, the patient also speaks of his regret at the subsequent appearance of the cavities he causes by, as Freud quotes him, '"constantly fiddling about with his hand"'.[23] 'Pressing out the content of the blackheads,' declares Freud, 'is clearly to him a substitute for masturbation', and interpreting the hole left in the skin as an analogy for the vagina, Freud concludes that it manifests the fantasy of the threat of castration brought about by masturbation.[24] Dalí's autoerotism and castration anxiety are well documented in his writings and paintings of the 1930s, and his exultant commendation in 'Aerodynamic Apparitions of "Beings-Objects"' of this 'intimate act', his intoxicated description of the slippery movement of the shaft of pus, and his recounting of a 'veritable "neurotic ceremonial"' duplicate the sexual neurosis of Freud's obsessive-compulsive patient.[25]

Surprisingly, Dalí goes on to say that his remorseless account of squeezing blackheads introduces a metaphor for all that is 'most vital and lyrical in contemporary moral, scientific, and artistic thought'.[26] This is the changing concept of space, which in Dalí's hands becomes a 'nutritive history' of the evolution of the notion, calling on Freud's 'oral tactility' from the text 'Infantile Sexuality', the second of the *Three Essays on the Theory of Sexuality* (1905).[27] Again, it is established in the scholarship that Dalí's work leans heavily on Freud's assertion that at the infant stage sexual desire and eating are not yet distinct but mutually identifiable, having 'the *incorporation* of the object' as their aim.[28] The iconography of cannibalism and narcissism of his paintings of the 1930s springs from his knowledge of Freud's itinerary of infantile 'perversion', particularly the oral and anal phases of sexual organisation.

Somewhat unexpectedly again (to say the least), Dalí's interpretation of the pregenital oral and sadistic-anal stages in 'Aerodynamic Apparitions of "Beings-Objects"' is refracted through his reading of Relativity physics (fig. 80). The original psychoanalytic data is used to express, in terms of the texture of food, the 'thickening' of space

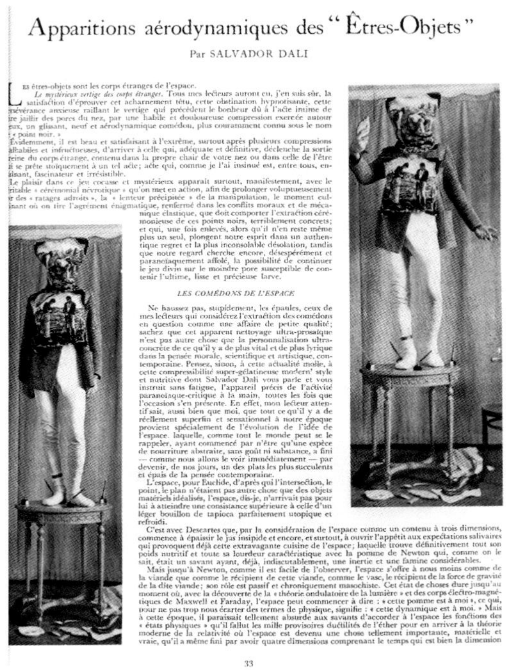

Apparitions aérodynamiques des "Êtres-Objets"

Par SALVADOR DALI

Les êtres-objets sont les corps étranges de l'espace.

Le mystérieux vertige des corps étranges. Tous mes lecteurs auront eu, j'en suis sûr, la satisfaction d'éprouver cet acharnement têtu, cette obstination hypnotisante, cette [...]rsévérance anxieuse raillant le vertige qui précèdent le bonheur dû à l'acte intime de [...]ire jaillir des pores du nez, par une habile et douloureuse compression exercée autour [...]eux, un glissant, neuf et aérodynamique comédon, plus couramment connu sous le nom [...] « point noir. »

Évidemment, il est beau et satisfaisant à l'extrême, surtout après plusieurs compressions [...]habiles et infructueuses, d'arriver à celle qui, adéquate et définitive, déclenche la sortie [...]eine du corps étrange, contenu dans la propre chair de votre nez ou dans celle de l'être [...]i se prête stoïquement à un tel acte; acte qui, comme je l'ai insinué est, entre tous, en[...]înant, fascinateur et irrésistible.

Le plaisir dans ce jeu cocasse et mystérieux apparait surtout, manifestement, avec le [...]ritable « cérémonial névrotique » qu'on met en action, afin de prolonger voluptueusement [...]r des « ratages adroits », la « lenteur précipitée » de la manipulation, le moment cul[...]inant où on tire l'agrément énigmatique, renfermé dans les conflits moraux et de mécanique élastique, que doit comporter l'extraction cérémonieuse de ces points noirs, terriblement concrets; et qui, une fois enlevés, alors qu'il n'en reste même plus un seul, plongent notre esprit dans un authentique regret et la plus inconsolable désolation, tandis que notre regard cherche encore, désespérément et paranoïaquement affolé, la possibilité de continuer le jeu divin sur le moindre pore susceptible de contenir l'ultime, lisse et précieuse larve.

LES COMÉDONS DE L'ESPACE

Ne haussez pas, stupidement, les épaules, ceux de mes lecteurs qui considérez l'extraction des comédons en question comme une affaire de petite qualité; sachez que cet apparent nettoyage ultra-prosaïque n'est pas autre chose que la personnalisation ultra-concrète de ce qu'il y a de plus vital et de plus lyrique dans la pensée morale, scientifique et artistique, contemporaine. Pensez, sinon, à cette actualité molle, à cette compressibilité super-gélatineuse modern' style et nutritive dont Salvador Dali vous parle et vous instruit sans fatigue, l'appareil précis de l'activité paranoïaque-critique à la main, toutes les fois que l'occasion s'en présente. En effet, mon lecteur attentif sait, aussi bien que moi, que tout ce qu'il y a de réellement superfin et sensationnel à notre époque provient spécialement de l'évolution de l'idée de l'espace. laquelle, comme tout le monde peut se le rappeler, ayant commencé par n'être qu'une espèce de nourriture abstraite, sans goût ni substance, a fini — comme nous allons le voir immédiatement — par devenir, de nos jours, un des plats les plus succulents et épais de la pensée contemporaine.

L'espace, pour Euclide, d'après qui l'intersection, le point, le plan n'étaient pas autre chose que des objets matériels idéalisés, l'espace, dis-je, n'arrivait pas pour lui à atteindre une consistance supérieure à celle d'un léger bouillon de tapioca parfaitement utopique et refroidi.

C'est avec Descartes que, par la considération de l'espace comme un contenu à trois dimensions, commence à épaissir le jus insipide et encore, et surtout, à ouvrir l'appétit aux expectations salivaires qui provoquent déjà cette extravagante cuisine de l'espace; laquelle trouve définitivement tout son poids nutritif et toute sa lourdeur caractéristique avec la pomme de Newton qui, comme on le sait, était un savant ayant, déjà, indiscutablement, une inertie et une famine considérables.

Mais jusqu'à Newton, comme il est facile de l'observer, l'espace s'offre à nous moins comme de la viande que comme le récipient de cette viande, comme le vase, le récipient de la force de gravité de la dite viande; son rôle est passif et chroniquement masochiste. Cet état de choses dure jusqu'au moment où, avec la découverte de la « théorie ondulatoire de la lumière » et des corps électro-magnétiques de Maxwell et Faraday, l'espace peut commencer à dire : « cette pomme est à moi », ce qui, pour ne pas trop nous écarter des termes de physique, signifie : « cette dynamique est à moi. » Mais à cette époque, il paraissait tellement absurde aux savants d'accorder à l'espace les fonctions des « états physiques » qu'il fallut les mille provisoires ductilités de l'éther pour en arriver à la théorie moderne de la relativité où l'espace est devenu une chose tellement importante, matérielle et vraie, qu'il a même fini par avoir quatre dimensions comprenant le temps qui est bien la dimension

33

80 Salvador Dalí, 'Apparitions aérodynamique des "Être-Objets"', *Minotaure*, no. 6, Winter 1934–5, 33–34.

he says took place over the period spanning Euclid and Einstein. According to Dalí, with the advance of classical physics, space underwent a change, becoming 'active' following its previous 'passivity'. With Einstein's Relativity, space took on a physical state, 'material and real',[29] having 'four dimensions, including time which is certainly the delirious [délirant] and Surrealist dimension par excellence'.[30] Dalí adds that 'modern' non-Euclidean space is, like meat/flesh, 'bloody and perfectly viscous', having a 'physicality' that encloses human beings and cars in individual pockets out of which they are squeezed in the same way that blackheads, he maintains, slip out the flesh of the nose.[31] The contemporary *époque de mou*, he declares, apparent in the 'soft' morphology of modern architecture, furniture and cars, was a product of the same fashionable tendency that heralded Einstein's Relativity and the soft structures of Salvador Dalí.[32]

Working with an obvious knowledge of the languages and concepts of Relativity and psychoanalysis, 'Aerodynamic Apparitions of "Beings-Objects"' also hilariously satirises their authority and fashionable nature in the services of Surrealist iconoclasm. Its scatological humour and modernist fantasies mark a pungent contrast with academic scholarship or even the writings of Georges Bataille, who, like most of the Surrealists, always took humour too seriously. Yet Dalí's intelligence precluded regression to Dada's puerile anti-intellectualism. His writings and paintings building upon and combining ideas taken from Relativity and psychoanalysis are peculiarly aligned with Surrealism, because they are fastened to the moment that Breton, already steeped in the work of Freud, factored Relativity into the foundations of Surrealist theory in the early 1920s. Indeed, Dalí might have had Breton's 'Max Ernst' in mind when he wrote in 1933 that Surrealism, 'from its beginnings, rejected mechanical materialism, confining itself to a wholly provisional relativistic idealism'.[33]

However, the accent placed upon psychoanalysis in the literature on Dalí has obscured until now the metaphorical and satirical investment in Relativity that his work made. This is not surprising, for both the 'divided self' of psychoanalysis and Relativity's multiple observers occupying separate frames of reference issued from, shared in, and contributed to a 'post-humanist' and yet subjectivist discourse that emphasised rhetorical tropes of fragmentation and decentralisation. In the wake of Dada, Surrealism was primed at the outset by the two dominant intellectual fashions of the day to co-ordinate an iconographic field of metamorphosis, juxtaposition and decontextualisation, which Dalí pressed into parodic service in 'Aerodynamic Apparitions of "Beings-Objects"' under its proper double rubric of psychoanalysis and Relativity.

The Use-Value of Modern Physics: Roger Caillois and Dalí

Emulating Breton's own harnessing of psychoanalysis and Relativity in his early theory of poetics, Dalí's art and writings lie safely within the boundary drawn by Breton's ongoing line of enquiry delimiting Surrealist activity. This was not the case for the contemporary writings of Roger Caillois, who would desert Surrealism in response to its lax attitude towards empirical verification. Subsequently, Caillois would use the question of scientific rigour to mark his distance from the group, taking Dalí's absurd manipulation of scientific concepts and metaphors, particularly, as an antithetical point of reference.

From the beginning, Caillois's position within Surrealism was fairly ambiguous, and where Dalí's sporadic love affair with the group lasted for ten years and spanned the 1930s, his sojourn with the Surrealists was brief. He joined them in April 1932 at the age of only nineteen, expecting to come into contact with serious experimental procedures, contributing in May 1933 his first sceptical, anti-literature (and already virtually anti-Surrealist) article 'Spécification de la poésie' to the fifth number of *Le Surréalisme au service de la Révolution* (SASDLR).[34] In February 1934, he was perhaps the most vocal among those Surrealists lobbying for Dalí's expulsion from the group for alleged 'Hitlerism'.[35] In May, his 'Systematization and Determination' was published in the '*Intervention surréaliste*' number of the Belgian *Documents 34*, calling on Émile Meyerson and the philosophy of mathematics with the idea of giving poetry an experimental role, which ends up sounding excruciatingly functional.[36] Breton duly complained about the article because, as Caillois remembered his remarks, it 'revealed an antilyrical position'.[37]

The future differences between the two were forecast in Breton's censure, coming fully out into the open in the course of the silly yet significant Mexican jumping beans incident of December 1934, well rehearsed in the literature on Surrealism. Witnessing their mysterious behaviour with Breton and Jacques Lacan (jumping beans were rare in Paris at the time), Caillois innocently suggested splitting them open to reveal the secret of their motion, leading to an average-sized tantrum from Breton who, Caillois claimed, wanted their riddle to remain intact.[38] Giving notice of his departure from Surrealism in a slightly pompous letter to Breton of 27 December 1934, he situates their disagreement between the two poles of 'research and poetry', expressing disappointment at Surrealism's preservation of 'poetic' mystery for its own sake, when, in his view, research into the irrational required a rigorous 'scientific' methodology.[39] Implying that Surrealism was tied to outmoded tenets and insisting that 'the point of contention is chiefly methodological',[40] Caillois declares that the *merveilleux*, so dear to Breton, no longer belongs to poetry but to new adventures in physics, which blazed a novel, radical epistemological trail:

For instance, modern atomic theory is at present an adventure into the dark: somewhat like children raised in boxes, who are amazed to discover ferns (in the words of a physicist). This does not involve the *distress* or *jouissance* brought about by a *beautiful picture* but instead involves a sense of utter confusion; utter confusion in the face of what I choose to call the *debacle of the evident*. For there is nothing left of the old intuitions, and any philosophy that cannot fit together with this new science of the *why not* is absurd.[41]

Caillois does not mention the source for his 'why not', but perhaps he thought Breton knew where he was heading, for he is quoting directly from Bachelard's recently published *New Scientific Spirit* in this passage.[42] Begun here, Caillois's search for a more stable methodology than he felt Surrealism could offer would lead him to investigate modern physics, Bachelard's epistemology and the related thought of the Vienna Circle.

Caillois's disillusioned report to Breton, insisting upon the value of an outlook on knowledge grounded in scientific (ir)rationalism, bears useful comparison with another letter composed at the same time. Confirming that different expectations were being voiced at that moment concerning the critical role that Surrealism should play and the place its theory and practice should assign modern physics, this was sent by Dalí to Paul Éluard from the US. Dalí now believed it imperative that Surrealism take cognisance of the physics of Relativity if it were to continue to advance:

I think more and more seriously that Surrealism has to evolve towards the foundation of a new materialist, anti-mystical religion, based on the progress of scientific knowledge (especially the new ideas of space, which were inaccessible not only to the Greeks but also to Christianity) a religion which would occupy the emotional imaginative void that the collapse of metaphysical ideas has produced in our era. I have some very concrete ideas on this subject.[43]

But Dalí's thoughts on the subject were not 'concrete' in the sense of the 'objective certitude' demanded by Caillois, whose sober defence of experimental rigour begins from here to look strikingly like an alternative position to Dalí's, precluding any 'delirious' talk of a Surrealist religion. This distance between Caillois and Dalí sets out the ground over which the debate on the value of modern physics to Surrealism, along with the question of the methodological regulation of Surrealism itself, was played out in the mid-1930s.[44] Their respective publications in 1935 demonstrate these divergences. Appearing within a few weeks of each other, Caillois's *Procès intellectuel de l'art* and Dalí's *Conquest of the Irrational* both draw upon elements of modern physics, turning them to radically divergent ends (figs 80 and 81).

Method and Madness: Procès intellectuel de l'art *and* Conquest of the Irrational

I looked first at Caillois's 1934 letter to Breton because it constitutes a kind of manifesto statement, established by its place at the front of his short book of 1935 in which

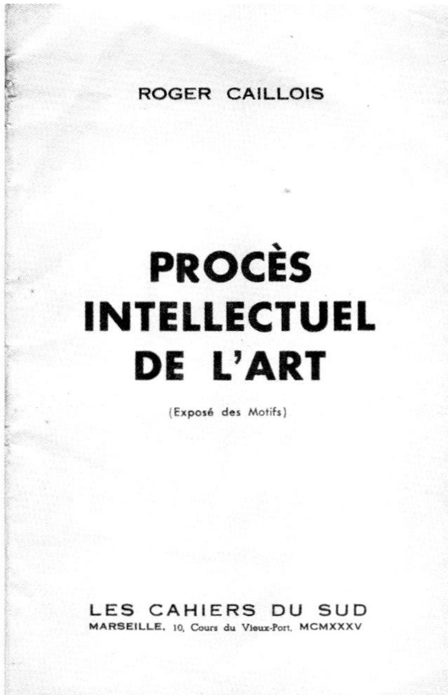

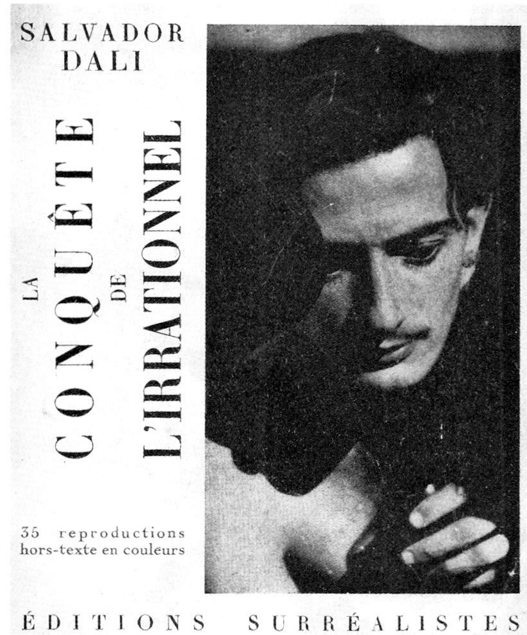

81 (left)
Roger Caillois,
Procés intellectuel de l'art (1935).

82 (right)
Salvador Dalí,
Conquest of the Irrational (1935).

he declares his independence from Surrealism, *Procès intellectuel de l'art* (*Intellectual Trial of Art*).[45] It is followed by three texts, on philosophy, art and literature, which I examine briefly here to give a flavour of Caillois's evaluation of the interpretive purchase of modern physics. On closer inspection, the book itself has something of a speculative manifesto-type 'aims and objectives' feel, sketching rather than fully detailing the means by which scientific rigour could be applied in the humanities.

In 'Décision préliminaire sur la métaphysique' ('Preliminary Decision on Metaphysics'), Caillois discusses the difficulty of posing the 'metaphysical problem', which, we saw, exercised many materialists at the time who were critical of Heidegger's influence.[46] Transposing a physical model onto the mental sphere, he argues that the failure of the Michelson–Morley experiment can be used to illustrate the impossibility, without fixed points of reference, of reaching reliable conclusions about the activity of the mind, declaring 'one can neither from the earth discover the movement of the earth, nor from consciousness the movement of consciousness'.[47] Apparently drawing on a further model, this one deployed by Einstein to illustrate the idea of a finite yet unbounded universe, Caillois compares humankind's embeddedness in, and therefore necessarily partial knowledge of its own consciousness, to 'the situation of curved, infinitely flat beings living on the surface of a gigantic sphere', who can conceive their world as finite, but who experience it, in their two-dimensional state, as infinite.[48] Concluding, Caillois's loyalties in his search for 'simple rigour'[49] outside the Surrealist group are spelled out in no uncertain terms: the 'concrete complexity' he prized 'could be arrived at only by the thankless patience of the scientist and not by the intuitions of the artist'.[50]

Caillois substantiates this view in the next essay of *Procès intellectuel de l'art*, 'Notice sur l'impureté dans l'art' ('Notes on Impurity in Art'), turning more specifically to the epistemology of science after making a distinction between 'two antagonistic elements' in art.[51] Broadly speaking, these are the tradition extending from Cézanne to Picasso, on the one hand, which Caillois regards as predominantly 'harmonic', and, on the other, that including Gustave Moreau and (significantly) Dalí, which Caillois claims is chiefly 'lyric' – a designation ironically echoing Breton's criticism of his own 'antilyricism.'[52] Where the former is objective, being constituted mainly of '*pure* elements', the latter is subjective, composed primarily of '*impure*' factors.[53] Again, Caillois seeks guidance from science, claiming that the appeal of 'pure' elements in art came not from any notion of transcendence but from their communication of the underlying (mathematical) order of the universe, to which humankind was also subject.[54]

Although Dalí is not singled out for any particularly rough treatment, he is obviously the target for Caillois's equivocal appraisal of the 'delirious [délirante] fantasising' of 'impure art', and Caillois's censure of the general aspects of that category call forth some of the main features of contemporary fashions in Surrealist painting.[55] When Caillois claims in a footnote that 'the results of the reflection on science (epistemology) are singularly more fertile and more solid than those of the reflection on art', he names Bachelard, Émile Meyerson, Hans Reichenbach, Rudolph Carnap and Philipp Frank as those whose work evinced 'new modes of thought demanded by the development of physical science'.[56] Signalling his disapproval of Dalí's exotic, 'impure' morphologies, Caillois also seeks to define the (rigorous) terms by which the methods and findings of the new physics could be put in play outside of their proper field.[57]

So far, so vague; which is also true of Caillois's hope, expressed in the final text of *Procès*, 'Crise de la littérature' ('Literature in Crisis'), of 'applying scientific method to the artistic domain, for scientific rigor here seems, in our opinion, more essential than elsewhere'.[58] Bemoaning the loss to literature of a younger generation more interested in politics and philosophy, Caillois projects onto his contemporaries his own dissatisfaction with woolly thinking within the humanities and his taste for the structure, stability and precision of science. While conceding 'it is nonsense to try to base psychological freedom on the indeterminacy relations of the subatomic world', Caillois entertains the possibility of mobilising the 'epistemological constructions required by the problems of contemporary scientific methodology', for an experimental, objective, systematic study of the mind.[59]

Like the rest of the book, 'Literature in Crisis' sounds provisional and preachy. Yet *Procès intellectuel de l'art* works well on the whole as youthful statement of intent, promising to deal with issues close to Surrealism from a positivistic experimental base. For a few years, Caillois continued to endorse the *logique nouvelle* of Bachelard and the Vienna Circle, avoiding the imprecise adoption of ideas and experiments from science simply as poetic ornament devoid of their theoretic weight, while seeming rather vague all the same about the exact nature of the new method he was advocating.

Bachelard's philosophy of science seems to have entered into Dalí's writings of 1935, too, when the 'very concrete ideas' he mentioned to Éluard, concerning an order of space unknown to previous times adding up to a new post-classical, post-Christian epistemology, found a wider audience in his short volume *Conquest of the Irrational*.[60] The book comprises thirty-three plates preceded by four brief texts, of which three continue Dalí's reflections on modern physics. Consistent with my earlier discussion of Dalí's writings since the inception of paranoia-criticism, they aim to unite physics and psychoanalysis/psychology in a single theory. Using the same alogical, broken narrative and 'comestible' imagery prevalent in 'Aerodynamic Apparitions of "Beings-Objects"', *Conquest of the Irrational* appeared only weeks after *Procès intellectuel de l'art*. When placed in dissonant relation with Caillois's plea for engagement with the new physics and its philosophy, Dalí's performative revision of physics to an irrational mode in *Conquest* brings into sharper focus the terms of the debate concerning the methodological direction that the Surrealist intervention should take.

Dalí's *Conquest of the Irrational* is a further acknowledgement of the artist's conviction that the description of nature offered by Relativity, confounding common sense predictions of the constitution of time, space and matter, could be mobilised in support of Dalí's project to engender systematic confusion in the 'objective' world, and cast doubt on rational expectation as it had been tutored by the senses. Accordingly, the first essay, 'The Waters in Which We Swim', praises advances made by science in 'the overwhelming repudiation of "logical intuition"'.[61] Dalí goes on to declare that modern science is the offspring of this logical intuition and Newtonian physics. On 'a cultural dining table' he finds:

> the cold, insubstantial leftovers of art and literature, and . . . the burning analytic precisions of the particular sciences, impervious, for the moment, to a nutritive synthesis owing to their disproportionate extension and specialisation and, in any case, totally inassimilable, apart from speculative cannibalisms.[62]

Dalí presents the Surrealists as a group of intellectual aristocrats charged with the task of bringing art and science together, yet they are as alien to those currents as Breton's 'soluble fish': 'for we, Surrealists . . . we are not exactly artists, any more than we are exactly true men of science', rather, 'we are carnivorous fish, which, as I have already implied, swim between two waters, the cold water of art, and the warm water of science'.[63] Though the means to accomplish this task are, characteristically, left obscure, neither the tone nor the intention of Dalí's book is programmatic like that of Caillois. Rather, the aim of remaking the world in the image of the irrational is carried in Dalí's metaphors and style – the disconnected thinking and writing performance itself – as previously counselled by Aragon in *Treatise on Style*.

Another section of *Conquest*, 'Abjection and Misery of Abstraction-Creation', supplements an attack on the 'model of mental debility called abstract art' and its 'neo-Kantian' progenitors with a eulogistic poem celebrating the art of Picasso.[64] There is clear evidence here that Dalí used the same sources as Caillois, for his criticism of the group Abstraction-Creation centres on what he saw as abstract art's outmoded and restrictive adoption of the Kantian notion that Euclidean geometry and Newtonian time and space are *a priori* particulars circumscribing metaphysical speculation, the objects of the imagination and rational behaviour.[65] Dalí's aversion to abstraction and admiration for Cubism, then, along with his attraction to the 'surreal' non-Euclidean geometry and fourth, temporal dimension of Relativity, are all rooted in the same revision of the realism and rationalism of Descartes, Newton and Kant, forced by the discoveries of modern physics, and characterised by Bachelard as an 'epistemological rupture.'[66] Wedded to the same sources or not, Caillois's steady pragmatism and the accent he placed on clarity and careful evaluation are a million miles away from Dalí's hyperbole and confusive rhetoric.

Typical, in this regard, is Dalí's aberrant assessment of Picasso's Cubism, which, he says, surpassed 'the experimental ambitions/of modern/mathematical physics'.[67] In a similar hyperbolic, rhetorical vein, he predicts a future philosophy of paranoia-criticism:

> there does not exist/these days/as far as I know/a physics/of psychopathology/a physics of paranoia/which could be considered/only/as/the experimental basis/of the imminent/philosophy/of/psychopathology/of the imminent/philosophy of 'paranoiac-critical' activity/which one day/I shall attempt to envisage polemically/if I have the time for it/and I am in the mood.[68]

This extract is Dalí's fullest acknowledgment that what he seeks to formulate is an amalgam of psychoanalysis and physics, while its irresistible, bathetic coda punctures any sense of portentousness such a claim might carry. While undoubtedly lacking in *rigueur*, Dalí's unique perspective on science is nevertheless finely calibrated: avoiding a refutation of scientific method, he allows a measure of the seriousness of scientific endeavour to bolster his claims for the importance of paranoia-criticism.

The final essay of *Conquest of the Irrational*, 'The Tears of Heraclitus', is a kind of coalition of the imagery of 'Aerodynamic Apparitions of "Beings-Objects"' and the concepts introduced in his letter to Éluard. Giving another abbreviated historiography of the concept of space, Dalí places the emphasis on its effect on contemporary artists. Further expanding the Bachelardian ideas imparted to Éluard, Dalí states that as the art of the Greeks was subject to their understanding of space as Euclid described it, he should have recourse to the modern concept of space, 'this harrowing and colossal question of Einsteinian space-time'.[69] Dalí's characterisation in the same book of the soft watches of his 1931 *Persistence of Memory* (fig. 83) as 'soft, extravagant, and solitary paranoiac-critical Camembert of time and space' insists again upon the context of Relativity that I have developed here.[70] More than some psychoanalytic concept, against the background of physics the painting evokes Einstein's notion of 'time dilation'; that is, the relative variation, depending upon position in space, of the rate of temporal flow – the function within Relativity of the fourth dimension, the 'delirious and Surrealist dimension par excellence'.[71]

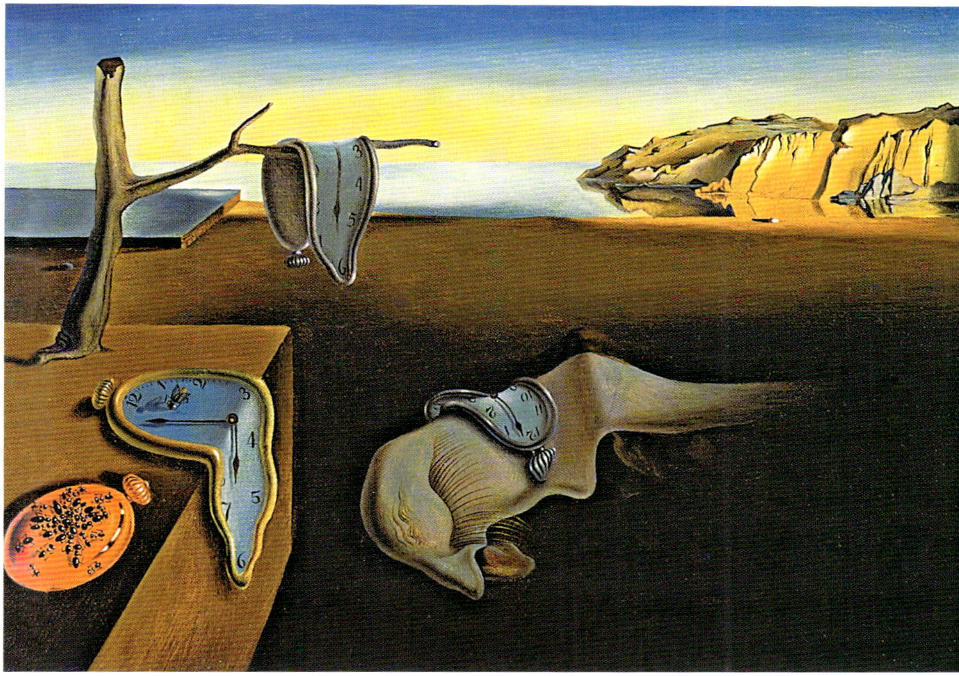

83 Salvador Dalí, *The Persistence of Memory* (1931). Oil on canvas, 24.1 × 33 cm. The Museum of Modern Art, New York.

Dalí pursued his off-beat accommodation of Relativity in his work up to the end of the 1930s, while Caillois continued to discuss the new physics in a manner implicitly critical of the more 'poetic' or 'metaphorical' interpretations of Dalí and Breton, rooted in 'Max Ernst'.[72] The original cultural context of Surrealism created the conditions whereby Caillois's writings from within its confines were ill-suited to the (poetic) aims of the movement, but also to those by which Dalí's excesses are allowed free reign. Neither a poet nor a scientist, Caillois argued that poetry and science investigate the world by means of entirely different methods – the speculative poetical and the empirical experimental – which should not be allowed to converge. Where he (rhetorically, at least) chose the sober certitudes of science over the ambiguities and shifting identities of poetry, Dalí and the Surrealists continued to swim between the waters of art and science. This uncertainty surrounding the nature of Surrealism was a provisional footing that Breton and Dalí were keen to maintain and develop, whereas Caillois saw merely 'art' and 'literature' in Surrealism.

When I looked at his writings in the political context, I argued that Caillois was sometimes torn between a disparagement of the 'abuses' of Relativity and QM in line with the Vienna Circle, and a 'poetic' treatment of aspects of those theories bordering on Breton's, presumably due to his residual Surrealism. Yet the *Inquisitions* article I looked at, 'For a Militant Orthodoxy', can also be viewed as a clarification of the earlier writings of *Procès intellectuel de l'art*. Arguing from a position that gives priority to science over poetry in the service of order, Caillois's version of Relativity exceeds, explains and rationalises poetry's 'irrationalism', organising it through a physico-mathematical philosophy designed to rein in some of the excesses of the Surrealists.

My discussion so far has shown this to be unequivocally antithetical to Dalí's 'conquest of the irrational', which seeks a systematic dissemination of confusion through a creative modification of the more uncanny aspects of the exact sciences towards chaotic ends. However, Caillois and Dalí cannot be positioned simply as polar opposites within Surrealism; both underwent theoretical dialogue with the early Lacan, and elements of Dalí's paranoia-criticism are comparable with Caillois's characterisation of the praying mantis as an 'ideogram'.[73] Yet their differences, which can seem subtle and significant, are, I feel, exposed best by the context of physics (never considered, by the way, in the endless scholarship that has accrued to Caillois's reflections on the praying mantis). Caillois's opposition, for instance, of the *objectivity* of the ideogram to the *subjectivity* and arbitrariness of Dalí's paranoia-criticism is closely related to his advocacy of a rigorous method derived from physics that could expand the boundaries of the real through the exercise of the imagination. Dalí, on the other hand, was more intent on appropriating Relativity pictorially or conceptually for delirium. In this way, Dalí's take on Relativity was close to the Rimbaldian 'intellectual hedonism' criticised by Caillois in 'For a Militant Orthodoxy' (demonstrating further that the dispute over the interpretation of physics within Surrealism had implications for Rimbaud's legacy).

For fairly explicit evidence that Caillois was rankled by Dalí's consistently immoderate art and writings concocted upon a very free interpretation of Relativity, we can look at his more pointed criticism of the Surrealists' 'abuses' of modern physics in the important essay 'L'alternative (*Naturphilosophie* ou *Wissenschaftlehre*)'. Written for a special 1937 issue of *Cahiers du Sud* dedicated to German Romanticism, Caillois's article ponders Romanticism's 'unhealthy' engagement with science at the end of the eighteenth century (with Novalis, an important precursor of Surrealism, singled out as chief culprit; fig. 84).[74] Caillois's criticism of the Romantics centres on their 'inappropriate' use of science as a pathway towards poetic reverie, for it was not as 'a method of investigation' that they approached the natural sciences but as 'solely a system of aesthetic relationships if not a reservoir of expressions'.[75]

Caillois's essay was probably intended as a partial response to a 1933 assessment by Breton of the German Romantics' fascination with physics.[76] Written as an introduction to a new French translation of the then-obscure Surrealist favourite Achim von Arnim's

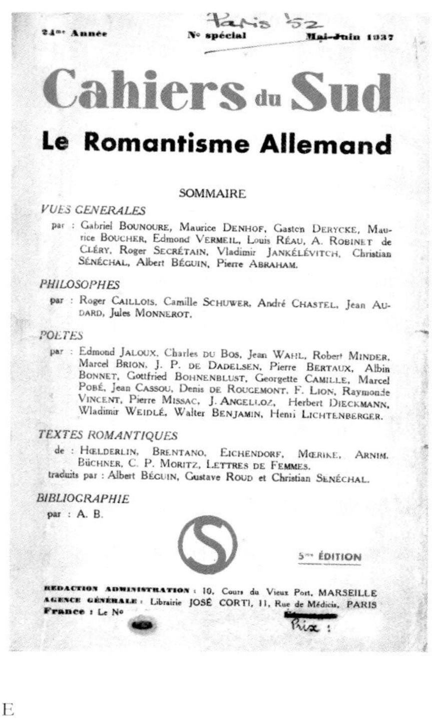

24me Année — No spécial — Mai-Juin 1937

Paris '52

Cahiers du Sud

Le Romantisme Allemand

SOMMAIRE

VUES GENERALES

par : Gabriel BOUNOURE, Maurice DENHOF, Gaston DERYCKE, Maurice BOUCHER, Edmond VERMEIL, Louis RÉAU, A. ROBINET de CLÉRY, Roger SECRÉTAIN, Vladimir JANKÉLÉVITCH, Christian SÉNÉCHAL, Albert BÉGUIN, Pierre ABRAHAM.

PHILOSOPHES

par : Roger CAILLOIS, Camille SCHUWER, André CHASTEL, Jean AUDARD, Jules MONNEROT.

POETES

par : Edmond JALOUX, Charles DU BOS, Jean WAHL, Robert MINDER, Marcel BRION, J. P. DE DADELSEN, Pierre BERTAUX, Albin BONNET, Gottfried BOHNENBLUST, Georgette CAMILLE, Marcel POBÉ, Jean CASSOU, Denis DE ROUGEMONT, F. LION, Raymonde VINCENT, Pierre MISSAC, J. ANGELLOZ, Herbert DIECKMANN, Wladimir WEIDLÉ, Walter BENJAMIN, Henri LICHTENBERGER.

TEXTES ROMANTIQUES

de : HŒLDERLIN, BRENTANO, EICHENDORF, MŒRIKE, ARNIM, BÜCHNER, C. P. MORITZ, LETTRES DE FEMMES.
traduits par : Albert BÉGUIN, Gustave ROUD et Christian SENÉCHAL.

BIBLIOGRAPHIE

par : A. B.

5me ÉDITION

RÉDACTION ADMINISTRATION : 10, Cours du Vieux Port, MARSEILLE
AGENCE GÉNÉRALE : Librairie JOSÉ CORTI, 11, Rue de Médicis, PARIS
France : Le No
Prix :

84 *Cahiers du Sud*, Year 24, May–June 1937.

Strange Tales, Breton's piece is consequently free of his usual polemicism. In its place, it combines a 'scholarly' style (replete with footnotes) with anecdotal content and a melodramatic romantic tone, as though its author had lately overindulged in midnight readings of Gothic novels. Within the historical preamble are Breton's observations of Arnim's investigations in physics, noting the 'experimental' as opposed to 'speculative' course taken by his scientific methodology.[77] This brought Arnim into contact with Johann-Wilhelm Ritter, Breton tells us, whose Tourette's-like improvisation wins him the compliment of '"Surrealist" before the fact',[78] just as Breton scooped up Arnim for Surrealism the following year, as a kind of Einstein before the fact, for being 'Surrealist absolutely, in space and time'.[79]

Close in theme to his 1934 letter to Breton complaining about the still unreconciled division in Surrealist activity between poetry and research, Caillois's far more sceptical take on German Romanticism is a development of that position. It makes much of what Caillois supposes to be the salon dilettantism of the *Naturphilosophie* of Schelling and Schlegel, observing the 'abyss' separating it from the hard-edged positivism of Carnap and Reichenbach's *Wissenschaftlehre*, which he had grown to admire by reading *Erkenntnis*.[80] By contrast, Breton places Arnim and Ritter on the experimental wing of German Romanticism without condemning the poetic speculation of Novalis, Schelling and the Schlegels. Asserting that 'a striking coalition . . . groups together poets, artists, and scientists', he already paves the way for the 'Bachelardism' of 'Crisis of the Object'.[81]

A great admirer of German Romanticism, Breton definitely read this special number of *Cahiers du Sud*, and would certainly have paid particular attention to Caillois's grumpy dismissal of *Naturphilosophie*, especially since Caillois appears to pick on Surrealism in the second half of this essay.[82] Turning from the convergence of poetry and science in German Romanticism to topical concerns, Caillois claims 'an analogous situation seems to be emerging today'.[83] Surveying the languages of Relativity and quantum theory first as poetic then as philosophic resources (specifically Heisenberg's ever present Uncertainty Principle and the picturesque imagery of alpha particles and electron 'spin'), Caillois shudders at the thought of the mistreatment those theories were about to undergo once processed by 'eccentric minds which for the moment still confine themselves to the symbolism of dreams.'[84] As we know, Dalí's interpretation of Freudian symbolism and autobiographical exercise of psychoanalytic categories had by that time been yoked to an equally free rendering of concepts derived from the new physics. Caillois's accompanying remark: 'Some fools will try to explain everything by means of the fourth dimension', confirms that Dalí's appropriations were the object of his ire, and will take on a peculiarly clairvoyant quality in the context of Breton's writing about Surrealist painting in the 1930s.[85]

Le Nouvel Esprit scientifique de la peinture

In the wake of his estrangement from Surrealism and his ongoing dispute with Breton and Dalí, Caillois probably allowed himself a degree of satisfaction when he noticed

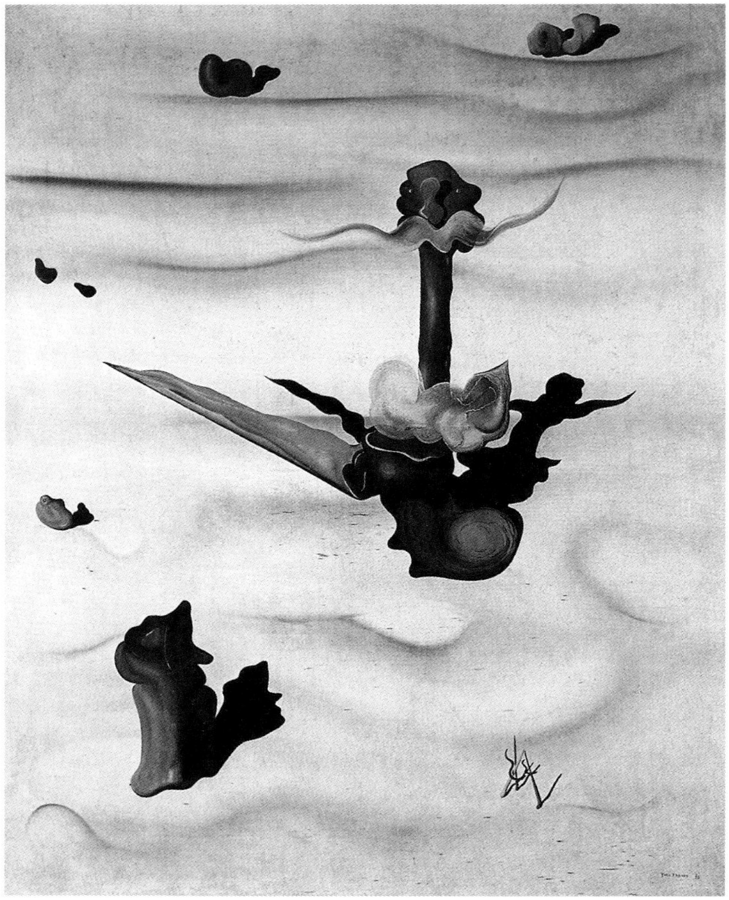

85 Yves Tanguy, *Man and Woman*, 1930.

in 1939 that Breton's thinking on Surrealist painting had turned towards a notion of four dimensionality. Sketching hesitantly a 'new' and yet confused and incomplete quasi-scientific paradigm for Surrealist painting, 'The Most Recent Tendencies in Surrealist Painting' introduced a youthful generation of Surrealist artists who Breton hoped would silence those who believed that Surrealism was a spent force at the close of the decade, and that Surrealist painting was at a dead end.[86] Ironically, a slating of the now distant figure of Dalí provides the impetus for Breton, who announces 'a very rapid decline' in his influence and in the paranoia-criticism Breton himself once lauded. Thematising a renewal by a return to first (pre-Dalínian) principles, Breton now sees a new importance vested in Yves Tanguy's work (fig. 85) by the latest recruits to Surrealism, such as Esteban Francès, Gordon Onslow Ford and Matta (fig. 86). The 'principled' spontaneity that Breton valued in Tanguy's work had, he says, brought about a general return to automatism, even heralding new automatic methods such as Oscar Domínguez's decalcomania and Wolfgang Paalen's *fumage*. Writing twenty years after the publication of *The Magnetic Fields*, Breton fixes upon that fundamental Surrealist practice to endorse the significance of these developments.

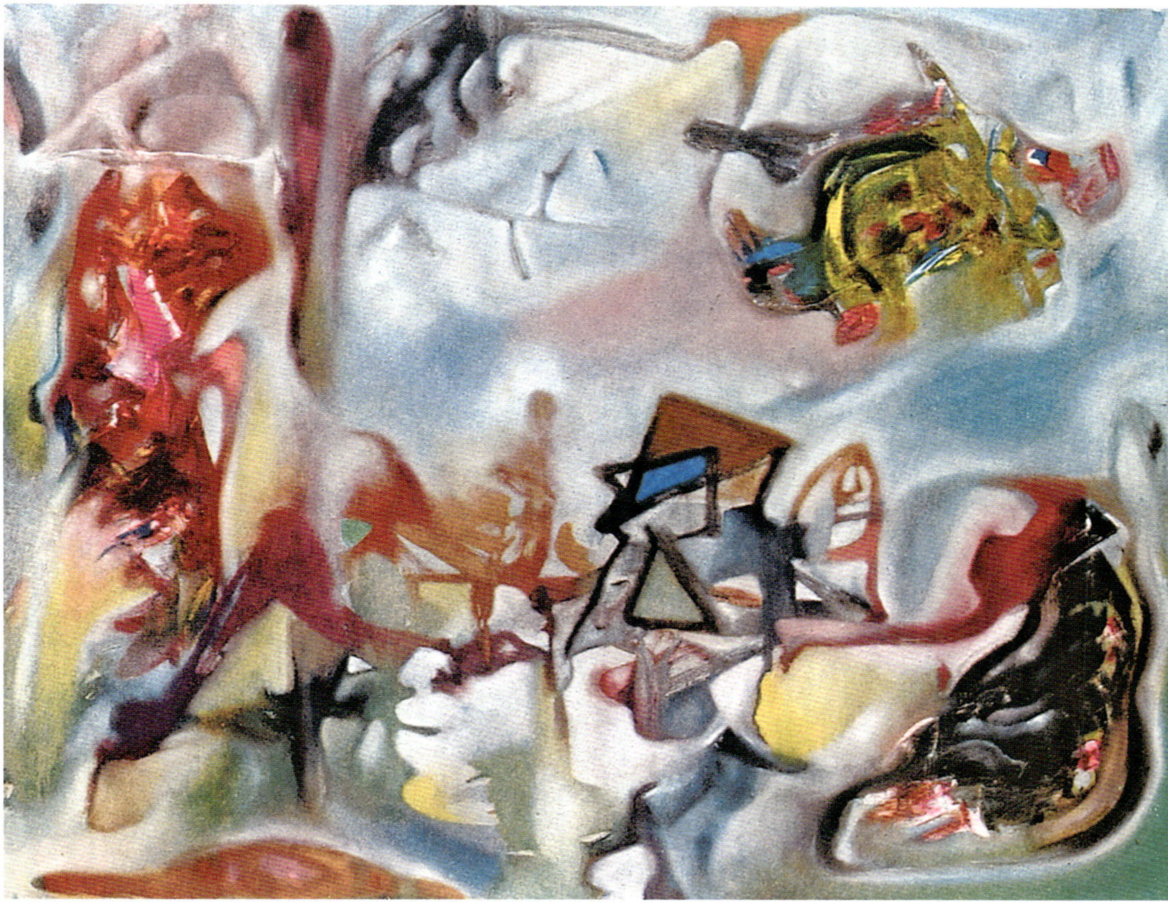

86 Matta, *Morphology of Desire* (1938). Oil on canvas, 73 × 92.1 cm. Former Gordon Onslow Ford Collection.

Perhaps less obviously, Breton's strategy of validating the new work by indicating a return to the basics of Surrealism is equally a feature of the second half of 'Most Recent Tendencies', where he gives notice of the entrance of these artists into 'the realm of science'.[87] Although he would have read Caillois's cutting remark about the interpretive overload of the fourth dimension in his essay in *Cahiers du Sud*, it did not prevent Breton asserting that the explorations of Surrealism's recent intake of painters revealed their desire 'to transcend the three-dimensional universe'.[88] Confirming his concern with the temporal sense of the fourth dimension as opposed to its mystical or *n*-dimensional connotations, Breton adds that these artists were inspired in their quest by 'Einstein's introduction into physics of the concept of the *space-time continuum*'.[89] The outcome of conversations with Onslow Ford and Matta, this was the first time since 'Max Ernst' of 1921 that he had equated the imagery of Surrealist art with the idea of space-time.[90]

Once this link is made with 'Max Ernst', Breton's underlying intentions in 'Most Recent Tendencies' become clearer. He goes on to declare that the 'need for a suggestive representation of the four-dimensional universe',[91] expressed through multiple hori-

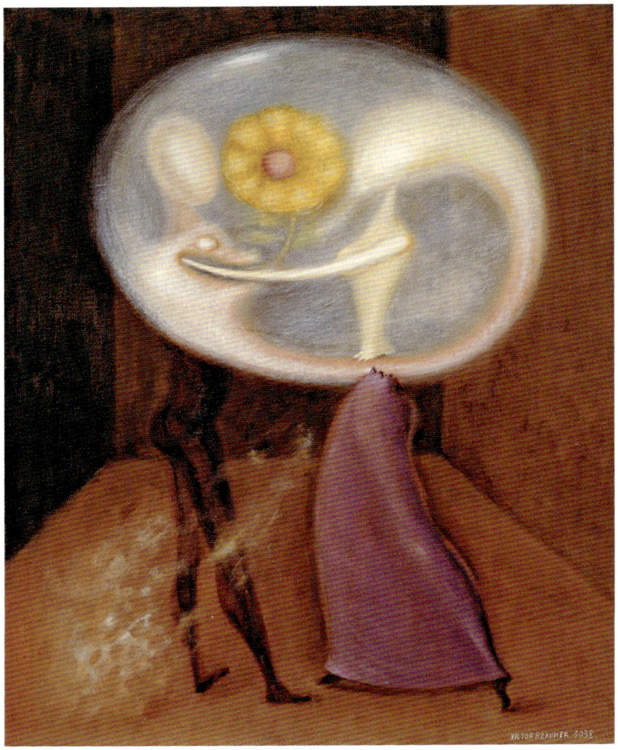

87 (above) Victor Brauner. *The Anemone* (1938). Oil on canvas, 80.6 × 64.8 cm. Collection of Francesco Pellizzi.

88 (facing page) Oscar Domínguez, untitled (Époque lithocronique) (1939). Oil on canvas, 98 × 79 cm. Private collection, Paris.

zons in Matta's art, and which 'describes a kind of world in which the last clear angles of Cubism are fragmented'[92] in Onslow Ford's, is of a different, scientific (temporal) order from the 'post-Euclidean' interpretation that provided 'one of the leitmotifs of Cubism in its heroic period'.[93] This characterisation of Surrealist art as a development of Cubism in the age of modern physics reiterates Breton's view, first given in 'Max Ernst', of Surrealism as an advance on Symbolism and Cubism through its refusal of the absolute nature of sensory experience and meaning, and its emphasis instead on the quality of the differences lying between phenomena.[94]

Expanding on its theme, 'Most Recent Tendencies' also deals with the 'operation' of the fourth dimension in the art of Victor Brauner and Oscar Domínguez. For Breton, as we observed earlier, 'the transition to the fourth dimension is undertaken on the psychic rather than the physical plane' in Brauner's paintings (fig. 87), a notion he elaborates with reference to the same Ducassian 'encounter' that he had married with Relativity in his discussion of Ernst's early collages in 'Max Ernst'.[95] If Breton was reaching a bit in trying to accommodate Brauner's painting to the concept of the fourth dimension (Brauner's work obviously has no truck whatsoever with the fourth dimension as temporality), then in the case of Domínguez he was on firmer ground. For Domínguez's investment in Relativity was clearly stated through his research into and theory of 'lithochronic surfaces' carried out with Ernesto Sábato (fig. 88). Quoting from their

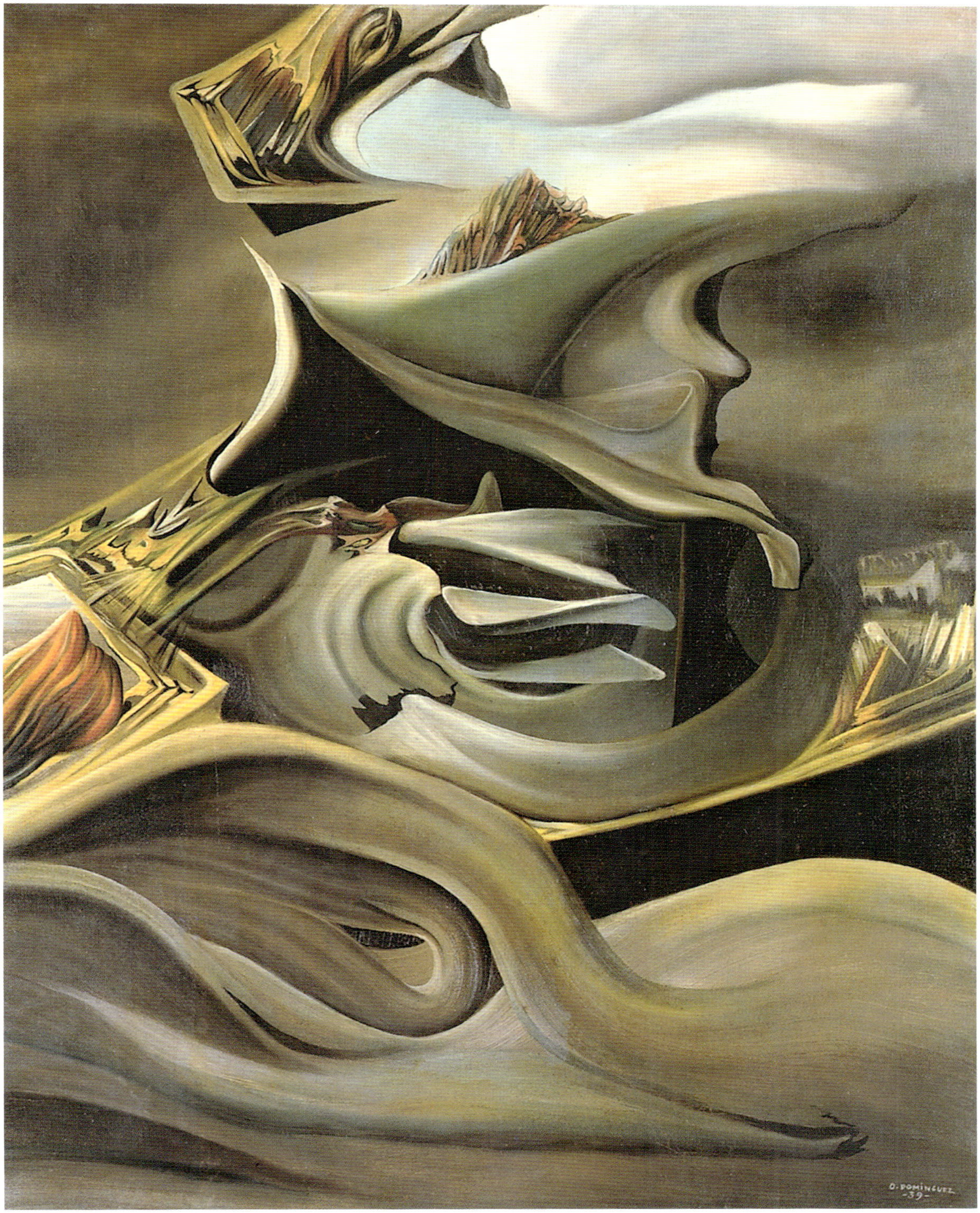
O. DOMINGUEZ
-39-

work, Breton airs a bland metaphor for Surrealist painting, consistently present in his own writing over the years, averring that Domínguez's canvases 'open a window upon the strange world of the fourth dimension, constituting a kind of solidification of time'.[96]

Yet as I have argued, Dalí's hybrids and his composite images using paranoia-criticism shared a similar purpose, insofar as Dalí strove to give an image to temporality (the 'Surrealist dimension par excellence') through the fixing of events to location by exploiting Einstein's knot of four-dimensional space-time. Breton's text is itself, then, a curious hybrid. It is openly critical of Dalí as a person and artist, endeavouring to leap back beyond his example to the potent source of Surrealism in automatism, whilst unwittingly complicit with Dalí's programme as set out over the previous ten years in its second 'scientific' part. This *had* to be the case, because Dalí's imagery, if not his method (like Ernst in 1921, he was no automatist), was created within a theoretical framework given partly by Relativity in accord with the terms set by Breton for the Surrealist definition of beauty in 'Max Ernst' (confirmed in the *Manifesto of Surrealism*). At the same time as he reinstated the importance of automatism over Dalínian paranoia-criticism in 'Most Recent Tendencies', Breton went back to the same (non-automatist) pseudo-scientific juncture that Dalí himself had taken as his point of departure.

It is debatable whether or not Breton was aware of this consonance as he wrote 'Most Recent Tendencies' in 1939, yet the text is undergirded by a strong rationale. As an art critic/theorist, he was on the wane by then (his last major statement of the kind remains 'Crisis of the Object'), as was his writing's symbiotic 'other', Surrealist painting itself, grasping at scant sources of creative inspiration. In this light, the content and contradictions of 'The Most Recent Tendencies in Surrealist Painting' can be viewed as the product of a failed attempt at triangulating a response to three crises. Anxious to replenish his own criticism, Surrealist painting, and the original impetus of the movement, Breton invoked and brought together two of Surrealism's founding textual moments: the automatism of *The Magnetic Fields*, the first Surrealist text, and the space-time model of 'Max Ernst', his first monograph on a contemporary artist.

In my chapter on Breton, I reviewed both of these at length to give an idea of what Surrealism owed Relativity at its birth, in the immediate aftermath of the First World War. With another war looming in 1939, Surrealism placed great faith in modern physics again, and I look at some of the detail of that now in a final section. By pursuing Surrealism's bad-tempered bantering with Dalí, I can recap some of the histories and arguments I have given in this book, and also record and explain Surrealism's drastic turn away from physics in the era of the Cold War.

CODA

NUCLEAR PHYSICS AND THE COLD WAR

Surrealism and Salvador Dalí

André Breton probably hoped he had heard the last of Dalí in 1939, but there was trouble brewing. We can track that now from the perspective of modern physics, for just as the scientific and philosophic discourses on the Theory of Relativity and QM in the 1920s and 1930s make greater sense of Dalí's Surrealist-period art and writings, they also help unravel the ill-natured 'dialogue' between Surrealism and post-Surrealist Dalí. This was transatlantic and impersonal for the most part (though not entirely, as we shall see), extending well into the 1960s. In this final section, I argue that the growth of nuclear physics and onset of the Cold War in the 1940s played as much a part in turning the Surrealists' interwar enthusiasm for the atom into scepticism as it did in transforming Dalí's into mysticism. I demonstrate this by means of a historical account of the rise of Big Science and the anti-nuclear protest that it met with in America and France, focusing ultimately on the late 1950s.

The almost caricatural differences that characterised relations between Breton and Dalí from the 1940s were already fermenting in the divergent views they expressed of physics in the 1930s. We heard Breton in the Tenerife *Indice* interview in 1935 testify to the high value he then placed in modern physics even more than in the findings of Freud, though at the same time he cautioned: '[l]et us restrain ourselves from contributing to the formation of a new religion which might be, paradoxically, the religion of science.'[1] Now look again at the opinion given by Dalí in his letter to Éluard a few months earlier, insisting that Surrealism 'has to evolve towards the foundation of a new, anti-mystical, materialist religion, based on the progress of scientific knowledge (especially the new ideas of space)'.[2] By the time he wrote those words, suspicions and recriminations were already circulating in Surrealism over Dalí's fascination with Hitler, leading to the artist's increased marginalisation and eventual departure from the group, probably in 1939 (although the actual cut-off point remains vague).

The Surrealists' misgivings about Dalí had reached the wider world by the time Breton went to New Haven to address the French students at Yale University on 10 December 1942, during the poet's wartime exile in America. Something of a blot on Breton's revolutionary CV, his lecture 'Situation of Surrealism between the Two Wars' was meant to sustain visibility for Surrealism in the New World, and was doubtless accompanied by the idea of gaining converts to Surrealism from the academic elite. Those who attended were lucky enough to get a taste of Breton's now-venomous hatred of Dalí, whose image and growing commitment to a classical style were hard to avoid in America at the time.[3] But they were also treated to Breton's prediction of 'a spectacular discovery that, as some unmistakable forerunners have already indicated, will take place in the sphere of *physics*'.[4] Breton avowed optimistically that this would act as the spark re-establishing communication between human beings, and that hopeful statement was given privileged status when the Yale lecture was published in the latest Surrealist journal *VVV*, illustrated with a drawing by Max Ernst (see fig. 89).[5]

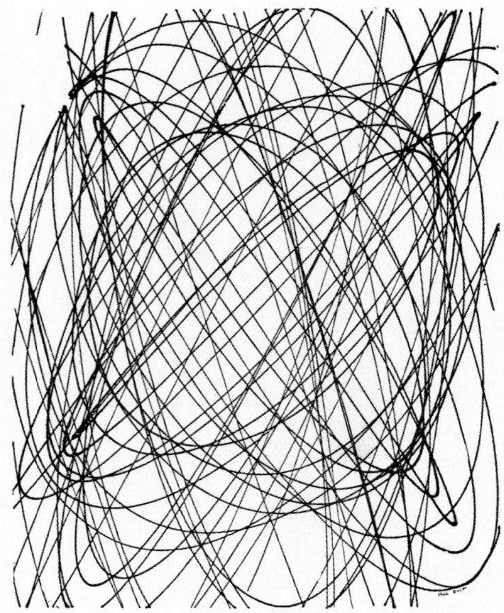
" . . . *sur le plan dé la* physique"

89 Max Ernst, '. . . sur le plan de la physique' (1943). Ink on paper, 63 × 49 cm. Private collection, Japan.

A few years later, in furtherance of Surrealism's theory of precognition or 'objective chance', Breton would claim prophetic status for his nebulous remark about physics in the Yale lecture.[6] The truth is far more banal: he had been reading the newspapers. With Rome and Berlin, Paris had been a leading player in nuclear physics in the 1930s. Although nuclear fission was discovered in Berlin in December 1938 and quickly replicated in laboratories around the world, its possibilities were explored chiefly at the Collège de France in Paris in 1939, by Frédéric Joliot-Curie and his team. These developments were widely reported in the French and international press, where speculation raged in 1939 about the future of atomic energy, from which electric power stations, submarines and bombs were expected soon.[7] According to the memoirs of Marie Curie's assistant Bertrand Goldschmidt, who escaped Vichy France for America on the same ship as André Masson within a few days of Breton, France would certainly have assembled the first atomic pile ahead of the US if she had not been invaded by Germany in 1940.[8] An atomic pile is the means by which fission creates a self-sustaining chain reaction, achieved for the first time in Chicago, in fact, by the Italian Enrico Fermi (another escapee from European fascism) on 2 December 1942, only eight days before Breton made his great 'prediction' to the students of Yale.[9]

Less a prediction, then, and more an informed statement that makes perfect sense in the context of contemporaneous general knowledge about research in nuclear physics. In the month before Breton's address, the US government bought a site for a nuclear weapons factory at Los Alamos, New Mexico, for what became known as the 'Manhattan Project', and the explosion of the first atomic device took place on 16 July 1945 in New Mexico.[10] Heralding the arrival of the Big Science that the Surrealists would come to despise, the new epoch it inaugurated would soon be celebrated by Dalí. Naturally, the nuclear test happened in great secrecy, yet by an odd coincidence for the story I am telling here, Breton had near-direct contact with the event. In blissful ignorance of the recent local occurrence of the first man-made nuclear explosion in history, he arrived in a radioactive New Mexico via Arizona a few weeks later on a visit to Zuni and Hopi reservations (fig. 90).[11]

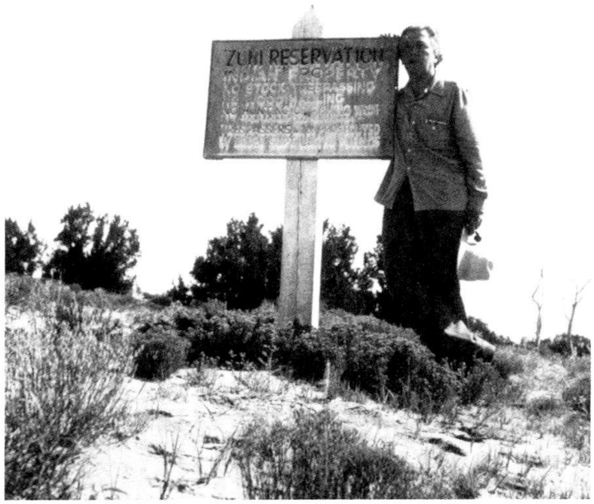

90 Elisa Breton, photograph of André Breton at Zuni Reservation, New Mexico, 1945. Courtesy Jean-Michel Goutier.

Meanwhile, the world was already dividing into pro- and anti-nuclear camps. In the run-up to August 1945, for instance, Chicago physicists were showing resistance to dropping the bomb on a city.[12] And in spite of the famous letter he wrote to Roosevelt in August 1939, identifying recent work in France and the US on the uranium nucleus and warning the President of scientific research and development in Germany, Einstein had been an extremely vocal advocate for peace since 1930.[13] Although Einstein did not play a crucial role in the initiation of the Manhattan Project and as a 'security risk' had no part in the creation of the first atomic bomb, his unmatched network of contacts in the world of science would have kept him informed of progress at Los Alamos. This accounts, perhaps, for the devastating fatalism with which he greeted news of the Hiroshima bomb: '*Oh, Weh*', he said, 'And that's that.'[14]

The responses of Breton and Dalí to the Bomb are far more representative of the widespread shock felt at the time at the instantaneous transition to a new age of military-industrial complexes (as they would be termed by Eisenhower in 1962), immensely powerful physics and hugely influential physicists. In an interview he gave in October 1946 where he spoke at length about Hiroshima, Breton referred metaphorically to the severe jolt that the explosion that destroyed that city gave to human affairs.[15] In the same interview, carried out four months after his return to Paris, Breton contrasts the relatively insouciant French reaction to the blast with the excitement it had caused in America, where 'atomic disintegration was widely held to be the most important discovery since fire'.[16] Commending Denis de Rougemont's recent *Lettres sur la bombe atomique* as a model of sobriety on the issue, Breton indicates the two paths along which nuclear energy could take the human race, while engineering the interview so that it seemed as though one of them was set out in the 'prediction' made in his Yale lecture four years earlier.[17]

This interview registers Breton and the Surrealists' hesitation in their formulation of a creative response to the onset of the nuclear age, and it bears stark comparison with Dalí's reaction to the news. Years later, Dalí would speak of how the detonation had shaken him 'seismically', and this is evident in the immediate change his work underwent in the aftermath of the Bomb, heralded by *Melancholic, Atomic, Uranic Idyll* of 1945 (fig. 91).[18] Evoking a mood of foreboding reminiscent of Magritte's remarkable, ominous 1937 canvas *The Black Flag* (fig. 92), Dalí's painting presents his familiar though increasingly Americanised jumble of enigmatic figures assembled in a shallow, box-like space against a black backdrop, holed through and torn at the top. The planked floor resembles here a stage, though the theatrical illusionism is compromised by the torn drop curtain revealing supplementary figures from Dalí's lexicon of fantasia in a separate, non-theatrical space beyond. The startled human face at bottom left looks on at the apocalyptic spectacle of the nuclear age, in which the melodramatic imaginary and dramatic real are conjoined as a deeply ironic, tragi-comic, utopian opera, indebted to Hieronymus Bosch.

91 Salvador Dalí, *Melancholic, Atomic, Uranic Idyll* (1945). Oil on canvas, 65 × 85 cm. Museo Nacional Centro de Arte Reina Sofía, Madrid.

If *Melancholic, Atomic, Uranic Idyll* gives a fair idea of Dalí's instant fascination with, and spectacularisation of, the atom, its nucleus and the Bomb, why were the Surrealists, regrouped in Paris around the recently returned Breton from 1946, so slow to reach their own definitive assessment of the new era? The fullest possible answer to this question would deal at length with the discursive nature of life in the Surrealist group, where extended debate played a considerable role in reaching collective agreement on its theoretical and political orientation. However, in tune with the theme of this part of my study, I shall answer this question by focusing solely on the historical context of the Cold War.

92 René Magritte, *The Black Flag* (1937). Oil on canvas, 54.2 × 73.7 cm. National Galleries of Scotland.

The historian David Pace has counted no less than seventy-three books, pamphlets or articles published in France between 1945 and 1949, 'in which the post-Hiroshima period was explicitly characterised as the atomic age or era'.[19] Of those, twenty-five appeared in 1945 and 1946, almost none of which 'seriously questioned whether atomic energy would one day offer great new possibilities for mankind'.[20] The day after the second, plutonium bomb devastated Nagasaki, the most distinguished physicist in France, Louis de Broglie, celebrated the arrival of the atomic age in the pages of *France-soir* in unequivocal terms:

Having succeeded in drawing from a few grams of matter more energy than can be derived from the richest coal mines, the most beautiful waterfalls, or the most abundant oil wells, human industry will find itself one day entirely transformed, and its possibilities will be multiplied to infinity. Theoretically, we know that even now we are scarcely aware of what treasures are hidden in the nucleus of the atom, but the discovery of atomic bombs shows us that the dream of yesterday is ready to become the reality of tomorrow. Everything leads us to predict that a new era is going to begin: the era of the use of atomic energies.[21]

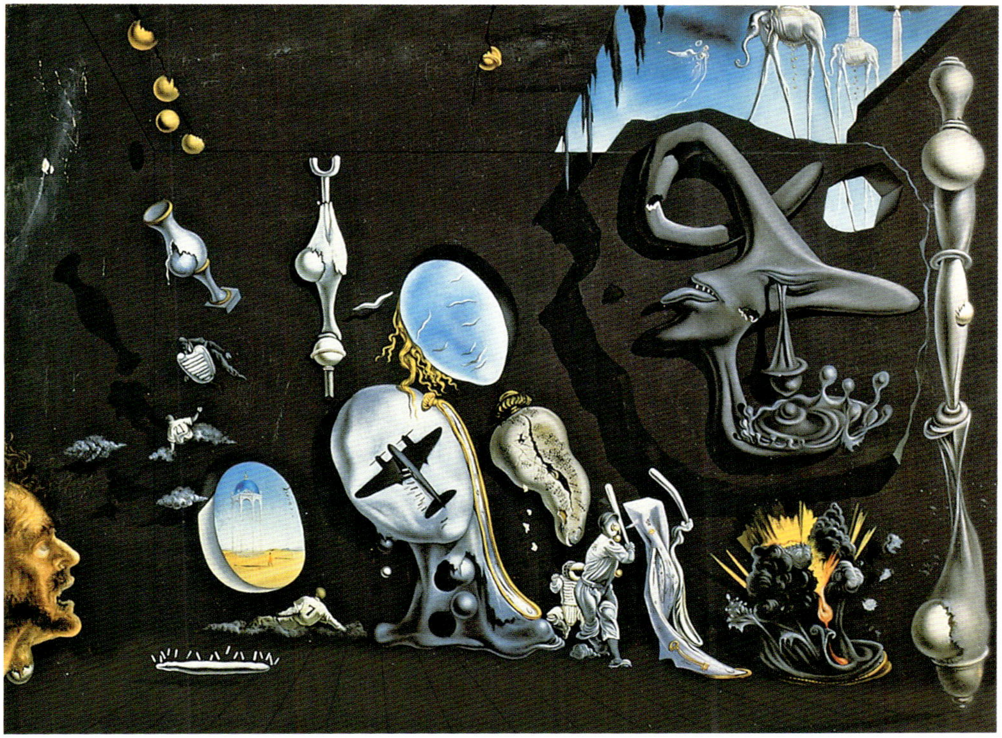

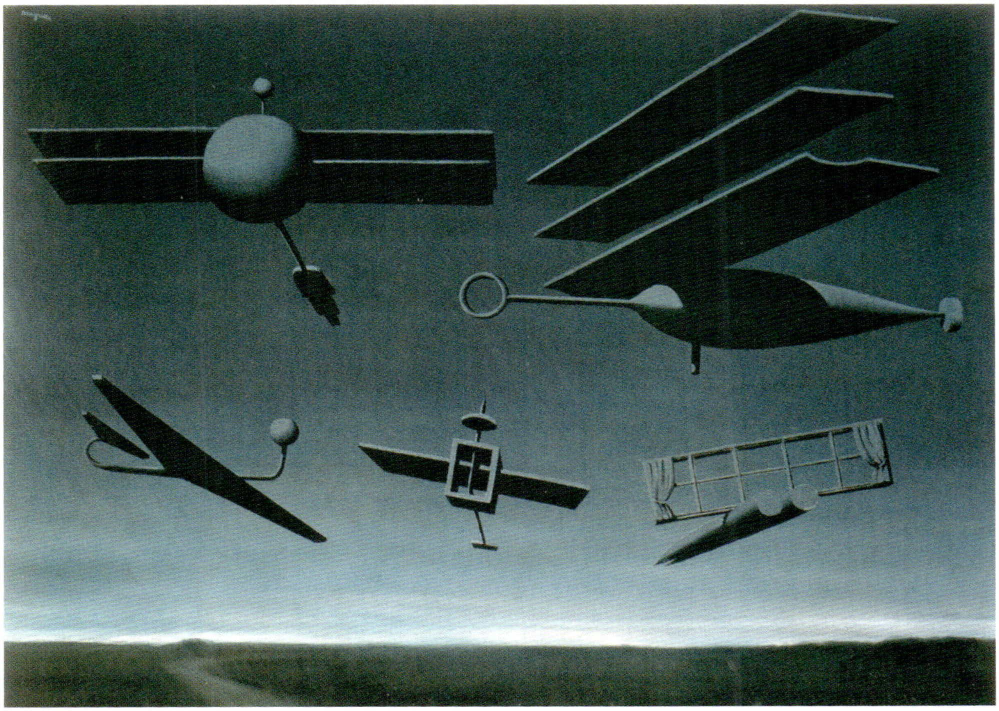

In the post-Chernobyl era it is easy to scoff at de Broglie and others who made it sound as though scientists had discovered El Dorado, but for some the scale of the achievement obscured the dangers it brought with it. In a Paris subject to postwar food shortages and blackouts, it is no wonder that the Surrealists – no strangers themselves to utopianism – were reluctant to condemn the advent of the nuclear age and foreclose on the riches it promised.

This reluctance is best seen among the contributions to the catalogue of the 1947 Exposition Internationale du Surréalisme, the show that confirmed the continued existence of Surrealism to a sceptical postwar Paris.[22] Many of the catalogue's thirty-eight short texts were written by youthful new recruits to the group, and even though five have as their main theme or sub-theme modern physics, its philosophy and their relations with Surrealism, not a single one takes the opportunity to condemn the Bomb, the rise of modern physics, the irresponsibility of the physicist, or industrialised, militarised Big Science.[23] Several of them did extend Surrealism's longstanding anti-clericalism, though, and I feel part of the explanation lies here. The Surrealists' unwillingness to denounce the nuclear industry stems, in part, it seems, from a desire to distinguish their purpose from the predictable criticism aimed at atomic power by the religious right after the war, which manifested a uniformly bleak view, counterbalancing the euphoria expressed in other quarters. Typical of the anti-scientific, anti-materialist religious statements of the period were the dire warnings of the commentator in the Catholic periodical *La Croix* in August 1945, roundly comparing 'the disintegration of the atom with the "disintegration" of the family, the working world, faith, and the French soul itself'.[24]

What a contrast with Dalí, who would soon celebrate in America this self-same disintegration as the herald of a new religious era. His attitude towards the Bomb was certainly affected by the American viewpoint on atomic energy as Breton had described it, where the explosions in Japan in all their monstrous magnitude were regarded alongside the inconceivable energy resource they harboured. This was reflected in the pages of *Scientific American*. A cursory reading of responses to the Bomb in its pages between 1945 and 1950 reveals a mood of curiosity, excitement and optimism.[25] By 1947, the nuclearphilia of the estranged Surrealist Dalí was closer to the French engineer and designer Louis Reard, who in July 1946 designated a new age in swimwear by adopting the name 'Bikini', the site of the atomic tests of that year.[26] Looking back now, following around fifty years of anti-nuclear protest, Reard's marketing decision seems rather ill-judged, only making sense when we put it in the social and cultural context of the immediate postwar period, where the words 'atomic' and 'bomb', together with those associated with them, quickly broke loose of their moorings in physics to designate by early 1946 anything astonishing, groundbreaking, new or sensational. Dalí swiftly hitched his post-Surrealist wagon to the Next Big Thing, and in the year after an article on the atomic tests at Bikini in *La Vie catholique illustrée* compared scientists to the sorcerer's apprentice – 'unleashing forces, which their will can no longer master' – Dalí gave mythical shape to those same forces in his conceptually questionable, clumsily realised landscape painting *The Three Sphinxes of Bikini* (fig. 93).[27]

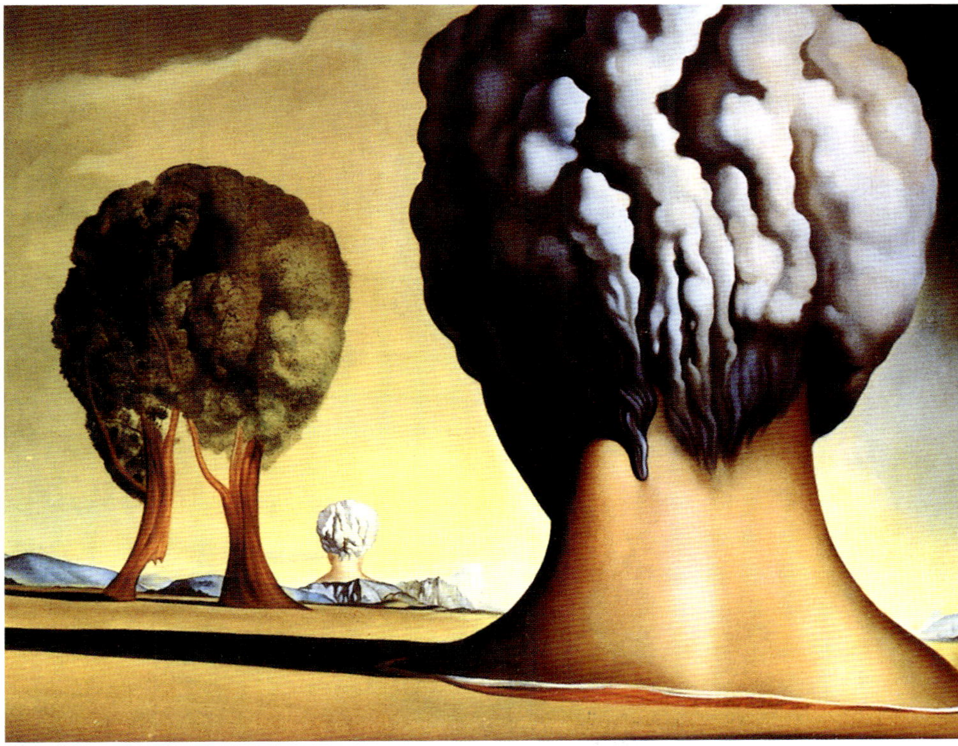

93 Salvador Dalí, *The Three Sphinxes of Bikini* (1947). Oil on canvas, 30 × 50 cm. Private Collection, Geneva.

Making a crude grab at the New, Dalí utilises in this painting the by then well-established 'method' of paranoia-criticism to extend an old theme. In a hollow echo of the creative mandate that the ambiguity and interpretability of clouds allowed the Surrealists (and, indeed, Shakespeare), Dalí envisions two enigmatic heads and a damaged tree in the mushroom clouds of Bikini.[28] Here Dalí reworked the 'visionary' aspects and Freudian and mythological themes of his earlier Surrealist-period paintings like *The Metamorphosis of Narcissus* (1937), aligning them with his new self-perception as a 'nuclear mystic' who could 'see and understand the forces and hidden laws of things' through his grasp of modern science.[29] In *Melancholic, Atomic, Uranic Idyll*, Dalí had isolated, transposed and incorporated one of the figures of *Couple with their Heads Full of Clouds* (1936), replacing the clouds in its head with an aircraft, probably a bomber. The shift in emphasis from his earlier equivocal, 'elevated' vision to alternative concerns is clear enough from these paintings of the mid-1940s: Dalí's hallucinations were moving away from a relatively arbitrary 'Freudian' symbolism of extensive scope, to solidify solely around the Bomb and its attendant imagery of atoms, nuclei, particles and so on.

Dalí took a further step in the makeover of his imagery at this time by moving towards a coalition of geometry, nuclear physics and Catholic mysticism, later theorised in his *Mystical Manifesto* of 1951.[30] What Dalí called there the 'paroxysmal crisis of Dalinian mysticism', founded upon the 'metaphysical spirituality of the substan-

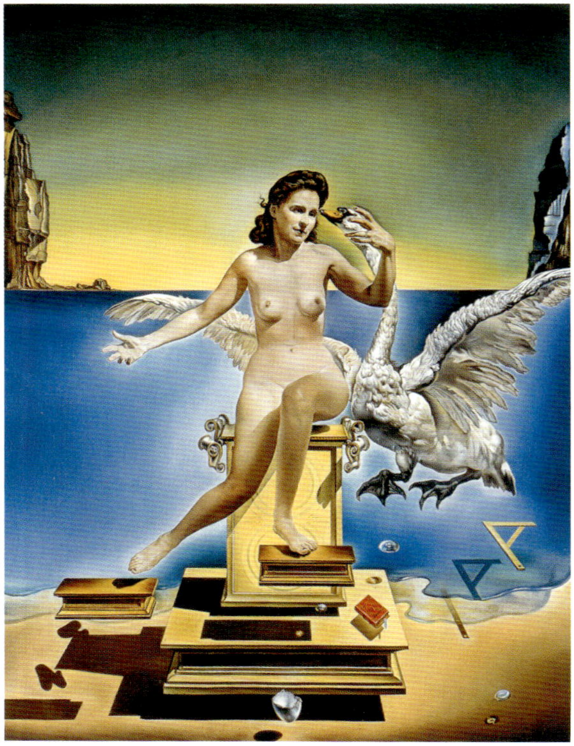

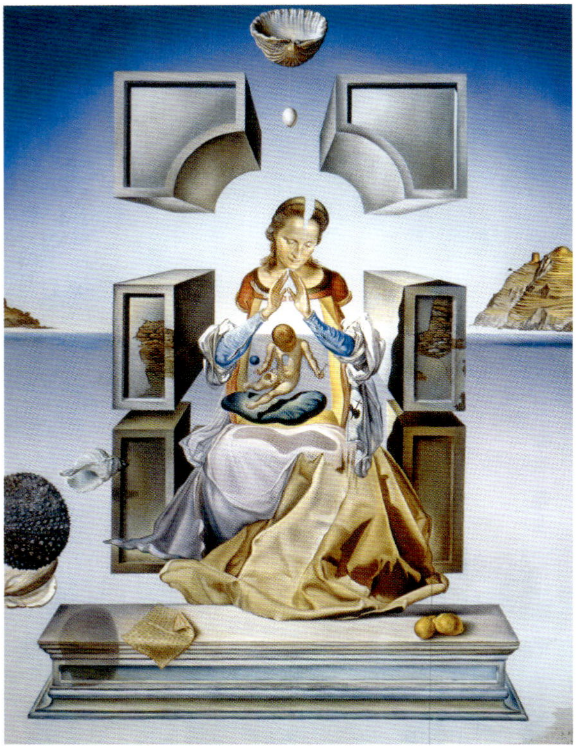

tiality of quantum physics', is piloted at first by a smooth, plasticky style that might be called 'neo-Renaissance kitsch', and can be viewed in his *Leda Atomica* (1947–9) and *Madonna of Port Lligat* (1949) (figs 94 and 95).[31] Alternatively, from 1951, in paintings like *Raphaelesque Head Exploding* (fig. 96), Dalí sought to give voice to his post-Surrealist theory by shattering these neo-Renaissance motifs. In this, he was inspired quite specifically by the picture of the physical world developed by writers like Eddington, whose 'two tables' (the common sense solid one and the scientific one swarming with atoms that open *The Nature of the Physical World*) Dalí alluded to in the 1950s and 1960s in his interviews and writings.[32] The resulting disintegrative style seems distantly related to Cubism – the Cubism that Carl Einstein insisted in 1934 was perfectly comprehensible only once it was understood that 'quantum theory had shattered the unitary continuity of the universe'.[33]

A bit closer to Dalí's histrionic, architectonic, fragmented head of rhinoceros horns and his daft 'immaculately corpuscular' Madonnas – though free of their religious and classicising tendencies – is Wolfgang Paalen's *Figure pandynamique* of 1940 (fig. 70).[34] A composite head in the manner of Archimboldo by whom Paalen owned a painting, we saw that *Figure Pandynamique* is indebted to the microcosmic flux of reality as revealed by the *wave* version of QM formulated by de Broglie and Erwin Schrödinger, rather than the *particle* quantum interactions described by Werner Heisenberg, which inspired Dalí in the 1950s. He could have seen *Figure pandynamique* and other 'wave'

94 (top left) Salvador Dalí, *Leda Atomica* (1947-49). Oil on canvas, 61 x 45 cm. Figueres, Fundació Gala-Salvador Dalí.

95 (top right) Salvador Dalí, *Madonna of Port Lligat* (1949). Oil on canvas, 49.5 x 38.3 cm. The Patrick and Beatrice Haggerty Museum of Art, Marquette University, Milwaukee, gift of Mr. and Mrs. Ira Haupt, 1959.

96 (facing page) Salvador Dalí, *Raphaelesque Head Exploding* (1951). Oil on canvas, 43 x 33 cm. Scottish National Gallery, Edinburgh, on permanent loan from Miss Stead-Ellis, Summerset.

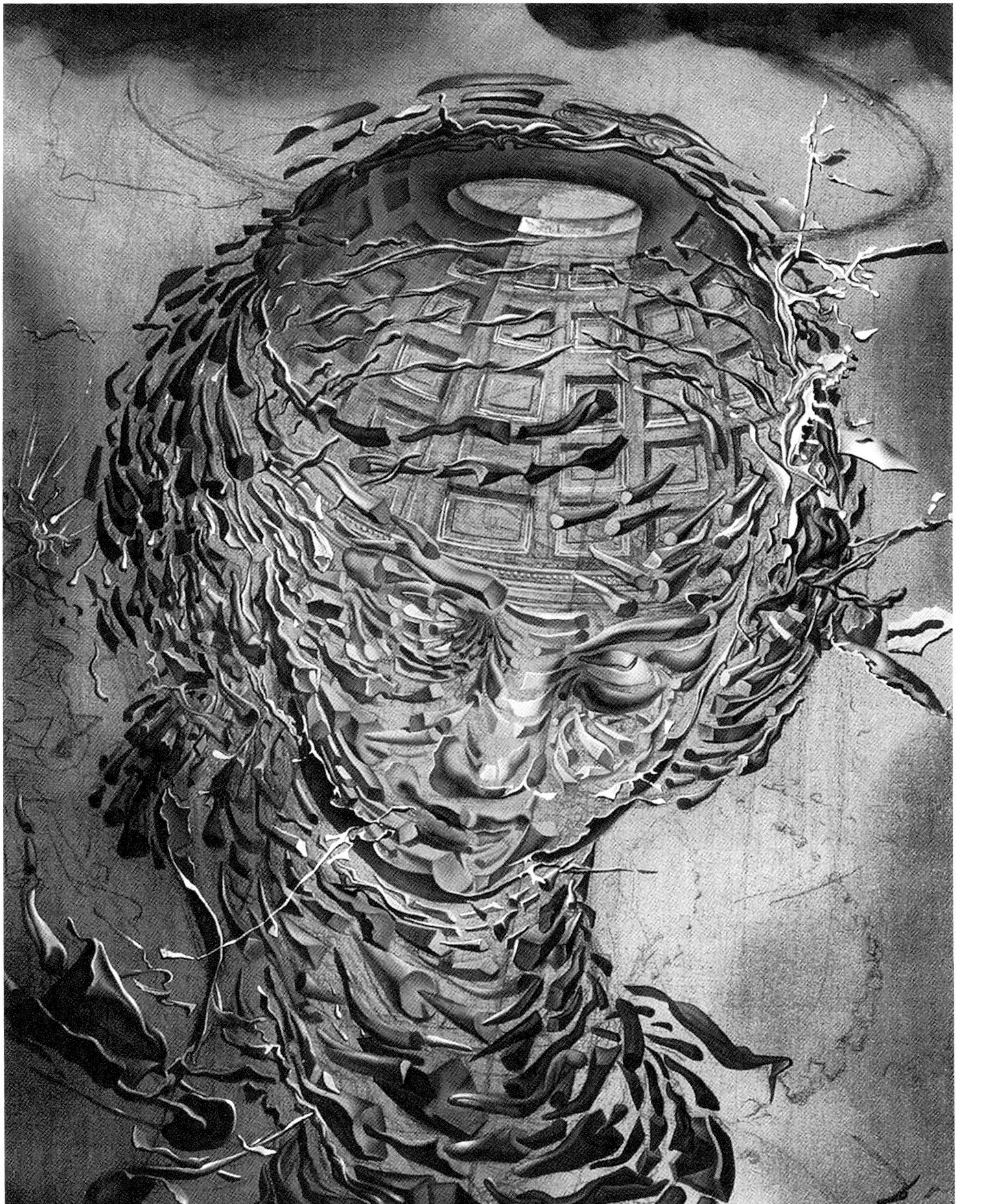

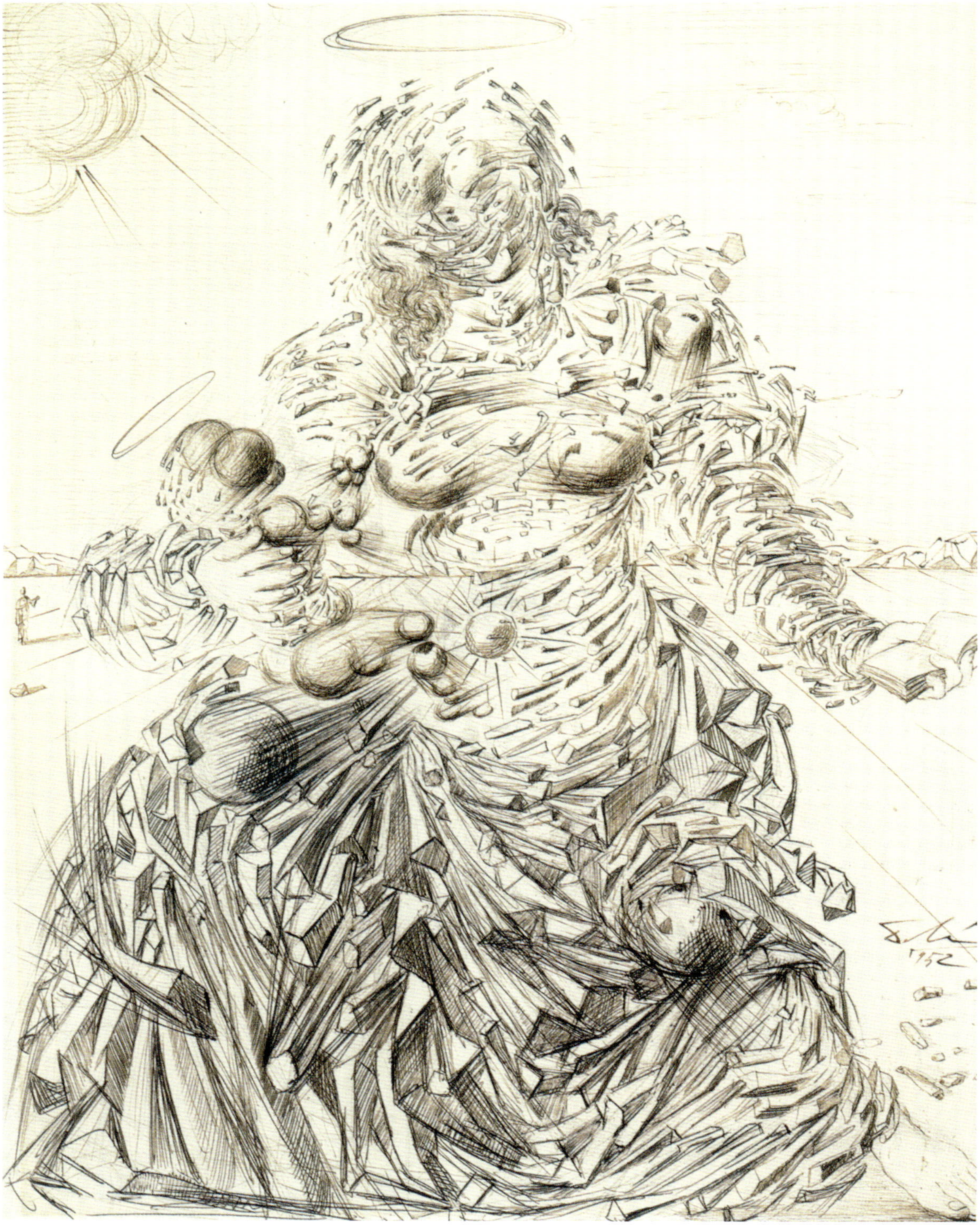

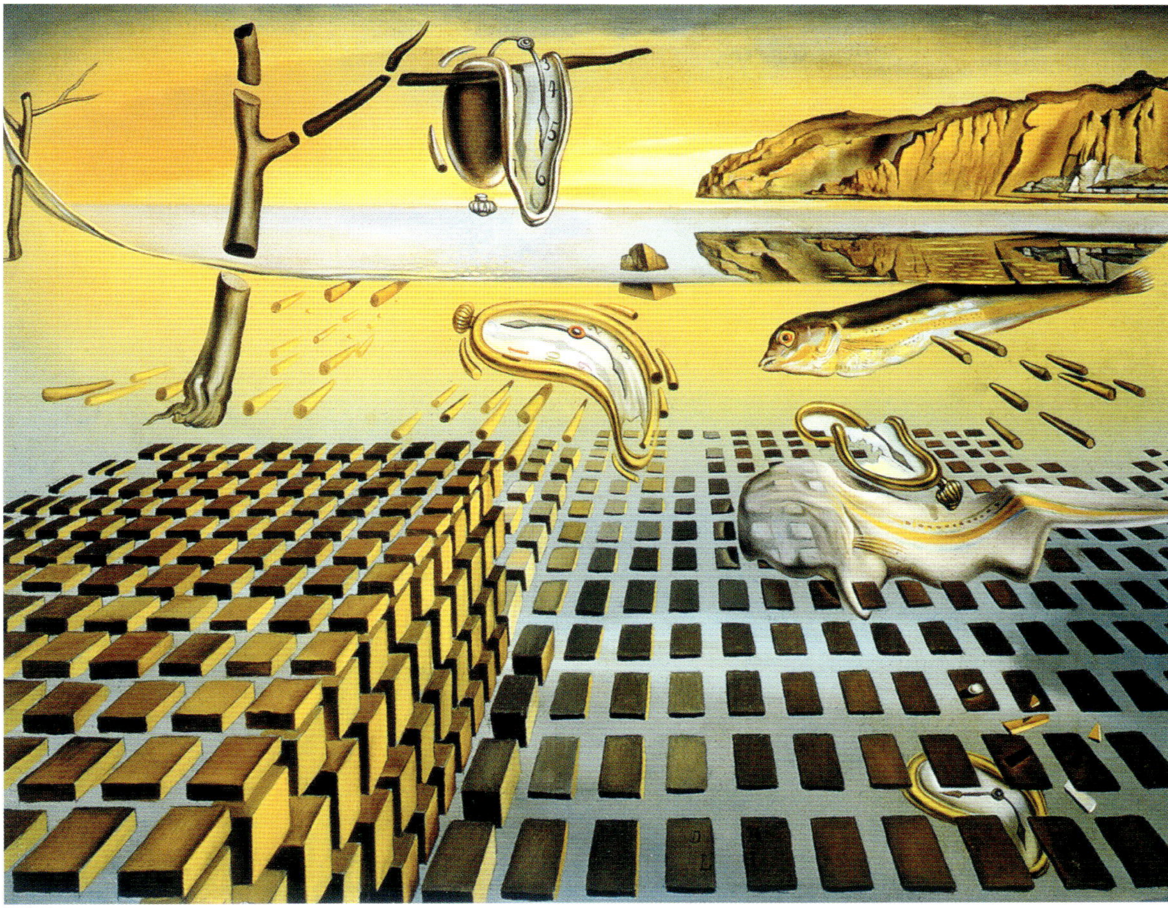

97 (facing page) Salvador Dalí, *Corpuscular Madonna* (1952). Pencil, sepia-colour and China ink, 55.8 × 43 cm. Birmingham Museum of Art, gift of Mr. And Mrs. Charles W. Ireland.

98 (above) Salvador Dalí, *The Disintegration of the Persistence of Memory* (1952). Oil on canvas, 25 × 33 cm. The Salvador Dalí Museum, St. Petersburg, Florida.

paintings reproduced in Paalen's journal *Dyn*, which, as I noted earlier, was published in Mexico from 1942, with a wide distribution during the war among artists and galleries in New York. Works like *Figure pandynamique* were shown in Paalen's exhibition *Dynaton*, held at the San Francisco Museum of Art in 1951, the very year Dalí initiated his corpuscular style, which he claimed, typically contrarily, 'show[ed] the spirit of reintegration' as opposed to what he characterised as Surrealism's 'disintegration' (fig. 97).[35]

Michael R. Taylor has detailed how in early 1952 Dalí spread the good news of nuclear mysticism through lectures given across America, receiving a predictably baffled response before the tour collapsed under the weight of Dalí's geographical naïvety (he thought America was about the same size as an average European country).[36] Undeterred, in his diary at the time he recorded that the style was the direct outcome of his Surrealist period, and the self-confessed opportunist linked the two for the usual purposes (publicity) in the disintegrating revision of his most celebrated Surrealist painting, *The Persistence of Memory* of 1931 (fig. 98).

By that year, 1952, the Surrealists had formulated a criticism of nuclear physics and the Bomb, which converged with their prewar, anti-authoritarian rhetoric. I do not believe this necessarily emerged from a deeper reflection on the Hiroshima and Nagasaki bombs, nor from subsequent statistics of radioactive poisoning and the longer term, multigenerational, human cost. Rather, I think it more likely that French nuclear policy and the intensification of the Cold War convinced the Surrealists that nuclear energy posed a threat to life on earth. The turning point in Surrealism was 1948. In that year, Breton wrote 'The Lamp in the Clock', which I regard as a transitional essay in Surrealism's interpretation of modern physics. Here, Breton attacks the continued opposition '[o]n the politico-philosophical level', of materialism and idealism, 'whose opposition is purely formal,' he writes, 'as the latest discoveries in physics should have sufficed to make clear'.[37] However, from this positive appraisal of the political, social and philosophical value of the epistemology of modern physics, Breton moves on to condemn weapons testing and stockpiling by making an analogy with a tradition of apocalyptic imagery in poetry and painting. Placing the world truly at risk, unlike the metaphorical, Biblical world endings imagined by Bosch and others, the scientist and dictator are cast by Breton in the same lethal mould:

> However much we try and guess what is being hatched under the locks of Professor Einstein or what is thriving behind the crew cut of the strange comrade Stalin, this final slaughter has nothing to do with what had been envisioned.[38]

The simultaneous approval of the epistemology of QM and criticism of research in experimental physics in 'The Lamp in the Clock' in 1948 mark the beginning of the end of Surrealism's love affair with modern physics – exactly the year, probably not coincidentally, that witnessed Dalí's nuclear epiphany.[39]

The next ten years saw the world separate into two mutually hostile camps, as well as massive increases in investment in nuclear arms. In 1949, the year after Breton's text appeared, NATO was established and Germany divided following the Berlin blockade. Also in that year, the Mouvement de la Paix (MDP) was founded in France, led by Frédéric Joliot-Curie and others, counting Picasso and former Surrealists Paul Éluard and Louis Aragon among its members. Ultimately, it became a subgroup of the French Communist Party, helping create an identification between Communism and the peace movement in France that lasted many years, even though the PCF had welcomed the Hiroshima bomb in the pages of *L'Humanité* (!) and the French Communist press greeted with equal enthusiasm the explosion of the first Soviet atomic device in 1949.[40] Naturally, the Surrealists' longstanding, hardened hostility to Stalinised Communism and growing opposition to atomic energy prevented them entertaining relations either with the MDP or PCF.

Joliot-Curie also led the Commissariat à l'energie atomique (CEA), which was set up by de Gaulle in 1945. Supposedly concerned with the peaceful use of nuclear power, it counted Georges Bataille's friend Georges Ambrosino among its employees. By 1950, the year North Korea heightened Cold War tension by launching a surprise attack on

South Korea, Bertrand Goldschmidt, back in France after his wartime exile in North America, had isolated 'the first French made plutonium' under its auspices.[41] Over the next ten years, the CEA grew rapidly due to huge funding increases from 1952 and the go-ahead given in 1956–7 by the Socialist government (which included François Mitterand among its most prominent members) for a five-year plan of nuclear experiments. As very powerful particle accelerators became more affordable, the Conseil Européen pour la Recherche Nucléaire (CERN) was established in 1952 by twelve member nations, including France, to compete with the US. The first French nuclear reactor was built in 1956. This swift expansion of French nuclear policy helps explain the persistent comment on, and critique of, science by Surrealism in the 1950s, visible chiefly in the pages of the Surrealist journal of that decade, *Médium*, and in a series of remarks made by Breton.

Naturally, exchanges within Surrealism on the subject of atomic physics were an adjunct to feverish speculation in the wider world on the subject. An invitation to Breton to participate in this broader debate came as early as January 1950, in a letter from Marcel Duchamp. Requesting an article for the journal planned by the American abstract artist Harry Holtzman, titled *Trans/formation: Arts, Communication, Environment*, Duchamp writes: 'if you don't have any preference for a particular subject, he would be pleased if you would agree to write something on a theme like "Science and Surrealism" or something similar'.[42] Duchamp ended up as a consulting editor on *Trans/formation* (alongside the American artist Stuart Davis), which in its three issues (1950–52) went on to become, in the words of Linda Dalrymple Henderson, 'the most avant-garde art magazine in New York', stating the belief that 'art, science, technology are interacting components of the *total human enterprise*', and declaring it would 'cut across the arts and sciences by treating them as a continuum'.[43] This would have been music to the ears of 1930s Surrealism, but by the 1950s the movement had definitively switched sides and would have been unsympathetic to the aims of *Trans/formation*. Consequently, Breton never wrote that essay on Surrealism and science.

In 1952, *Trans/formation* gave over its third and final number entirely to science, carrying Heisenberg's modestly titled 'The Quantum Theory: A Formula Which Changed the World', together with an essay on 'The Transformatory Implications of Einstein's Totalism'. In the same year, Breton was interviewed on the subject of art and science, and took the opportunity to lay into scientists like Heisenberg and Einstein for what he saw as their collusion in the life-threatening aspect of scientific discovery. In that interview, Breton claims 'the poets and artists I've known seem to have adopted a dissident attitude toward [science],'[44] adding that artists had 'unanimously . . . been uninterested in quantum mechanics and Heisenbergian physics'.[45] Even putting to one side his choice of Heisenberg's 'unvisualisable' matrix mechanics to 'prove' his point, Breton's Cold War strategy to rescue art in general and Surrealism in particular from what he now saw as the destructive force of science and technology is unconvincing. Ignoring a mountain of conflicting evidence, much of which I have assembled in this book, he insists that '[l]a technique scientifique' had not been 'a source of artistic enrichment: quite the opposite.'[46] With the danger the scientist's research now posed to humanity, Breton maintains that the artist, as much as the layperson, 'more than ever has the right to glower at the scientist'.[47] And quoting the judgment passed on mathematicians by a 'physics professor', his old favourite the German Romantic Georg Christoph Lichtenberg – 'their heads are most often filled with junk' – Breton concludes: '[w]hat more could we say about *atomic scientists*?'[48]

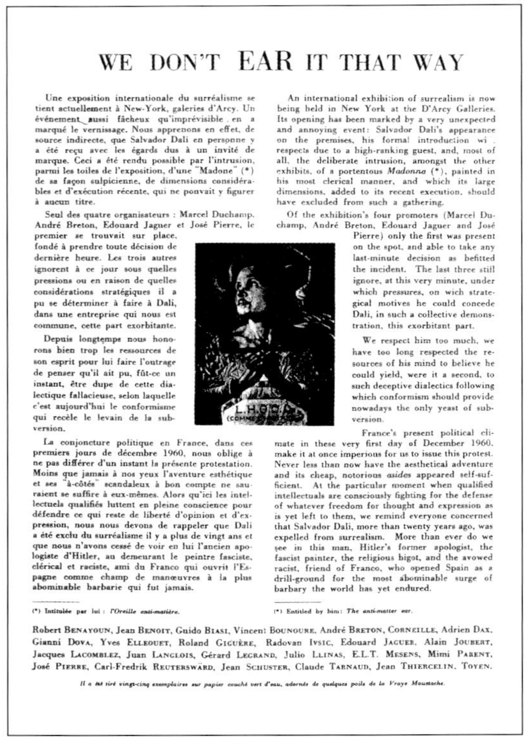

WE DON'T EAR IT THAT WAY

Une exposition internationale du surréalisme se tient actuellement à New-York, galeries d'Arcy. Un événement aussi fâcheux qu'imprévisible en a marqué le vernissage. Nous apprenons en effet, de source indirecte, que Salvador Dali en personne y a été reçu avec les égards dus à un invité de marque. Ceci a été rendu possible par l'intrusion, parmi les toiles de l'exposition, d'une "Madone" (*) de sa façon sulpicienne, de dimensions considérables et d'exécution récente, qui ne pouvait y figurer à aucun titre.

Seul des quatre organisateurs : Marcel Duchamp, André Breton, Edouard Jaguer et José Pierre, le premier se trouvait sur place, fondé à prendre toute décision de dernière heure. Les trois autres ignorent à ce jour sous quelles pressions ou en raison de quelles considérations stratégiques il a pu se déterminer à faire à Dali, dans une entreprise qui nous est commune, cette part exorbitante.

Depuis longtemps nous honorons bien trop les ressources de son esprit pour lui faire l'outrage de penser qu'il ait pu, fût-ce un instant, être dupe de cette dialectique fallacieuse, selon laquelle c'est aujourd'hui le conformisme qui recèle le levain de la subversion.

La conjoncture politique en France, dans ces premiers jours de décembre 1960, nous oblige à ne pas différer d'un instant la présente protestation. Moins que jamais à nos yeux l'aventure esthétique et ses "à-côtés" scandaleux à bon compte ne sauraient se suffire à eux-mêmes. Alors qu'ici les intellectuels qualifiés luttent en pleine conscience pour défendre ce qui reste de liberté d'opinion et d'expression, nous nous devons de rappeler que Dali a été exclu du surréalisme il y a plus de vingt ans et que nous n'avons cessé de voir en lui l'ancien apologiste d'Hitler, au demeurant le peintre fasciste, clérical et raciste, ami du Franco qui ouvrit l'Espagne comme champ de manœuvres à la plus abominable barbarie qui fut jamais.

(*) Intitulée par lui : *l'Oreille anti-matière.*

An international exhibition of surrealism is now being held in New York at the D'Arcy Galleries. Its opening has been marked by a very unexpected and annoying event: Salvador Dali's appearance on the premises, his formal introduction wi respects due to a high-ranking guest, and, most of all, the deliberate intrusion, amongst the other exhibits, of a portentous *Madonna* (*), painted in his most clerical manner, and which its large dimensions, added to its recent execution, should have excluded from such a gathering.

Of the exhibition's four promoters (Marcel Duchamp, André Breton, Edouard Jaguer and José Pierre) only the first was present on the spot, and able to take any last-minute decision as befitted the incident. The last three still ignore, at this very minute, under which pressures, on wich strategical motives he could concede Dali, in such a collective demonstration, this exorbitant part.

We respect him too much, we have too long respected the resources of his mind to believe he could yield, were it a second, to such deceptive dialectics following which conformism should provide nowadays the only yeast of subversion.

France's present political climate in these very first day of December 1960, make it at once imperious for us to issue this protest. Never less than now have the aesthetical adventure and its cheap, notorious *asides* appeared self-sufficient. At the particular moment when qualified intellectuals are consciously fighting for the defense of whatever freedom for thought and expression as is yet left to them, we remind everyone concerned that Salvador Dali, more than twenty years ago, was expelled from surrealism. More than ever do we see in this man, Hitler's former apologist, the fascist painter, the religious bigot, and the avowed racist, friend of Franco, who opened Spain as a drill-ground for the most abominable surge of barbary the world has yet endured.

(*) Entitled by him: *The anti-matter ear.*

Robert BENAYOUN, Jean BENOIT, Guido BIASI, Vincent BOUNOURE, André BRETON, CORNEILLE, Adrien DAX, Gianni DOVA, Yves ELLEOUET, Roland GIGUÈRE, Radovan IVSIC, Edouard JAGUER, Alain JOUBERT, Jacques LACOMBLEZ, Juan LANGLOIS, Gérard LEGRAND, Julio LLINAS, E.L.T. MESENS, Mimi PARENT, José PIERRE, Carl-Fredrik REUTERSWÄRD, Jean SCHUSTER, Claude TARNAUD, Jean THIERCELIN, TOYEN.

Il a été tiré vingt-cinq exemplaires sur papier couché vert d'eau, adornés de quelques poils de la Vraye Moustache.

99 (left) The Surrealist Group, 'We Don't Ear it that Way' (December 1958).

100 (facing page) Salvador Dalí, *The Sistine Madonna* (1958). Oil on canvas, 224.6 × 191.3 cm. The Metropolitan Museum of Art, New York, H. J. Heinz Collection.

Finally, in 1958, the contradictory positions taken over nuclear power by Dalí and the Surrealists, around which I have defined their conflicting postwar attitudes in the final part of this book, reached their extreme point. On the very same day – 18 February 1958 – Dalí's latest exhibition opened in at the M. Knoedler & Co. Gallery in New York, including the series of diffusive drawings and watercolours titled 'Anti-Matter Archangels and Celestial Visions', whilst in Paris the Surrealists launched their broadside 'Expose the Physicists, Empty the Laboratories' ('Démasquez les physiciens, Videz les laboratoires').[49] Defining their hostility towards atomic energy in the era of the Cold War, the Surrealists' text denounced the irresponsibility and lack of ethics evinced in what they saw as nuclear physicists' complacent research into, and development of lethal weapons at the behest of governments. Whereas Breton had spoken of modern physics as the most pioneering field of thought in his 1935 Tenerife interview, now in 1958 the Surrealists spoke caustically of the same revolution: 'Modern physics has . . . promised, has maintained and still promises tangible results, in the form of heaps of corpses.'[50] The note of caution sounded by Breton in 1935, on the danger of science becoming a new religion, was now transformed into a polemical statement by the Surrealists, aimed at a wayward science threatening to become the new 'opium of the people', creating a 'theology of the Bomb' through its apocalyptic menace.[51] The name of Salvador Dalí or even 'Avida Dollars' stubbornly refuses to appear in the Surrealists' anti-nuclear statement, probably because Dalí's weird mix of nuclear physics and Catholic mysticism had by then made him a *persona non grata*, beyond the scope of even the Surrealists' far-reaching radar of contempt.

Dalí's *mélange* of science and religion received another outing later in 1958, when his 'Anti-Matter Manifesto' appeared as the preface to the catalogue of another exhi-

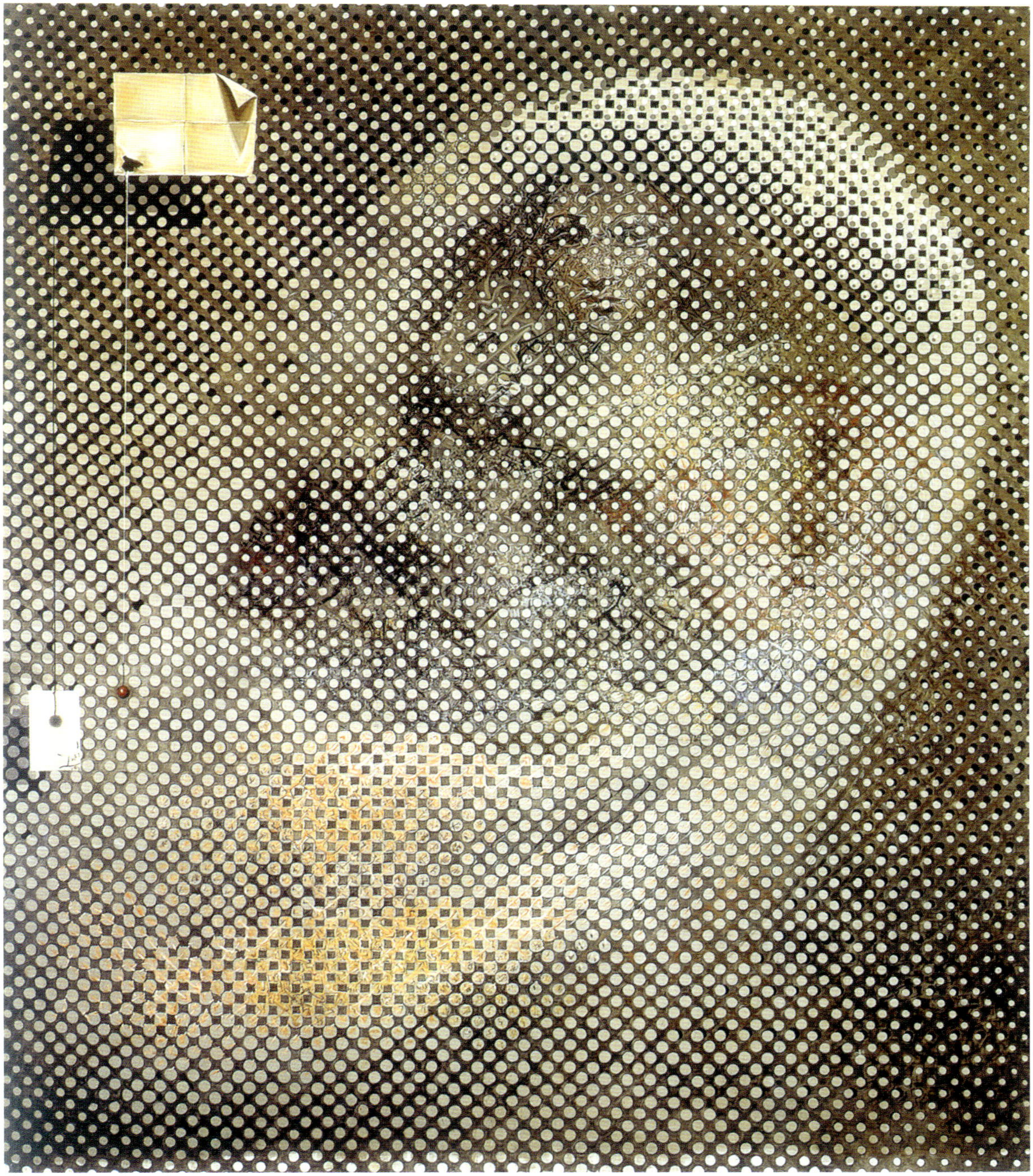

bition of his work, at the Carstairs Gallery in New York.[52] Speaking of his love for physics, Dalí made his now well-known post-Surrealist conversion statement, redrawing his intellectual genealogy along lines laid down by QM rather than psychoanalysis:

> In the Surrealist period I wanted to create the iconography of the interior world – the world of the marvellous, of my father Freud. I succeeded in doing it.
>
> Today the exterior world – that of physics – has transcended the one of psychology. My father today is Dr. Heisenberg.
>
> It is with pi-mesons and the most gelatinous and indeterminate neutrinos that I want to paint the beauty of the angels and of reality.[53]

In the postwar period, the Surrealists did not really need another reason to hate Dalí, but he was more than happy to give them one. It is difficult to conceive of a starker contrast than the one lying between the Surrealists' support for the Comité de Lutte Anti-Nucléaire and call to 'boycott conferences devoted to the exaltation of the atom', and Dalí's exactly contemporary, provocative celebration and mythification of its powers.[54] Unsurprisingly, his ridiculous particle angels found no takers among the Surrealists.

I believe that the reverential obeisance Dalí paid Werner Heisenberg in his 1958 'Anti-Matter Manifesto' – continued in a lecture on the great physicist at the Théâtre de l'Étoile in Paris in May of that year – is also countered by the Surrealists' anti-nuclear tract. In a passage denouncing '[p]rotests against the arms race that some physicists today put on a show of signing', the Surrealists declare such protestations comparable to those made by apologists for Christianity who, they maintain, similarly presume to condemn a state of affairs in the social sphere for which they are largely to blame.[55] This swipe at peace protesting physicists sounds to me like a direct reference to the so-called Göttingen Manifesto, first published in the German press in April 1957. From a besieged Federal Republic of Germany, rearming from the early 1950s (with a nuclear weapons programme from 1955), criss-crossed with American nuclear devices and increasingly seen as the buffer and potential battleground between East and West, a refusal to participate in 'the production, testing, or use of atomic weapons' was signed by eighteen of the most prominent members of the German Physics Society, among whom Heisenberg was the best-known signatory.[56] Taking umbrage at what the Surrealists saw as barefaced hypocrisy, 'Expose the Physicists, Empty the Laboratories' could well have been prompted by the Göttingen Manifesto, which was widely circulated and debated for nearly a year. If so, it extends the tussle between Dalí and the Surrealists over Heisenberg's legacy in the arts.[57]

Deaf, as ever, at the height of the Cold War, to the humanitarian and political concerns that preoccupied the Surrealists, Dalí's exhibition at the Carstairs Gallery opened on 6 December 1958, including among its works the large-scale canvas completed that year, *The Sistine Madonna* (fig. 100). The day before, Breton had spoken in France at an event in support of conscientious objection to the Algerian War, where he used religion again as a context within which to attack those physicists who colluded in the build-up of nuclear arsenals.[58] There is, then, vicious irony in Dalí's 'anti-matter ear', *The Sistine Madonna*, finding its way soon after into the Surrealists' international exhibition co-directed by Breton, *Surrealist Intrusion in the Enchanters' Domain*. This happened courtesy of unusual indelicacy on the part of fellow organiser Marcel Duchamp, who over the years had pulled off the unique trick of juggling friendships with Dalí and Breton simultaneously. Opening at the d'Arcy Galleries in New York in November 1960, in the year a new surge of anti-nuclear and peace activity greeted France's detonation of her first atomic bomb, that exhibition saw the temporary 'reinsinuation' of Dalí into the Surrealist group, coinciding with the Surrealists' most extreme Dalíphobia, anti-clericalism and opposition to nuclear physics.

This is the background against which we should view the Surrealists' vitriolic response to the 'accidental' inclusion of their former comrade's *Sistine Madonna* in their show, in the tract 'We Don't Ear it that Way' (fig. 99).[59] For there, Dalí is impugned in similar terms to those used by Breton in his 1942 Yale lecture from which I quoted at the beginning of this final section of my history of Surrealism and modern physics, as 'Hitler's former apologist, the fascist painter, the religious bigot, and the avowed racist, friend of Franco, who opened Spain as a drill-ground for the most abominable surge of barbary [*sic*] the world has yet endured.'[60] It is ironic that the trigger for that vilification was what is plainly one of Dalí's best post-Surrealist paintings, but the Surrealists had lost all capacity for looking favourably on the work of their one-time friend by then.[61] We can be certain that if the despised Dalí's Cold War defence of nuclear physics did not exactly engender the Surrealists' opposition to it, then it certainly confirmed it towards the end of the 1950s, by which time Surrealism's love affair with modern physics was well and truly over.

NOTES

Introduction

1 Michel Carrouges, *André Breton and the Basic Concepts of Surrealism*, trans. Maura Prendergast, Alabama: University of Alabama Press, 1974, 230; '[o]n n'a pas assez souligné le fait que le surréalisme est né à l'époque de cette grande révolution des sciences physiques et mathématiques opérée par Einstein, Niels Bohr, Riemann, Heisenberg, de Broglie, et tant d'autres', Michel Carrouges, *André Breton et les données fondamentales du surréalisme*, Paris: Gallimard, 1950, 277.

2 Carrouges, *André Breton et les données fondamentales du surréalisme*, 265.

3 Carrouges, *André Breton and the Basic Concepts of Surrealism*, 268–9; Carrouges, *André Breton et les données fondamentales du surréalisme*, 320–21.

4 Linda Dalrymple Henderson, *The Fourth Dimension and Non-Euclidean Geometry in Modern Art*, Princeton: Princeton University Press, 1983. Also see Linda Dalrymple Henderson, Editor's Introduction: I. 'Writing Modern Art and Science – An Overview'; II. 'Cubism, Futurism, and Ether Physics in the Early Twentieth Century', *Science in Context* (sp. issue, 'Modern Art and Science'), vol. 17, no. 4, December 2004, 423–66.

5 Henderson, *The Fourth Dimension and Non-Euclidean Geometry in Modern Art*, 345–9.

6 Virginia Parrott Williams, *Surrealism, Quantum Philosophy, and World War I*, New York and London: Garland Publishing Inc., 1987.

7 Williams, *Surrealism, Quantum Philosophy, and World War I*, iii.

8 C. P. Snow, *The Two Cultures* [1959], Cambridge: Cambridge University Press, 1993.

9 Snow, *The Two Cultures*, 16–17.

10 Snow, *The Two Cultures*, 16.

11 Lingua Franca (eds), *The Sokal Hoax: The Sham that Shook the Academy*, Lincoln: University of Nebraska Press, 2000.

12 Alan Sokal, 'Transgressing the Boundaries: Toward a Transformative Hermeneutics of Quantum Gravity', *Social Text*, no. 46/47, Spring–Summer 1996, 217–52. Quotations are from Alan Sokal and Jean Bricmont, *Intellectual Impostures*, London: Profile, 1998, 199–240. Sokal's enjoyable exploit gained extra appeal from its appearance in a special number of *Social Text* meant as a response to criticism from the scientific community of contemporary critical theory.

13 Sokal and Bricmont, *Intellectual Impostures*, 199.

14 Sokal and Bricmont, *Intellectual Impostures*, 200.

15 Sokal and Bricmont, *Intellectual Impostures*, 210.

16 Sokal and Bricmont, *Intellectual Impostures*, 249.

17 Sokal and Bricmont, *Intellectual Impostures*, 186, 187.

18 'les années 1930 furent caractérisées par une nouvelle aggravation de la crise du rationalisme qui, de plus en plus menacé, essayait cependant de maintenir ses positions avec la fondation de l'*Union rationaliste* par Bayet, Boll et Langevin et avec la publication des *Ages de l'intelligence* et des *Progrès de la conscience dans la philosophie occidentale* de Léon Brunschvicg. Ces sursauts n'empêchèrent pas le rationalisme, pourtant maître encore de beaucoup de chaires universitaires, d'être de plus en plus battu en brêche. Bergson, après plusieurs années de méditation silencieuse, fit paraître en 1932 son dernier grand livre *Les Deux Sources de la morale et de la religion* qui élargit encore son audience. Les progrès de la physique quantique, representée en France par Louis de Broglie, incitèrent savants et philosophes à remettre en cause les notions de causalité et de déterminisme. S'inspirant des recherches freudiennes sur l'inconscient, la pensée philosophique tendit à s'élargir en s'ouvrant à des explorations plus subtiles des marges mystérieuses et des zones profondes de la conscience avec, par exemple, les travaux de Bachelard élaborant une méthode de connaissance totale et une critique explicative du symbolisme des images', Jean-Louis Loubet del Bayle, *Les non-conformistes des années 30. Une tentative de renouvellement de la pensée politique française*, Paris: Éditions du seuil, 1969, 24. Symptomatic is Albert Bayet's discussion of the various rationalisms, religious, metaphysical and scientific, in his *Qu'est-ce que le rationalisme?* Paris: Éditions Rationalistes, 1939.

19 Sokal and Bricmont, *Intellectual Impostures*, 248.

1 Modern Physics and its Philosophy: *The Reception in France*

1 In this book, the terms 'modern physics' and 'the new physics' are meant to refer to Relativity and quantum physics, though I accept that the scholarship is not consistent on this issue, usually regarding Relativity as the last gasp of classical physics, and often equating only the quantum with the 'modern age'.

2 From a vast industry of scholarship, I have found the following useful in developing some understanding of Relativity: Mara Beller, Robert S. Cohen, and Jürgen Renn (eds), *Einstein in Context*, Cambridge: Cambridge University Press, 1993; Jeremy Bernstein, *Einstein*, Bungay, Suffolk: Fontana/Collins, 1973; Albert Einstein, *Relativity: The Special and the General Theory* [1920], London: Methuen, 1960; Albert Einstein et al, *The Principle of Relativity: A Collection of Original Papers on the Special and General Theory of Relativity*, trans. W. Perrett and G.B. Jeffery [1923], New York: Dover, 1952; Philipp Frank, *Einstein: His Life and Times*, London: Jonathan Cape, 1948; Peter Galison, *Einstein's Clocks, Poincaré's Time: Empires of Time*, London: Sceptre, 2003; Peter Galison, Michael Gordin, David Kaiser (eds), *Science and Society: The History of Modern Physical Science in the Twentieth Century*, vol. 2, 'Making General Relativity', New York and London: Routledge, 2001; Stanley Goldberg, *Understanding Relativity: Origin and Impact of a Scientific Revolution*, Oxford: Clarendon Press, 1984; Gerald Holton, *Thematic Origins of Scientific Thought: Kepler to Einstein*, Cambridge, Mass.: Harvard University Press, 1974; Don Howard and John Stachel (eds), *Einstein and the History of General Relativity*, Einstein Studies, vol. 1, Boston, Basel, Berlin: Birkhäuser: 1989; Abraham Pais, *'Subtle is the Lord . . .' The Science and the Life of Albert Einstein*, Oxford: Oxford University Press, 1982; L. Pearce Williams (ed.), *Relativity Theory: Its Origins and Impact on Modern Thought*, New York: John Wiley and Sons, 1968.

3 Albert Einstein, 'Autobiographical Notes', Paul Arthur Schlipp (ed.), *Albert Einstein: Philosopher-Scientist*, Illinois: Open Court, 1949, 2–94, 17.

4 Quoted in Bernstein, *Einstein*, 33.

5 Einstein, 'Autobiographical Notes', 53.

6 Einstein, 'Autobiographical Notes', 53.

7 Albert Einstein, 'On the Electrodynamics of Moving Bodies', *The Principle of Relativity*, 37–65, ['Zur Elektrodynamik bewegter Körper', *Annalen der Physik*, vol. 17, no. 10, 26 September 1905, 891–921].

8 Albert Einstein, 'Über einen die Erzeugung und Verwandlung des Lichtes betreffenden heuristischen Gesichtspunkt', *Annalen der Physik*, vol. 17, no. 6, 9 June 1905, 132–48, and Albert Einstein, 'Über die von der molekularkinetischen Theorie der Wärme geforderte Bewegung von in ruhenden Flüssigkeiten suspendierten Teilchen', *Annalen der Physik*, vol. 17, no. 7, 18 July 1905, 549–60.

9 Quoted in Albrecht Fölsing, *Albert Einstein*, trans. Ewald Osers, London: Penguin, 1997, 245.

10 Albert Einstein, 'The Foundation of the General Theory of Relativity', *The Principle of Relativity*, 109–64, ['Die Grundlage der allgemeinen Relativitätstheorie', *Annalen der Physik*, vol. 49, no. 7, 11 May 1916, 769–822].

11 Bernstein, *Einstein*, 100. The principle of equivalence first appears in Albert Einstein, 'On the Influence of Gravitation on the Propagation of Light', *The Principle of Relativity*, 97–108, ['Über den Einfluß der Schwerkraft auf die Ausbreitung des Lichtes', *Annalen der Physik*, vol. 35, no. 10, 1 September 1911, 898–908].

12 Philipp Frank, *Einstein: His Life and Times*, London: Jonathan Cape, 1948, 158.

13 For a detailed account of the expedition, see John Earman and Clark Glymour, 'Relativity and Eclipses: The British Expeditions of 1919 and Their Predecessors', *Historical Studies in the Physical Sciences*, vol. 11, 1980, 49–85. For an excellent interpretation of Eddington's intentions, see Matthew Stanley, '"An Expedition to Heal the Wounds of War": The 1919 Eclipse and Eddington as Quaker Adventurer', *ISIS*, no. 94, 2003, 57–89.

14 I have been as rigorous as a book of this kind can allow in my use of the scientific terminology, though I am aware that there are precise meanings ascribed to and overlaps between 'quantum physics', 'quantum theory' and 'quantum mechanics' that I have probably not always been able to observe. In gaining an art historian's imperfect understanding of the concepts, philosophy and early history of QM, I have been helped by the following, plucked from an intimidating mountain of scholarly work: David Z. Albert, *Quantum Mechanics and Experience*, Cambridge, Mass. and London: Harvard University Press, 1992; Niels Bohr, *Atomic Theory and the Description of Nature*, Cambridge: Cambridge University Press, 1934; Louis de Broglie, *Matter and Light: The New Physics*, trans. W.H. Johnstone, London: George Allen and Unwin Ltd., 1939; Paul Dirac, *The Development of Quantum Theory*, New York: Gordon and Breach, 1971; Paul Dirac, *Directions in Physics*, New York: John Wiley and Sons, 1978; Richard P. Feynman, *Six Easy Pieces: The Fundamentals of Physics Explained*, Harmondsworth: Penguin, 1998; George Gamow, *Thirty Years that Shook Physics*, New York: Doubleday, 1966; John Gribbin, *In Search of Schrödinger's Cat: Quantum Physics and Reality* [1984], London: Black Swan, 1998; Werner Heisenberg, *The Physical Principles of the Quantum Theory*, trans. Carl Eckart and Frank C. Hoyt, Chicago: University of Chicago Press, 1930; Armin Hermann, *The Genesis of Quantum Theory (1899–1913)*, trans. Claude W. Nash, Cambridge, Mass. and London: MIT, 1971; Max Jammer, *The Conceptual Development of Quantum Mechanics*, New York: McGraw-Hill, 1966; Max Jammer, *The Philosophy of Quantum Mechanics: The Interpretations of Quantum Mechanics in Historical Perspec-*

tive, New York: John Wiley and Sons, 1974; Helge Kragh, *Quantum Generations: A History of Physics in the Twentieth Century*, Princeton: Princeton University Press, 1999; Edward M. MacKinnon, *Scientific Explanation and Atomic Physics*, Chicago and London: University of Chicago Press, 1982; Tim Maudlin, *Quantum Non-Locality and Relativity: Metaphysical Intimations of Modern Physics*, Oxford: Blackwell, 1994; Jagdish Mehra and Helmut Rechenberg, *The Historical Development of Quantum Theory: The Quantum Theory of Planck, Einstein, Bohr and Sommerfeld: Its Foundations and the Rise of its Difficulties* 1900–1925, vol. 1, part 1, New York, Heidelberg, Berlin: Springer-Verlag, 1982; Alastair I. M. Rae, *Quantum Physics: Illusion or Reality?* Cambridge: Cambridge University Press, 1986; Erwin Schrödinger, *Science and the Human Temperament*, trans. James Murphy, London: George Allen and Unwin Ltd., 1935; Gary Zukav, *The Dancing Wu Li Masters: An Overview of the New Physics* [1979], New York: Bantam Books, 1986. These publications have their own history and reception, of course, ranging from the mathematically intimidating to the philosophically demanding and the scientifically disreputable.

15 Max Planck, 'Zur Theorie des Gesetzes der Energieverteilung im Normalspectrum', *Verhandlungen der Deutschen Physikalischen Gesellschaft*, vol. 2, no. 17, 14 December 1900, 237–45; Max Planck, 'Ueber das Gesetz der Energieverteilung im Normalspektrum', *Annalen der Physik*, vol. 4, no. 3, 1 March 1901, 553–63.

16 h = 0.0000000000000000000000000006626 joule-seconds.

17 Max Jammer, *The Conceptual Development of Quantum Mechanics*, New York: McGraw-Hill, 1966, 22.

18 Albert Einstein, 'Über einen die Erzeugung und Verwandlung des Lichtes betreffenden heuristischen Gesichtspunkt', *Annalen der Physik*, vol. 17, no. 6, 4 July 1905, 132–48.

19 Einstein, 'Autobiographical Notes', 45.

20 Einstein, 'Autobiographical Notes', 45.

21 Niels Bohr, 'On the Constitution of Atoms and Molecules', *The Philosophical Magazine and Journal of Science*, vol. 26, no. 151, July 1913, 1–25; 'On the Constitution of Atoms and Molecules', *The Philosophical Magazine and Journal of Science*, vol. 26, no. 153, September 1913, 476–502; 'On the Constitution of Atoms and Molecules', *The Philosophical Magazine and Journal of Science*, vol. 26, no. 155, November 1913, 857–75. These classic papers of 1913 received a fiftieth anniversary reprint as Niels Bohr, *On the Constitution of Atoms and Molecules*, Copenhagen: Munksgaard, 1963.

22 Bohr, 'On the Constitution of Atoms and Molecules', *The Philosophical Magazine and Journal of Science*, vol. 26, no. 151, July 1913, 24–5.

23 A. H. Compton, 'A Quantum Theory of the Scattering of X-Rays by Light Elements', *The Physical Review*, vol. 21, Series 2, no. 5, May 1923, 483–502.

24 Compton, 'Quantum Theory of the Scattering of X-Rays', *Physical Review*, vol. 21, 501.

25 Jammer, *Conceptual Development of Quantum Mechanics*, 248–9.

26 Louis de Broglie, 'Sur les interférences et la théorie des quanta de lumière', *Comptes Rendus*, vol. 175, 1922, 811–13; 'Rayonnement noir et quanta de lumière', *Journal de Physique*, vol. 3, November 1922, 422–8; 'Ondes et quanta', *Comptes Rendus*, vol. 177, 1923, 507–10; 'Quanta de lumière, diffraction et interférences', *Comptes Rendus*, vol. 177, 1923, 548–50; 'Les quanta, la théorie cinétique des gaz et le principe de Fermat', *Comptes Rendus*, vol. 177, 1923, 630–32. The three brief *Comptes Rendus* papers were summarised and translated as Louis de Broglie, 'A Tentative Theory of Light Quanta', *The Philosophical Magazine and Journal of Science*, vol. 47, no. 278, February 1924, 446–58.

27 Louis de Broglie, 'Recherches sur la théorie des quanta', *Annales de Physique*, vol. 3/4, January–February 1925, 22–128.

28 André Breton, 'Manifesto of Surrealism', *Manifestoes of Surrealism*, trans. Richard Seaver and Helen R. Lane, Michigan: University of Michigan Press, 1972, 1–47, 5; 'La seule imagination me rend compte de ce qui *peut être* . . .', Breton, 'Manifeste du surréalisme', *Oeuvres complètes* [OC], vol 1, Paris: Gallimard, 1988, 309–46, 312.

29 Wolfgang Pauli, *Exclusion Principle and Quantum Mechanics*, Neuchatel: Griffon, 1947, 13. His classic paper formulating the electron's new two-valued quantum number is Wolfgang Pauli, 'Über den Einfluß der Geschwindigkeitsabhängigkeit der Elektronenmasse auf den Zeemaneffekt', *Zeitschrift für Physik*, vol. 31, 1925, 373–85.

30 Pauli, *Exclusion Principle and Quantum Mechanics*, 13.

31 George Uhlenbeck and Samuel Goudsmit, 'Ersetzung der Hypothese vom unmechanischen Zwang durch eine Forderung bezüglich des inneren Verhaltens jedes einzelnen Elektrons', *Die Naturwissenschaften*, vol. 13, 20 November 1925, 953–4, 954. The term 'spin' does not appear in Uhlenbeck and Goudsmit's text, though it does in its English-language version, George Uhlenbeck and Samuel Goudsmit, 'Spinning Electrons and the Structure of Spectra', *Nature*, vol. 117, no. 2938, 20 February 1926, 264–5.

32 Wolfgang Pauli, 'Über den Zusammenhang des Abschlusses der Elektronengruppen im Atom mit der Komplexstruktur der Spektren', *Zeitschrift für Physik*, vol. 31, 1925, 765–83.

33 Einstein, 'Autobiographical Notes', 47.

34 Pauli, *Exclusion Principle and Quantum Mechanics*, 9.

35 'There never exist two or more equivalent electrons in an atom which, in strong fields, agree in all quantum numbers . . . If there exists in the atom an electron for which these quantum numbers (in the external field) have definite values, this state is "occupied"', Pauli quoted in Jammer, *Conceptual Development of Quantum Mechanics*, 144.

36 Werner Heisenberg, *Encounters with Einstein and Other Essays on People, Particles, and Places* [1983], Princeton, New Jersey: Princeton University Press, 1989, 38.

37 Quoted in David. C. Cassidy, *Uncertainty: The Life and Science of Werner Heisenberg*, New York: W.H. Freeman and Company, 1992, 197.

38 Paul Dirac, *The Development of Quantum Theory*, New York: Gordon and Breach, 1971, 19.

39 Werner Heisenberg, 'Über quantentheoretische Umdeutung kinematischer und mechanischer Beziehungen', *Zeitschrift für Physik*, vol. 33, 1925, 879–93.

40 Jammer, *Conceptual Development of Quantum Mechanics*, 209.

41 Max Born and Pascual Jordan, 'Zur Quantenmechanik', *Zeitschrift für Physik*, vol. 34, 1925, 858–88.

42 'die "verschärfte Quantenbedingung"', Born and Jordan quoted in Jammer, *Conceptual Development of Quantum Mechanics*, 211. The famous equation makes its first appearance on p. 871 of the Born-Jordan paper.

43 Max Born, Werner Heisenberg and Pascual Jordan, 'Zur Quantenmechanik. II', *Zeitschrift für Physik*, vol. 35, 4 February 1926, 557–615.

44 Jammer, *Conceptual Development of Quantum Mechanics*, 212.

45 Dirac, *Development of Quantum Theory*, 23.

46 Dirac, *Development of Quantum Theory*, 29.

47 P. A. M. Dirac, 'The Fundamental Equations of Quantum Mechanics', *Proceedings of the Royal Society of London*, vol. 109, no. A752, 1 December 1925, 642–53, 642.

48 P. A. M. Dirac, 'Quantum Mechanics and a Preliminary Investigation of the Hydrogen Atom', *Proceedings of the Royal Society of London*, vol. 110, no. A755, 1 March 1926, 561–79; 'The Elimination of the Nodes in Quantum Mechanics', *Proceedings of the Royal Society of London*, vol. 111, no. A757, 1 May 1926, 281–305; 'Relativity Quantum Mechanics with an Application to Compton Scattering', *Proceedings of the Royal Society of London*, vol. 111, no. A758, 2 June 1926, 405–23; 'On Quantum Algebra', *Proceedings of the Cambridge Philosophical Society*, vol. 23, no. 4, October 1926, 412–18. Each of these papers emphasises the importance of non-commutation in its introduction.

49 Referring to de Broglie's work, Einstein wrote: 'I shall discuss this interpretation in greater detail because I believe it involves more than merely an analogy', quoted in Jammer, *Conceptual Development of Quantum Mechanics*, 249, noting Albert Einstein, 'Quantentheorie des einatomigen idealen Gases', *Berliner Berichte* 1925, 3–14, 9.

50 Quoted in Walter Moore, *A Life of Erwin Schrödinger*, Cambridge: Cambridge University Press, 1994, 138.

51 Erwin Schrödinger, 'Quantisierung als Eigenwertproblem', *Annalen der Physik*, vol. 79, no, 4, 1926, 361–76. The other classic papers of 1926 are 'Quantisierung als Eigenwertproblem', *Annalen der Physik*, vol. 79, no. 6, 1926, 489–527; 'Über das Verhältnis der Heisenberg-Born-Jordanschen Quantenmechanik zu der meinen' ['On the Relation between the Quantum Mechanics of Heisenberg, Born, and Jordan, and that of Schrödinger'], *Annalen der Physik*, vol. 79, no. 8, 1926, 734–56; 'Quantisierung als Eigenwertproblem', *Annalen der Physik*, vol. 80, no. 13, 1926, 437–90; 'Quantisierung als Eigenwertproblem', *Annalen der Physik*, vol. 81, no. 18, 1926, 109–39. These papers are collected and supplemented with other key statements in Erwin Schrödinger, *Collected Papers on Wave Mechanics* [1927], trans. J.F. Shearer, New York: Chelsea Publishing Co., 1982.

52 For an authoritative and entertaining account of the place of Schrödinger and his equation in the history of quantum theory, see Arthur I. Miller, 'Erotica, Aesthetics and Schrödinger's Equation', Graham Farmelo (ed.), *It Must be Beautiful: Great Equations of Modern Science*, London and New York: Granta, 2002, 81–101.

53 Heisenberg, *Encounters with Einstein*, 50.

54 Schrödinger, *Collected Papers on Wave Mechanics*, 46, n. 1 (trans. slightly modified); Schrödinger, 'Über das Verhältnis der Heisenberg-Born-Jordanschen Quantenmechanik zu der meinen', 735, n. 2.

55 Quoted in Cassidy, *Uncertainty*, 215.

56 Werner Heisenberg, 'Über den anschaulichen Inhalt der quantentheoretischen Kinematik und Mechanik', *Zeitschrift für Physik*, vol. 43, 31 May 1927, 172–98.

57 Quoted in Arthur I. Miller, *Insights of Genius: Imagery and Creativity in Science and Art*, New York: Copernicus, 1996, 120; 'so wird durch Quantenmechanik die Ungültigkeit des Kausalgesetzes definitiv festgestellt', Heisenberg, 'Über den anschaulichen Inhalt der quantentheoretischen Kinematik und Mechanik', 197.

58 Although Bohr's ideas on Complementarity were first made public in 1927, the two chief statements were published in the following year as Niels Bohr, 'Das Quantenpostulat und die neuere Entwicklung der Atomistik', *Die Naturwissenschaften*, vol. 16, 13 April 1928, 245–7, and Niels Bohr, 'Wirkungsquantum und Naturbeschreibung', *Die Naturwissenschaften*, vol. 17, 28 June 1929, 483–6. The first was translated as Niels Bohr 'The Quantum Postulate and the Recent Development of Atomic Theory', *Nature*, vol. 121, no. 3050, 14 April 1928, 580–90, and both are reprinted in Niels Bohr, *Atomic Theory and the Description of Nature*, Cambridge: Cambridge University Press, 1934, 52–91 and 92–101.

59 Bohr, *Atomic Theory and the Description of Nature*, 54.

60 John L. Heilbron, 'The Earliest Missionaries of the Copenhagen Spirit', Edna Ullman-Margalit (ed.), *Science in Reflection*, Dordrecht: Kluwer Academic Publishers, 1988, 201–33, 202. Apart from Bohr's writings and Jammer, *Conceptual Development of Quantum Mechanics*, 345–61, the two historical

essays I have found most helpful on the concept, philosophy and context of Complementarity are Heilbron's and Gerald Holton, 'The Roots of Complementarity', *Thematic Origins of Scientific Thought*, 115–61.

61 Werner Heisenberg, *Physics and Philosophy: The Revolution in Modern Science*, London: Penguin, 1990, 37.

62 Michel Paty, 'The Scientific Reception of Relativity in France', trans. Wayne K. Ishikawa, Thomas F. Glick (ed.), *The Comparative Reception of Relativity*, Dordrecht: D. Reidel, 1987, 113–67, 115. The indispensable study is Dominique Pestre, *Physique et physiciens en France*, 1918–1940, Paris: Archives Contemporaines, 1984.

63 Paty, 'The Scientific Reception of Relativity in France', 116.

64 The standard history of the reception and development of psychoanalysis in France, introduced with an assessment of the role of Surrealism, is Elisabeth Roudinesco, *A History of Psychoanalysis in France, 1925–1985*, trans. Jeffrey Mehlman, London: Free Association Books, 1990, 3–34.

65 Michel Biezunski, 'Einstein's Reception in Paris in 1922', Glick (ed.), *The Comparative Reception of Relativity*, 169–88, 174–5.

66 The larger canvas I am frantically trimming under the usual pressure of limited space includes the vast achievements of Poincaré and his supposed discovery of a version of Relativity before Einstein. A lifelong inspiration for Marcel Duchamp with his reduction of Euclidean geometry to a linguistic convention and other revolutionary-sounding postulates (though reform was his aim), Poincaré trumped Einstein, rather, by rejecting absolute time and space in 1902 in *La Science et l'hypothèse*. For the whole story of Einstein, Poincaré and France, see Galison, *Einstein's Clocks, Poincaré's Time: Empires of Time*; Stanley Goldberg, 'Henri Poincaré and Einstein's Theory of Relativity', *American Journal of Physics*, vol. 35 (ii), no. 10, October 1967, 934–44; Stanley Goldberg, 'Poincaré's Silence and Einstein's Relativity', *The British Journal for the History of Science*, vol. 5, part 1, no. 17, June 1970, 73–84; Holton, 'Poincaré and Relativity', *Thematic Origins of Scientific Thought*, 185–95; Paty, 'The Scientific Reception of Relativity in France', 116–23. For Duchamp and Poincaré see Linda Dalrymple Henderson, *Duchamp in Context: Science and Technology in the* Large Glass *and Related Works*, Princeton: Princeton University Press, 1998.

67 Paty, 'The Scientific Reception of Relativity in France', 150.

68 Paul Langevin, 'L'évolution de l'espace et du temps', *Scientia*, vol. 10, no. XIX, 1911, 31–54. Langevin invited Einstein to speak in France as early as 1914 but the war prevented his visit.

69 Paty, 'The Scientific Reception of Relativity in France', 116.

70 Breton, *Manifestoes of Surrealism*, 9–10; 'les procédés logiques, de nos jours, ne s'appliquent plus qu'à la résolution de problèmes d'intérêt secondaire. Le rationalisme absolu qui reste de mode ne permet de considérer que des faits relevant étroitement de notre expérience. Les fins logiques, par contre, nous échappent. Inutile d'ajouter que l'expérience même s'est vu assigner des limites. Elle tourne dans une cage d'où il est de plus en plus difficile de la faire sortir. Elle s'appuie, elle aussi, sur l'utilité immédiate, et elle est gardée par le bon sense', Breton, *OC*, vol. 1, 316.

71 Breton, *Manifestoes of Surrealism*, 6; 'l'attitude réaliste, inspirée du positivisme, de saint Thomas à Anatole France, m'a bien l'air hostile à tout essor intellectuel et moral. Je l'ai en horreur, car elle est faite de médiocrité, de haine et de plate suffisance. C'est elle qui engendre aujourd'hui ces livres ridicules, ces pièces insultantes. Elle se fortifie sans cesse dans les journaux et fait échec à la science, à l'art, en s'appliquant à flattter l'opinion dans ses goûts les plus bas . . .', Breton, *OC*, vol. 1, 313.

72 Biezunski, 'Einstein's Reception in Paris in 1922', 180.

73 Michel Biezunski, 'Popularisation and Scientific Controversy: The Case of the Theory of Relativity in France', Terry Shinn and Richard Whitley (eds), *Expository Science: Forms and Functions of Popularisation*, Dordrecht: D. Reidel, 1985, 183–94. Einstein was received in Paris by Xavier Léon's Société française de Philosophie as well as the Collège de France but not by the Académie française or Société française de Physique, Paty, 'The Scientific Reception of Relativity in France', 137. Einstein decided not to attend the meeting arranged for him at the Académie des sciences when he heard that thirty members planned to walk out on his arrival in support of the boycott of German scientists.

74 Robert L. Weber, *Pioneers of Science: Nobel Prize Winners in Physics*, Bristol and London: The Institute of Physics, 1980, 91.

75 Kragh, *Quantum Generations*, 189.

76 Kragh, *Quantum Generations*, 20–22, 150.

77 Kragh, *Quantum Generations*, 144.

78 Kragh, *Quantum Generations*, 189.

79 Léon Brunschvicg, 'Science et prise de conscience', *Scientia*, vol. 55, no. CCLXV, May 1934, 329–40.

80 Gaston Bachelard, 'La Richesse d'inférence de la physique mathématique', *Scientia*, vol. 50, no. CCXXXII, August 1931, 79–88.

81 Hans Reichenbach, 'Crise de la causalité', *Documents*, year 1, no. 2, May 1929, 105–8. I am grateful to David Lomas for mentioning this to me.

82 In 1913, at the request of the editor of *Scientia*, Freud wrote 'The Claims of Psychoanalysis to Scientific Interest', *Historical and Expository Works on Psychoanalysis*, Pelican Freud Library, vol. 15, Harmondsworth: Penguin, 1986, 27–56.

83 'un lieu de convergence de la spéculation proprement philosophique et de la critique scientifique et religieuse', Louis Lavelle, *La Philosophie française entre les deux guerres*, Paris: Éditions Montaigne, 1942, 231.

84 Émile Meyerson, *La Déduction relativiste*, Paris:

Payot, 1925; *Réel et déterminisme dans la physique quantique*, Paris: Hermann, 1933. Léon Brunschvicg, *L'Expérience humaine et la causalité physique*, Paris: Alcan, 1922; *La Physique du vingtième siècle et la philosophie*, Paris: Hermann, 1936.

85 Émile Meyerson, 'Le relativisme, théorie du réel', *Revue de Métaphysique et de Morale* [*RMM*], year 31, no. 1, January–March 1924, 29–48.

86 'un renouvellement de la psychologie de l'intelligence', Léon Brunschvicg, 'Le Temps et la causalité', *RMM*, year 29, no. 1, January–March 1922, 1–33, 1.

87 For evidence of a recent new wave of interest in Meyerson, see Frédéric Fruteau de Laclos, *La philosophie de l'intellect d'Émile Meyerson. De l'épistémologie à la psychologie*, unpublished dissertation submitted for the degree of Ph.D., Université de Paris X-Nanterre, 2004.

88 Louis de Broglie, 'La physique moderne et l'œuvre de Fresnel', *RMM*, year 34, no. 4, October–December 1927, 421–40.

89 Further evidence of de Broglie's crossover appeal is his presentation to the Société française de Philosophie (also founded by Xavier Léon) on 12 November 1929, 'Déterminisme et causalité dans la physique contemporaine', *Bulletin de la Société française de Philosophie*, year 29, no. 5, October–December 1929, 141–60. Einstein, Langevin, Perrin, Émile Borel and the gifted amateur Paul Valéry took part in the discussion that followed. Einstein had not made a special trip, but was passing through Paris at the time of de Broglie's lecture. Consequently he was not on hand (though Valéry was) when the Société staged Brunschvicg's similarly titled *séance* in March the following year, 'Le déterminisme et la causalité dans la physique contemporaine', *Bulletin de la Société française de Philosophie*, year 30, no. 2, April–June 1930, 50–78.

90 As ever, lack of space limits my discussion here, but the *Revue de Synthèse* (before 1931 the *Revue de Synthèse Historique*) which published Bachelard, de Broglie, Langevin, Abel Rey and Philipp Frank deserves greater English-language treatment than it has received hitherto in the scholarship on French culture and the avant-garde.

91 Eugenio Rignano, 'La psychologie dans ses rapports avec la philosophie et avec la science', *RMM*, year 33, no. 4, October–December 1926, 455–67. Rignano was also one of the founding editors of *Scientia*.

92 'quantité effarante des publications de la physique nouvelle', Michel Souriau, 'G. Bachelard, *Le Nouvel Esprit scientifique*' (book review), *Recherches philosophiques*, no. 4, 1934–5, 440–2, 440.

93 'entrepris de faire connaître en France les travaux du Cercle de Vienne', Lucien Brunet, 'La Logique de la science et l'école de Vienne' (book review), *Revue de Synthèse*, Tome XII, no. 1, April 1936, 113–15, 113.

94 Léon Brunschvicg was active at Hermann too, overseeing the sub-section 'Le Progrès de l'esprit'.

2 Relativity and Epistemology: *André Breton and Gaston Bachelard*

1 André Breton, 'Crisis of the Object', *Surrealism and Painting*, trans. Simon Watson Taylor, London: Macdonald and Co., 1972, 275–80; Breton, 'Crise de l'objet', *Le Surréalisme et la peinture*, Paris: Gallimard, 1965, 275–80. Gaston Bachelard, *The New Scientific Spirit*, trans. Arthur Goldhammer, Boston: Beacon Press, 1984; Gaston Bachelard, *Le Nouvel Esprit scientifique*, Paris: Alcan, 1934.

2 Judith Robinson-Valéry, 'The Fascination of Science', Paul Gifford and Brian Stimpson (eds), *Reading Paul Valéry: Universe in Mind*, Cambridge: Cambridge University Press, 1998, 77–84, 77. Fluent in English, Valéry was also familiar with classical physics, reading Faraday and Maxwell in 1893. In that year, too, he read Lord Kelvin's *Popular Lectures and Addresses* (1889) from which Alfred Jarry would later borrow elements for chapters 37 and 38 of *Gestes et Opinions du Docteur Faustroll* (1898).

3 We glimpsed Valéry in my last chapter, at the 1929 and 1930 talks given by de Broglie and Brunschvicg to the Société française de Philosophie. Shortly after, he published Valéry, 'Les Sciences de l'esprit sont-elles essentiellement différentes des Sciences de la nature?' *Revue de Synthèse*, no. 3, October 1931, 9–11. Worth noting in passing is another well-connected amateur, the Cubist painter Albert Gleizes whose friendships with Émile Borel, Langevin and René Allendy led to scientific speculation in his art and theory, such as his attempt to attain a kind of Complementarity in his paintings, and in his theoretical discussion of the possibility of depicting space and time (with fast and loose reference to Heisenberg's Uncertainty Principle) in his survey 'Art and Science' [1931], *Art and Religion, Art and Science, Art and Production*, trans. Peter Brooke, London: Francis Boutle, 1999, 60–95, 89.

4 'il pouvait trouver dans les developpements de la physique moderne tout ce qui plaisait à son esprit: la netteté des concepts et des mots, les vastes horizons intellectuels brusquement découvertes, la fine subtilité des théories qui, en nous révélant la relativité de l'espace et du temps et les incertitudes du monde atomique, ont ouvert à la pensée humaine tant d'étranges perspectives nouvelles'; Louis de Broglie, 'Le Penchant de Paul Valéry pour la physique', *Cahiers du Sud*, 'Paul Valéry vivant', Paris, 1946, 109–10, 109.

5 'triomphante conquête de l'inconnu', 'heureuses métaphores', de Broglie, 'Le Penchant de Paul Valéry pour la physique', 109. On this subject, see Reino Virtanen, *L'Imagerie scientifique de Paul Valéry*, Paris: Librairie Philosophique J. Vrin, 1975, and Christine M. Crow, *Paul Valéry and Maxwell's Demon: Natural Order and Human Possibility*, Hull: University of Hull, 1972.

6 Robinson-Valéry, 'The Fascination of Science', Gifford and Stimpson (eds), *Reading Paul Valéry*, 77.

7 Valéry's aptitude in the sciences, and particularly his

knowledge of physics, only became known to a wide audience with the publication of his *Cahiers*, the notebooks he began shortly after the commencement of his twenty-five year 'silence' from poetry: Paul Valéry, *Cahiers* (facsimile edition of twenty-nine volumes), Paris: Centre National de la Recherche Scientifique, 1957–61. The 1946 tribute by de Broglie serves as a preface to the facsimile edition, underlining the high scientific content of the 257 notebooks. The page references I give are from the later edition, *Cahiers* (two vols), Paris: Gallimard, 1973. Valéry's reflections on mathematics and science are collected in vol. 2, respectively 775–830 and 831–920. Also see Judith Robinson-Valéry (ed.), *Fonctions de l'esprit: Treize savants redécouvrent Paul Valéry*, Paris: Hermann, 1983.

8 'Ce que je cherche dans la physique (ex[emple]: p[rincipe] de la relativité) c'est de quoi aborder les problèmes du moi', Valéry, *Cahiers*, vol. 2, 845. Linda Dalrymple Henderson has suggested to me that Valéry is, in fact, responding in this note to Henri Poincaré's pre-Einstein reference to a 'principe de relativité', Poincaré, *La Valeur de la Science*, Paris: Ernest Flammarion, 1904, 196.

9 Paul Valéry, *Ideé fixe*, trans. David Paul, London: Routledge and Kegan Paul, 1965, 120.

10 Quoted in Matthew Stanley, '"An Expedition to Heal the Wounds of War": The 1919 Eclipse and Eddington as Quaker Adventurer', *ISIS*, no. 94, 2003, 57–89, 74.

11 André Breton, 'Manifesto of Surrealism', *Manifestoes of Surrealism*, trans. Richard Seaver and Helen R. Lane, Michigan: University of Michigan Press, 1972, 1–47, 35; Breton, 'Manifeste du surréalisme', *Oeuvres complètes* [OC], vol. 1, Paris: Gallimard, 1988, 309–46, 327.

12 André Breton and Philippe Soupault, *The Magnetic Fields*, trans. David Gascoyne, London: Atlas, 1985, 41; 'La mer tourmentée illuminait ces régions; une végétation instantanée disparut et des agglomérations de vapeurs découvrirent les astres. Activité céleste pour la première fois explorée. Les planètes s'approchaient à pas de loup et des silences obscurs peuplaient les étoiles. Les collines s'entourent des moindres lenteurs. Il ne reste sur les marais que les souvenirs des vols. La nécessité des absurdités mathématiques n'est pas démontrée', Breton, 'Éclipses', *OC*, vol. 1, 62–7, 64.

13 Breton and Soupault, *Magnetic Fields*, 39; 'ville du nord-est', 'montagnes de sable et de fossiles', 'la mer change de couleur', '[s]ur ces plages de galets tachés de sang, on peut entendre les tendres murmures des astres', Breton, *OC*, vol. 1, 62.

14 Breton and Soupault, *Magnetic Fields*, 40; '[l]es nuages sentimentaux accouraient. C'était l'heure exacte et prévue', '[a]u contact des étoiles filantes, les yeux anxieux des femmes se sont fermés pour plusieurs années. Elles ne verront plus que les tapisseries du ciel de juin et des hautes mer; mais il y a les bruits magnifiques des catastrophes verticales et des événements historiques', Breton, *OC*, vol. 1, 63, 63–4.

15 Breton and Soupault, *Magnetic Fields*, 44; 'la tête en bas, les élégantes partent pour un voyage au centre de la terre', Breton, *OC*, vol. 1, 67.

16 Breton and Soupault, *Magnetic Fields*, 42; 'les derniers adeptes de ces théories prennent place sur la colline devant les cafés qui ferment', Breton, *OC*, vol. 1, 65.

17 Breton and Soupault, *Magnetic Fields*, 39; [l]orsque l'on tourne le dos à cette plaine, on aperçoit de vastes incendies', Breton, *OC*, vol. 1, 62.

18 Breton and Soupault, *Magnetic Fields*, 40; '[l]es richesses chimiques importées brûlaient aussi lourdement que l'encens', Breton, *OC*, vol. 1, 63.

19 Breton and Soupault, *Magnetic Fields*, 41; '[l]es balles de coton arrivaient à donner naissance au soleil vomi comme sur les affiches. Ce qui précède a trait aux singularités chimiques, à ces beaux précipités certains', Breton, *OC*, vol. 1, 65.

20 André Breton, 'Max Ernst' [1921], *The Lost Steps*, trans. Mark Polizzotti, Lincoln and London: University of Nebraska Press, 1996, 60–61, 60; 'une véritable photographie de la pensée', *OC*, vol. 1, 245–6, 245.

21 Breton, *Manifestoes of Surrealism*, 28; 'les modestes *appareils enregistreurs*', Breton, *OC*, vol. 1, 330.

22 Breton and Soupault, *Magnetic Fields*, 39; 'un trou sans profondeur connue qui attire tous nos regards', Breton, *OC*, vol. 1, 62.

23 'Joint Eclipse Meeting of the Royal Society and the Royal Astronomical Society', *The Observatory*, vol. XLII, no. 545, November 1919, 389–98.

24 In 1921 Einstein himself thought the response to him 'seemed psychopathological', quoted in Gerald Holton, *Einstein, History, and Other Passions: The Rebellion against Science at the End of the Twentieth Century*, New York: Addison-Wesley, 1996, 128. For a close reading of contemporary reaction in Germany and an interpretation of the origins of the Einstein cult see Lewis Elton, 'Einstein, General Relativity, and the German Press, 1919–1920', *ISIS*, vol. 77, no. 1, March 1986, 95–103.

25 Paul Valéry, 'Un article de *L'Athenaeum*', *NRF*, vol. 18, year 7, no. 75, 1 December 1919, 1118–22; 'X', 'Einstein's Theory of Gravitation', *The Athenaeum*, no. 4672, 14 November 1919, 1189. Valéry wrote two 'Letters from France' columns for *The Athenaeum* earlier that year. Michael Whitworth has established that the original *Athenaeum* article was authored by none other than Bertrand Russell, and I am very grateful to him for sharing this information with me.

26 Breton, *OC*, vol. 2, Paris: Gallimard, 1992, 1224, n.1.

27 Julien Gracq, *André Breton. Quelques aspects de l'écrivain*, Paris: Librairie José Corti, 1948.

28 Gracq, *André Breton*, 47–87.

29 'il est légitime de dire qu'un fil unique, un fil conducteur par excellence par l'intermédiaire du subconscient de Breton traverse le surréalisme, l'hégélianisme et le freudisme . . .', Gracq, *André Breton*, 82.

30 'une des découvertes de la physique moderne qui seraient les plus capables d'élucider par analogie les relations les moins exprimables d'un 'moi' à l'autre paraît bien être celle des phénomènes d'*induction*', Gracq, *André Breton*, 52.

31 'série d'essais sur l'imagination de la matière', Gracq, *André Breton*, 57. Gracq is alluding here to the four books of 1942–8 dealing with the four elements, beginning with Bachelard, *Water and Dreams: An Essay on the Imagination of Matter*, trans. Edith R. Farrell, Dallas: Pegasus, 1999; Bachelard, *L'eau et les rêves: Essai sur l'imagination de la matière*, Paris: José Corti, 1942. Gracq notes these 'remarquables ouvrages' in a text he published alongside the Surrealists, 'Les yeux bien ouverts', Breton et al, *Farouche à quatre feuilles*, Paris: Éditions Bernard Grasset, 1954, 75–116, 115.

32 Bachelard, *New Scientific Spirit*, 6; 'la véritable pensée scientifique est métaphysiquement inductive', Bachelard, *Nouvel Esprit*, 6.

33 'Avec la puissante monotonie d'un leit-motiv, toute l'oeuvre, d'un bout à l'autre, apparaît traversée de ce thème qui la galvanise . . .', Gracq, *André Breton*, 62.

34 For more on Valéry's metaphors of electromagnetism, see the section 'Electricité et Magnétisme', Virtanen, *L'Imagerie scientifique de Paul Valéry*, 74–83. I should point out that Gracq was writing before the publication of Valéry's *Cahiers*.

35 'De ces réceptions quelque peu cénaculaires, des échos transpiraient naturellement dans la société parisienne. Bientôt il n'y eut plus un salon, plus une réunion de jeunes intellectuels, presque plus un café où il ne fût question de la psychanalyse; c'était devenu le sujet de conversation à la mode, comme l'avait été quelques mois auparavant la théorie de la relativité de M. Einstein', Angelo Hesnard and Édouard Pichon, 'Aperçu historique du mouvement psychanalytique français', *Revue de Psychologie Concrète*, no. 1, 1 February 1929, 101–20, 111. Breton gave this short-lived Marxist review a mauling in the *Second Manifesto of Surrealism* [1929–30], Breton, *Manifestoes of Surrealism*, 117–87, 145; Breton, 'Seconde manifeste du surréalisme', *OC*, vol. 1, 775–828, 797. For more on the French reception of Freud–Einstein, see Marie-Odile Briot, '*L'Esprit Nouveau*: son regard sur les sciences', *Léger et l'esprit moderne*, 1918–1931, Paris: Musée d'art moderne de la ville de Paris, 1982, 27–59.

36 Jules Romains, 'A Sketch of Psychoanalysis', trans. Blanche A. Price, Justin O'Brien (ed.), *The Most Significant Writings from the* Nouvelle Revue française, 1919–1940, London: Eyre & Spottiswoode, 1958, 240–55, 240.

37 Jacques Rivière, 'Reconnaisance à dada', *NRF*, year 7, no. 83, 1 August 1920, 216–37. Rivière not only confirmed Dada's radicalism and legitimated its authority by noting its departure from the 'positivism' of the 'classics' and linking its experiments with language to a distinguished (and Breton-friendly) literary heritage including Rimbaud, Lautréamont and Mallarmé, he also predicted its future by connecting it, in a rare, early citation, to the '*surréalisme*' of Apollinaire, 221. Rivière's appreciation was a kind of response to the supportive yet arch assessment of Dada's efforts in the April issue by André Gide, 'Dada', trans. Justin O'Brien, O'Brien (ed.), *Most Significant Writings from the* Nouvelle Revue française, 12–15.

38 Harriett Ann Watts, *Chance: A Perspective on Dada*, Ann Arbor, Michigan: UMI Research Press, 1980, 97–111, 107.

39 A great artist, Arp's protestations of the epistemological relevance of his artistic practice in terms of quantum theory are nevertheless pretty unconvincing. Looking back in 1950 on his 1927–48 works, in which he exploits chance as an artistic resource, he evokes obliquely 'Planck's constant' claiming, 'J'OBÉISSAIS INCONSCIEMMENT À UNE LOI QUI AUJOURD'HUI EST DEVENUE / UNE LOI SUPRÊME / JE DONNAIS CE NOM "SELON LES LOIS DU HASARD" NAÏVEMENT / SANS AVOIR QUE C'ÉTAIT UNE LOI QUI ENGLOBAIT LE LOI DE CAUSE ET EFFET SELON PLANCK', Arp, 'Formes', *Jours effeuillés: Poèmes, essais, souvenirs*, 1920–1965, Paris: Gallimard, 1966, 360–61, 360. From beginning to end, Arp's work is a hymn to nature, frequently given in the face of its destruction by civilisation, making him an advocate of human frailty and the 'insufficiency' of quantum theory expressed as Heisenberg's Uncertainty Principle, '[c]e n'est pas une fable qu'en science on est incapable de fixer en même temps le lieu et la vitesse d'une particule. La présomption, la vanité démesurée de l'homme me semblent pour cela d'un grotesque indescriptible', Arp, 'Danger de mort' [1954], *Jours effeuillés*, 403–4, 403.

40 Lewis S. Feuer, *Einstein and the Generations of Science* (second ed.), New Brunswick and London: Transaction, 1982.

41 Feuer, *Einstein and the Generations of Science*, 158–99, 170.

42 Feuer, *Einstein and the Generations of Science*, 187. The controversial quotation is from Breton, 'Second Manifesto of Surrealism', *Manifestoes of Surrealism*, 125; Breton, 'Seconde manifeste du surréalisme', *OC*, vol. 1, 782–3.

43 I am extremely grateful to my friend Dr. Simone Perks for several of the references I give here and elsewhere on Dada, Surrealism and chance. Simone's doctorate promises to be a significant contribution to a field of study close to my own, Perks, *The Dice-Box of Chance: Chance and Determinism in Surrealism, Science, and the Occult*, unpublished dissertation submitted for the degree of Ph.D., University of Manchester, 2007.

44 Richard Sheppard, *Modernism–Dada–Postmodernism*, Evanston, Illinois: Northwestern University Press, 2000.

45 Sheppard, *Modernism–Dada–Postmodernism*, 31–70, 199.

46 As the Dada who, more than any other, privileged chance in his work yet relied primarily on Henri Poincaré and had barely any knowledge of Relativ-

ity or QM, Marcel Duchamp presents an inconvenient test case for those arguing for parallel practices in modern art and physics. A model of sobriety on this is Linda Dalrymple Henderson, *Duchamp in Context: Science and Technology in the* Large Glass *and Related Works*, Princeton: Princeton University Press, 1998.

47 Paul Neuhuys, 'Quelques Poètes', *Ça Ira!*, no. 14, 1921, 53–67.

48 'investigations transcendantes de l'esprit humain', Neuhuys, 'Quelques Poètes', 56.

49 'Le dadaïsme est, un résultat de la philosophie intuitive', 'se libérer des concepts relatifs de la raison humaine', Neuhuys, 'Quelques Poètes', 57.

50 'Einstein a définitivement établi sa philosophie du Relativisme', Clément Pansaers, 'Dada et moi', *Ça Ira!*, no. 16 (sp. issue, 'Dada: sa naissance, sa vie, sa mort'), November 1921, 111–15, 111.

51 'Poincaré, Freud, Einstein représentant les étapes successives de cette idée', Louis Aragon and André Breton, 'Projet pour la bibliothèque de Jacques Doucet' [February 1922], Breton, *OC*, vol. 1, 631–6, 631.

52 Michel Biezunski, 'Einstein's Reception in Paris in 1922', Thomas F. Glick (ed.), *The Comparative Reception of Relativity*, Dordrecht: D. Reidel, 1987, 169–88, 183. Threats of violence against Einstein by Charles Maurras's Royalist, Catholic, nationalist group Action française would have only increased sympathy towards the physicist from intellectuals like Breton with emerging leftist political views.

53 André Breton, 'The Disdainful Confession', *The Lost Steps*, trans. Mark Polizzotti, Lincoln and London: University of Nebraska Press, 1996, 1–11; Breton, 'La Confession dédaigneuse', *OC*, vol. 1, 193–202. Breton dispensed with chronological ordering to make this the introductory text of *Les Pas perdus*, the first volume of his collected writings, suggesting he viewed it as an important breakaway step in the direction of the new poetic consciousness that would be Surrealism, Breton, 'La Confession dédaigneuse', *Les Pas perdus*, Paris: Gallimard, 1924, 7–24.

54 Breton confirmed to André Parinaud that his relationship with Valéry unravelled from 1921, Breton, *Conversations: The Autobiography of Surrealism*, trans. Mark Polizzotti, New York: Paragon House, 1993, 10; Breton, *OC*, vol. 3, Paris: Gallimard, 1999, 434.

55 Breton, *The Lost Steps*, 1 (trans. modified); 'un certain déplacement de son centre de gravité', Breton, *OC*, vol. 1, 193.

56 Breton, *The Lost Steps*, 1; 'équilibre tout relatif', Breton, *OC*, vol. 1, 193.

57 Albert Einstein, *La théorie de la Relativité: Restreinte et Généralisée*, trans. J. Rouvière, Paris: Gauthier-Villars et Cie, 1921.

58 Breton, *The Lost Steps*, 3 (trans. modified); '"Un événement ne peut être la cause d'un autre que si on peut les réaliser tous deux au même point de l'espace," nous apprend Einstein. C'est ce que j'ai toujours grossièrement pensé. Je nie tant que je touche terre, j'aime à une certaine altitude, plus haut que ferai-je? Encore dans l'un quelconque de ces états ne repassai-je jamais par le même point et disant: je touche terre, à une certaine altitude, plus haut, ne suis-je pas dupe de mes images', Breton, *OC*, vol. 1, 195.

59 'Breton partage le contresens général des intellectuels quant à la relativité des physiciens', Gérard Legrand, 'Breton et l'inauguration philosophique du surréalisme', Christian Deschamps (ed.), *Surréalisme et philosophie*, Paris: Centre Georges Pompidou, 1992, 13–19, 13.

60 'Je prétends quelquefois qu'on lui doit une figure plastique conforme aux nouveaux modes des réprésentation mises en vigueur par Einstein', letter of 3 October 1921 quoted in Werner Spies, *Max Ernst, Collages: The Invention of the Surrealist Universe*, trans. John William Gabriel, New York: Harry N. Williams, 1991, 256, n. 333.

61 Breton, 'Max Ernst', *The Lost Steps*, 60–61; Breton, 'Max Ernst', *OC*, vol. 1, 245–6.

62 Breton, *The Lost Steps*, 60–61 (trans. modified); 'La croyance en un temps et un espace absolus semble prête à disparaître. Dada ne se donne pas pour moderne. Il juge inutile, aussi, de se soumettre aux lois d'une perspective donné . . . Mais la faculté merveilleuse, sans sortir du champ de notre expérience, d'atteindre deux réalités distantes et de leur rapprochement de tirer une étincelle; de mettre à la portée de nos sens des figures abstraites appelées à la même intensité, au même relief que les autres; et, en nous privant de système de référence, de nous dépayser en notre propre souvenir, voilà qui provisoirement le retient', Breton, *OC*, vol. 1, 245–6.

63 'It is, as it were, from the fortuitous juxtaposition of the two terms that a particular light has sprung, *the light of the image*, to which we are infinitely sensitive. The value of the image depends upon the beauty of the spark obtained; it is, consequently, a function of the difference of potential between the two conductors', Breton, 'Manifesto of Surrealism', *Manifestoes of Surrealism*, 36; 'C'est du rapprochement un quelque sorte fortuit des deux termes qu'a jailli une lumière particulière, *lumière de l'image*, à laquelle nous nous montrons infiniment sensibles. La valeur de l'image dépend de la beauté de l'étincelle obtenue; elle est, par conséquent, fonction de la différence de potentiel entre les deux conducteurs', Breton, 'Manifeste du surréalisme', *OC*, vol. 1, 337–8. Bonnet writes: 'Il est remarquable que, pour la première fois avec ce texte ['Max Ernst'], vienne sous sa plume une formule très proche de celle qui définira l'image dans le *Manifeste*: "la faculté merveilleuse, sans sortir du champ de notre expérience, d'atteindre deux réalités distants et de leur rapprochement tirer une étincelle"', Marguerite Bonnet, *André Breton: Naissance de l'aventure surréaliste*, Paris: José Corti, 1975, 238–9. While Gracq interprets Breton's famous quotation, inspired by Pierre Reverdy, through its second outing in the *Manifesto of Surrealism* as an example of the *classical*, inductive nature of Breton's theory of

poetry, the context of my study reveals that Breton originally drew an analogy between the *modern* physics of Relativity and the new poetic imagination in 'Max Ernst', Gracq, *André Breton*, 63.

64 Breton, *The Lost Steps*, 61; 'on pressent avec une émotion extrême ce que peuvent être ces temps locaux dont on entend parler', Breton, *OC*, vol. 1, 246.

65 Breton, *The Lost Steps*, 61; 'quelque jour à échapper au principe d'identité', Breton, *OC*, vol. 1, 246.

66 Breton, *The Lost Steps*, 61; Breton, *OC*, vol. 1, 246.

67 'Les notions développées par M. Thirring sont aujourd'hui très connues . . .', Anonymous, review of H. Thirring, *L'idée de la théorie de la Relativité*, *Revue de Métaphysique et de Morale*, year 31, no. 4, October–December 1924, supplement, 3–5, 3. It is worth noting that Thirring was a contemporary and friend of Erwin Schrödinger at the University of Vienna in the years leading up to the First World War, and later taught Karl Popper and self-styled 'flippant Dadaist' Paul Feyerabend.

68 Shortly after the appearance of the *Manifesto of Surrealism*, Breton alluded to Relativity briefly in his 'Introduction to the Discourse on the Paucity of Reality', a key early statement about the role of language in Surrealism. Insisting language belonged at the leading edge of thought in the service of exploration and not of description, Breton writes that poetic language should henceforth reach outside of the known. For this, he seems to commend the languages of science, unencumbered by the deadening hand of common sense description: '[d]oes every man of today, eager to conform to the directions of his time, feel he could describe the latest biological discoveries, for example, or the Theory of Relativity?' André Breton, 'Introduction to the Discourse on the Paucity of Reality' [1924–5], *Break of Day*, trans. Mark Polizzotti and Mary Ann Caws, Lincoln and London: University of Nebraska Press, 1999, 3–20, 14; 'Tout homme d'aujourd'hui, soucieux de se conformer aux directions de son époch, se sent-ils, par exemple, en mesure de faire la part dans son langage des dernières découvertes biologiques, ou de la théorie de la relativité?' Breton, 'Introduction au discours sur la peu de réalité', *OC*, vol. 2, 265–80, 275.

69 Louis Aragon, *Treatise on Style*, trans. Alyson Waters, Lincoln and London: University of Nebraska Press, 1991; Aragon, *Traité du style*, Paris: Gallimard, 1928.

70 Aragon, *Treatise on Style*, 70 (trans. slightly amended); 'justement ces années dernières, dans leur flux d'insanités et d'événements, ont apporté deux énormes statues de spécialistes, ce qui est l'espèce de statue le plus propice à l'inscription du nom des garnements qui veulent à tout prix lier leur sort à celui de quelque personne considérable . . . le malin qui puise chez le cordonnier la confirmation de ses calculs pour l'établissement des verres de lunettes, alliant au flair de la mode et à l'astuce du camouflage l'ignorance bénie de la carpe, n'a pas de peine à se duper avant les autres et croit dur comme fer à la solidité de la ficelle et de la poudre aux yeux. Je jette dans la balance ces deux épées, Freud, Einstein', Aragon, *Traité du style*, 141, 142.

71 Aragon, *Treatise on Style*, 74; '[v]oici le Mauvais Horloger, le Calendrier Malhonnête, le Généalogiste Bègue, le Mécanicien des Trajets inconciliables, le Mahomet mathématicien, Einstein auréolé de couscous et de roulettes! Il jette le désarroi dans les familles avec de petits problèmes qui laissent des doutes sur l'âge du Papa, Maman se met à courir pour rester jeune, les Bébés ne tètent plus à l'heure dite, enfin c'est l'anarchie. Einstein serait-il l'Antéchrist, ce qui simplifierait sa future canonification dialectique . . . [?]' Aragon, *Traité du style*, 150.

72 Aragon, *Treatise on Style*, 75; 'ces nageurs traversant la voie lactée, ces voyageurs sur boulets de canon, ces bateaux longs comme un jour sans pain', Aragon, *Traité du style*, 151.

73 Aragon, *Treatise on Style*, 75; 'à la propagation de notions physico-mathématiques nouvelles', Aragon, *Traité du style*, 151. Aragon was obviously reading Lewis's recent condemnation of the mania for philosophies of time in English-language writing, in Lewis, *Time and Western Man*, London: Chatto and Windus, 1927. Aragon wrote *Traité du style* in 1927, but Valéry and André Gide, shocked by its content and tone, used their influence at the *NRF* to delay its publication for a year.

74 Aragon, *Treatise on Style*, 75; 'aucune différence ici entre le sort des images einsteiniennes et celui des images rimbaldiennes', Aragon, *Traité du style*, 152. For an extension of Aragon's discussion of image and metaphor to politics, see Walter Benjamin, 'Surrealism: The Last Snapshot of the European Intelligentsia' [1929], *One Way Street and Other Writings*, trans. Edmund Jephcott and Kingsley Shorter, London and New York: Verso, 2000, 225–39, 238–9.

75 Aragon, *Treatise on Style*, 76; 'étude soignée', 'en apprendrait long sur le mauvais emploi des images', Aragon, *Traité du style*, 153.

76 Lucien Fabre, *Les Théories d'Einstein*, Paris: Payot & Cie, 1921.

77 Aragon, *Treatise on Style*, 79; 'le déroutant usage quotidien des seules pensées respectables dans le monde moderne', Aragon, *Traité du style*, 160. Lucien Fabre, *Connaissance de la déesse*, Paris: Éditions *NRF*, 1924.

78 Breton, *Manifestoes of Surrealism*, 143; 'comme eût aussi bien pu le faire celle de "relativistes" pour des einsteiniens, de "psychanalystes" pour des freudiens', Breton, *OC*, vol. 1, 795.

79 As I demonstrated earlier, there was obviously interest in Einstein, shown by the favourable mark of 15 given him by Breton in the 'referendum' held on various individuals by the Dadas and carried by *Littérature*, year 3, no. 18, March 1921, 1–7. Further research is required on Dada and Relativity and could be set on strong empirical foundations (Tristan Tzara's knowledge of mathematics is one obvious basis). The larger literary context has not been examined to my knowledge, apart from some (unsourced) remarks by Balakian, '[b]oth Saint-Pol-Roux and Apollinaire refer to Einstein . . . For them Einstein is

the supreme poet of their era', Anna Balakian, *Surrealism: The Road to the Absolute* (revised edition), London: Unwin, 1970, 89.

80 Exceptions are Henry Okun, *The Surrealist Object*, Ph.D., New York University, 1981, 90–94, 153, and J. H. Matthews, *Languages of Surrealism*, Columbia: University of Missouri Press, 1986, 177–85. Neither of these books investigates the larger cultural context and both have dated badly. The best book on the Surrealist object is also a groundbreaking study of Surrealism's politics, Steven Harris, *Surrealist Art and Thought in the 1930s: Art, Politics, and the Psyche*, Cambridge: Cambridge University Press, 2004.

81 Teresa Castelao, 'Gaston Bachelard et le milieu scientifique et intellectuel français', Pascal Nouvel (ed.), *Actualité et Postérités de Gaston Bachelard*, Paris: PUF, 1997, 101–15.

82 'une partie de l'avant-garde intellectuelle française, incluant des savants et des philosophes, a reçu avec enthousiasme les progrès de la physique théorique allemande. Les travaux épistémologique de Bachelard appartiennent à ce genre de pensée, décidément en marge de la philosophie institutionelle', Castelao, *Actualité et Postérités de Gaston Bachelard*, 105.

83 Bachelard, *Nouvel Esprit*, 1.

84 Bachelard, *New Scientific Spirit*, 3–4; '[s]*i elle expérimente, il faut raisonner; si elle raisonne, il faut expérimenter*', Bachelard, *Nouvel Esprit*, 3.

85 Bachelard, *Nouvel Esprit*, 4.

86 Bachelard, *New Scientific Spirit*, 4; 'le sens du *vecteur* épistémologique nous paraît bien net. Il va sûrement du rationnel au réel . . .', Bachelard, *Nouvel Esprit*, 4.

87 Stephen Kern, *The Culture of Time and Space, 1880–1918* [1983], Cambridge, Mass. and London: Harvard University Press, 2003, 133. For a friendly yet rigorous introduction, see Rudolf v. B. Rucker, *Geometry, Relativity and the Fourth Dimension*, New York: Dover, 1977.

88 Bachelard, *New Scientific Spirit*, 8; 'la totalisation . . . l'absorption dans une pangéométrie', Bachelard, *Nouvel Esprit*, 8.

89 French mathematician Guillaume-Jules Houel quoted in Bachelard, *New Scientific Spirit*, 27; '"géométrie généralisée"', Bachelard, *Nouvel Esprit*, 26.

90 Bachelard, *New Scientific Spirit*, 3 (trans. modified); 'ne peut plus être un rationalisme fermé', Bachelard, *Nouvel Esprit*, 3.

91 '*C'est au moment où un concept change de sens qu'il a le plus de sens*', Bachelard, *Nouvel Esprit*, 52. Bachelard's introduction is titled 'La Complexité essentielle de la philosophie scientifique', signalling the theme of the complication, revision or *généralisation* of concepts, which we will find of particular interest to Roger Caillois, and is central to Bachelard's argument, *Nouvel Esprit*, 45–53, 57, 59–61, 70–75, 83, 86–7, 140–42, 148–9.

92 Bachelard, *New Scientific Spirit*, 13; 'renforce ce qui transparaît derrière ce qui apparaît', Bachelard, *Nouvel Esprit*, 13.

93 Bachelard, *New Scientific Spirit*, 14; 'substantif substantialisé', Bachelard, *Nouvel Esprit*, 13. Bachelard derived this piece of arcana from the now almost forgotten Charles Renouvier, *Les dilemmes de la Métaphysique pure*, Paris: Alcan, 1901, 248–9. However, an unhappy typographical error in Bachelard's text transforms Renouvier's 'sujet logique de qualités et de relations définissables' into a 'sujet logique de qualités et de relations indéfinissables', which serves to erase the dialectical sense of Renouvier's argument. The error was still present in the fifteenth edition of *Le Nouvel Esprit scientifique* in 1983 and is also reproduced in Arthur Goldhammer's English translation, *New Scientific Spirit*, 13–14.

94 Bachelard, *New Scientific Spirit*, 7; '[c]omme le dit Nietzsche: tout ce qui est décisif ne naît que *malgré*', Bachelard, *Nouvel Esprit*, 7. '[e]verything decisive comes about "in spite of"', Friedrich Nietzsche, *Ecce Homo*, trans. R. J. Hollingdale, London: Penguin, 1980, 100.

95 Bachelard, *New Scientific Spirit*, 7–8; 'Nous insisterons donc sur la valeur dilemmatique des nouvelles doctrines comme la géometrie non-euclidienne, la mesure non-archimédienne, la mécanique non-newtonienne avec Einstein, la physique non-maxwellienne avec Bohr, l'arithmétique aux opérations non-commutatives qu'on pourrait désigner comme non-pythagoricienne', Bachelard, *Nouvel Esprit*, 7.

96 Bachelard, *New Scientific Spirit*, 39, 40; 'Il a suffi pour cela que le physicien aborde un nouveau domaine en toute indépendance d'esprit, après une psychanalyse des entraînements euclidiens . . . Psychologiquement, le physicien contemporain se rende compte que les habitudes rationelles nées dans la connaissance immédiate et dans l'action utilitaire sont autant d'ankyloses dont il faut triompher pour retrouver le mouvement spirituel de la découverte', Bachelard, *Nouvel Esprit*, 38, 39.

97 Bachelard, *New Scientific Spirit*, 58; 'beaux symboles mathématiques où s'allient le possible et le réel', Bachelard, *Nouvel Esprit*, 56.

98 Bachelard, *New Scientific Spirit*, 60 (trans. significantly modified); '*enveloppement* des anciennes pensées par les nouvelles. Les générations spirituelles procèdent par emboîtements successifs', Bachelard, *Nouvel Esprit*, 58.

99 Bachelard, *New Scientific Spirit*, 133–4, 134; 'Dès lors, il nous semble que dans l'intervalle qui sépare l'evanouissement d'un objet scientifique et la constitution d'une nouvelle réalité, il y a place pour une pensée non réaliste, pour une pensée se faisant un appui de son mouvement. Instant éphémère, dira-t-on, qui ne peut guère compter quand on le compare aux périodes de science acquise, assise, expliquée, enseignée. C'est pourtant là, sur ce bref instant de la découverte, qu'on doit saisir l'inflexion décisive de la pensée scientifique . . . C'est alors que se produisent les brusques contradictions expérimentales, les doutes sur l'évidence des axiomes, ces synthèses *a priori* qui, comme la synthèse géniale de M. Louis de Broglie, viennent doubler le réel . . .', Bachelard, *Nouvel Esprit*, 133, 134.

100 Bachelard, *New Scientific Spirit*, 146–7; '[a] cette crise réelle et utile de l'évolution intellectuelle humaine ne correspond plus une crise réelle de la culture intellectuelle', Bachelard, *Nouvel Esprit*, 147.

101 Bachelard, *New Scientific Spirit*, 173 (trans. modified); Bachelard, *Nouvel Esprit*, 175.

102 Gaston Bachelard, 'Le Surrationalisme', *Inquisitions*, no. 1, June 1936, 1–6; Bachelard, *L'Engagement rationaliste*, Paris: PUF, 1972, 7–12. For some reason, Bachelard's text was translated almost immediately by gallerist Julien Levy and slotted into his book on Surrealism – meant to coincide with the Museum of Modern Art exhibition of that year, *Fantastic Art, Dada, Surrealism* – which I use here, Bachelard, 'Surrationalism', Julian Levy (ed.), *Surrealism*, New York: The Black Sun Press, 1936, 186–9.

103 'Partout et toujours nous trouvons le même fait fondamental: la séparation de l'instinct et de la raison est fausse . . . La vie humaine ne se comprend que comme une coopération intime de l'instinct et de l'intelligence . . . nous proposons d'appeler "surrationalisme" une telle philosophie', Richard Müller-Freienfels, 'Rationalisme et irrationalisme', *Recherches philosophiques*, no. 1, 1931–2, 125–37, 137.

104 Michel Souriau, 'La Matière et le concret', *Recherches philosophiques*, no. 2, 1932–3, 81–111, 83.

105 For a different opinion, see Harris, *Surrealist Art and Thought in the 1930s*, 283, n. 62.

106 Bachelard, 'Surrationalism', Levy (ed.), *Surrealism*, 186; 'une révolution spirituelle', Bachelard, 'Le Surrationalisme', *Inquisitions*, 1.

107 Levy (ed.), *Surrealism*, 189 (trans. modified); 'rationalisme sclérosé', Bachelard, *Inquisitions*, 3.

108 Bachelard, *New Scientific Spirit*, 21; Bachelard, *Nouvel Esprit*, 20–21.

109 Levy (ed.), *Surrealism*, 187; '[il] a dialectisé la notion de parallèle', Bachelard, *Inquisitions*, 2.

110 Levy (ed.), *Surrealism*, 187; 'la liberté de la raison', Bachelard, *Inquisitions*, 3.

111 Levy (ed.), *Surrealism*, 186; 'pourra être mis en rapport avec le surréalisme . . .', Bachelard, *Inquisitions*, 1.

112 Levy (ed.), *Surrealism*, 189; '[d]e cette liberté . . . on n'a malheuresement pas fait un usage positif, réel, surréaliste', Bachelard, *Inquisitions*, 3.

113 Levy (ed.), *Surrealism*, 187; '*révolution de la raison*', Bachelard, *Inquisitions*, 3.

114 Levy (ed.), *Surrealism*, 189; 'Nietzsche a reconnu à la fois le caractère tardif et le caractère méthodologique des saines transmutations, "Les vues les plus précieuses se trouvent les dernières; mais les vues les plus précieuses sont les méthodes"', Bachelard, *Inquisitions*, 5. 'The most valuable insights are the last to be discovered; but the most valuable insights are *methods*', Friedrich Nietzsche, *Twilight of the Idols and The Anti-Christ*, trans. R. J. Hollingdale, London: Penguin, 1968, 123.

115 Levy (ed.), *Surrealism*, 189 (trans. slightly modified); 'rationalisme fermé fait place le rationalisme ouvert', Bachelard, *Inquisitions*, 6.

116 André Breton, 'What is Surrealism?', *What is Surrealism? Selected Writings*, Franklin Rosemont (ed.), London: Pluto, 1978, 112–41, 112; '*poétiquement* il nous est impossible aujourd'hui d'apercevoir dans le passé une époque aussi riche, aussi victorieuse, aussi révolutionnaire, aussi chargée de sens lointain que celle . . .', Breton, 'Qu'est-ce que le surréalisme?', *OC*, vol. 2, 223–62, 225–6.

117 Breton, 'What is Surrealism?', 112; '"[à] l'heure que j'écris, de nouveaux frissons parcourent l'atmosphère intellectuelle . . .,"' Breton, *OC*, vol. 2, 225.

118 Breton, 'What is Surrealism?', 114; 'le problème de leur interprétation et de leur propagation est resté le même et nous entendons qu'il reste posé', Breton, *OC*, vol. 2, 228.

119 Breton, *OC*, vol. 2, 258.

120 André Breton, 'Surrealist Situation of the Object: Situation of the Surrealist Object', *Manifestoes of Surrealism*, 255–78; Breton, 'Situation surréaliste de l'objet. Situation de l'objet surréaliste', *OC*, vol. 2, 472–96.

121 Breton, 'Crisis of the Object', *Surrealism and Painting*, 275 (trans. modified); 'au cours de ce dernier siècle, des idées scientifiques, d'une part, poétiques et artistiques, d'autre part. Je prendrai pour repères deux dates, littérairement des plus significatives: 1830, à quoi l'on fixe l'apogée du mouvement romantique, 1870 d'où partent, avec Isidore Ducasse et Arthur Rimbaud, les "nouveaux frissons" qui vont être ressentis de plus en plus profondément jusqu'à nous', Breton, 'Crise de l'objet', *Le Surréalisme et la peinture*, 275.

122 Breton, *Surrealism and Painting*, 275 (trans. modified); '"ouvre", comme on a fort bien dit, le rationalisme', Breton, *Le Surréalisme et la peinture*, 275.

123 Breton, *Surrealism and Painting*, 275 (trans. modified); '"ouverture" du rationalisme', 'ouverture du réalisme . . . sous la pression des idées romantiques proprement dites . . .', Breton, *Le Surréalisme et la peinture*, 275.

124 Breton, *Surrealism and Painting*, 275; 'une "géometrie généralisé" qui intègre à un système d'ensemble, au même titre que tout autre, la géometrie euclidienne . . ., Breton, *Le Surréalisme et la peinture*, 275.

125 Breton, *Surrealism and Painting*, 275 (trans. modified); '*contradiction surmontée* du même type que celle que Ducasse et Rimbaud, dans un autre domaine, prennent alors pour tremplin, dans le dessein de provoquer le bouleversement total de la sensibilité . . .', Breton, *Le Surréalisme et la peinture*, 275.

126 Breton, *Surrealism and Painting*, 276; '[l]a pensée scientifique et la pensée artistique modernes présentent bien à cet égard la même structure . . .', 276. Without reference to Bachelard, Foucault and so on, Umberto Eco makes similar claims about art and science, perhaps due to his indebtedness to a comparable 'structuralist' epistemology (whose intellectual genealogy, nevertheless, lies at several removes from the Bachelardian one): 'In every century the way that artistic forms are structured reflects the way in which science or contemporary culture views reality', Eco,

The Role of the Reader: Explorations in the Semiotics of Texts, Bloomington and London: Indiana University Press, 1979, 57. However, some of the examples used by Eco to sketch a cross-disciplinary *epistème* are miles wide of the mark (Mallarmé and Bohr?), 58–9.

127 The following also prefigures aspects of Jacques Derrida's deconstruction: 'Ainsi des principes épistémologiques vraiment nouveaux nous semblent devoir s'introduire dans la philosophie scientifique contemporaine. Telle serait, par exemple, l'idée que les caractères complémentaires doivent être inscrits dans l'essence de l'être, en rupture avec cette tacite croyance que l'être est toujours le signe de l'unité. En effet, si l'être en soi est un principe qui se communique à l'esprit . . . il ne saurait être le symbole d'une unité. Il conviendrait donc de fonder une ontologie du complémentaire moins âprement dialectique que la métaphysique du contradictoire', Bachelard, *Nouvel Esprit*, 15–16.

128 Breton, *Surrealism and Painting*. 276; '"Tout ce qui est réel est rationnel, et tout ce qui est rationnel est réel"', 'la raison d'aujourd'hui ne se propose rien tant que l'assimilation continue de l'irrationnel . . .', Breton, *Le Surréalisme et la peinture*, 276.

129 Breton, *Surrealism and Painting*, 276 (trans. modified); 'C'est en ce sens qu il faut admettre que le *surréalisme* s'accompagne nécessairement d'un *surrationalisme* qui le double et le mesure. L'introduction récente, par M. Gaston Bachelard, dans le vocabulaire scientifique du mot *surrationalisme* qui aspire à rendre compte de toute une méthode de pensée, prête un surcroît d'actualité et de vigueur au mot "surréalisme", dont l'acceptation jusqu'ici était restée strictement artistique. Encore une fois l'un des deux termes vérifie l'autre: cette constatation suffit à mettre en évidence l'esprit commun, fondamental, qui anime de nos jours les recherches de l'homme, qu'il s'agisse du poète, du peintre ou du savant', Breton, *Le Surréalisme et la peinture*, 276. Note that Breton uses the word 'savant', meaning 'scientist', in 'Crise de l'objet', as does Bachelard throughout *Le Nouvel Esprit scientifique*. One of the reasons why English-speaking critics have been slow to recognise fully the parallels made by Breton between the creative approaches of poet, artist and scientist in 'Crise de l'objet' is that its two best-known English translations render 'savant' and 'savants' as 'scholarship' and 'scholars' instead of 'scientist(s)', which is obviously what Breton is getting at, Breton, 'Crisis of the Object', Lucy R. Lippard (ed.), *Surrealists on Art*, New Jersey: Spectrum, 1975, 51–5. For evidence of this, see the barely known version of the essay correctly translated by Angus Malcolm, which was the only one approved by Breton, 'Crise de l'objet', *Art and Artists*, vol. 1, no. 4, July 1966, 12–15.

130 Breton, *Surrealism and Painting*, 277 (trans. modified); 'pensée en rupture avec la pensée millénaire', Breton, *Le Surréalisme et la peinture*, 277.

131 See especially Georges Canguilhem, *The Normal and the Pathological* [1966], trans. Carolyn R. Fawcett with Robert S. Cohen, New York: Zone, 1991. Foucault's introduction emphasises the importance to Canguilhem of both Bachelard and Alexandre Koyré, 7–24.

132 Louis Althusser, *Pour Marx*, Paris: Librairie François Maspero, 1965; Althusser, 'A Letter to the Translator', *For Marx*, trans. Ben Brewster, London: Penguin, 1969, 256–7, 256.

133 Michel Foucault, *The Archaeology of Knowledge*, trans. A. M. Sheridan Smith, London and New York: Routledge, 1989, 4–5. Given the presence in the pages of Georges Bataille's journal *Critique* of Foucault, Barthes and sundry (post)structuralists, a significant juncture in the postwar interpretation of Bachelard might have been the extensive review of five of his books of 1929–48, where his vast erudition is set out by the mathematician Paul Jaffard, 'Philosophie, science et psychanalyse', *Critique*, vol. 7, no. 46, 15 March 1951, 219–32.

134 Foucault, *Archaeology of Knowledge*, 4, 5.

135 As well as Foucault's introductions to his own *Archaeology of Knowledge* and to Canguilhem's *Normal and the Pathological*, I refer the reader to Dominique Lecourt, *Marxism and Epistemology: Bachelard, Canguilhem and Foucault*, trans. Ben Brewster, London: NLB, 1975; Lecourt, *L'épistémologie historique de Gaston Bachelard*, Paris: Vrin, 1974. See also Walter Privitera, *Problems of Style: Michel Foucault's Epistemology* [1990], trans. Jean Keller, Albany, New York: State University of New York Press, 1995, and Cristina Chimisso, *Gaston Bachelard: Critic of Science and the Imagination*, London and New York: Routledge, 2001. The gap (so to speak) in the history of the concept of the epistemological break that I am filling here is really a symptom of the widespread anxiety in postwar French art and writing at the intimidating cultural impact of Surrealism, which led to the denial and even expurgation of its influence by writers from Barthes and Foucault onwards, and is only now being fully understood and recounted in the scholarship on art and intellectual history.

136 Breton, *Surrealism and Painting*, 277 (trans. modified); 'non plus réductive mais indéfiniment inductive et extensive', Breton, *Le Surréalisme et la peinture*, 277. Bachelard, *New Scientific Spirit*, 6; 'la véritable pensée scientifique est métaphysiquement inductive', Bachelard, *Nouvel Esprit*, 6.

137 I am aware that the word 'inductive' has philosophical properties, too, but I am using it prominently in this chapter in its poetico-scientific meaning since it occurs frequently in key passages of Bachelard's book in a manner conducive to Breton's theory: '[q]uand on voudra mesurer la valeur épistémologique d'une idée fondamentale, c'est toujours du côté de l'induction et de la synthèse qu'il faudra se tourner', *Nouvel Esprit*, 145–6.

138 Breton, *Surrealism and Painting*, 277; 'se caractérise essentiellement par le fait qu'y préside une *volonté d'objectivation* sans précédent. Que l'on comprenne bien, en effet, que les 'objets' mathématiques, au même titre que les 'objets' poétiques, se recomman-

dent de tout autre chose, aux yeux de ceux qui les ont construits, que de leurs qualités plastiques et que si, d'aventure, ils satisfont à certaines exigences esthétiques, ce n'en serait pas moins une erreur que de chercher à les apprécier sous ce rapport', Breton, *Le Surréalisme et la peinture*, 277.

139 Bachelard, *New Scientific Spirit*, 4; 'le sens du *vecteur* épistémologique nous paraît bien net. Il va sûrement du rationnel au réel . . .', Bachelard, *Nouvel Esprit*, 4.

140 Breton, 'What is Surrealism?', *What is Surrealism?* 116; 'une époque purement *intuitive*', 'une époque *raisonnante*', Breton, 'Qu'est-ce que le surréalisme?', *OC*, vol. 2, 232.

141 Breton, *What is Surrealism?* 116 (trans. modified); '*la primauté de la pensée sur la matière*', Breton, *OC*, vol. 2, 232.

142 Breton, 'Introduction to the Discourse on the Paucity of Reality', *Break of Day*, 16; 'certains de ces objets qu'on n'approche qu'en rêve', Breton, 'Introduction au discours sur le peu de réalité', *OC*, vol. 2, 277.

143 Breton, *Surrealism and Painting*, 277; 'véritables désirs solidifiés', 'rien moins que l'objectivation de l'activité de rêve . . .', Breton, *Le Surréalisme et la peinture*, 277.

144 Breton, *Surrealism and Painting*, 277–9; 'Il n'est pas de raison qui puisse se tenir durablement pour acquise et négliger, de ce fait, la contradiction qu'est toujours prête à lui apporter l'expérience. C'est avant tout la poursuite de l'experience qui importe . . .', Breton, *Le Surréalisme et la peinture*, 277–9.

145 Bachelard, *New Scientific Spirit*, 173–4 (trans. modified); 'un état de surprise effective devant les suggestions de la pensée théorique. Comme le dit très bien M. Juvet: "C'est dans la surprise créée par une nouvelle image ou par une nouvelle association d'images, qu'il faut voir le plus important élément du progrès des sciences physiques, puisque c'est l'étonnement qui excite la logique, toujours assez froide, et qui l'oblige à établir de nouvelles coordinations, mais la cause même de ce progrès, la raison même de la surprise, il faut la chercher au sein des champs de forces créés dans l'imagination par les nouvelles associations d'images, dont la puissance mesure le bonheur du savant qui a su les assembler"', Bachelard, *Nouvel Esprit*, 175. Gustave Juvet, *La Structure des nouvelles théories physiques*, Paris: Alcan, 1933, 105. Bachelard quotes from Juvet's book several times in *Le Nouvel Esprit*. The same passage by Juvet is quoted in a book I look at in my next chapter, Armand Petitjean, *Imagination et réalisation*, Paris: Denöel et Steele, 1936, 26, n.

146 Juvet was a friend of Jean Piaget and collaborated with him on at least one essay. Juvet's early death in his fortieth year in 1936 is marked by Piaget's 'La philosophie de Gustave Juvet', *A la mémoire de Gustave Juvet*, 1896–1936, Lausanne: Université de Lausanne, 1937, 37–52. Piaget's work on child psychology itself bears the impress of Relativity and was significantly motivated by Einstein's probing questions about temporality, Holton, *Einstein, History, and Other Passions*, 139–40.

147 Breton, *Surrealism and Painting*, 279 (trans. modified); 'Les poètes, les artistes se rencontrent avec les savants au sein de ces "champs de force" créés dans l'imagination par le rapprochement de deux images différentes', Breton, *Le Surréalisme et la peinture*, 279. Breton uses part of the Juvet quotation again in *L'Amour fou*, published a year later, this time mentioning the physicist by name, Breton, *Mad Love* [1937], trans. Mary Ann Caws, Lincoln and London, University of Nebraska Press, 1987, 83–4; *OC*, vol. 2, 751.

148 Breton, *Surrealism and Painting*, 279–80; 'Cette faculté de rapprochement des deux images leur permet de s'élever au-dessus de la considération de la vie manifeste de l'objet, qui constitue généralement une borne. Sous leurs yeux, au contraire, cet objet, tout achevé qu'il est, retourne à une suite ininterrompue de *latences* qui ne lui sont pas particulières et appellent sa transformation . . . "Qu'est-ce, écrit M. Bachelard, que la croyance à la réalité, qu'est-ce que l'idée de réalité, quelle est la fonction métaphysique primordiale du réel? C'est essentiellement la conviction qu'une entité dépasse son donné immediat, ou, pour parler plus clairement, c'est la conviction que . . . *l'on trouvera plus dans le réel caché que dans le donné immédiat*"', Breton, *Le Surréalisme et la peinture*, 279–80. Bachelard's quotation, which Breton gives slightly amended (ending with '*immédiat*' where Bachelard uses 'évident') and with his own emphasis, is from *Nouvel Esprit*, 30; Bachelard, *New Scientific Spirit*, 31–2.

149 Breton's transferral of the dreamwork of condensation to everday life is best seen in his narrative of the 'slipper-spoon' in his *Mad Love*, 25–38; *L'Amour fou*, 33–46; *OC*, vol. 2, 697–709. For Breton, that object underwent the same process as words undergo in psychoanalysis where, according to Freud, 'a single word . . . takes over the representation of a whole train of thought', Sigmund Freud, 'The Unconscious', *On Metapsychology*, Penguin Freud Library, vol. 11, Harmondsworth: Penguin, 1991, 204. It is more likely that Breton read Freud's closely related remark, 'the intensity of a whole train of thought may eventually be concentrated in a single ideational element', *The Interpretation of Dreams*, Pelican Freud Library, vol. 4, London: Penguin, 1978, 753.

150 Breton, *Surrealism and Painting*, 280; '*révolution totale de l'objet*', '*mutation de rôle*', Breton, *Le Surréalisme et la peinture*, 280.

151 Breton, *Surrealism and Painting*, 280; Breton, *Le Surréalisme et la peinture*, 280.

152 André Breton, 'Crise de l'objet', *Cahiers d'Art*, year 11, May 1936, 21–6.

153 Isabelle Fortuné, 'Man Ray et les objets mathématiques', *Études photographiques*, no. 6, May 1999, 100–113, 113, n. 47. Thanks to Kelly Wilder for directing me to Fortuné's essay which offers a good introduction to the subject. My discussion in this chapter picks up where hers leaves off by giving the larger scientific context for the Surrealists' appreciation of mathematical models. The only extensive discussion of the models' impact on Surrealism, which

focuses on three artists and refers in part to the scientific context, is Gabriele Werner, *Mathematik im Surrealismus: Man Ray – Max Ernst – Dorothea Tanning*, Marburg: Jonas Verlag, 2002. Evidence of the continuing fascination of mathematical models for artists can be found in the recent series of photographs by Hiroshi Sugimoto exhibited with reference to Duchamp and Dada rather than Man Ray and Surrealism, Paris: Fondation Cartier pour l'art contemporain, *Hiroshi Sugimoto: Conceptual Forms*, 2004.

154 The most informed history, historiography and technical analysis I know of mathematical models in English is by Angela G. Vierling, http://math.harvard.edu/~angelavc/models.

155 Gerd Fischer, *Mathematische Modelle/Mathematical Models* (two vols), Braunschweig/Wiesbaden: Friedr. Vieweg und Sohn, 1986, vol. 2, v.

156 For those involved in the modelling of surfaces from the beginning of the nineteenth century, see Vierling, 'Cast of Characters', http://math.harvard.edu/~angelavc/models.

157 In this connection, see the publication based on the collection in the South Kensington Museum, C. W. Merrifield, *A Catalogue of a Collection of Models of Ruled Surfaces*, London: George E. Eyre and William Spottiswoode, 1872.

158 Fischer, *Mathematische Modelle*, p. ix. Fischer refers to the *Catalog mathematischer Modelle für den höheren mathematischen Unterricht*, Leipzig: Martin Schilling, 1911. Its uses in Ernst's work are extensively detailed by Werner, *Mathematik im Surrealismus*.

159 Man Ray, *Self Portrait*, London: Andre Deutsch, 1963, 368.

160 Twenty-four were shown at the exhibition 'Man Ray: Mathematical Objects' at the Robert Miller Gallery in New York, May–June 1983. I am grateful to Samantha Lackey for this information.

161 Christian Zervos, 'Mathématiques et art abstrait', *Cahiers d'Art*, year 11, 4–10.

162 Zervos, 'Mathématiques et art abstrait', 6. One of the conflicting stories about the discovery of the models has Zervos originally requesting Man Ray to photograph them, Marcel Jean, *The History of Surrealist Painting*, trans. Simon Watson Taylor, New York: Grove Press, 1960, 251.

163 'Procède par calcul et raisonnement, elle s'astreint à une soumission à des règles formelles, à un ensemble d'opérations déductives. Dans la science de notre temps, qui présente des points communs avec la science grecque d'avant l'ère platonicienne, plus de poèsie se révèle. Les précisions excessives y sont supprimées comme à charge. Les formules abstraites de la science contemporaine s'enrichissent d'inquiétude et de sens humaine, car la passion qui est à la base de notre vie s'incorpore aussi bien dans la pensée abstraite que dans le domaine esthétique', Zervos, 'Mathématiques et art abstrait', 8.

164 Breton, 'Crisis of the Object', *Surrealism and Painting*, 280; 'par simple *mutation de rôle*', Breton, 'Crise de l'objet', *Le Surréalisme et la peinture*, 280.

165 Briony Fer, David Batchelor, Paul Wood, *Realism, Rationalism, Surrealism: Art between the Wars*, New Haven and London: Yale University Press and the Open University, 1993, 223.

166 Bachelard, 'Surrationalism', Levy (ed.), *Surrealism*, 187; 'ces géométries multiples, laissées dans l'ombre par l'enseignement officiel et pragmatique', Bachelard, 'Le Surrationalisme', *Inquisitions*, 3.

167 Breton's titling exercise takes place in the final paragraph of 'Crisis of the Object' as it appears in *Cahiers d'Art*, 26. It was excised from the text when it was reprinted without Man Ray's photographs in the expanded editions of *Le Surréalisme et la peinture* in 1945 and 1965.

168 Breton might have had in mind the collection of poems published by Max Jacob, *Les Pénitents en maillots roses*, Paris: Éditions du Sagittaire, 1925.

169 Breton detaches Surrealism from the modernist lineage by gleefully emphasising its link with 'abstract' mathematical models and their non-artistic heritage in his 1941 retrospective, 'Artistic Genesis and Perspective of Surrealism', *Surrealism and Painting*, 51–82, 72; Breton, 'Genèse et perspective artistiques du surréalisme', *Le Surréalisme et la peinture*, 49–82, 72.

170 Madrid: Palacio de Velázquez, *Henry Moore: Sculptures, Drawings, Graphics* 1921–1981, 1981, 81.

171 'The similarity that exists between the mathematical object and the sculptor's morphological research cannot long be ignored . . .', Gordon Onslow Ford, 'The Wooden Giantess of Henry Moore', *London Bulletin*, no. 18/19/20, June 1940, 10.

172 Jean, *History of Surrealist Painting*, 269. My thanks to Jo Episcopo for this illuminating translation.

173 Although I have been unable to write at length on Domínguez here, his interest in geometry and Relativity in the 1930s and 1940s is, of course, extremely relevant to this study. See José Pierre, 'Óscar Domínguez o el Triunfo del Fantasma', Emmanuel Guigon, 'Nostalgia del Espacio', and Nilo Palenzuela, 'Óscar Domínguez: Paisajes del Deseo', Las Palmas de Gran Canaria: Centro Atlántico de Arte Moderno, *Oscar Domínguez*, 1926–1957, 1995, 17–32, 33–43 and 45–60.

174 Werner Spies, *Max Ernst, Oeuvre-Katalog: Werke 1939–1953*, Houston, Texas: Menil Foundation, 1987, 133.

175 The sources of these elements are established in the best account by Werner, *Mathematik im Surrealismus*, 164–5.

176 Marcel Duchamp, *Marchand du sel, écrits de Marcel Duchamp*, ed. Michel Sanouillet, Paris: Le Terrain Vague, 1958, 144.

177 William S. Rubin, *Dada, Surrealism, and Their Heritage*, New York: The Museum of Modern Art, 1968, 160.

178 '[J]e suis passé à une autre famille de formes irrégulières, ininterrompues, géodésiques', Matta, 'Novembre 1936', Germana Ferrari (ed.), *Matta: Entretiens Morphologiques: Notebook no. 1, 1936–44*, London: Sistan, 1987, 36.

179 Gordon Onslow Ford, 'Notes on Matta and Painting (1937–41)', Ferrari (ed.), *Matta*, 23–5, 23.

180 Josefina Alix, 'Transparent Mountains, Luminous Lands', Santiago de Compostela: Fundación Eugenio Granell, *Gordon Onslow Ford: Mirando en lo profundo/Seeing in Depth*, 1998, 199–210, 200.

181 Martica Sawin, *Surrealism in Exile and the Beginning of the New York School*, Cambridge, Mass.: MIT, 262.

182 Édouard Monod-Herzen, *Principes de Morphologie générale* (two vols), Paris: Gauthier-Villars et Cie, 1927.

183 Ferrari (ed.), *Matta*, 36, n. 1. Dalí draws an analogy between the necks of Pre-Raphaelite women, the '"geodesic lines" of Europe's structural legend', and the 'chainette', a morphological shape he derives (including diagram) from Monod-Herzen's 'étonnante étude', Salvador Dalí, 'The Spectral Surrealism of the Pre-Raphaelite Eternal Feminine', *The Collected Writings of Salvador Dalí*, ed. and trans. Haim Finkelstein, Cambridge: Cambridge University Press, 1998, 310–14, 311, 313, 312; Dalí, 'Le Surréalisme spectrale de l'éternel féminin préraphaélite', *Minotaure*, no. 8, 1936, 46–9, 46, 47, 48. See Monod-Herzen, *Principes de Morphologie générale*, vol. 1, 29. My thanks to Amy Winter for information on Paalen.

184 Édouard Monod-Herzen, 'Science et Intuition', Marc Thivolet (ed.), *Le Grand Jeu: les textes parus dans les trois numéros publiées de la revue*, Paris: Éditions de l'Herne, 1968, 77. Nicolas Calas, *Foyers d'incendie*, Paris: Denöel, 1938. See principally Jagadis Chunder Bose, *Life Movements in Plants*, London: Longmans, Green, and Co., 1918–31, and the classic Bose, *Plant Response*, London: Longmans, Green, and Co., 1906. Also see Visvapriya Mukherji, *Jagadis Chandra Bose*, New Delhi: Ministry of Information and Broadcasting, Government of India, 1983.

185 'ces lois où se forge le nouvel esprit esthétique', Calas, *Foyers d'incendie*, 34. Bachelard probably knew it at the time of Calas's writing but only gets round to mentioning Monod-Herzen's 'beau livre' in his writings twenty years later, Gaston Bachelard, *The Poetics of Space*, trans. Maria Jolas, Boston: Beacon Press, 1969, 105; Bachelard, *La Poétique de l'espace*, Paris: PUF, 1957, 105.

186 'une vue synthétique de recherches poursuivies dans trois directions différentes: dans le domaine de l'expérience physique par les recherches géniales de Bose, dans le domaine de l'expérience psychique par les recherches non moins géniales de Freud sur la symbolisme du rêve, et dans le domaine de la spéculation scientifique où la morphologie prend une place nouvelle grâce à la conception physico-mathématique due aux recherches de Bachelard. Le nouvel esprit scientifique est triangulaire . . .', Calas, *Foyers d'incendie*, 48–9.

187 'Il est grand temps que nous abordions enfin l'esthétique comme nous abordons la science, avec un esprit nouveau', Calas, *Foyers d'incendie*, 77. For a good account of the origins of Surrealism and Purism in the *esprit nouveau* defined by Apollinaire, see Fer et al, '"This Liberty and this Order,": Art in France after the First World War', *Realism, Rationalism, Surrealism*, 3–86.

3 Epistemology and Politics: *Gaston Bachelard and Surrealism*

1 'un peu distante', Gerard Legrand, *André Breton et son temps*, Paris: Le Soleil Noir, 1976, 50.

2 Steven Harris, *Surrealist Art and Thought in the 1930s: Art, Politics, and the Psyche*, Cambridge: Cambridge University Press, 2004, and Raymond Spiteri and Donald LaCoss (eds), *Surrealism, Politics and Culture*, London: Ashgate, 2003.

3 Aragon's famous provocative phrase appeared in *Clarté* in December 1924. For the full story and context, see José Pierre (ed.). *Tracts surréalistes et déclarations collectives*, 1922–1969, Tome I: 1922–1939, Paris: Le Terrain Vague, 1980, 375–9. My thanks to Steven Harris for clarification on this point.

4 André Breton, 'Léon Trotsky: *Lénine*', *La Révolution surréaliste*, no. 5, 15 October 1925, 29.

5 Quoted in Mark Polizzotti, *Revolution of the Mind: The Life of André Breton*, New York: Farrar, Straus and Giroux, 1995, 346.

6 In a postwar tract placing Surrealism firmly in the past, Tristan Tzara regards Kharkov as the original 'crise profonde du Surréalisme', Tzara, *Le Surréalisme et l'Après-Guerre*, Paris: Les Éditions Nagel, 1948, 31.

7 The names of Breton, René Crevel, Paul Éluard, Georges Hugnet, Marcel Jean, Jehan Mayoux, and Benjamin Péret appeared on the CVIA's manifesto. Breton and Éluard can be seen protesting under the banner of the CVIA in a photograph in Paris: Centre Georges Pompidou, Musée National d'Art Moderne, *André Breton: La Beauté convulsive*, 1991, 214. For more information on the CVIA, see the essay by Nicole Racine-Furland and interview with François Walter in Anne Roche and Christian Tarting (eds), *Des Années trente: groupes et ruptures*, Paris: Centre national de la recherche scientifique, 1985, 57–68, 69–72. For the CVIA and the Surrealists, see Robert Short, *The Political History of the Surrealist Movement*, unpublished dissertation submitted for the degree of Ph.D., University of Sussex, 1965, 442–4.

8 For Contre-Attaque, see Robert Stuart Short, 'Contre-attaque', Ferdinand Alquié (ed.), *Entretiens sur le surréalisme*, Paris: Mouton, 1968, 144–76. A closer look at Contre-Attaque in relation with the PCF and Popular Front is given in Harris, *Surrealist Art and Thought in the 1930s*, 149–52.

9 André Breton, 'Interview with *Indice*', *What is Surrealism? Selected Writings*, ed. Franklin Rosemont, London: Pluto, 1978, 144–7, 147; '[q]uelle est l'attitude du surréalisme à l'égard des thèses les plus importantes du matérialisme dialectique et de la psychologie contemporaine?' Breton, 'Interview d'*Indice*', *Oeuvres complètes* [OC], vol. 2, Paris: Gallimard, 1992, 445–50, 449.

10 Breton, *What is Surrealism?* 147; 'la plus grande

partie de la philosophie de Freud', Breton, *OC*, vol. 2, 449–50.

11 Breton, *What is Surrealism?* 147; '[n]ous souvenant que trois grandes découvertes scientifiques (celles de la cellule, de la transformation de l'énergie, la découverte darwinienne) ont permis l'édification du système matérialiste dialectique de la nature, et aidé corréalitivement à la compréhension des lois générales du développement de la société, nous estimons que la considération du mouvement scientifique actuel est plus profitable que celle du mouvement psychologique, toujours très en retard sur le précédent. Entre toutes les sciences, la physique moderne semble devoir retenir particulièrement l'attention', Breton, *OC*, vol. 2, 450.

12 Harris, *Surrealist Art and Thought in the* 1930s, 118.

13 André Breton, 'On the Time the Surrealists were Right', *Manifestoes of Surrealism*, trans. Richard Seaver and Helen R. Lane, Michigan: University of Michigan Press, 1972, 243–53; Breton, 'Du temps que les surréalistes avaient raison', *OC*, vol. 2, 460–71.

14 Harris, *Surrealist Art and Thought in the* 1930s, 132.

15 Harry W. Paul, *From Knowledge to Power: The Rise of the Science Empire in France*, 1860–1939, Cambridge: Cambridge University Press, 1985, 342.

16 Mary Jo Nye, 'Science and Socialism: The Case of Jean Perrin in the Third Republic', *French Historical Studies*, vol. IX, no. 1, Spring 1975, 141–69, 163.

17 Pascal Ory, *La Belle Illusion: Culture et politique sous le signe du Front populaire*, 1935–38, Paris: Plon, 1994, 481.

18 Matta worked on the Spanish pavilion where Picasso's *Guernica* received its first showing. An increasingly detached Dalí showed at the exhibition held at the Jeu de Paume, 'Origines et developpement de l'art international indépendant', which ran July–October 1937. Breton was not amused by its Dalí bias (eight paintings) over others he felt artistically and morally superior.

19 Paul, *From Knowledge to Power*, 352.

20 Perrin quoted in Nye, 'Science and Socialism', 152, 150.

21 Jacques Solomon, 'À propos des entretiens de physique à Varsovie', *Cahiers du Bolchevisme*, year 15, no. 7, July 1938, 322–5.

22 Solomon, 'À propos des entretiens de physique à Varsovie', 323–4.

23 For a critical synopsis and a bit of background to Lenin's inquiry, see Stephen Kern, *The Culture of Time and Space*, 1880–1918 [1983], Cambridge, Mass. and London: Harvard University Press, 2003, 134–6. Although it does not touch on the reception of physics in specifically French Marxism, there is some valuable information on Lenin, modern physics and Mach in David Joravsky, *Soviet Marxism and Natural Science*, 1917–1932, London: Routledge and Kegan Paul, 1961, 275–95. Lenin's flawed study was translated into French in 1928 and had a currency in Surrealism which has still not been given its full due by historians. I cannot really go into it on this occasion, except to say that the rock hard materialism of *Materialism and Empirio-Criticism* is noted by Maxime Alexandre, 'A propos de morale', *La Révolution surréaliste*, no. 12, 15 December 1925, 49–50, 50, n. 2, and that Breton was introduced to it by Georges Sadoul in 1931, and was reading it at the time he wrote *The Communicating Vessels* (1932) according to Marguerite Bonnet, Breton, *OC*, vol. 2, 1359–62.

Charles Renouvier receives a drubbing from Lenin for his perceived closeness to Mach in *Materialism and Empirio-Criticism*, and Lenin's book was quoted in the now-lost polemic of Soviet physicists Blochinzev and Alexandrov aiming to assimilate the philosophical content of the Copenhagen Interpretation of QM to dialectical materialism, contested but not fully referenced by Werner Heisenberg, *Physics and Philosophy: The Revolution in Modern Science* [1958], London: Penguin, 1990, 123–7.

24 'Langevin et Perrin démontraient brillamment *la réalité des atomes*', Solomon, 'À propos des entretiens de physique à Varsovie', 323. As we will see, the modern physics–alchemy analogy pops up with some frequency from the 1930s on.

25 Solomon, 'À propos des entretiens de physique à Varsovie', 324.

26 'Il faut leur montrer que le matérialisme dialectique de Marx, d'Engels, Lénine et Staline représente la forme moderne de la conception scientifique du monde. Elle est liée à la science en mouvement, son contenu s'adaptant, se modifiant avec les progrès et les transformations de la science', Solomon, 'À propos des entretiens de physique à Varsovie', 325.

27 Solomon also contributed articles on physics to *Europe* and *L'Humanité* during the period of the Popular Front, Ory, *La Belle Illusion*, 936, n. 29.

28 Jules Monnerot, *La Poésie moderne et le sacré*, Paris: Gallimard, 1945. The Martiniquan Monnerot's sympathies took a bizarre turn much later when he stood unsuccessfully as a European candidate for Jean-Marie Le Pen's National Front party, Denis Hollier, *Absent Without Leave*, trans. Catherine Porter, Cambridge, Mass. and London: Harvard University Press, 1997, 76–93.

29 Tzara's farewell note is buried discreetly in 'Chroniques', *Cahiers du Sud*, year 22, no. 170, March 1935, 226–56, 231–2. According to Caillois, the political differences of the editorial board of *Inquisitions* were responsible for its premature demise, Caillois, *Approches de l'imaginaire*, Paris: Gallimard, 1974, 57–8.

30 Aragon, 'Entre nous', *Inquisitions*, no. 1, June 1936, 43–4. Aragon affirms the centrality of politics in the formation of *Inquisitions*, writing 'l'expérience de cette revue se poursuit strictement dans les limites du Front Populaire', 42. My own view that individual and group alignments during the period had a measure of flexibility and were not entirely determined politically conflicts somewhat with Harris's, even though I maintain his is the best historical analysis of Surrealism's politics.

31 '– trouver le courant d'idées théorique aussi bien qu'affectif susceptible d'intégrer les problèmes que se

posent les sciences humaines', '– ne pas se limiter à l'aspect littéraire de cette problématique et dégager sur la plan intellectuel, une perspective englobant toutes les sciences au service de l'homme, à l'instar du Front populaire sur la plan politique . . .', Henri Béhar (ed.), *Inquisitions: Du surréalisme au Front populaire*, Paris: Centre national de la recherche scientifique, 1990, 11. This is a facsimile reprint of the journal with added documentary material. For more on *Inquisitions*, see Henri Béhar, '*Inquisitions*: le surrationalisme, la poésie et l'actualité', Roche and Tarting (eds), *Des Années trente*, 225–36.

32 'La revue "INQUISITIONS" se propose de réunir et de confronter les résultats de ceux des travaux scientifiques ou para-scientifiques modernes qui, dans quelque domaine qu'ils aient été entrepris, se sont avérés susceptibles de concourir à une cristallisation capable de répondre aux principales des questions présentement agitées . . .', '[l]a science moderne, dans la plupart de ses domaines, offre des foyers d'intérêt infiniment plus estimables et plus prometteurs que les apports des préoccupations artistiques, par exemple. On concevrait difficilement une tâche plus urgente que la coordination ne saurait être que l'aspect directement humain de ces recherches, incompatibles avec toutes sortes de tendances rétrogrades', Béhar (ed.), *Inquisitions*, 115.

33 The volumes edited by psychologist and member of the CVIA, Henri Wallon (ed.), *À la lumière du Marxisme*, Paris: ÉSI, 1935–7, are representative of a fair amount of publishing activity on the part of a PCF bent on bringing Marxism together with modern physics.

34 Spitz, 'La Théorie Quantique et le problème de la connaissance', *Inquisitions*, 20-28.

35 As a hinge figure in the still-untold story of Surrealism, science fiction and the comic book, Spitz deserves some attention from historians of the movement. Published by Gallimard in the same 'Éditions NRF' series as Aragon's *Treatise on Style*, his fantastic novels appeared under Wellsian titles like *Les Évadés de l'an* 4000, *L'homme Élastique* and *L'expérience du Docteur Mops*, which give a pretty good idea of their style and content. His *L'agonie du Globe* was translated into English in 1936 as *Sever the Earth*.

36 Jacques Spitz, '*Lautréamont*, par Léon-Pierre Quint' (book review), *NRF*, year 18, no. 203, 1 August 1930, 273–4; Spitz, 'Les Quanta et l'individu', *NRF*, year 26, no. 303, 1 December 1938, 998–1007.

37 'l'impossibilité de considérer les phénoménes comme régis par un déterminisme strict', Spitz, *Inquisitions*, 21.

38 'purement physiologique (purement objective), ou . . . purement psychique (purement subjective)', Spitz, *Inquisitions*, 26.

39 *Inquisitions*, 70–72. The discussion following each *séance* is published towards the back of the periodical.

40 Caillois, 'Pour une orthodoxie militante: les taches immédiates de la pensée moderne', *Inquisitions*, 6–14; Caillois, 'For a Militant Orthodoxy: The Immediate Tasks of Modern Thought', *The Edge of Surrealism: A Roger Caillois Reader*, ed. Claudine Frank, trans. Claudine Frank and Camille Naish, Durham and London: Duke University Press, 2003, 132–8.

41 Caillois, *Edge of Surrealism*, 132; 'activités à demiesthétiques', Caillois, *Inquisitions*, 7. Caillois's criticism of Surrealism's attachment to the art world is ironic given his own contributions to *Minotaure* and the equally sumptuous art review *Verve*. I return to Caillois's quarrel with Surrealism over its lack of experimental rigour in my final chapter.

42 Caillois, *Edge of Surrealism*, 134; '[i]l est temps de rompre avec cet hédonisme intellectuel', Caillois, *Inquisitions*, 9.

43 Caillois, *Edge of Surrealism*, 133, n. 1 (trans. modified); '[i]l est significatif en tout cas que les tentatives les plus audacieuses et les mieux menées de destruction des *formes de la sensibilité* et des *anticipations de la perception* soient venues non de la poésie, mais de la science, spécialement de la physique relativiste, qui fournit ainsi la plus importante contribution à la mise en œuvre du *long, immense et raisonné dérèglement de tous les sens* réclamé par Rimbaud', Caillois, *Inquisitions*, 7, n. 1.

44 In his *NRF* review Wahl wrote of Caillois, '[i]l pense que l'œuvre de transvaluation de Nietzsche, l'œuvre de règlement de Rimbaud, doivent se continuer par la science', Jean Wahl, reviews of *Inquisitions* and *Acéphale*, *NRF*, year 24, no. 275, 1 August 1936, 402–3, 403.

45 Caillois, *Edge of Surrealism*, 136; Caillois, *Inquisitions*, 12.

46 Published in 1934 in the Hermann series 'Actualités Scientifiques et Industrielles', Frank's book was reviewed at length by Italian mathematician Federigo Enriques, 'Philipp Frank, *Théorie de la connaissance et physique moderne*' (book review), *Scientia*, vol. 57, no. CCLXXV, March 1935, 227–9.

47 Caillois, *Edge of Surrealism*, 135; 'par lequel la géométrie de Riemann a résorbé celle d'Euclide et la physique relativiste celle de Newton en les admettant comme cas particuliers d'une synthèse plus compréhensive . . .', Caillois, *Inquisitions*, 10. Caillois's use of the term *synthèse* is a bit misleading. There is *no* synthesis of classical and quantum physics, but rather a switching back and forth where circumstances demand between convenient classical concepts and quantum reality.

48 'en septembre 1934, que Bachelard a vraiment connu les poètes et le mouvement surréalistes . . .', Vincent Therrien, *La Révolution de Gaston Bachelard en critique littéraire*, Paris: Klincksieck, 1970, 44, n. 4.

49 Béhar (ed.), 'Correspondance', *Inquisitions*, 152. To Caillois he writes: 'Votre bel article des *Recherches philosophiques* m'a beaucoup frappé', and to Tzara, '[e]n lisant p. 271 *Grains et Issues* [Tzara's recently published theoretical text] j'ai pense à une nouvelle doctrine de la substantialisation . . .', Béhar (ed.), *Inquisitions*, 152, 155.

50 Bachelard, 'Le Surrationalisme', *Inquisitions*, 1–6; Bachelard, 'Surrationalism', Julien Levy (ed.), *Surrealism*, New York: The Black Sun Press, 1936, 186–9.

51 Levy (ed.), *Surrealism*, 186; 'la sensibilité et la raison seront rendues . . . à leur fluidité', Bachelard, *Inquisitions*, 1.

52 Caillois used the word 'manifeste' later on to describe both his own essay and Bachelard's, *Approches de l'imaginaire*, 57. If we take this term to designate the laying of a theoretical foundation, then it is obvious to what extent far-reaching statements and new beginnings were deemed necessary during the period.

53 Levy (ed.), *Surrealism*, 188; '[s]*i, dans une expérience, on ne joue pas sa raison, cette expérience ne vaut pas la peine d'être tentée*', Bachelard, *Inquisitions*, 5.

54 Duchamp's 'relativist' critique of *The New Scientific Spirit* took place in 1945, and is recorded in Denis de Rougemont, 'Marcel Duchamp mine de rien', *Preuves*, year 18, no. 204, February 1968, 43–7, 44. A passage was underlined by de Rougemont for Duchamp's perusal, Bachelard, *The New Scientific Spirit*, trans. Arthur Goldhammer, Boston: Beacon Press, 1984, 69–70; Bachelard, *Le Nouvel Esprit scientifique*, Paris: Alcan 1934, 69–70.

55 In my opinion, further consideration of Bachelard's writings in relation with Surrealism, and of Surrealism's negotiation with (mainstream) philosophical rationalism itself, must take place within the context of the work of Meyerson and Bergson, and the wide-ranging 1930s debate on the rational and irrational, typified by the (Surrealism-free) 'Symposium sur l'irrationel', *Recherches philosophiques*, no. 1, 1931–2, 125–77.

56 Bachelard, *New Scientific Spirit*, 4; 'l'application de la pensée scientifique nous paraît essentiellement réalisante', 'montrer au cours de cet ouvrage ce que nous appellerons la réalisation du rationnel ou plus généralement la réalisation du mathématique', Bachelard, *Nouvel Esprit*, 4.

57 André Breton, 'Crisis of the Object', *Surrealism and Painting*, trans. Simon Watson Taylor, London: Macdonald and Co., 1972, 275–80, 277; '[t]out le pathétique de la vie intellectuelle d'aujourd'hui tient dans cette volonté d'objectivation', Breton, 'Crise de l'objet', *Le Surréalisme et la peinture*, Paris: Gallimard, 1965, 275–80, 277.

58 Salvador Dalí, 'The Rotting Donkey', *The Collected Writings of Salvador Dalí*, ed. and trans. Haim Finkelstein, Cambridge: Cambridge University Press, 1998, 223–6, 225 (trans. modified); 'vraies réalisations de désirs soldifiés', 'L'âne pourri', *Le Surréalisme au Service de la Révolution*, no. 1, July 1930, 9–12, 12. In the same vein Dalí elaborated his '"concret délirant"', 'hyper-matérialisme', '[r]éalisation de désirs solidifiés' (again), '[a]spirations-concrètes' and 'côtelette concrète grillé et savoureuse que le merveilleux matérialisme dialectique a placé, comme l'authentique Guillaume Tell, sur la tête même de la politique' in 'De la beauté terrifiante et comestible, de l'architecture modern style', *Minotaure*, no. 3/4, December 1933, 69–76, 69, 72, 73. I look at Éluard below. Unsurprisingly given its title, the Danish art journal of the 1930s, *Konkretion*, gave particular attention to Surrealism.

59 Albert Spaïer, *La Pensée concrète*, Paris: Alcan, 1927.

60 Jean Wahl, *Vers le concret. Études d'histoire de la philosophie contemporaine*, Paris: Librairie Philosophique, 1932. In the course of his long exploration of the thought of A. N. Whitehead, Wahl discusses his 'sens de l'espace comme ensemble concret', 9. Bataille read this book, and Dalí might have known Whitehead's work too, possibly through Wahl.

61 Armand Petitjean, *Imagination et réalisation*, Paris: Denöel et Steele, 1936.

62 For more information on Petitjean and *Imagination et réalisation*, see Claudine Frank-Elster, 'Les Retournements Petitjeans: The Return of Armand Petitjean', *MODERNISM/modernity*, vol. 3, no. 1, 2001, 87–116.

63 Frank-Elster, 'Les Retournements Petitjeans: The Return of Armand Petitjean', 89. Also see Armand Petitjean, 'Analyse spectrale du singe', *Minotaure*, no. 7, June 1935, 58–9. Petitjean responded to 'L'enquête sur la rencontre', *Minotaure*, no. 3/4, 112.

64 Breton, *OC*, vol. 2, 868, n. 2.

65 'à la physique même de la pensée', Petitjean, *Imagination et réalisation*, 11.

66 'l'intermédiaire de la psycho-physiologique de l'oeil . . . permettrait d'aperçevoir en esprit les grandes lois qui régissent à la fois le monde et l'esprit', Petitjean, 'L'oeil', *Mesures*, year 2, no. 4, 15 October 1936, 59.

67 Further evidence of Petitjean's interest in physics lies in his contribution to the translation of Rogers D. Rusk, *Les Atomes, les hommes et les étoiles*, trans. L. Hervé and A.-M. Petitjean, Paris: Gallimard, 1939.

68 Frank-Elster, 'Les Retournements Petitjeans: The Return of Armand Petitjean', 89.

69 Petitjean was an imitator of Bachelard in Jean Wahl's opinion, 'La Philosophie française en 1939', *Renaissance*, vol. 2/3, 1944–5, 336–9.

70 Bachelard, '*Imagination et réalisation*, par Armand Petitjean', *NRF*, year 25, no. 282, 1 March 1937, 455–6.

71 Gaston Bachelard, *Lautréamont*, trans. James Hillman and Robert S. Dupree, Dallas, Texas: The Pegasus Foundation, 1986, 84; 'la poésie biologique', Bachelard, *Lautréamont* [1939], Paris: Librairie José Corti, 1956, 144. Petitjean is an important presence in Bachelard's study as is Roger Caillois. The intellectual exchange between Lacan and Petitjean has yet to be explored at length, though it receives some notice in the excellent primer by Frank-Elster, 'Les Retournements Petitjeans: The Return of Armand Petitjean', 97.

72 Bachelard, *Lautréamont*, 66; 'l'énergie est une esthétique', Bachelard, *Lautréamont*, 114.

73 Bachelard, *Lautréamont*, 28; 'lyrisme musculaire', Bachelard, *Lautréamont*, 52.

74 Bachelard, *Lautréamont*, 23; 'images de la motricité', Bachelard, *Lautréamont*, 44.

75 Bachelard, *Lautréamont*, 25; 'hiérarchie des vitesses', Bachelard, *Lautréamont*, 46.

76 Bachelard, *Lautréamont*, 56; 'bons surréalistes', Bachelard, *Lautréamont*, 98.

77 Bachelard, *Lautréamont*, 49 (trans. modified); '[a]u début de l'ère relativiste, pour prouver la solidité des nouvelles doctrines, Painlevé dit aux ignorants, en parlant des cinquante mathématiciens réunis autour d'Einstein: "Regardez, on les *voit se comprendre*". Il faut dire de même à ceux qui s'affolent des libertés des surréalistes: "Regardez, on les *voit* comprendre Lautréamont"'. Bachelard, *Lautréamont*, 88.

78 Bachelard, *Lautréamont*, 90; Bachelard, *Lautréamont*, 154.

79 Bachelard, *Lautréamont*, 90; 'd'éveiller des dialectiques au niveau des principes ducassiens les plus féconds', Bachelard, *Lautréamont*, 154.

80 Breton owned a copy, but was unimpressed. In a letter to Pierre Mabille of April 1940, he gives Bachelard's book a minor trampling, probably because it contained too much Jung and Gide for his liking: '[j]e n'en dirai certes pas autant du *Lautréamont* de M. Bachelard qui est une pure saloperie doublée d'une triste rigolade', quoted in Remy Laville, *Pierre Mabille: Un Compagnon du Surréalisme*, Faculté des Lettres et Sciences humaines de l'Université de Clermont-Ferrand II, 1983, 86, n. 86. Breton was miffed in any case about Bachelard's decision not to get involved with the Fédération internationale de l'art révolutionnaire indépendant (F.I.A.R.I) to which Breton himself was deeply committed, see Breton, 'Et vous?', *OC*, vol. 2, 1222–32, 1223–4.

81 Nicolas Calas, *Foyers d'incendie*, Paris: Denöel, 1938.

82 'volonté d'objectivation qui s'exprime en art par le surréalisme et en philosophie par le surrationalisme . . .', Calas, *Foyers d'incendie*, 11. Without specifying its title, Calas praises on his first page Breton's 'Crisis of the Object', as 'tout à fait remarquable' (Calas's source is confirmed by his choice of Breton's term 'volonté d'objectivation'). Breton returned the compliment, signalling his approval of *Foyers d'incendie* in the enthusiastic *prière d'insérer* he penned for it, Breton, *OC*, vol. 2, 1221–2. Compare the bruising it got from Raymond Queneau, 'Minotaurisme et monogamie', *Volontés*, no. 15, March 1939.

83 'par la grâce du "nouvel esprit scientifique"', Calas, *Foyers d'incendie*, 22. Although Caillois and Calas use the same terminology there is some conceptual slippage between them. Caillois describes the art of Cézanne and Picasso as 'abstraits' and 'objectifs' and Dalí's works as 'subjectifs', while Calas characterises abstract, subjective art as that containing no subject matter (Mondrian and the Constructivists, for example) insisting Dalí as a Surrealist, 'a engendré une nouvelle objectivité', *Foyers d'incendie*, 29.

84 'nous éloigne trop du concret . . .', Calas, *Foyers d'incendie*, 25.

85 'un tableau de Tanguy est tragique, une expérience scientifique est tragique, la pomme de Newton est tragique', Calas, *Foyers d'incendie*, 103.

86 Closely tied to debates now obscure in their detail, neither *Imagination et réalisation* nor *Foyers d'incendie* has ever been reprinted. For a rare gloss of *Foyers d'incendie* and some subsidiary information, see Gérard Durozoi, *History of the Surrealist Movement* [1997], trans. Alison Anderson, Chicago and London: University of Chicago Press, 2002, 354–5, 735, n. 26.

87 René Crevel, 'Notes en vue d'une psycho-dialectique', *SASDLR*, no. 5/6, 15 May 1933, 48–52.

88 'très concrets désirs', '[r]etour de l'abstrait au concret', Crevel, 'Notes en vue d'une psycho-dialectique', 50, 48.

89 'Grace à la dialectique les matérialistes reprennent contact avec la vie', 'la possibilité d'agir sur son univers', Crevel, 'Notes en vue d'une psycho-dialectique', 48.

90 'les immenses progrès des sciences naturelles', Crevel, 'Notes en vue d'une psycho-dialectique', 48. A potential source for Crevel's interest in physics was psychoanalyst, art collector and doctor to Juan Gris, René Allendy. By the time Crevel underwent an ill-fated analysis at his hands in 1929, Allendy was already a well-known figure, having published in both academic and avant-garde reviews. Two articles by him appeared between 1920 and 1925 in Ozenfant and Le Corbusier's *L'Esprit nouveau*, both titled 'La Constitution de la matière'. The first quotes from Jean Perrin's 1913 scientific bestseller *Les Atomes*, whilst the second includes a diagram of Perrin's interpretation of the structure of the atom among several alternatives. More information is in Marie-Odile Briot, '*L'Esprit Nouveau*: son regard sur les sciences', Paris: Musée d'art moderne de la ville de Paris, *Léger et l'esprit moderne*, 1918–1931, 1982, 27–59. Crevel might have been prompted to reflect on the relation between mind and matter by Allendy, who construed their relationship in such a way that he hypothesised to his analysand Anaïs Nin that tuberculosis was a symptom of neurosis! Crevel's botched psychoanalysis led to his rancorous satirical portrait of Allendy in *Êtes-vous fous?* Paris: Éditions NRF, 1929. Georges Bataille knew Allendy well enough to found in 1938, with him, Leiris, and others, something called the Société de psychologie collective, which seems to have sunk without trace.

91 René Crevel, 'Au carrefour de l'amour, la poésie, la science et la révolution', *L'Esprit contre la raison*, Paris: Tchou, 1969, 125–46.

92 Paul Éluard, *La Rose publique*, Paris: Gallimard, 1934.

93 'la psychologie contemporaine décèle, explore, conquiert les montagnes de mouvant désirs, cueille les geysers inavoués, les coraux de l'inconscient', 'torrents d'invisibles mais non moins réelles énergies', 'percer les mystères physico-chimiques pour en connaître, dominer, utiliser les élans . . .', Crevel, *L'Esprit contre la raison*, 130.

94 'finit par perdre son utilité, sa signification même', Crevel, *L'Esprit contre la raison*, 137. Crevel misquotes Perrin while retaining the meaning of the

passage, which is from Jean Perrin et al., *L'Orientation Actuelle des Sciences*, Paris: Alcan, 1930, 27.

95 'Socialisme scientifique. Le socialisme *par* et *pour* la science. La science *par* et *pour* la socialisme', Crevel, *L'Esprit contre la raison*, 133.

96 Crevel, *L'Esprit contre la raison*, 136. Two months after Crevel wrote his text, Carnap's book was reviewed in the *Revue de Métaphysique et de Morale*, year 42, no. 2, April 1935, supplement, 2–3.

97 'calcul des probabilités sentimentales, l'algébra des nuances . . .', Crevel, *L'Esprit contre la raison*, 136.

98 'Comment, pourquoi opposer encore science et poésie, dès l'instant que poètes et savants aussi bien par les éclairs d'une pensée non dirigée que par les déductions expérimentales et logiques d'une pensée dirigée, illustrent, affirment, prouvent la loi d'universelle réciprocité, son action sur la vie de l'âme et du corps et la nécessité de son triomphe', Crevel, *L'Esprit contre la raison*, 145.

99 'La science moderne a corrigé les vues de l'astronomie newtonienne qui, selon Gaston Bachelard (*Le Nouvel Esprit scientifique*), donna sa rigueur à la doctrine des catégories kantiennes et son absolu aux formes *a priori* de l'espace et du temps. La rose d'Éluard est aussi différente de la rose de Condillac que le matérialisme dialectique du vingtième siècle l'est du matérialisme mécanique du dix-huitième', Crevel, *L'Esprit contre la raison*, 144. Crevel does not give a reference for the quotation he uses by rationalist (!) philosopher André Lalande in this same passage, but it is actually plucked from Bachelard, *New Scientific Spirit*, 11; Bachelard, *Nouvel Esprit*, 11.

100 The French commentator on Surrealism Henri Béhar makes an abbreviated claim comparable to Crevel's, which I would file under the heading 'Bad Intellectual History' for its delusive simplicity and generality: 'le surréalisme représent une seconde crise de la conscience européenne, semblable à celle qui marqua le XVIIIe siècle en ce qu'elle conteste les valeurs établies, integré les discours scientifiques de son temps (relations d'incertitude d'Heisenberg, physique relativiste d'Einstein, psychanalyse, marxisme . . .)', Henri Béhar and Michel Carassou, *Le Surréalisme: Textes et débats*, Paris: Librairie Générale Française, 1984, 9.

101 A dim view of Surrealism emerges from his correspondence with Tristan Tzara between August 1934 and March 1935, Crevel, 'Lettres à Tristan Tzara', *Les Pieds dans le plat*, Paris: Jean-Jacques Pauvert, 1974, 306–16.

102 Paul Éluard, *Lettres à Gala*, 1924–1948, Paris: Gallimard, 1984, 249, letter 204.

103 Paul Éluard, 'Physique de la poésie', *Minotaure*, no. 6, 1935, 6–12. By dropping the female/male definite article, Éluard puns on physics/physique.

104 'Que le langage se concrétise!', Éluard, 'Physique de la poésie', 6.

105 'On ne voit ce qu'on veut que les yeux fermés . . .', Éluard, 'Physique de la poésie', 12. My thanks to Anna Dezeuze for translation help. Éluard's talk at the congress on aesthetics in August 1937 under the aegis of the Exposition Internationale is essentially a rerun of his *Minotaure* article, though it deals only with Picasso, Ernst and Dalí, Paul Éluard, *OC*, vol. 1, Paris: Gallimard, 1968, 1554–8. It was also used for his *Donner à voir* (1939), *OC*, vol. 1, 936–9. Caillois spoke at the congress too on 'Le Mythe et l'art: Nature de leur opposition'.

106 Paul Éluard, 'Premières vues anciennes', *Minotaure*, no. 10, Winter 1937, 49–50.

107 Tristan Tzara, 'Note IV: Le problème du langage en tant qu'attitude mentale', *Grains et issues: rêve expérimental*, Paris: Denoël et Steele, 1935, 263–76; *OC*, vol. 3, Paris: Flammarion, 1979, 7–145, 122–7.

108 Tzara's actual 1935 farewell note to Surrealism in *Cahiers du Sud* received a response from Breton in his text of the same year, 'Position politique de l'art d'aujourd'hui', Breton, *OC*, vol. 2, 416–40; Breton, 'Political Position of Today's Art', *Manifestoes of Surrealism*, 212–33. It lacked the bitterness of the splits with Aragon and later Éluard, and it had been obvious that Breton and Tzara were heading in different directions since about mid-1934.

109 'dont les revendications idéologiques mises en avant ont préparé cette même révolution', Tzara, *OC*, vol. 3, 123.

110 'les notions récentes des corps à plus de trois dimensions, du temps-espace, de la matière-rayonnement, de l'indétermination', Tzara, *OC*, vol. 3, 123.

111 's'est de longue date intéressé aux progrès scientifiques, particulièrement dans le domaine de la physique . . .', Tzara, *OC*, vol. 3, 547, n. 3.

112 Elmer Theodore Peterson, *Tristan Tzara: Dada and Surrational Theorist*, New Brunswick, New Jersey: Rutgers University Press, 1971, 106.

113 Harris, *Surrealist Art and Thought in the 1930s*, 127–32.

114 Tzara, 'Le Poète dans la société', and 'De la nécessité en poésie', *Inquisitions*, 29–35, 45–9.

115 'un nouveau courant philosophique se dessine en France', 'G. Bachelard à élargi ce principe en une sorte de psychologie de savant moderne, fondement du nouvel esprit scientifique', Zdenko Reich, 'Propos sur une prétendue orthodoxie militante', *Europe*, no. 171, 15 March 1937, 385–98, 385.

116 Bachelard, *New Scientific Spirit*, 7–8; 'la géometrie non-euclidienne, la mesure non-archimédienne, la mécanique non-newtonienne avec Einstein, la physique non-maxwellienne avec Bohr, l'arithmétique aux opérations non-commutatives qu'on pourrait désigner comme non-pythagoricienne', Bachelard, *Nouvel Esprit*, 7.

117 Breton, 'Nonnational Boundaries of Surrealism', *Free Rein*, trans. Michel Parmentier and Jacqueline d'Amboise, Lincoln and London: University of Nebraska Press, 1995, 7–18; Breton, 'Limites non-frontières du surréalisme', *OC*, vol. 3, 659–71.

118 Breton, *Free Rein*, 10; 'qui définit la position actuelle des savants (par suite de la conception de la géométrie non-euclidienne, puis d'une géométrie généralisée, de la mécanique non-newtonienne, de la physique non-maxwellienne, etc.)', Breton, *OC*, vol. 3, 663.

119 Breton, 'L'Oeuvre d'art et la pensée philosophique', *OC*, vol. 2, 1267–75.

120 'se mettre à jour par rapport aux aquisitions de la science, ce que pratiquement elle a omis de faire depuis plus de cinquante ans', Breton, *OC*, vol. 2, 1274.

121 'modifié grandement depuis un demi-siècle la position philosophique des savants', Breton, *OC*, vol. 2, 1274.

122 '*signe des temps*', 'le *surréalisme* poétique s'accompagne d'un *surrationalisme* scientifique qui est le double et le mesure', Breton, *OC*, vol. 2, 1274–5. Breton's approving remarks in 'Crisis of the Object' were reprinted when Bachelard's *surrationalisme* received an entry in 1938 in the *Dictionnaire abrégé du surréalisme*, Breton, *OC*, vol. 2, 787–862, 846.

123 Gaston Bachelard, *La Philosophie du non: Essai d'une philosophie du nouvel esprit scientifique*, Paris: PUF, 1940.

124 The Surrealist Group, 'Rupture inaugurale', José Pierre (ed.), *Tracts surréalistes et déclarations collectives, 1922–1969,* Tome II: 1940–1969, Paris: Le Terrain Vague, 1982, 30–36, 316–20. For some translated excerpts, see Breton, *What is Surrealism?* 340–43.

125 Henri Jones, *Le Surréalisme ignoré*, Paris: Naaman Dilif, 1969, 102–3.

126 The Surrealist Group, 'Rupture inaugurale', José Pierre (ed.), *Tracts surréalistes et déclarations collectives*, 36. Probably influenced by this broadside, a couple of years later came another of those rare attempts to link Surrealism and QM (lacking the necessary historical spadework, but at least positing a connection) in the overly credulous pocket survey of Surrealism by Yves Duplessis, *Le Surréalisme*, Paris: PUF, 1950. Under the rubric of humour and *hasard objectif*, the author writes that Breton's more expansive notion of reason 's'accorde avec les découvertes de la science moderne qui mettent en échec le principe du déterminisme. Ainsi la mécanique ondulatoire a ruiné les prétentions orgueilleuses du mécanisme classique à la prevision rigoureuse des phénomènes, puisque dans le monde de la microphysique on ne peut connaître à la fois, avec précision la position et la vitesse d'un corpuscule à un moment donné. Heisenberg a été jusqu'à parler du "principe d'incertitude"', Duplessis, *Le Surréalisme*, 26–7. Duplessis goes on to rope in Louis de Broglie to support the semi-digested notion of science's 'retreat from objectivity', though the familiar 'non-' list appended to the discussion fixes it closely to debates of the 1930s, and specifically Breton's 'Crisis of the Object'. Many thanks, once again, to Simone Perks who discovered this quotation and generously passed it on to me. In my introduction, I noted comparable remarks to Duplessis's made in the same year by Michel Carrouges, 'Le Surréalisme et la science', *André Breton et les données fondamentales du surréalisme*, Paris: Gallimard, 1950, 264–326.

127 'En cela, sur un plan qui n'avait rien de systématique, nous nous trouvions d'accord avec ceux des hommes de science qui, vers la même époque, avec une rigueur minutieuse, reprenaient les expériences les plus communes à la physique, en constataient les déficiences et construisaient à partir de là cet édifice monumental qu'est aujourd'hui la physique modern', Tzara, *Le Surréalisme et l'Après-Guerre*, 17–18.

128 Isidore Isou, 'Qu'est-ce que le Lettrisme?', *Fontaine*, vol. 11, no. 62, October 1947, 529–50.

129 '*la poésie lettrique, qui est la poésie atomique et représente notre siécle*', Isou, 'Qu'est-ce que le Lettrisme?', vol. 11, 535, 544.

130 Laville, *Pierre Mabille*, 25.

131 'Selon Jean-Pierre, troisième enfant de Pierre Mabille, Bachelard aurait été en effet le seul homme devant lequel son père "se sentait redevenir petit garçon"', Laville, *Pierre Mabille*, 25.

132 Mabille, '*La Formation de l'esprit scientifique: contribution à une psychanalyse de la connaissance objective* par Gaston Bachelard', *Traversées de Nuit*, Paris: Plasma, 1981, 87–92. Gaston Bachelard, *The Formation of the Scientific Mind: A Contribution to a Psychoanalysis of Objective Knowledge*, trans. Mary McAllester Jones, Manchester: Clinamen, 2002; Bachelard, *La Formation de l'Esprit Scientifique*, Paris: Librairie Philosophique J. Vrin, 1938. Breton owned a copy of this book, which carried a standard dedication by Bachelard: 'A André Breton avec toutes les sympathies de l'esprit, Bachelard.'

133 'une révolution s'est produite dans la sensibilité artistique et dans les modes d'expression de celle-ci', Mabille, *Traversées de Nuit*, 87.

134 'nécessité d'une totale mutation de son être', Mabille, *Traversées de Nuit*, 88. This idea is extended in both title and contents of Mabille's first work, *La Construction de l'homme*, Paris: Jean Flory, 1936. This book is yet another attempt to bring together psychology and physics in the Surrealist quest to renew humankind. Here, Mabille voices the belief he shared with Breton that the total liberation of humanity is the central concern of all thought and behaviour and that revolutionary action must take place on several planes at once. In defiance of idealism and metaphysics, Mabille believes modern science and modern psychology can light the way for a new synthesis of natural world and the human being. An anonymous reviewer in *Cahiers d'Art* (Zervos?) emphasises the pivotal importance of modern physics to Mabille's thesis: '[l]a Physique moderne vient de parler un coup mortel aux subtilités métaphysiques en affirmant que matière et énergie ne sont qu'un et on le prouvant (les grains d'énergie), Pierre Mabille tente la même jonction entre le corporel et le mental', 'Les Livres', *Cahiers d'Art*, year 11, May 1936, 67.

135 Pierre Mabille, *Mirror of the Marvelous* [1940], trans. Jody Gladding, Rochester, Vermont: Inner Traditions, 1998, 58; '[l]a réelle tradition humaine n'est pas faite d'acceptation mais de révolte, elle oppose l'homme à Dieu, elle le dresse contre le ciel, dont il lui faut capter le feu pour l'asservir à son propre volonté. La route va de Prométhée aux Hermétistes,

d'eux aux physiciens actuels. Ceux-ci usent de méthodes plus précises, ils possèdent une instrumentation plus puissante. Ils sont néanmoins les successeurs légitimes de la tradition merveilleuse', Mabille, *Le Miroir du Merveilleux*, Paris: Sagittaire, 1940, 81.

136 'l'alchimiste moderne, écrira Pierre Mabille en 1945 . . . s'enferme dans son laboratoire avec les instruments les plus avancés de la recherche scientifique', Laville, *Pierre Mabille*, 34.

137 Louis de Broglie, *Matter and Light: The New Physics*, trans. W. H. Johnstone, London: George Allen and Unwin Ltd., 1939, 32. This remark by de Broglie was probably prompted by the book which updated the public on research in elementary particles, by Ernest Rutherford, *The Newer Alchemy*, Cambridge: Cambridge University Press, 1937. For more on the popular analogy between alchemy and the new physics, see Karen A. Bearor, *Irene Rice Pereira: Her Paintings and Philosophy*, Austin, Texas: University of Texas Press, 1993, 144–50.

138 'La science moderne demeure le premier point de référence, elle se met au service d'un esprit de synthèse venu de la Renaissance qui se born à inspirer le choix d'un chemin. Le fil de la Tradition oriente l'esprit, et celui-ci, rationellement, découvre souvent que les géométries non euclidiennes ou la théorie des *quanta* ne sont pas étrangères à la pensée pré-positiviste. Si l'esprit, calcule, selon les règles de la science moderne et de la raison, le fil de la Tradition fonctionne sur un tout autre mode, l'homologie . . . L'homologie, que d'autres nommeraient analogie, est à la fois une constatation scientifique et un ressort poétique essentiel . . . le terme final en est une conception humaniste originale, une forme de la synthèse art-science que Pierre Mabille estime nécessaire pour l'équilibre d'une civilisation', Laville, *Pierre Mabille*, 26–7.

139 'Mabille affirme qu'une nouvelle conception du monde est près de naître . . . il est temps, nous dit-il, de participer à une représentation nouvelle dont l'artiste et le scientifique doivent être les premiers hérauts . . . le monisme est à l'ordre du jour: Ame/matière – corps/esprit sont les dualités d'une ère écoulée. Le 'nouvel esprit scientifique' est sa morale, faire naître l'âge d'or est son but', Laville, *Pierre Mabille*, 27.

140 Mabille, 'Del Nuevo Mundo', *Cuadernos Americanos*, vol. 11, no. 5, September–October 1943, 94–110. I refer, of course, to an entertaining literary genre that drives most physicists mad, and includes Fritjof Capra, *The Tao of Physics: An Exploration of the Parallels between Modern Physics and Eastern Mysticism* [1975], London: Flamingo, 1983, and Gary Zukav, *The Dancing Wu Li Masters: An Overview of the New Physics*, New York: William Morrow & Company, Inc., 1979. In my opinion (and at the risk of annoying the physicists), the less reputable extra-curricular enthusiasms of so many of the major figures of QM – Wolfgang Pauli's interest in Jungian archetypes, Erwin Schrödinger's Vedantic 'view of the world', even Pascual Jordan's take on psychoanalysis – are evidence of the same kind of imaginative leaping (connecting those 'systems' to QM) that brought about the bold jigsawing of ideas in search of an underlying unity in physics. Their ideas outside of science are not aberrant, I am saying, but *inseparable from* the unifying aims of physical science.

141 André Breton, 'Des tendances les plus récentes de la peinture surréaliste', *Minotaure*, no. 12/13, May 1939, 16–17. Page references are from André Breton, 'The Most Recent Tendencies in Surrealist Painting', *Surrealism and Painting*, trans. Simon Watson Taylor, New York: Harper and Row, 1972, 145–50; Breton, 'Des tendances les plus récentes de la peinture surréaliste', *Le Surréalisme et la peinture*, Paris: Gallimard, 1965, 145–50.

142 Breton, *Surrealism and Painting*, 149; 'le passage à la quatrième dimension tend à s'opérer sur le plan non plus physique, mais psychique', Breton, *Le Surréalisme et la peinture*, 149.

143 Pierre Mabille, 'L'oeil du peintre', *Minotaure*, no. 12/13, 53–6.

144 's'élève au-delà des dimensions euclydiennes visuelles', Mabille, 'L'oeil du peintre', 55.

145 Bachelard, *New Scientific Spirit*, 3; '[l]a science crée en effet de la philosophie', Bachelard, *Nouvel Esprit*, 3.

146 'Ces modes de pensée, les sciences abstraites les utilisent déjà, mais la timidité, l'inertie craintive nous empêchent de les mettre au service de la connaissance de la vie humaine quotidienne. Il est clair que psychologie, philosophie, morale stagnent par rapport aux conquêtes de la science, la volonté rétrograde de maintenir les choses de l'homme dans les cadres d'un cartésianisme dépassé par ailleurs est une des raisons profondes du drame contemporain', Mabille, 'L'oeil du peintre', 55.

147 Raymond Queneau, *Le Voyage en Grèce*, Paris: Gallimard, 1973, 220–21, and Queneau, *OC*, vol. 2, Paris: Gallimard, 2002, 1442–3. Queneau abandoned his translation but uses Dunne's time-travelling 'observer' in his *Le Chiendent*, Paris: Gallimard, 1933; Queneau, *The Bark Tree*, trans. Barbara Wright, London: Calder and Boyers, 1968.

148 Petitjean, *Imagination et réalisation*, 59, n. 1.

149 Duchamp mentions Dunne's *Serialism* (1934) in the same note, Marcel Duchamp, *Duchamp du signe: Écrits*, ed. Michel Sanouillet and Elmer Peterson, Paris: Flammarion, 1975, 141, n. 1; Duchamp, *Salt Seller: The Essential Writings of Marcel Duchamp*, ed. Michel Sanouillet and Elmer Peterson, London: Thames and Hudson, 1975, 101, n.

150 J. W. Dunne, *Le Temps et le Rêve*, trans. Eugène de Veauce, Paris: Éditions du Seuil, 1948.

151 Michel Souriau, 'J. W. Dunne, *An Experiment with Time*' (book review), *Recherches philosophiques*, no. 6, 1936–7, 399–400, 399.

152 J.W. Dunne, *An Experiment with Time*, London: A. & C. Black, 1927, 34. Oddly, a glass that melted in the lava flow of this very eruption appeared as a found object in the Surrealist-controlled special

number of *Cahiers d'Art*, year 11, May 1936.

153 André Breton, *Conversations: The Autobiography of Surrealism*, trans. Mark Polizzotti, New York: Paragon House, 1993, 241; 'Quels sont vos projets les plus immédiats?' 'revenir au problème du rêve sur lequel l'ouvrage de John W. Dunne: *Le Temps et le Rêve* me semble avoir apporté de nouvelles lueurs', Breton, *OC*, vol. 3, 627.

154 'cette *autre dimension*, inconnue, dont John W. Dunne nous invite à chercher le secret dans le *temps sériel*', André Breton, 'Le Pont Suspendu' [1955], *Perspective cavalière*, Paris: Gallimard, 1970, 93–5, 94. See also Breton's lengthier reference to Dunne and his 'captivant ouvrage', together with biographical information he took from the dust jacket of the French translation of *An Experiment with Time*, in Breton et al, *Farouche à quatre feuilles*, Paris: Éditions Bernard Grasset, 1954, 17–18. Contemporary with high Surrealism, Dunne's ideas about time and the mind in *The Serial Universe* (1934), *The New Immortality* (1938), and *Nothing Dies* (1940), are frequently concerned with philosophical problems close to Surrealism (though without knowledge of it as far as I can see), such as the 'Magrittean' reflections of *Nothing Dies*, London: Faber and Faber, 1940, 24–30. Breton owned a copy of *Le Temps et le Rêve*, so we can assume his enthusiasm is based chiefly on that first study.

155 Breton, *Conversations*, 241; 'j'ai trop sacrifié au "matérialisme" de l'heure', Breton, *OC*, vol. 3, 627.

4 Astrophysics and Mysticism: *Georges Bataille and Arthur Eddington*

1 Denis Hollier (ed.), *The College of Sociology (1937–39)*, trans. Betsy Wing, Minneapolis: University of Minnesota Press, 1988, 103.

2 Hollier (ed.), *College of Sociology*, 111.

3 The latter part of the passage should have been translated into English as 'a physicist cited by Bataille', the French original reading 'Eddington . . . (pour citer un physicien, et un physicien que cite Bataille) . . .', Denis Hollier (ed.), *Le Collège de Sociologie (1937–1939)*, Paris: Gallimard, 1979, 188.

4 '[S]ans bien connaître le tout des sciences de son temps, loin de là, Bataille en aura deviné et entrevu un peu plus que beaucoup d'autres, certainement mieux informés', Francis Marmande, *Georges Bataille politique*, Lyon: Presses Universitaires de Lyon, 1985, 133. Marmande plays the dangerous game of seeking out broad analogies between Bataille's thought and themes in modern physics (of disorder, chance, paradox, wave picture of reality, and so on), 131–5. For a consideration of Bataille's critique of science that does not look at the scientific detail or historical and philosophical context, see Jean-Louis Baudry, 'Bataille and Science: Introduction to Inner Experience', Leslie Anne Boldt-Irons (ed.), *On Bataille: Critical Essays*, New York: SUNY, 1995, 265–81.

5 Details in English of Bataille's early life and the trajectory of his thought can be found in Georges Bataille, 'Autobiographical Note' [1958?], *October*, no. 36, Spring 1986, 107–110; John Brenkman, 'Introduction to Bataille', *New German Critique*, no. 16, Winter 1979, 59–63; Georges Bataille, *Visions of Excess: Selected Writings, 1927–1939*, trans. and ed. Alan Stoekl, Minneapolis: University of Minnesota Press, 1985, ix–xxv; Denis Hollier, *Against Architecture: The Writings of Georges Bataille*, trans. Betsy Wing, Cambridge, Mass. and London: MIT, 1992; Michel Surya, *Georges Bataille: An Intellectual Biography*, trans. Krzysztof Fijalkowski and Michael Richardson, London and New York: Verso, 2002.

6 [Georges Bataille], 'Fatrasies', *La Révolution surréaliste*, no. 6, 1 March 1926, 2–3.

7 Bataille, 'Autobiographical Note', 108.

8 André Breton, *Manifestoes of Surrealism*, trans. Richard Seaver and Helen R. Lane, University of Michigan Press, 1972, 185, 186; '*des philosophes-excréments*', 'un "assis" de bibliothèque', Breton, *Oeuvres complètes* [*OC*], vol. 1, Paris: Gallimard, 1988, 826, 827.

9 Bataille, 'The "Old Mole" and the Prefix *Sur* in the Words *Surhomme* and *Surrealist*' [1929–30], *Visions of Excess*, 32–44, 42; 'l'illumination icarienne', '[b]ien qu'il s'agisse de l'élévation d'esprit quand la voûte céleste aveuglante est devenue une obsession psychologique . . .', Bataille, 'La "vieille taupe" et le préfixe *sur* dans les mots *surhomme* et *surréaliste*', *OC*, vol. 2, Paris: Gallimard, 1970, 93–109, 107.

10 André Breton, *Surrealism and Painting*, trans, Simon Watson Taylor, New York: Harper and Row, 1972, 35; '"chimie de l'intelligence"', 'la science de Masson', André Breton, *Le Surréalisme et la peinture*, Paris: Gallimard, 1965, 35. Poe used the term 'chemistry of the intellect' on more than one occasion, to designate 'the admixture of two elements', the final time in a passage on beauty and the imagination remarkable for its proto-Surrealism, Edgar Allan Poe, 'Marginalia', *Southern Literary Messenger*, vol. XV, no. 5, May 1849, 292–6, 295.

11 '[F]ont une brèche dans l'ordre des processus mécaniques; elles introduisent des blocs d'"a-causalité" dans cette réalité que l'on s'était donné absurdément comme une', Carl Einstein, 'André Masson, étude ethnologique', *Documents*, year 1, no. 2, May 1929, 93–102, 95.

12 Hans Reichenbach, 'Crise de la causalité', *Documents*, year 1, no. 2, May 1929, 105–8. The full story of how Reichenbach came to publish in *Documents* has never been uncovered. It might have come about through a connection with Carl Einstein and/or have something to do with concerns about the excess of articles appearing on causality in science journals at the time, since Arnold Berliner was turning down essays on the subject submitted to his periodical *Die Naturwissenschaften* in 1931 by the likes of Reichenbach and Pascual Jordan, Finn Aaserud, *Redirecting Science: Niels Bohr, Philanthropy, and the Rise of*

Nuclear Physics, Cambridge: Cambridge University Press, 1990, 87.

13 '[S]ignifie une rupture radicale avec la représentation causale que se fait du monde la physique classique . . .', Reichenbach, 'Crise de la causalité', 108.

14 Émile Meyerson, 'Le physicien et le primitif', *Revue philosophique*, year 55, no. 5/6, May–June 1930, 319–58. The essay was an extract from Meyerson's forthcoming *Du cheminement de la pensée*, read by Bataille at the beginning of 1935 according to the list of books he borrowed from the Bibliothèque Nationale, Bataille, *OC*, vol. 12, Paris: Gallimard, 1988, 599.

15 See especially Michel Leiris, 'Civilisation', *Documents*, year 1, no. 4, September 1929, 221–2.

16 '[L]es récents travaux de M. L. de Broglie'; 'la théorie des quanta, qui réintroduit le discontinu et l'irrationnel dans la science', Arnaud Dandieu, 'Émile Meyerson, "Le Physicien et le Primitif"' (article review), *Documents*, year 2, no. 5, 1930, 312. Later that year appeared Dandieu, 'Le conflit du réel et du rationnel dans la psychologie du temps et de l'espace', *Revue philosophique*, year 55, no. 11/12, November–December 1930, 448–61, immediately preceded by Meyerson acolyte André Metz's lengthy, glowing review of Eddington's *La nature du Monde physique*, 418–47.

17 See also the doctrinaire bible of the personalist movement of Ordre Nouveau (reissued by Place in 1993), Robert Aron and Arnaud Dandieu, *La Révolution nécessaire*, Paris: Bernard Grasset, 1933. Its origins can be found in *Bifur*, a review not too distant from Surrealism (though described by Surrealism's leader in the *Second Manifesto* as a 'remarkable dustbin', Breton, *Manifestoes of Surrealism*, 166 (trans. altered); 'remarquable poubelle', Breton, *OC*, vol. 1, 812), Robert Aron and Arnaud Dandieu, 'Le Plan Young: Un phénomène de névrose sociale', *Bifur*, no. 3, 30 September 1929, 169–82. The other site upon which Aron and Dandieu aired the ideas that would reach fruition in *La Révolution nécessaire* was Romain Rolland's review *Europe*. I am deeply indebted to Christian Roy for the generous advice and guidance he offered on Dandieu and Ordre Nouveau after reading an earlier version of this chapter.

18 Claude Chevalley and Alexandre Marc, 'Dialectique de l'action (En marge de la pensée d'Arnaud Dandieu)', *Recherches philosophiques* [*RP*], no. 6, 1936–7, 373–83.

19 Chevalley and Marc, 'Dialectique de l'action', *RP*, no. 6, 373, 374. This cannot be read as an unbiased view of Dandieu's thought given the Catholic (since 1933) Marc's leadership of Ordre Nouveau, the nonconformist movement Dandieu and Aron helped establish. Fellow member of Ordre Nouveau and Dandieu's collaborator on several essays (especially on the philosophy of science), Claude Chevalley was the mathematician son of Abel and Marguerite Chevalley, co-authors of the *Oxford Concise French Dictionary*, and in 1935, age 26, he became the youngest of the founders of the avant-garde mathematics group Bourbaki. See Jean-Louis Loubet del Bayle, *Les non-conformistes des années 30: Une tentative de renouvellement de la pensée politique française*, Paris: Éditions du Seuil, 1969 and (more cautiously) John Hellman, *The Communitarian Third Way: Alexandre Marc's Ordre Nouveau, 1930–2000*, Montreal and Kingston: McGill-Queen's University Press, 2002. Marc was a co-founder of Ordre Nouveau, whose ideology helped lay the foundaton for European federalism and create the early impetus for the postwar congresses that created the first European institutions (surviving in the network of training centres on European integration, minus its ideological origins and revolutionary tendencies since the death of Marc in 2000). In spite of other links such as *College of Sociology* fellow traveller Denis de Rougemont's membership of Ordre Nouveau and Aron's involvement in setting up the Théâtre Alfred-Jarry with Antonin Artaud and Roger Vitrac, barely anything has been said in the English language on either Bataille's or Surrealism's dealings with the movement, apart from the note by Surya, *Georges Bataille: An Intellectual Biography*, 522, n. 9. See Christian Roy, *Alexandre Marc et la Jeune Europe, 1904–1934: l'Ordre Nouveau aux origins de personnalisme*, Nice: Presses d'Europe, 1998, and Christian Roy, 'Arnaud Dandieu's Epistemological Contributions to *Documents*: Evidence of Common Ground between Surrealism and "Nietzschean" Personalism', paper delivered at the Fourth Annual Ph.D. Symposium, Tate Modern, London, 2 June 2006; Thomas Keller, *Deutsch-französische Dritte-Weg-Diskurse: Personalistische Intellektuellendebatten der Zwischenkriegszeit*, Munich: Wilhelm Fink, 2001; and the memoir by ON member and co-founder with Bataille of *Critique*, Pierre Prévost, *Rencontre Georges Bataille*, Paris: Jean-Michel Place, 1987.

20 Bataille's view of the same subject, antagonistic to academic philosophy, precedes Dandieu, 'Espace', *Documents*, year 2, no. 1, 1930, 41–4. Dandieu's is an abbreviated version of his important *Revue philosophique* essay I noted above, which compares the 'primitive' time and space recorded by Hubert and Mauss, the Einsteinian epistemological space–time reported by Meyerson, Bergson's *durée*, and Eugène Minkowski's characterisation of schizophrenic space and time.

21 'paraissant entièrement tributaire des sciences'; 'union nouvelle'; 'dans laquelle la philosophie jouait le rôle de serviteur zélé', Bataille, 'Lorsque M. Meyerson . . .', *OC*, vol. 2, Paris: Gallimard, 1970, 137–9, 139, 138.

22 As early as December 1928 Bataille had read Meyerson's *Les Travaux de M. Charles Henry sur une théorie mathématique de l'expression*, Bataille, *OC*, vol. 12, 571. For further evidence of Meyerson's importance in these circles, see Arnaud Dandieu, 'La philosophie d'Émile Meyerson et l'avenir du rationalisme', *Europe*, vol. XXIX, no. 116, 15 August 1932, 633–41.

23 Émile Durkheim, *Les Formes élémentaires de la vie religieuse*, Paris: Alcan, 1912; Marcel Mauss, 'Essai sur le Don: Forme et raison de l'échange dans les sociétés archaïques', *L'Année Sociologique*, New Series, vol. 1, 1923–1924, 30–186; Marcel Mauss, *The Gift: The Form and Reason for Exchange in Archaic Societies* [1925], trans. W. D. Halls, London: Routledge, 1990. For the Maussian component in the thought of Dandieu, see Christian Roy, 'Transpositions of Mauss' Theory of the Gift in the Personalist Social Critique of Arnaud Dandieu (1897–1933)', Antoon Vandevelde (ed.), *Gifts and Interests*, Leuven: Peeters, 2000, 177–89.

24 Georges Bataille, *The Accursed Share: An Essay on General Economy. Vol. 1: Consumption* [1949], New York: Zone, 1991, 22–3; '[l]'activité économique, envisagée comme un ensemble, est conçue sur le mode de l'opération particulière, dont la fin est limitée. L'esprit généralise en composant l'ensemble des opérations: la science économique se contente de généraliser la situation isolée, elle borne son objet aux opérations faites en vue d'une fin limitée, celle de l'homme économique; elle ne prend pas en considération un jeu de l'énergie qu'aucune fin particulière ne limite: le jeu de *la matière vivante en général*, prise dans le mouvement de la lumière dont elle est l'effet. A la surface du globe, pour *la matière vivante en général*, l'énergie est toujours en excès, la question est toujours posée en termes de luxe . . .', Bataille, *OC*, vol. 7, Paris: Gallimard, 1976, 30–31. Bataille regarded this as his most important book, and it expands upon the ideas contained in his key statement of 1933, 'The Notion of Expenditure', *Visions of Excess*, 116–29; Bataille, 'La Notion de dépense', *OC*, vol. 1, Paris: Gallimard, 1970, 302–20. For more on Bataille's general economy, see Jacques Derrida, 'From Restricted to General Economy: A Hegelianism without Reserve' [1967], *Writing and Difference*, trans. Alan Bass, Chicago: University of Chicago Press, 1978, 251–77; Michèle H. Richman, *Reading Georges Bataille: Beyond the Gift*, Baltimore and London: The John Hopkins University Press, 1982; Jean Piel, 'Bataille and the World: From "The Notion of Expenditure" to *The Accursed Share*', Boldt-Irons (ed.), *On Bataille*, 95–106.

25 Arkady Plotnitsky, *Complementarity: Anti-Epistemology after Bohr and Derrida*, Durham and London: Duke University Press, 1994, 17.

26 Plotnitsky, *Complementarity*, 271, n. 1. Bataille's work should be examined extensively in the light of Plotnitsky's statement. Obvious space limitations mean I can give only a brief review here of certain of Bataille's writings against the historical and cultural background I set out in earlier chapters.

27 André Breton, 'Interview with *Indice*', *What is Surrealism? Selected Writings*, ed. Franklin Rosemont, London: Pluto, 1978, 144–7; Breton, 'Interview d'*Indice*', *OC*, vol. 2, Paris: Gallimard, 1992, 445–50.

28 Bataille, 'Materialism', *Visions of Excess*, 15–16; 'les faits psychologiques ou sociaux et non sur des abstractions telles que les phénomènes physiques artificiellement isolés. Ainsi c'est à Freud, entre autres, – plutôt qu'à des physiciens depuis longtemps décédés et dont les conceptions sont aujourd'hui hors de cause – qu'il faut emprunter une représentation de la matière', Bataille, 'Matérialisme', *OC*, vol. 1, 179–80.

29 Breton, *Manifestoes of Surrealism*, 183; 'un retour offensive du vieux matérialisme antidialectique qui tente, cette fois, de se frayer gratuitement un chemin à travers Freud', Breton, *OC*, vol. 1, 825.

30 Bataille, *Visions of Excess*, 105–15; Bataille, 'La Critique des fondements de la dialectique hégélienne', *OC*, vol. 1, 277–90. The broader background to this essay has been discussed by Christian Roy, as a response to the controversy about Nicolai Hartmann's reading of Hegel in *Gegner*, a revival of a Dada review that was Ordre Nouveau's main ally in Germany, Roy, 'Transpositions of Mauss' Theory of the Gift', 184, n. 36. Shortly before his death, Dandieu responded to Bataille's essay in Dandieu, 'La philosophie social marxiste', *Demain?*, no. 16/17, July–August 1933, 24–6.

31 Details of Queneau's book lists are taken from Raymond Queneau, *Journaux* 1914–1965, Paris: Gallimard, 1996, 295–336.

32 Jacques Guicharnaud, *Raymond Queneau*, trans. June Guicharnaud, New York and London: Columbia University Press, 1965, 14.

33 Raymond Queneau, review of Russian mineralogist and geochemist Vladimir Vernadsky's 'L'étude de la vie et la nouvelle physique', *La Critique sociale*, no. 3, October 1931, 133–4. Also see Queneau, 'Les mathématiques dans la classification des sciences', *Bâtons, chiffres et lettres*, Paris: Gallimard, 1950, 240–47.

34 'Il est avéré que Breton . . . a fréquenté les célèbres cours de Kojève à l'Institut des Hautes Études entre 1935 et 1938', Gérard Legrand, 'Breton et l'inauguration philosophique du surréalisme', Christian Deschamps (ed.), *Surréalisme et philosophie*, Paris: Centre Georges Pompidou, 1992, 13–19, 17.

35 See Robert Stuart Short, 'Contre-attaque', Ferdinand Alquié (ed.), *Entretiens sur le surréalisme*, Paris: Mouton, 1968, 144–76.

36 The complete quotation runs as follows: 'De 33 (je pense) à 39 je suivis le cours qu'Alexandre Kojève consacra à l'explication de la *Phénoménologie de l'Esprit* (explication géniale, à la mesure du livre: combien de fois Queneau et moi sortîmes suffoqués de la petite salle – suffoqués, cloués)./A la même époque, par d'innombrables lectures, j'étais au courant du mouvement des sciences', Bataille, *OC*, vol. 6, Paris: Gallimard, 1973, 416.

37 I referred earlier to Bataille's reading of the sociology of Durkheim and Mauss, and by 'du mouvement des sciences' he could of course be referring to the human and social sciences rather than specifically physics. But the larger scientific and cultural context for those, never traced in the literature on Bataille, connects closely to physics and therefore alters the per-

spective on Bataille's interest in sociology and elucidates his attraction to physics. Further study is required to attune Bataille's readings in sociology with his interest in modern physics through an examination of their shared epistemology, concepts, and cultural and intellectual context. The essay I noted earlier by Meyerson, 'Le physicien et le primitif', reviewed by Dandieu, linking the writings of Lévy-Bruhl, Mauss and Alexandre Koyré to QM (with reference to Eddington's *The Nature of the Physical World*), is a fascinating contemporary manifestation of this, and must have been read by Bataille. Recent, historical perspectives on the links between physics, sociology and anthropology in the period in question are given by Robin Horton, 'Lévy-Bruhl, Durkheim and the Scientific Revolution', Robin Horton and Ruth Finnegan (eds), *Modes of Thought: Essays on Thinking in Western and Non-Western Societies*, London: Faber & Faber, 1973, 249–305, E. A. Tiryakian, 'L'École Durkheimienne à la recherche de la société perdue: La sociologie naissante et son milieu culturel', *Cahiers Internationaux de sociologie*, vol. LXVI, year 26, January–June 1979, 97–114, and Michèle Richman, 'Anthropology and Modernism in France: From Durkheim to the *Collège de sociologie*', Marc Manganaro (ed.), *Modernist Anthropology: From Fieldwork to Text*, Princeton, New Jersey: Princeton University Press, 1990, 183–214.

38 Alexandre Kojève, 'La dialectique du réel et la méthode phénoménologique chez Hegel (texte intégral des Conférences 6 à 9 du Cours de l'année 1934–5)', *Introduction à la lecture de Hegel*, ed. Raymond Queneau, Paris: Gallimard, 1947, 445–526. Translations are from Alexandre Kojève, 'The Dialectic of the Real and the Phenomenological Method in Hegel: Complete Text of the Sixth through Ninth Lectures of the Academic Year 1934–1935', *Introduction to the Reading of Hegel: Lectures on the Phenomenology of Spirit*, trans. James H. Nichols, Jr., ed. Raymond Queneau and Allan Bloom, Ithaca and London: Cornell University Press, 1980, 169–259.

39 Dominique Auffret, *Alexandre Kojève: la philosophie, l'État, la fin de l'Histoire*, Paris: Grasset & Fasquelle, 1990, 159.

40 Alexandre Kojève, *L'Idée du déterminisme dans la physique classique et dans la physique moderne*, Paris: Librairie Générale Française, 1990.

41 'forment alors l'ébauche concrète du système de philosophie qui va occuper Kojève toute sa vie', Auffret, *Alexandre Kojève*, 205.

42 Kojève, *Introduction to the Reading of Hegel*, 172; Kojève, *Introduction à la lecture de Hegel*, 448.

43 Kojève, *Introduction to the Reading of Hegel*, 172; '"le Tout (Ganze) simple-ou-indivis qui-a-un-aperçu de soi-même (sich übersehende) . . ."', Kojève, *Introduction à la lecture de Hegel*, 448.

44 Kojève, *Introduction to the Reading of Hegel*, 172; 'pseudo-philosophie' and 'pseudo-science (newtonienne), qui réfléchit *sur* le Réel en ce situant *hors* de lui . . .', Kojève, *Introduction à la lecture de Hegel*, 448.

45 Kojève, *Introduction to the Reading of Hegel*, 176; Kojève 'la "méthode" du savant hégélien consiste à ne pas avoir de methode . . .', *Introduction à la lecture de Hegel*, 451.

46 '"Transformer le monde," a dit Marx; "changer la vie," a dit Rimbaud: ces deux mots d'ordre pour nous n'en font qu'un', Breton, 'Discours au Congrès des écrivains' [June 1935], *OC*, vol. 2, 459.

47 Kojève, *Introduction to the Reading of Hegel*, 176; 'un homme vivant au sein de la Nature et indissolublement lié à elle, mais aussi opposé à elle et voulant la transformer: la science naît du désir de transformer le Monde en fonction de l'Homme', Kojève, *Introduction à la lecture de Hegel*, 451.

48 Kojève, *Introduction to the Reading of Hegel*, 177; '[l]'expérience scientifique perturbe l'Objet en raison de l'intervention active du sujet, qui applique à l'Objet une *méthode* d'investigation qui lui est propre et à laquelle rien ne correspond dans l'Objet lui-même', Kojève, *Introduction à la lecture de Hegel*, 452.

49 Kojève, *Introduction to the Reading of Hegel*, 177; 'ce réel autonome, cette "chose en soi" de Kant–Newton . . .', Kojève, *Introduction à la lecture de Hegel*, 452.

50 Kojève, *Introduction to the Reading of Hegel*, 177; 'non pas au réel concret, mais à une *abstraction*', Kojève, *Introduction à la lecture de Hegel*, 452.

51 Kojève, *Introduction to the Reading of Hegel*, 177, n. 2; '[c]ette interprétation de la science, sur laquelle Hegel a beaucoup insisté, est actuellement admise par la science elle-même. Dans la physique quantique elle est exprimée sous forme mathématique par les relations d'incertitude de Heisenberg. Ces relations montrent d'une part que l'expérience physique n'est jamais parfaite, puisqu'elle ne peut pas aboutir à une description à la fois complète et adéquate (précise) du "réel physique". D'autre part il en découle le fameux principe des "notions complémentaires", formulé par Bohr: celles d'onde et de corpuscule, par exemple', Kojève, *Introduction à la lecture de Hegel*, 452, n. 1. This note in the text could well have been made by Queneau, who was uniquely well placed to record these reflections, but my larger point holds about Bataille's (and perhaps Breton's) exposure to the philosophy of QM in the mid-1930s (mainly, in this case, the altered relationship between subject and object) because they are so evident in Kojève's explication.

52 Kojève, *Introduction to the Reading of Hegel*, 178, n. 2; 'il est seul à se rapporter au Réel *concret*, c'est-à-dire à la *totalité* de la réalité de l'Être', Kojève, *Introduction à la lecture de Hegel*, 452, n. 1. Kojève's claim that Hegel's phenomenology predicts the findings of quantum physics is rare, for as one writer has rightly stated, 'Hegel is conspicuous by his absence in most accounts of the philosophical genesis of quantum mechanics . . .', Arkady Plotnitsky, *In the Shadow of Hegel: Complementarity, History, and the Unconscious*, Florida: University Press of Florida, 1993, 72.

53 Breton's interest in Hegel can be traced as far back as 1922, when he gave the philosopher an ambivalent mention in his post-Dada call to arms, 'Lachez tout', Breton, *OC*, vol. 1, 262–3, 262. Further back still, in the previous year Hegel figures in the Dadas' *Littérature* 'referendum' receiving a rock-bottom mark of −25 from Tzara next to Breton's favourable 15 (the highest score Hegel received), *Littérature*, year 3, no. 18, March 1921, 1–7. For a compact introduction to Kojève's Hegel and its implications for French philosophy in the twentieth century, see Vincent Descombes, *Modern French Philosophy*, trans. L. Scott-Fox and J. M. Harding, Cambridge: Cambridge University Press, 1980, 9–54. More extensive accounts are given in the indispensable Michael S. Roth, *Knowing and History: Appropriations of Hegel in Twentieth-Century France*, Ithaca: Cornell University Press, 1988, and by Bruce Baugh, *French Hegel: From Surrealism to Postmodernism*, New York and London: Routledge, 2003.

54 Records show Bataille borrowed volumes of the *Revue de Métaphysique et de Morale* from the Bibliothèque Nationale and he mentions the journal in his 1932 'La Critique des fondements de la dialectique hégélienne', *OC*, vol. 1, 277.

55 'Fait pénétrer en France, et en langue française, beaucoup des doctrines allemandes', Raymond Aron, reviews of *Recherches philosophiques* and Gaston Bachelard, *La dialectique de la durée*, *Zeitschrift für Sozialforschung*, year 6, no. 2, 1937, 417–19, 417.

56 For a critical overview, coinciding with the moment I am describing here, of the Kantian–Cartesian bias in French philosophy and the academic rejection of Hegel, see Alexandre Koyré, 'Rapport sur l'état des études hégéliennes en France' [1930–31], *Études d'histoire de la pensée philosophique*, Paris: Librairie Armand Colin, 1961, 205–27.

57 For some rare biographical information, see Joseph Campbell (ed.), *Man and Time*, Papers from the Eranos Yearbooks, vol. 3, New York: Princeton University Press, 1957, 356.

58 Henri-Charles Puech, 'Picasso et la représentation', *Documents*, no. 3 (sp. issue, 'Hommage à Picasso'), 1930, 118–22; 'Les "Prisons" de Jean-Baptiste Piranèse', *Documents*, no. 4, 1930, 199–204; 'Le dieu Bésa et la magie hellénistique', *Documents*, no. 7, 1930, 415–25. Puech remains for now a marginal character even though Bataille must have read his work and seems to have continued contact with him after the war, see Georges Bataille, *Choix de lettres, 1917–1962*, Paris: Gallimard, 1997, 329–30.

59 Henri-Charles Puech, 'Signification et représentation', *Minotaure*, no. 6, Winter 1934–5, 52. See chapter five for my in-depth discussion of this article in the light of Matta, Paalen and particle physics.

60 Spaïer's chief publications are *La Pensée concrète: Essai sur le symbolisme intellectuel*, Paris: Alcan, 1927, and *La Pensée et la quantité: Essai sur la signification et la réalité des grandeurs*, Paris: Alcan, 1927. In keeping with the temper of the times, the Sorbonne sociologist Charles Lalo wrote in his obituary of Spaïer that these books revealed 'les principes d'un rationalisme élargi, qui se refuse à laisser abdiquer l'esprit humain devant des réalités prétendues irrationelles', *RP*, no. 3, 1933–4, vii–viii, vii.

61 Alexandre Koyré, *Essai sur l'idée de Dieu et les preuves de son existence chez Descartes*, Paris: Éditions Ernest Leroux, 1922; *La Philosophie de Jakob Boehme*, Paris: Librairie Philosophique J. Vrin, 1929.

62 Among Koyré's most important works are his *Études galiléennes*, Paris: Hermann, 1940, and especially his widely read study of seventeenth-century astronomy, *From the Closed World to the Infinite Universe*, Baltimore: The John Hopkins University Press, 1957.

63 The texts by Bachelard are 'Noumène et microphysique', *RP*, no. 1, 1931–2, 55–66; 'Le Monde comme caprice et miniature', *RP*, no. 3, 1933–4, 306–20; 'Idéalisme discursif', *RP*, no. 4, 1934–5, 21–9.

64 Émile Meyerson, 'La notion de l'identique', *RP*, no. 3, 1–17.

65 R. Johan, 'La raison et l'irrationnel chez M. Meyerson', *RP*, no. 1, 138–65.

66 Albert Spaïer, 'Pensée et étendue', *RP*, no. 1, 21–54.

67 Maurice Janet, 'La finalité en mathématiques et en physique', *RP*, no. 2, 1932–33, 1–17.

68 H. J. Jordan, 'L'indeterminisme vital et le dynamisme des structures causales', *RP*, no. 2, 18–47.

69 Michel Souriau, 'La Matière et le concret', *RP*, no. 2, 81–111. Arthur Eddington, *The Nature of the Physical World*, Cambridge: Cambridge University Press, 1928, xi–xix, xiii. Souriau resumed this research in 'Matière, Lettre et Verbe', *RP*, no. 3, 89–124. It was in 'La Matière et le concret' that Souriau employed the term 'surrationalisme', created in the previous number of the journal by Richard Müller-Freienfels and soon taken up by Bachelard.

70 Henri Ey, review of Jacques Lacan, *De la psychose paranoïaque dans ses rapports avec la personalité*, *RP*, no. 3, 457–60. For more on Lacan's thesis and its circulation, see Steven Harris, *Surrealist Art and Thought in the 1930s: Art, Politics, and the Psyche*, Cambridge: Cambridge University Press, 2004, 121–2, 270, n. 85. Around the same time, Lacan entered the orbit of the Surrealists with his 'Le problème du style et les formes paranoïaque de l'expérience', *Minotaure*, no. 1, February 1933, 68–9. There is also a later contribution by Lacan to *Recherches*, his review of Eugène Minkowski, *Le temps vécu*, *RP*, no. 5, 1935–6, 424–31.

71 Eugène Minkowski, 'Le problème du temps en psychopathologie', *RP*, no. 2, 231–56. Louis Lavelle indicates the three main themes of *Recherches philosophiques* were 'le *thème du temps et de l'éternité*', 'l'*épaisseur du réel*', and 'l'*existence personelle, unique et subjective*', Lavelle, *La Philosophie française entre les deux guerres*, Paris: Éditions Montaigne, 1942, 253.

72 '[L]'idée d'extériorité a nécessairement deux sens très différents suivant que l'on se place au point de vue du psychologue ou au point de vue du physicien', Raymond Ruyer, 'Sur quelques arguments nouveaux

contre le réalisme', *RP*, no. 4, 30–50, 30. Ruyer followed a similar line of enquiry in his articles for the *Revue de Métaphysique et de Morale*, see his 'Microphysique et Micro-spiritualisme', *RMM*, year 45, no. 4, October 1938, 527–41.

73 René Poirier, 'Temps spirituels et temps matériel', *RP*, no. 5, 1935–6, 1–40.

74 Roger Caillois, 'Analyse et commentaire d'un exemple d'association libre d'idées', *RP*, no. 4, 321–36.

75 Pierre Klossowski, 'Le mal et la négation d'autrui dans la philosophie de D.A.F. de Sade', *RP*, no. 4, 268–93.

76 René Daumal, 'Les limites du langage philosophique et les savoirs traditionnels', *RP*, no. 4, 209–31.

77 Pierre Klossowski, 'Temps et agressivité: Contribution à l'étude du temps subjectif', *RP*, no. 5, 100–111; Roger Caillois, 'Le Mythe et l'homme', *RP*, no. 5, 252–63.

78 'Jamais nous n'en avons rencontré un exposé aussi original, amusant et amathématique que chez Sir A. Eddington', Alexandre Koyré, review of Arthur Eddington, *New Pathways in Science*, *RP*, no. 5, 455–7, 457. Koyré could be referring to Hermann Weyl or André Weil, here. Both were distinguished mathematicians and their names are often confused.

79 Records reveal Bataille borrowed Reichenbach's *La Philosophie scientifique: Vues nouvelles sur ses buts et ses methodes* in November 1934, Dirac's *Les Principes de la mécanique quantique* in December, and Bohr's *La Théorie atomique et la description des phénomènes* in August 1935, Bataille, OC, vol. 12, 597, 598, 602.

80 Bataille, 'Le Labyrinthe', *RP*, no. 5, 364–72. Quotations are from Bataille, 'The Labyrinth', *Visions of Excess*, 171–7; Bataille, 'Le labyrinthe', *OC*, vol. 1, 433–41.

81 Bataille, *Visions of Excess*, 174; 'l'atome compose son unité d'électrons variables', Bataille, *OC*, vol. 1, 437. Whilst acknowledging Bataille's importance to poststructuralism, commentary on philosophy's 'post-Heideggerian' (or poststructuralist) dismantling of origins, ends, ontology, 'presence', individual agency and identity, which culminated in Jacques Derrida's deconstruction, has not fully taken account of the role played in this by the new philosophy heralded by modern physics. That is an underlying argument of this chapter.

82 Bataille, *Visions of Excess*, 177, n. 1; Bataille, *OC*, vol. 1, 435, n.

83 Bataille, *Visions of Excess*, 175; 'L'être particulier ne se comporte pas seulement comme élément d'un ensemble informe sans structure . . . mais aussi comme élément périphérique gravitant autour d'un noyau où l'être se durcit', '[d]ans le cas où plusieurs villes abdiquent leur fonction de centre au profit d'une seule, un empire se forme autour d'une capitale où se concentre la souveraineté et les dieux', *Bataille*, *OC*, vol. 1, 438, 439.

84 Koyré, review of Paul Langevin, *La notion de corpuscules et d'atomes*, *RP*, no. 4, 436–8, 436.

85 Jean Ullmo, 'L'évolution de la notion de corpuscule d'après M. Langevin', *Scientia*, vol. 55, no. CCLXII, February 1934, 103–17.

86 'Eddington, qui est un grand savant, dans son livre intitulé "La nature du monde physique," s'exprime ainsi: "On pourra peut-être dire, comme conclusion à tirer de ces arguments fournis par la science moderne, que la religion est devenue acceptable pour un esprit scientifique raisonnable à partir de 1927 . . . Si notre prévision se conforme que 1927 aura vu l'élimination définitive de la causalité stricte par Heisenberg, Bohr, Born et d'autres, cette année representera certainement l'une des plus grandes époques dans le développement de la pensée scientifique"', Paul Langevin, *La notion de corpuscules et d'atomes*, Paris: Hermann, 1934, 33–4. Quotation is from Eddington, *Nature of the Physical World*, 350.

87 Obviously, I can give only an incomplete account of Bataille's complex text from within the limits imposed by my inquiry. For a short analysis of 'The Labyrinth' from the viewpoint of science, general economy and deconstruction, see Arkady Plotnitsky, *Reconfigurations: Critical Theory and General Economy*, Florida: University Press of Florida, 1993, 301–2, and for a non-scientific gloss, Joseph Libertson, 'Bataille and Communication: *Savoir*, *Non-Savoir*, *Glissement*, *Rire*', Boldt-Irons (ed.), *On Bataille*, 209–30.

88 Arnaud Dandieu and Denis de Rougemont, 'L'acte comme point de départ', *RP*, no. 6, 55–64.

89 Jean Wahl, 'Heidegger et Kierkegaard', *RP*, no. 2, 349–70; Martin Heidegger, 'De la nature de la cause', *RP*, no. 1, 83–124.

90 Jean-Paul Sartre, 'La transcendance de l'Ego. Esquisse d'une description phénoménologique', *RP*, no. 6, 85–123; Sartre, *The Transcendence of the Ego: An Existentialist Theory of Consciousness*, trans. Forrest Williams and Robert Kirkpatrick, New York: Noonday, 1957. Elsewhere, Sartre aggressively opposes the 'intentionality' of the consciousness of Husserlian phenomenology to the assimilating, identifying consciousness of the empirio-criticism and neo-Kantianism of Brunschvicg, Lalande, and Meyerson ('philosophie alimentaire!'), 'Une idée fondamentale de la phénoménologie de Husserl: *L'Intentionnalité*' [1939], *Situations*, vol. 1, Paris: Gallimard, 1947, 31–3, 31.

91 'la renaissance de l'intérêt philosophique chez les physiciens modernes n'est pas due au hasard, mais que c'est là une conséquence naturelle du caractère même de ces théories', Alexandre Kojevnikoff, book reviews, *RP*, no. 3, 464–6, 464.

92 'utile par la masse de richesses qu'il enferme en un petit volume, par la netteté et l'exactitude de ses conclusions', Michel Souriau, 'G. Bachelard, *Le Nouvel Esprit scientifique*', book review, *RP*, no. 4, 440–42, 440. Bachelard's book received plenty of attention: reviewed by 'G.S.' as 'parfois un peu vague', *Scientia*, vol. 57, no. CCLXXVI, July 1935, 318–19, 319; anonymously recommended for its 'perspicacité philosophique', *RMM*, year 42, no. 1, January 1935,

supplement, 8; and praised as 'une chose vivante et pleine d'enseignement', in a lengthy review essay by Sorbonne professor Lucien Brunet, '*Le Nouvel Esprit scientifique* d'après G. Bachelard', *Revue de Synthèse*, Tome VIII, no. 2, October 1934, 250–52, 252.

93 Space restrictions limit the detail I can allow in this section, but all of Eddington's books in English and in French translation were reviewed in the journals I discuss. For details of Eddington's articles in *Scientia* and French, German, Italian, Polish, Hungarian and other translations of his books, see A. Vibert Douglas, *Arthur Stanley Eddington*, London: Thomas Nelson and Sons Ltd, 1957, 193–8. Worth noting again given Eddington's near-obscurity today is the lengthy review of *La nature du Monde physique*, Paris: Payot, 1929, by André Metz in *Revue Philosophique*, year 55, no. 11/12, November–December 1930, 418–47. This helped set the stage for Jules Sageret, *Le Système du Monde de Pythagore à Eddington*, Paris: Payot, 1931. Although much has been published on popular science as a genre, to my knowledge the French (or even international) reception of Eddington's writings has not been assessed, although there is some excellent material on the British context. For an introduction, including a discussion of Alfred North Whitehead and James Jeans, see Michael Whitworth, 'The Clothbound Universe: Popular Physics Books, 1919–1939', *Publishing History*, vol. 40, 1996, 53–82. Also see Gillian Beer, 'Eddington and the Idiom of Modernism', Henry Krips, J. E. McGuire, Trevor Melia (eds), *Science, Reason, and Rhetoric*, Pittsburgh, P.A.: University of Pittsburgh, 1995, 295–315; Kate Price, *Poetry and Physics: Interchange within the Writings of Arthur Eddington, I.A. Richards, William Empson and Laura Riding*, unpublished dissertation submitted for the degree of Ph.D., Cambridge University, 2002; and Michael H. Whitworth, *Einstein's Wake: Relativity, Metaphor, and Modernist Literature*, Oxford: Oxford University Press, 2001.

In 1938, Eddington's *Nature of the Physical World* was published in Spanish translation by *Ediciones SUR*, as was Breton's epic poem 'Fata Morgana' four years later following its ban under Vichy. The Argentine literary periodical *SUR* was a significant site for the convergence of Surrealism and modern physics (not to mention its ties with Ordre Nouveau), carrying its first Surrealist text (by Breton) in May 1937, then the following month beginning a regular column on modern science, 'Cuestiones científicas de nuestro tiempo', constantly featuring Eddington's work. Although Octavio Paz was introduced to Surrealism through *SUR* (which carried a Surrealism section in one number), and Surrealism's immense impact on Latin American culture is well known, the full story of Surrealism and *SUR* has yet to be told, and is only touched on in the one English-language study of the journal by John King, Sur*: A Study of the Argentine Literary Journal and its Role in the Development of Culture,* 1931–1970, Cambridge: Cambridge University Press, 1986.

94 Meyerson, 'Le relativisme, théorie du réel', *RMM*, year 31, no. 1, January–March 1924, 29–48. The same set of sources (Meyerson/Eddington/Langevin) was immediately available to the Surrealists as it is unpacked by Bachelard, *New Scientific Spirit*, 130; Bachelard, *Nouvel Esprit*, 130, and Eddington's *New Pathways in Science* receives significant mention in Bachelard, *The Philosophy of No: A Philosophy of the New Scientific Mind*, trans. G. C. Waterston, New York: The Orion Press, 1968, 118; Bachelard, *La Philosophie du non: Essai d'une philosophie du nouvel esprit scientifique*, Paris: PUF, 1940, 139.

95 There is much to say about René Daumal's interest in modern physics and mysticism, which probably deserves a chapter of its own. The appendix on Daumal in my Ph.D. can act as a starting point, Gavin Parkinson, *Physique du Surréalisme: Surrealism, Modern Physics, and Epistemology*, dissertation submitted for the degree of Ph.D., Courtauld Institute of Art, University of London, 2000, 343–7.

96 Anonymous review of Eddington, *L'univers en expansion*, *RMM*, year 42, no. 1, January 1935, supplement, 1–3, 1. Elsewhere, the book is said to deal with its subject 'avec peu de formules mathématiques, d'une manière brillante et agréable', by the Italian astronomer Giuseppe Armellini, review of Arthur Eddington, *The Expanding Universe*, *Scientia*, vol. 57, no. CCLXXV, May 1935, 234.

97 S. Chandrasekhar, *Eddington: The Most Distinguished Astrophysicist of his Time*, Cambridge: Cambridge University Press, 1983, 4. Compare Marcel Duchamp, 'one is obliged to break open several large clocks to obtain a drawer of tender years', *Salt Seller: The Essential Writings of Marcel Duchamp*, ed. Michel Sanouillet and Elmer Peterson, London: Thames and Hudson, 1975, 174; and Surrealist poet Benjamin Péret, 'the fan-headed trout/dragging heavy chariots of champagne bubbles/which blacken your hair of fortified castle', *Death to the Pigs: Selected Writings*, trans. and ed. Rachel Stella, London: Atlas, 1988, 44.

98 André Breton, 'Des tendances les plus récentes de la peinture surréaliste', *Minotaure*, no. 12/13, May 1939, 16–17, 17. André Breton, 'The Most Recent Tendencies in Surrealist Painting', *Surrealism and Painting*, trans. Simon Watson Taylor, New York: Harper and Row, 1972, 145–50, 149; Breton, 'Des tendances les plus récentes de la peinture surréaliste', *Le Surréalisme et la peinture*, Paris: Gallimard, 1965, 145–50, 149.

99 Harley D. Oberhelman, *Ernesto Sábato*, New York: Twayne, 1970, 20. In 1940, Sábato returned to Argentina to take the chair of theoretical physics at the National University of La Plata. He became one of Argentina's most important intellectuals and novelists, working in an existentialist rather than a Surrealist mode.

100 Oscar Domínguez, 'Le Pétrification du Temps' [1938], *La Conquête du Monde par l'Image*, 1942, 27.

101 See Arthur Eddington, *Space Time and Gravitation:*

An Outline of the General Relativity Theory, Cambridge: Cambridge University Press, 1920, 32.

102 Eddington, *Space Time and Gravitation*, 57.

103 Bataille and Roger Caillois, 'Sacred Sociology and the Relationships between "Society," "Organism," and "Being"', Hollier (ed.), *College of Sociology*, 73–84; Bataille and Caillois, 'La sociologie sacrée et les rapports entre "société," "organisme" et "être"', Hollier (ed.), *Collège de Sociologie*, 139–63. Hollier's introductory note to this lecture indicates its eclecticism, adding that it carried echoes of 'Nietzsche et de Durkheim, d'un certain Belot et de l'alternative entre discours corpusculaire et discours ondulatoire devant laquelle les traveaux de de Broglie avaient récemment placé les physiciens . . .', *Collège de Sociologie*, 139.

104 Émile Belot, 'Le rôle capital de l'Astrophysique dans la cosmogonie', *Scientia*, vol. 62, no. CCIV, July 1937, 74–82.

105 Bataille, *OC*, vol. 12, 601.

106 Whitworth, 'The Clothbound Universe', 53.

107 A. N. Whitehead, *Process and Reality*, Cambridge: Cambridge University Press, 1929, 150.

108 Whitehead, *Process and Reality*, 150.

109 Jean Wahl, 'Vers le concret', *RP*, no. 1, 1–20; Jean Wahl, *Vers le concret. Études d'histoire de la philosophie contemporaine*, Paris: Librairie Philosophique J. Vrin, 1932. Bataille had pored over the premier issue of *Recherches philosophiques* on numerous occasions by the time he borrowed Wahl's book from the Bibliothèque Nationale in January 1935, Bataille, *OC*, vol. 12, 599. Also see the review of Wahl's book by Dandieu, 'Jean Wahl – *Vers le concret*' [book review], *Europe*, vol. 30, no. 120, 15 December 1932, 630–34.

110 Arthur Eddington, *Stars and Atoms*, Oxford: Clarendon Press, 1927, 10.

111 'qui cimente le macroscopique avec le microscopique', Anonymous, review of Arthur Eddington, *Étoiles et Atomes*, *RMM*, year 38, no. 3, July–September 1931, supplement, 5–6, 5.

112 Arthur Eddington, *The Expanding Universe*, Cambridge: Cambridge University Press, 1933. For a typically Dalínian interpretation of Eddington's idea, see Salvador Dalí, *Oui: La Révolution paranoïaque-critique; L'Archangélisme scientifique*, Paris: Denoël, 2004, 329.

113 Arthur Eddington, *The Rotation of the Galaxy*, Oxford: Clarendon Press, 1930, 6.

114 Eddington, *Rotation of the Galaxy*, 6.

115 Eddington, *Rotation of the Galaxy*, 20–21.

116 Eddington, *Rotation of the Galaxy*, 30.

117 Eddington, *Rotation of the Galaxy*, 30.

118 Bataille, 'Formless' [1929], *Visions of Excess*, 3; Bataille, 'Informe', *OC*, vol. I, 217.

119 Martin Jay, *Downcast Eyes: The Denigration of Vision in Twentieth-Century French Thought*, Berkeley, Los Angeles, and London: University of California Press, 1993, 228.

120 Jay, *Downcast Eyes*, 227.

121 Bataille, 'Rotten Sun' [1930], *Visions of Excess*, 57; 'poétiquement le sens de la sérénité mathématique et de l'élévation d'esprit', 'esprit de celui qui doit l'émasculer nécessairement par suite de l'incapacité des yeux', 'à l'éjaculation mentale, à l'écume aux lèvres et à la crise d'épilepsie', Bataille, 'Soleil pourri', *OC*, vol. 1, 231–2, 231.

122 Jay, *Downcast Eyes*, 222–3.

123 Jonathan Strauss, 'The Inverted Icarus', *Yale French Studies*, no. 78 ('On Bataille'), 1990, 106–23, 115.

124 Strauss, 'The Inverted Icarus', 115.

125 Strauss, 'The Inverted Icarus', 115.

126 Bataille, 'The Obelisk' [1938], *Visions of Excess*, 213–222, 214; 'horreur, violence, haine, sanglot, crime, dégoût et rire et amour humain', Bataille, 'L'obélisque', *OC*, vol. 1, 501–13, 503.

127 Eddington, *Nature of the Physical World*, 327.

128 Bataille, 'The "Old Mole"', *Visions of Excess*, 42; 'naïveté icarienne', Bataille, *OC*, vol. 2, 107.

129 Bataille, 'The "Old Mole"', *Visions of Excess*, 43; 'les *garde-à-vous* du ciel commandant aujourd'hui à l'esprit de n'importe quel homme la plus imbécile élévation', Bataille, *OC*, vol. 2, 109.

130 Arthur Eddington, 'The Internal Constitution of the Stars', *The Observatory*, vol. XLIII, October 1920, no. 557, 341–58, 358.

131 Arthur Stanley Eddington, *Science and the Unseen World*, London: George Allen & Unwin, 1929, 16.

132 Bataille, 'The Solar Anus', *Visions of Excess*, 5–9, 9; 'immonde parodie du soleil torride et aveuglant', Bataille, 'L'anus solaire', *OC*, vol. 1, 79–86, 86.

133 Bataille, 'The "Old Mole"', *Visions of Excess*, 35; 'creuse des galeries dans un sol décomposé et répugnant pour le nez délicat des utopistes', Bataille, *OC*, vol. 2, 96.

134 Bataille, 'Corps célestes', *Verve*, vol. 1, no. 2, March–June 1938, 97–100 and *OC*, vol. 1, 514–20. Translated quotations are from Bataille, 'Celestial Bodies', *October*, no. 36, Spring 1986, 75–8. Bataille's text and its accompanying illustrations might have helped prompt Miró's remarkable *Contellations* series of gouaches begun the following year, though this connection is not made in the main text on Bataille–Miró by Rosalind Krauss, '"Michel, Bataille et moi"', *October*, no. 68, Spring 1994, 3–20.

135 Given the 1930 appearance of *The Rotation of the Galaxy* and the peculiar 1927–31 dating of Bataille's first major statement, *The Solar Anus*, it is perfectly plausible to assume that the overriding motif of *terrestrial* rotation which drives the imagery, analogies and ideas of *The Solar Anus* was inspired by an earlier reading of Eddington's book.

136 Eddington, *The Rotation of the Galaxy*, 7.

137 Bataille, 'Celestial Bodies', 75; Bataille, *OC*, vol. 1, 514.

138 Bataille, 'Celestial Bodies', 75; '[s]i le soleil était animé d'un mouvement de gravitation stable autour d'un centre, l'univers auquel appartient la Terre pourrait être tenu dans son ensemble pour analogue au système fermé de nos révolutions planétaires (seul l'ordre des grandeurs aurait changé). Alors que le mouvement décrit par le Soleil et par les planètes

autour du centre galactique s'ouvre en quelque sorte à travers le ciel. En effet, quand nous apercevons ce mouvement dans l'immensité – animant des univers semblables au nôtre – il n'a nullement l'aspect des anneaux stables et géométriques de Saturne: il semble jeté dans l'espace comme une explosion qui tournoie', Bataille, *OC*, vol. 1, 514.

139 Bataille, 'Celestial Bodies', 76; 'soleils tournoyants de nos fêtes de lumières', 'l'explosion éblouissante d'un feu d'artifice qui se perd en quelques instants dans la nuit', Bataille, *OC*, vol. 1, 515

140 Bataille, 'Celestial Bodies', 76; '[m]ais si nous tenons compte du jugement d'Eddington (*The Rotation of the Galaxy*, Oxford, 1930), la précarité réelle des univers semblables au nôtre répondrait à leur aspect . . .', Bataille, *OC*, vol. 1, 515.

141 Bataille, 'Celestial Bodies', 76; 'Dans le tourbillon même de la Galaxie s'est formé un domaine qui s'est constitué en apparence l'autonomie la plus entière, échappant à la réalité de l'existence mouvementée de l'Univers à ce point que, là où toute chose tourne avec une rapidité folle, une parfaite immobilité semble s'étendre. Et même s'il est vraie que l'existence humaine découvre maintenant l'univers qui la porte, cette existence, à moins de se nier elle-même, doit reconnaître cet univers comme un spectacle extérieur: comment pourrait-elle, en effet, prétendre s'identifier à l'ivresse du ciel, admettre qu'elle est le spectacle se regardant lui-même, quand le fait de regarder demande que celui qui regarde ait échappé en quelque façon au mouvement ivre de l'univers. Toute la vérité que reconnaît l'homme est nécessairement liée à l'erreur que représente le "sol immobile"', Bataille, *OC*, vol. 1, 516.

142 Denis Hollier, 'The Dualist Materialism of Georges Bataille', Fred Botting and Scott Wilson (eds), *Bataille: A Critical Reader*, Oxford: Blackwell, 1998, 59–73, 67.

143 Bataille, 'Celestial Bodies', 78; 'Le véritable don de soi, l'extase – qui ne lie pas seulement l'homme à la femme – n'en marque pas moins la limite de l'avidité, la possibilité d'échapper au mouvement froid et de retrouver la fête des soleils et des spirales', Bataille, *OC*, vol. 1, 519.

144 Susan Rubin Suleiman, 'Transgression and the Avant-Garde: Bataille's *Histoire de l'oeil*', Boldt-Irons (ed.), *On Bataille*, 313–33, 317.

145 Bataille, 'Celestial Bodies', 78; '[l]es hommes peuvent retrouver . . . le mouvement libre de l'univers, ils peuvent danser et tournoyer avec une ivresse aussi délivrante que celle des grands essaims d'étoiles', Bataille, *OC*, vol. 1, 520.

146 Bataille, 'Autobiographical Note', 109.

147 Bataille, *OC*, vol. 12, Paris: Gallimard, 1988, 604. Bataille, 'Nietzsche and the Fascists' and 'The Obelisk', *Visions of Excess*, 182–96, 213–22; Bataille, *OC*, vol. 1, 'Nietzsche et les fascistes' and 'L'obélisque', 447–65, 501–13.

148 Friedrich Nietzsche, *The Gay Science*, trans. Walter Kaufmann, New York: Vintage Books, 1974, 7.

149 The much misunderstood phrase appears in Nietzsche, *Gay Science*, 167, 181, 279. The first 1882 edition of the book carried an epigraph by Emerson, 'To the poet and sage, all things are friendly and hallowed, all experiences profitable, all days holy, all men divine.'

150 Nietzsche, *Gay Science*, 183.

151 Nietzsche, *Gay Science*, 230–31.

152 Nietzsche, *Gay Science*, 240–241.

153 Bataille, 'Autobiographical Note', 110.

154 Bataille, 'Autobiographical Note', 110. A brief argument suggestively tying Nietzsche's indeterminism to Bataille's general economy and Bohr's complementarity is made in Plotnitsky, *Complementarity*, 253–4.

155 Discussed as a formative text in Hollier, *Against Architecture*, 14–46.

156 Jean-Paul Sartre, 'Un nouveau mystique' [1943], *Situations*, vol. 1, 143–88. Sartre's review appeared originally in three parts in *Cahiers de Sud*, and as I write remains untranslated into English.

157 Georges Bataille, *Inner Experience*, trans. Leslie Anne Boldt, New York, SUNY, 1988; Bataille, *L'Expérience intérieure*, Paris: Gallimard, 1943; Bataille, *OC*, vol. 5, Paris: Gallimard, 1973, 7–189. Without reference to science, recent scholarship has given greater attention to this aspect of Bataille's work, Peter Tracey Connor, *Georges Bataille and the Mysticism of Sin*, Baltimore and London: John Hopkins University Press, 2000, and Andrew Hussey, *The Inner Scar: The Mysticism of Georges Bataille*, Amsterdam: Editions Rodopi, 2000.

158 Bataille, *Inner Experience*, 103; 'est déjà dans le néant de Son insatiabilité', Bataille, 'L'expérience intérieure', *OC*, vol. 5, 121.

159 Bataille, *Inner Experience*, 81–92; Bataille, *OC*, vol. 5, 97–109.

160 In this chapter particularly, the nature of my study of Surrealism and physics requires emphasis on an empirical and collative approach over a theoretical one. An extension of my inquiry into Bataille, Eddington and mysticism would gain immense leverage from Jacques Derrida's remarkable 'Différance', which notes the resemblance of its approach to that of negative theology, and is deeply indebted to Koyré and especially Bataille, Derrida, 'Différance' [1968], *Margins of Philosophy*, trans. Alan Bass, New York: Harvester Wheatsheaf, 1982, 1–27.

161 Douglas, *Arthur Stanley Eddington*, 1.

162 Arthur Eddington, *Science and Religion*, London: The Friends Home Service Committee, 1931, 5.

163 Eddington, *Science and Religion*, 16.

164 Eddington, *Science and Religion*, 8.

165 Eddington, *Science and Religion*, 10.

166 Eddington, *Nature of the Physical World*, 320.

167 Eddington, *Nature of the Physical World*, 335.

168 Eddington, *Nature of the Physical World*, 322.

169 Bataille, 'Un-Knowing: Laughter and Tears' [1953], *October*, no. 36, Spring 1986, 89–102, 95.

170 Eddington, *Nature of the Physical World*, 330.

171 Arthur Eddington, *New Pathways in Science*, Cambridge: Cambridge University Press, 1935, 322.

172 Eddington, *Nature of the Physical World*, 342.

173 Walter Benjamin, 'Max Brod's Book on Kafka and Some of My Own Reflections' [1938], *Illuminations*, trans. Harry Zorn, London: Collins/Fontana, 1973, 141–8, 145. Benjamin quotes from the 1928 German translation of *The Nature of the Physical World*.

174 Benjamin emphasises the importance of Heidegger to the journal, review of *Recherches philosophiques*, *Zeitschrift für Sozialforschung*, year 6, no. 1, 1937, 173–4. Given the dates of these two texts, we can assume Benjamin was drawn to modern physics in the late 1930s, though this has yet to be acknowledged in the voluminous literature on the writer.

175 Benjamin, *Illuminations*, 146.

176 Robert Benayoun, 'Le Chandail d'Einstein', *Médium: Communication Surréaliste*, Series 2, no. 4, January 1955, 17; Charles Fort, 'Les Livres des damnés', *Médium*, Series 2, no. 4, January 1955, 18–19.

177 Benayoun, 'Le Chandail d'Einstein', 17. Eddington, *Nature of the Physical World*, 291.

178 Fort, 'Les Livres des damnés', *Médium*, 19.

179 The case for science-as-farce is furthered through the figure of Professor Gottlieb – 'an unlikely cross between Einstein, Gurdjieff and Groucho Marx' – in Robert Benayoun, *La Science met bas*, Paris: Jean-Jacques Pauvert, 1959, 30.

180 Georges Bataille, *La Part maudite: Essai d'économie générale. La Consomation*, Paris: Minuit, 1949.

181 'Par Ambrosino, par Queneau, par Caillois, Bataille fait appel, fut-ce de manière "poétique," aux ressources contradictoires de la science moderne et à ses lois d'indétermination', Marmande, *Georges Bataille politique*, 133.

182 For some rare biographical information on the still-shadowy figure of Ambrosino, see Georges Bataille (ed.), *Encyclopaedia Acephalica*, trans. Iain White, London: Atlas Press, 1995, 158.

183 Georges Ambrosino, 'La Structure de l'atome: Le noyau selon L. de Broglie', *Critique*, no. 1, June 1946, 91–6.

184 Bataille, *The Accursed Share*, 191 n. 2; 'est aussi par une part importante l'œuvre d'Ambrosino', 'des mouvements de l'énergie à la surface du globe', Bataille, *OC*, vol. 7, 23, n. 2. Surya cites a letter from Ambrosino to Bataille of 28 November 1945 in support of his claim that Ambrosino 'inspired and corrected the text', Surya, *Georges Bataille: An Intellectual Biography*, 561, n. 15. Bataille's book deserves a new reading in the light of that evidence and what I have written in this chapter.

185 C. P. Snow, *The Two Cultures*, Cambridge: Cambridge University Press, 1993, 16.

186 Hollier (ed.), *Collège de Sociologie*, 188.

5 Quantum Mechanics and Particle Physics: *Matta, Wolfgang Paalen, Max Ernst*

1 C. P. Snow, *The Two Cultures*, Cambridge University Press: Cambridge, 1993, 16–17; Alan Sokal and Jean Bricmont, *Intellectual Impostures*, London: Profile, 1998, 248.

2 Carl Einstein, *Georges Braque*, trans. E. Zipruth, Paris: Chroniques du jour, 1934.

3 'ne sont qu'une sorte de résistance, une barrière, dans le flot d'un monde fonctionnel', Einstein, *Georges Braque*, 70.

4 'une scission entre l'homme et la conception conventionnelle du monde, une reconstruction de la structure de la vue', Einstein, *Georges Braque*, 70.

5 'cette rupture de la continuité'; 'brisent la chaîne fatale de la causalité unitaire', Einstein, *Georges Braque*, 47.

6 'Si l'on voulait arriver à une nouvelle conception et à une nouvelle forme de l'homme et de l'univers, il importait avant tout de briser le cliché du réel, de se briser soi-même ainsi que l'histoire conventionnelle. Cette entreprise fut osée par quelques poètes et quelques hommes de science, ainsi que par deux cubists, dont l'audace équivaut à un fait historique de première importance', Einstein, *Georges Braque*, 57. The 'deux cubists' are Picasso and Braque, of course.

7 'Il est à la fois pénible et humiliant de voir les historiens d'art considérer les faits de l'histoire de l'art comme des phénomènes isolés: traitement qui fait de l'art un monde monstrueux, adoré comme un miracle divin. On devrait une fois pour toutes s'incorporer à l'ensemble de l'histoire; se rappeler par exemple que les Riemann et Lobatschewski avaient déjà, depuis longtemps, élargi la notion de l'espace par-delà les limites de la description unilatérale, que le truquage avec les 'substances' et la perspective statique de l'âme n'ont plus cours, et que la grammaire, ne suffisant plus avec ses déclinaisons et conjugaisons aux courants psychiques et à la structure sociale, est désormais déchue de son règne. La théorie des quantums a déchiré la continuité unitaire de l'univers, et l'on a fini par mettre sérieusement en doute la causalité, âme de toute l'ancienne science. Le moi conscient n'est plus aujourd'hui qu'une façade, prête à sombrer dans l'acte spontané . . . Le soi-disant cubisme est un phénomène parfaitement normal dans l'ensemble d'une conception radicalement transformée . . .', Einstein, *Georges Braque*, 63–4.

8 'la fin du sujet "stable" et déterminé équivaut à une liquidation de l'attitude anthropocentrique', Einstein, *Georges Braque*, 70.

9 'la croyance à un moi hermétique, à la figuration définitive, devait être abandonée', Einstein, *Georges Braque*, 76.

10 Carl Einstein's interest in the new space and concepts of modern physics was closely linked to his attraction to *both* Cubism and tribal objects. According to one author, the work of Cubists and tribal artisans convinced Einstein it was necessary for art 'non pour reproduire mais 'inventer', inventer une nouvelle vision, créer l'espace . . .', Liliane Meffre, 'Georges Bataille et Carl Einstein', Dominique Lecoq (ed.), *Georges Bataille et la pensée allemande*, Paris: Les Amis de Georges Bataille, 1986, 95–9, 98.

11 All quotations are from Puech, 'Signification et représentation', *Minotaure*, no. 6, Winter 1934–5, 52.

12 'cherchant à se situer dans un ensemble qui le *comprenne* et qu'il *comprendrait* et ne parvenant à se représenter à soi-même que dans la mesure où il se représente l'univers. De toute façon, il n'y avait pas divorce entre la réalité et l'homme, mais correspondance entre un microcosme et un macrocosme ou identité de l'Être et de la Représentation, communauté du monde objectif et du sujet au sein de quelque chose qui se révélerait en fin de compte capable d'être représentée.'

13 'en fonction d'un arrière-plan immuable qui, bloquant la vision et formant écran, permettrait la représentation.'

14 'Second échec plus grave, et proche de la faillite: la toute récente promotion de la physique semble interdire à la représentation de retrouver un 'invariant' *en deçà* de la réalité apparente, dans l'élément du *contenu*. Le solide se dissout et se résout en fluide, la matière en rayonnement, le corpuscule en 'paquet d'ondes.' L'enserrement exact du réel fait place à une plus vague considération d'ensembles plus ou moins arbitrairement cernés: le détérminisme strict s'évanouit devant une supputation de probabilités qui s'efforce à composer avec le hasard. Désormais en deçà des points de repère – maintenant abolis – que l'atome fournissait à la représentation dernière des choses, il n'y a plus – et c'est encore s'accrocher à un lambeau illusoire d'image – que la houle lente et indéterminée des ondulations de l'espace ou de quelque chose qui pourrait encore ressembler à l'espace pur, à l'ultime expression du réel.'

15 'd'événements, d'accidents, d'associations singulières qu'il traîne avec soi.'

16 'L'individu est une *histoire* présente en lui et *qui n'est que pour lui-même . . .*'

17 'les philosophies réalistes ou "existentielles" substituent à la représentation fondée sur l'opposition d'un objet-spectacle et d'un sujet-spectateur, l'idée – si c'est encore une idée – d'une *présentation* réciproque du réel et de l'homme, d'une *présence* éprouvée, mais rebelle à toute projection objective, détachée de qui l'éprouve.' Existentialism had arrived in France in the form of Heidegger's *Sein und Zeit*, published in 1927, the same year Alexandre Koyré translated a text by Heidegger for the marginally Surrealist journal *Bifur* (Koyré also ensured his work appeared in *Recherches philosophiques*).

18 'par des moyens extrêmement compliqués, l'exprimer que par des symboles techniques de plus en plus abstraits.'

19 'il semble qu'il y ait rupture entre l'homme vivant et l'homme savant, entre la représentation que l'homme se fait quotidiennement de lui-même et de l'univers, pratique et familier, qui l'entoure, et, d'autre part, la réalité où il arrive tout au plus à se figurer qu'il est *quelque chose* et où la science ne lui procure plus qu'une gloire abstraite ou qu'une facilité de maniement technique.'

20 'manifestation ou système de forces *significatives*.'

21 'une copie plus ou moins idéalisée, mais à un objet magique ou à un *symbole* mental.'

22 '*réalisme* et symbole.'

23 'au sens banal du "symbolisme,"' 'à *imposer* et à agir.'

24 'qui, d'ailleurs, ne peut plus être une vision.'

25 Puech points out features shared by modern art and modern science as early as 1930, in remarks made in an article on Picasso, when he writes that the world, 'répugne de plus en plus à toute *représentation* . . .. Il faudra bien un jour mesurer la portée de cette remarque que vérifient toutes les manifestations de la science ou de l'art modernes, nous rendre un compte plus exact que la ruine de l'idée de *représentation* comme le caractère essentiel de notre époque . . .', Puech, 'Picasso et la répresentation', *Documents*, no. 3 (special number, 'Hommage à Picasso'), 1930, 118–22, 118. Carl Einstein contributed an essay to this number of *Documents* on Picasso, too. He wrote *Georges Braque* in 1931–2.

26 'revision des valeurs plastiques', Christian Zervos, 'ENQUÊTE', *Cahiers d'Art*, year 10, 1935, 5–72, 5.

27 'Ne serait-il pas temps d'intégrer la forme dans le fonde, suivant l'example de la philosophie qui commence déjà à élargir le concept du monde en intégrant petit à petit l'expérience scientifique dans la pensée?' Zervos, *Cahiers d'Art*, year 10, 6.

28 'c'est-à-dire forcer les spectateurs à une contemplation qui les branche bon gré mal gré sur le grand courant commun, qui remonte plus loin que toute notre histoire connue et qui les mène à un concept du monde inaccessible à l'outil grossier du déterminisme scientifique', Paalen, *Cahiers d'Art*, year 10, 6.

29 'communauté spirituelle qui peut prétendre à l'universalité n'existe que dans la domaine scientifique', Paalen, *Cahiers d'Art*, year 10, 62.

30 Nicolas Calas, *Matta: A Totemic World*, New York: Andrew Crispo Gallery, 1975, unpaginated.

31 '"Transformer le monde," a dit Marx; "changer la vie," a dit Rimbaud: ces deux mots d'ordre pour nous n'en font qu'un', Breton, 'Discours au Congrès des écrivains' [June 1935], *Oeuvres complètes* [OC], vol. 2, Paris: Gallimard, 1992, 451–9, 459.

32 Michel Leiris, 'Notes sur deux figures microcosmiques des XIV et XV siècles', *Documents*, no. 1, April 1929, 48–52.

33 See André Masson, *Anatomy of My Universe*, New York: Curt Valentin, 1943. Matta believed Masson was an 'over-literary artist', quoted in Romy Golan, 'R. E. Matta, 1937–1947: The Last Phase of Surrealism', M.A. report, Courtauld Institute of Art, 1982, 7, n. 6.

34 Calas, *Matta: A Totemic World*.

35 Similarly, though inspired more by Jung than Freud, in his 1940 lectures on Surrealism given at the New School for Social Research in New York, Matta's friend Gordon Onslow Ford, 'launched into his own theories on combining symbols from the collective unconscious with a time-space continuum that dissolved the artificial constructs of temporality and

Euclidean geometry', Martica Sawin, 'Onslow Ford in New York, June 1940 to June 1941', *Gordon Onslow Ford: Mirando en lo profundo/Seeing in Depth*, Santiago de Compostela: Fundación Eugenio Granell, 1998, 193–7, 195.

36 Sidney Simon, 'An Interview with Robert Motherwell Conducted by Sidney Simon in New York in January 1967', *Art International*, vol. XI, no. 6, Summer 1967, 20–23, 21.

37 R. Malbert, 'An Introduction to the Drawings', Plymouth: City Museum and Art Gallery, *Matta: The Logic of Hallucination*, 1984, 10–11, 10.

38 James Thrall Soby, *Contemporary Painters*, New York: Museum of Modern Art, 1948, 62.

39 Soby, *Contemporary Painters*, 62.

40 Matta, "Novembre 1936", *Matta: Entretiens Morphologiques*, Germana Ferrari (ed.), London: Sistan, 1987, 36.

41 Simon, 'An Interview with Robert Motherwell', *Art International*, no. 6, 62. The German exhibit at the 1893 World's Fair in Chicago displayed many mathematical models some of which were purchased and shown by American universities and colleges, according to Angela Vierling, 'The "Botanical" Study of Algebraic Surfaces in the 19th Century', http://math.harvard.edu/~angelavc/models.

42 Sidney Simon, 'Concerning the Beginnings of the New York School: 1939–1943. An Interview with Peter Busa and Matta Conducted by Sidney Simon in Minneapolis in December 1966', *Art International*, vol. XI, no. 6, Summer 1967, 17–20, 18.

43 William Rubin, *Matta*, New York: Museum of Modern Art, 1957, 3.

44 'Since Descartes', Matta said in 1955, 'we have only seen man isolated, the thing divorced. Now we must see the thing – and describe it – without abstracting it, see the thing in the world, *with* the world. We must paint man and his daily galaxy', quoted in Stockholm: Moderna Museet, *Sebastian Matta: 15 former av trivel/15 forms of doubting*, 1959, 32.

45 Breton, 'Matta: The Pearl is Marred in my Eyes . . .', *Surrealism and Painting*, trans. Simon Watson Taylor, London: Macdonald and Co., 1972, 183–8; 'il nous convie sans cesse à un *nouvel espace*, en rupture délibérée avec l'ancien, puisque ce dernier n'a de sens qu'autant qu'il est distributif de corps élémentaires et fermés. Peu importent les appuis momentanés que cette manière de voir et de faire voir s'est cherchés successivement dans les aperçus scientifiques de la morphologie psychologique, de la Gestalttheorie, de l'astrophysique, de l'histologie et de la physique moléculaire. Le besoin de tels appuis n'exprime autre chose que l'aspiration à étendre, en s'aidant des ressources les plus modernes, le champ du visible . . .', Breton, 'Matta: La perle est gâtée à mes yeux . . .', *Le Surréalisme et la peinture*, Paris: Gallimard, 1965, 183–8, 187. Recall that Breton spoke similarly of *surrationalisme* and Surrealism as evidence of a 'pensée en rupture avec la pensée millénaire', 'Crise de l'objet', *Le Surréalisme et la peinture*, 275–80, 277.

46 Marcel Duchamp, *Salt Seller: The Essential Writings of Marcel Duchamp*, ed. Michel Sanouillet and Elmer Peterson, London: Thames and Hudson, 1975, 154.

47 Duchamp, *Salt Seller*, 154. 'I made some puns', said Duchamp of the brief Société Anonyme writings, Pierre Cabanne, *Dialogues with Marcel Duchamp*, trans. Ron Padgett, London: Thames and Hudson, 1971, 84. Matta was dismissed by Greenberg in 1945 as 'that prince of the comic-strippers', Clement Greenberg, *The Collected Essays and Criticism*, vol. II: *Arrogant Purpose, 1945–1949*, Chicago and London: University of Chicago Press, 1986, 13. Of course, my argument that Matta called upon modern physics to attain an art we might call abstract, years ahead of the New York School, goes against the grain of Greenberg's idea that forces solely inclusive to art itself brought about the development of abstract art, but the scope of my project means I can sometimes refer to its 'local' historical and theoretical implications only in passing.

48 'la limitation de la vue à l'image rétinienne ou à son souvenir . . .', Alain Jouffroy, 'Matta: Ulysse passe-partout', Paris: Centre Georges Pompidou, *Matta*, 1985, 53–65, 56.

49 Matta, 'Auto-Elasto-Infra-Biography', Ferrari (ed), *Matta*, 12–19, 16.

50 Octavio Paz, 'Vestibule', Valerie Fletcher (ed.), *Cross-currents of Modernism: Four Latin American Painters*, Washington DC: Smithsonian Institution Press, 1992, 274–81, 277.

51 Soby, *Contemporary Painters*, 66.

52 Rosamund Frost, 'Matta's Third Surrealist Manifesto', *ARTNews*, vol. XLIII, no. 1, 15–29 February 1944, 18.

53 Max Kozloff, 'Interview with Matta by Max Kozloff', *Artforum*, vol. IV, no. 1, September 1965, 23–6, 26.

54 Kozloff, 'Interview with Matta', *Artforum*, vol. IV, no. 1, 26.

55 For a coalition of the discourses of physics and psychoanalysis in the postwar period to add to the ones I note here from Matta to Dalí and beyond, see Gherasim Luca, *Les Orgies des quanta*, Bucharest: Presses de l'imprimerie independenta, 1946.

56 Gustav Regler, *Wolfgang Paalen*, New York: Nierendorf Editions, 1946, 41.

57 'un nouvel espace, celui de la quatrième dimension, si j'ai bien compris . . .', Pierre Mabille, *Messages de l'étranger*, Paris: Plasma, 1981, 230.

58 However, Paalen was certainly interested in the fourth dimension and owned copies of most of the standard works on the subject, including Maeterlinck's *La Quatrième Dimension* (1928), Charles Howard Hinton's *The Fourth Dimension* (1904) and *An Episode in Flatland* (1907), Claude Bragdon's *A Primer of Higher Space (The Fourth Dimension)* (1913), and Ouspensky's *Tertium Organum* (1920).

59 Louis de Broglie, *Matter and Light: The New Physics*, trans. W. H. Johnstone, London: George Allen and Unwin Ltd., 1939, 29.

60 Amy Winter, *Wolfgang Paalen*, Dyn, *and the Ameri-*

can Avant-Garde of the 1940s, dissertation submitted for the degree of Ph.D., City University of New York, 1995, 513.

61 'cette solitude qui fait la beauté et la grandeur de *Dyn*', José Pierre, *Domaine de Paalen*, Paris: Éditions Galanis, 1970, 42.

62 Sawin reports that Pollock owned a complete run of *Dyn*, Martica Sawin, *Surrealism in Exile and the Beginning of the New York School*, Cambridge, Mass.: MIT, 1995, 267.

63 'les préoccupations dont *Dyn* est le reflet sont tout autant surréalistes que celles de *VVV*', Pierre, *Domaine de Paalen*, 41.

64 Breton, 'Wolfgang Paalen: Non plus le diamant au chapeau . . .', *Le Surréalisme et la peinture*, 136–140, 140.

65 'In my opinion it has been modern physics rather than Mexican landscape that has been Paalen's inspiration . . .', letter from Onslow Ford to Peggy Guggenheim (late 1944 or early 1945), quoted in Sawin, *Surrealism in Exile*, 283.

66 Wolfgang Paalen, *Form and Sense*, New York: Wittenborn and Co., 1945. The best English-language monograph on Paalen lays the stress on anthropology, his other main enthusiasm, Amy Winter, *Wolfgang Paalen: Artist and Theorist of the Avant-Garde*, Westport, Connecticut: Praeger, 2003. Paalen's friend José Pierre points out it was 'paradoxalement les arts primitifs qui l'ont persuadé de la nécessité d'aboutir à une relation intime entre . . . la science et l'art', Pierre, *Domaine de Paalen*, 11. Paalen's apparently divergent interests in quantum theory and 'primitive' art square around the theme of *rhythm*, which he sees underlying wave mechanics and perceives in the lyrical forms of tribal artifacts created by craftspeople 'attuned to nature.'

67 Paalen, 'During the Eclipse', *Dyn*, no. 6, 1944, 20.

68 Paalen, 'On the Meaning of Cubism Today', *Dyn*, no. 6, 1944, 4–8, 6. Linda Dalrymple Henderson notes that Paalen probably read Apollinaire's *The Cubist Painters*, which was republished in 1944, Henderson, 'Four-Dimensional Space or Space-Time?: The Emergence of the Cubism-Relativity Myth in New York in the 1940s', Michele Emmer (ed.), *The Visual Mind ii*, Cambridge, Mass: MIT, 2005, 349–97.

69 Paalen, 'On the Meaning of Cubism Today', *Dyn*, no. 6, 6.

70 Paalen, 'On the Meaning of Cubism Today', *Dyn*, no. 6, 7.

71 Paalen, 'On the Meaning of Cubism Today', *Dyn*, no. 6, 7.

72 Paalen, 'Art and Science', *Dyn*, no. 3, 1942, 4–9; Paalen, 'Le Grand Malentendu (Art et Science)', 22–6.

73 Paalen, 'Art and Science', *Dyn*, no. 3, 4.

74 Paalen, 'Art and Science', *Dyn*, no. 3, 4.

75 Paalen, 'Art and Science', *Dyn*, no. 3, 5.

76 Paalen, 'Art and Science', *Dyn*, no. 3, 4.

77 Paalen, 'Art and Science', *Dyn*, no. 3, 7.

78 See Gerald Holton, 'The Roots of Complementarity', *Thematic Origins of Scientific Thought: Kepler to Einstein*, Cambridge, Mass.: Harvard University Press, 1974, 115–61.

79 Werner Heisenberg, 'Die Gothesche und die Newtonsche Farbenlehre im Lichte der modernen Physik', *Geist der Zeit*, no. 19, 1941, 261–75; Werner Heisenberg, *Philosophic Problems of Nuclear Science*, trans. F. C. Hayes, London: Faber and Faber, 1952, 60–76.

80 Heisenberg, *Philosophic Problems of Nuclear Science*, 62.

81 Heisenberg, *Philosophic Problems of Nuclear Science*, 66.

82 Heisenberg, *Philosophic Problems of Nuclear Science*, 65.

83 Heisenberg, *Philosophic Problems of Nuclear Science*, 68.

84 Heisenberg, *Philosophic Problems of Nuclear Science*, 71.

85 Paalen, 'Art and Science', *Dyn*, no. 3, 6.

86 Heisenberg, *Philosophic Problems of Nuclear Science*, 74.

87 Paalen, 'Art and Science', *Dyn*, no. 3, 7.

88 Paalen, 'Art and Science', *Dyn*, no. 3, 5–6.

89 Werner Heisenberg, 'Zur Geschichte der physikalischen Naturerklärung', *Wandlungen in den Grundlagen der Naturwissenschaft*, Leipzig: Verlag von S. Hirzel, 1935, 27–45; Heisenberg, *Philosophic Problems of Nuclear Science*, 27–40.

90 Werner Heisenberg, 'Fundamental Problems of Present-day Atomic Physics', *Philosophic Problems of Nuclear Science*, 95–108, 97; Erwin Schrödinger, 'The Atomists', *Nature and the Greeks and Science and Humanism*, Cambridge: Cambridge University Press, 1996, 75–89, 89. Schrödinger writes that the parallels between the thought of Leucippus and Democritus and the theory of modern physics had been noted by several philosophers, including Bertrand Russell, so there were a number of avenues along which Paalen could have received the same information. Recall remarks made by Christian Zervos in 1936, to the effect that modern physics 'holds some points in common with Presocratic science', Christian Zervos, 'Mathématiques et art abstrait', *Cahiers d'Art*, year 11, 4–10, 8.

91 Paalen knew well the book by Erwin Schrödinger, *Science and the Human Temperament*, trans. James Murphy, London: George Allen and Unwin Ltd, 1935, and in *Dyn* he quotes from it on one occasion and gives a mild critique of its contents on another, Paalen, 'Art and Science', *Dyn*, no. 3, 8; Paalen, 'Nicolas Calas, *Confound the Wise*' (book review), *Dyn*, no. 3, 1942, 45. An exact contemporary of Marcel Duchamp, Schrödinger was an important figure for Paalen, who might even have attended lectures in Vienna given by Schrödinger, Winter, *Wolfgang Paalen,* Dyn, *and the American Avant-Garde of the* 1940s, 201, n. 42, 581.

92 Paalen, 'Art and Science', *Dyn*, no. 3, 8.

93 Paalen, 'Nicolas Calas', *Dyn*, no. 3, 45.

94 Paalen, 'Nicolas Calas', *Dyn*, no. 3, 45. Bataille, Paalen and Caillois respectively read, owned and quoted either from Niels Bohr's *La théorie atomique*

et la description des phénomènes, or its English translation *Atomic Theory and the Description of Nature*, Cambridge: Cambridge University Press, 1934, which provided a useful source for attacks made by all three, with recourse to modern physics, on Surrealism's alleged 'idealism' or 'metaphysics.'

95 Paalen, 'Hans Reichenbach, *From Copernicus to Einstein*' (book review), *Dyn*, no. 3, 1942, 44. Paalen owned a copy of Dean of the Vienna Circle, Rudolph Carnap's *Foundations of Logic and Mathematics*, Chicago: University of Chicago Press, 1937. He must have been attracted to the subject matter of his *La Science et la Métaphysique devant l'analyse logique du langage*, too, because he picks up on the same 'nonsensical' passage of Heidegger's highlighted there by Carnap, 'das Nichts nichtet' (also quoted from Carnap by René Crevel), without acknowledging Carnap when he (Paalen) criticises Heidegger's metaphysics in the later essay 'Metaplastic', San Francisco: San Francisco Museum of Art, *Dynaton*, 1951, 1–27, 3. Given this, we can assume Paalen read Crevel's 'Au carrefour de l'amour, la poésie, la science et la révolution' and/or Carnap's book, or (given Paalen's rendering of the passage in German) the original publication of the book as a much-debated article 'Über Windung der Metaphysik durch logische Analyse der Sprache', *Erkenntnis*, no. 2, 1931, 219–41, probably Carnap's best-known piece of writing. I draw out Surrealism's weirdly upbeat reception of Vienna Circle positivism in a moment, with reference to Max Ernst.

96 Paalen, 'Hans Reichenbach', *Dyn*, no. 3, 44. Reichenbach was widely read in France and Paalen might have seen the French translation of *Atom und Kosmos* anonymously reviewed in the *Revue de Métaphysique et de Morale*, year 42, no. 2, April 1935, supplement, 1–2.

97 'Charles Givors', reviews of *Fata Morgana* and *VVV*, *Dyn*, no. 4/5, 1943, 80–83, 82.

98 'Charles Givors', reviews of *Fata Morgana* and *VVV*, *Dyn*, no. 4/5, 81.

99 '"[D]écouverte spectaculaire . . . se produira sur la plan de la *physique*"', 'Charles Givors', reviews of *Fata Morgana* and *VVV*, *Dyn*, no. 4/5, 82. André Breton, 'Situation du surréalisme entre les deux guerres', *VVV*, no. 2/3, March 1943, 44–53, 49; André Breton, 'Situation of Surrealism between the Two Wars', *Free Rein*, trans. Michel Parmentier and Jacqueline d'Amboise, Lincoln and London: University of Nebraska Press, 1995, 51–67, 60; Breton, 'Situation du surréalisme entre les deux guerres', *OC*, vol. 3, Paris: Gallimard, 1999, 709–25, 718.

100 'Charles Givors', reviews of *Fata Morgana* and *VVV*, *Dyn*, no. 4/5, 82.

101 Paalen, 'Farewell au Surréalisme' *Dyn*, no. 1, 1942, 26.

102 'conquêtes et les méthodes de la science', Paalen, 'Farewell au Surréalisme', *Dyn*, no. 1, 26.

103 Paalen, 'Seeing and Showing', *Dyn*, no. 1, 1942, 27.

104 Paul Feyerabend, *Farewell to Reason*, London and New York: Verso, 1987, 81. As I show below, a more obvious example of pictorial Complementarity is given by Dalí's paranoia-critical method.

105 Matta, *Entretiens Morphologiques*, Ferrari, 228.

106 Richard von Mises, 'Ernst Mach and the Empiricist Conception of Science', Robert S. Cohen and Raymond J. Seeger (eds), *Ernst Mach: Physicist and Philosopher*, Dordrecht: D. Reidel, 1970, 245–70, 252.

107 Philipp Frank, 'Ernst Mach and the Unity of Science', Cohen and Seeger (eds), *Ernst Mach*, 235–44, 239.

108 Liliane Meffre, 'Aspects de la théorie de l'art de Carl Einstein', *Cahiers du Musée National d'Art Moderne*, no. 1, July–September 1979, 14–16, 14. There is no discussion of Carl Einstein's interest in modern physics or of his Machism in the recent, useful number of *October* no. 107, Winter 2004 (Carl Einstein special number).

109 Meffre, 'Aspects de la théorie', no. 1, 15.

110 'Sur le plan théorique, c'est peut-être Mach qui m'est le plus proche', quoted in Jean Laude, 'Un portrait', *Cahiers du Musée National d'Art Moderne*, no. 1, 10–13, 13.

111 'Pour Mach, qui abolit toute barrière entre le physique et le psychique, il n'existe aucune opposition entre le monde intérieur du sujet et le monde extérieur, formés tous deux d'éléments, l'espace, le temps, les couleurs, les sons. La "chose en soi" n'existe pas, "c'est des sensations et de leurs combinaisons que sortent nos concepts." . . . Mach remplace les concepts de cause et d'effet par la notion de fonction qui permet de bien mieux représenter les relations des éléments entre eux. Cette notion de fonction est à la base de la théorie de l'art d'Einstein', Meffre, 'Aspects de la théorie', no. 1, 15.

112 'It was Ernst Mach who, in his *History of Mechanics*, shook this dogmatic faith [in mechanics]; this book exercised a profound influence upon me in this regard while I was still a student. I see Mach's greatness in his incorruptible scepticism and independence; in my younger years, however, Mach's epistemological position also influenced me very greatly, a position which today appears to me to be essentially untenable', Albert Einstein, 'Autobiographical Notes', Paul Arthur Schlipp (ed.), *Albert Einstein: Philosopher-Scientist*, Illinois: Open Court, 1949, 2–94, 21. See Holton, 'Mach, Einstein, and the Search for Reality', *Thematic Origins of Scientific Thought*, 219–59.

113 John T. Blackmore, *Ernst Mach: His Work, Life, and Influence*, Berkeley: University of California Press, 1972, 315. For more detail, see the chapter 'Mach's Influence on Early Logical Positivism and on Quantum Theory', 300–318.

114 Mara Beller, *Quantum Dialogue: The Making of a Revolution*, Chicago and London: University of Chicago Press, 1999, 284.

115 Blackmore, *Ernst Mach*, 315. For more on Schrödinger and Mach, see Henk W. de Regt, 'Erwin Schrödinger', J. Blackmore, R. Itagaki and S. Tanaka (eds), *Ernst Mach's Vienna, 1895–1930, or Phenom-*

enalism as Philosophy of Science, Dordrecht: Kluwer Academic Publishers, 2001, 85–104.

116 Blackmore, *Ernst Mach*, 316.

117 Niels Bohr, 'Wirkungsquantum und Naturbeschreibung', *Die Naturwissenschaften*, vol. 17, 28 June 1929, 483–6; Niels Bohr, 'The Quantum of Action and the Description of Nature', *Atomic Theory and the Description of Nature*, Cambridge: Cambridge University Press, 1934, 92–101.

118 Unpublished manuscript translated and quoted in Finn Aaserud, *Redirecting Science: Niels Bohr, Philanthropy, and the Rise of Nuclear Physics*, Cambridge: Cambridge University Press, 1990, 84. Among other things, Aaserud offers an invaluable reconstruction of key debates on QM, biology and psychoanalysis in the 1920s and 1930s.

119 Max Ernst, 'What is Surrealism?' [1934], trans. Werner Spies, London: Tate Gallery, *Max Ernst: A Retrospective*, 1991, 309–10, 310.

120 Ernst, 'What is Surrealism?', *Max Ernst*, 310. I am very grateful (yet again) to Simone Perks, who alerted me to Ernst's text quoting Jordan. Ernst quotes the final sentence from the following passage by Jordan: 'Diese Erschütterung der klassisch-philosophischen Vorstellung des von der Beobachtung unberührten Objektes bedingt übrigens eine gewisse Abmilderung eines Gegensatzes, der von fast allen früheren Philosophen als ein *absoluter*, grundsätzlicher und vollkommener betrachtet wurde; wir meinen den Gegnesatz zwischen objektiver Beobachtung und rein subjektivem Erleben. Man ist gewohnt, von der Betrachtung und Beobachtung der sog. "Außenwelt" die Beobachtung der eigenen subjektiven "Innenwelt" grundsätzlich zu unterscheiden. Aber dieser Unterscheidung wird eine Hauptstütze entzogen mit der experimentellen Widerlegung der Vorstellung, daß in der "Außenwelt" Tatbestände vorlägen, welche unabhängig vom Beobachtungprozeß ein "objektives" Dasein besitzen', Pascual Jordan, 'Die Quantenmechanik und die Grundprobleme der Biologie und Psychologie', *Die Naturwissenschaften*, vol. 20, 4 November 1932, 815–21, 818.

121 Ernst, 'What is Surrealism?' *Max Ernst*, 310.

122 Aaserud, *Redirecting Science*, 93.

123 For some informed, sympathetic and detailed contemporary commentary, see Jean Cavaillès, 'L'École de Vienne au Congrès de Prague', *Revue de Métaphysique et de Morale*, year 42, no. 1, January 1935, 137–49.

124 See Aaserud, *Redirecting Science*, 96, 305. Bohr's lecture was published as Niels Bohr, 'Kausalität und Komplementarität', *Erkenntnis*, no. 6, April 1937, 293–303 and translated as 'Causality and Complementarity', *Philosophy of Science*, no. 4, July 1937, 289–98.

125 An important development in the debate I am sketching here was the paper given by Bohr at the Second International Light Congress held in Copenhagen in August 1932, published as Niels Bohr, 'Licht und Leben', *Die Naturwissenschaften*, vol. 21, 31 March 1933, 245–50, and translated in two parts as 'Light and Life', *Nature*, vol. 131, no. 3308, 25 March 1933, 421–3 and *Nature*, vol. 131, no. 3309, 1 April 1933, 457–9. This article and the issues I am raising are discussed by a writer we have met who admired Bachelard and was close to the Surrealists: Armand Petitjean, *Imagination et réalisation*, Paris: Denoël et Steele, 1936, 235–55. Amazingly, Petitjean's overview of problems in observation raised by QM not only refers the reader to Jordan's 1932 *Die Naturwissenschaften* article (misdated 1933 by Petitjean), but quotes exactly the same passage Max Ernst used in 'Was ist Surrealismus?', Petitjean, *Imagination et réalisation*, 247–8, n. 3.

126 Pascual Jordan, 'Quantenphysikalische Bemerkungen zur Biologie und Psychologie', *Erkenntnis*, no. 4, 1934, 215–52. For responses from Edgar Zilsel, Hans Reichenbach, Otto Neurath, Moritz Schlick and Philipp Frank, see *Erkenntnis*, no. 5, 1935, 56–64 and 178–84, and see Jordan's reply in the same volume, 'Ergänzende Bemerkungen über Biologie und Quantenmechanik', 348–52. This is the published record of the debate begun at the 1934 Prague conference where Zilsel, who was close to the Vienna Circle, chaired a session on the views aired by Jordan in *Erkenntnis* earlier that year, bringing dismissive responses from all but Reichenbach. Jordan continued to promote the 'amplifier theory' in Pascual Jordan, 'Die Verstärkertheorie der Organismen in ihrem gegenwärtigen Stand', *Die Naturwissenschaften*, vol. 26, 19 August 1938, 537–45, and in a later book *Verdrängung und Komplementarität: Eine philosophische Untersuchung*, Hamburg-Bergedorf, 1947, where he attempts to relate Freud's concept of inhibition to Bohr's Complementarity. See Aaserud, *Redirecting Science*, 82.

127 Hans Reichenbach, 'Crise de la causalité', *Documents*, no. 2, May 1929, 105–8. In a 1931 letter to Bohr, Jordan mentions difficulties encountered by himself and Reichenbach in getting their papers accepted by *Die Naturwissenschaften*, as Berliner felt too much had already been published on the subject of causality, Aaserud, *Redirecting Science*, 87. However, in that year appeared Hans Reichenbach, 'Das Kausalproblem in der Physik', *Die Naturwissenschaften*, vol. 19, 21 Aug 1931, 713–22.

128 Paalen owned a copy of Mach's *Analysis of Sensations*, and his Viennese origins make these intellectual connections obvious ones to explore, though this has not yet been carried through thoroughly in the scholarship outside some speculations by Winter, *Wolfgang Paalen, Dyn, and the American Avant-Garde of the 1940s*, 10–11, 88, 177.

129 Frank, 'Ernst Mach and the Unity of Science', Cohen and Seeger (eds), *Ernst Mach*, 235.

130 Rudolph Carnap, *The Unity of Science*, trans. Max Black, London: Kegan Paul, 1934, 28. Carnap is referring here to Heidegger.

131 Quoted in Victor Kraft, *The Vienna Circle: The Origin of Neo-Positivism*, trans. Arthur Pap, New York: Philosophical Library, 1953, 165.

132 Quoted in Kraft, *The Vienna Circle*, 165.

133 Not surprisingly, Freudian psychoanalysis was raised in discussion groups in Vienna that gave rise to the group, and Hans Reichenbach admired Freud. But I found few references to psychoanalysis either in writings by the Vienna Circle or in the main history, Friedrich Stadler, *The Vienna Circle: Studies in the Origins, Development, and Influence of Logical Empiricism* [1997], trans. Camilla Nielson, Joel Golb, Sabine Schmidt, Thomas Ernst, New York: Springer, 2001.

134 André Breton, 'Manifesto of Surrealism', *Manifestoes of Surrealism*, trans. Richard Seaver and Helen R. Lane, Michigan: University of Michigan Press, 1972, 1–47, 6; 'l'attitude réaliste, inspirée du positivisme, de saint Thomas à Anatole France', Breton, 'Manifeste du surréalisme', *OC*, vol. 1, Paris: Gallimard, 1988, 309–46, 313.

135 André Breton, 'Interview with *Indice*', *What is Surrealism? Selected Writings*, ed. Franklin Rosemont, London: Pluto, 1978, 144–7, 147; 'la plus grande partie de la philosophie de Freud comme *métaphysique*', Breton, 'Interview d'*Indice*', *OC*, vol. 2, 445–50, 449–50.

136 As I said, Bachelard took an interest in the Vienna Circle, which in turn left its mark on Caillois, see Jean Leroux, 'Bachelard et le cercle de Vienne', *Cahiers Gaston Bachelard*, no. 5, 2003, 107–27. An extension of this discussion would take in a book that has been influential on the methodology of art history, has been compared with Bachelard's writings, acknowledges a significant debt to Alexandre Koyré and Émile Meyerson, and was published under the aegis of the Vienna Circle, Thomas S. Kuhn, *The Structure of Scientific Revolutions* [1962], Chicago: University of Chicago Press, 1970.

137 'les conditions, la portée et les limites du problème de la connaissance', André Burnier, 'Néo-réalisme et idéalisme contemporain', *Revue de Métaphysique et de Morale*, year 46, no. 3, July 1939, 425–41, 425.

138 'Sans doute, l'histoire de la philosophie nous offre semblable expérience et chacun de nous en a profité; mais elle est aujourd'hui, sinon épuisée, du moins dépassé par les incessants progrès de la philosophie alliée à la science', Burnier, 'Néo-réalisme et idéalisme contemporain', 425.

139 Berge wrote of a freshly minted Surrealism in 1925, '[c]ette tendance artistique est d'ailleurs en rapport étroit avec certaines tendances philosophiques contemporaines: M. Meyerson, en effet, a déjà tenté de nous prouver que l'apparence logique et rationnelle de la Nature était une vaine création de l'esprit humain. Je crois cette théorie conforme à la plus pure doctrine surréaliste', André Berge quoted in Henri Béhar and Michel Carassou, *Le Surréalisme: Textes et débats*, Paris: Librairie Générale Française, 1984, 164.

140 Marie-Antoinette Tonnelat, 'La notion de réalité physique et l'œuvre de M. Louis de Broglie', André George (ed.), *Louis de Broglie: Physicien et penseur*, Paris: Éditions Albin Michel, 1953, 145–52.

141 'On arrive ainsi à ce résultat assez paradoxal: un positivisme de stricte observance que, par un mauvais jeu de mot, nous pourrions qualifier de *surréaliste*, ne peut atteindre une réalité physique permanente et autonome. En physique comme en peintre, le surréalisme nie la possibilité d'une description qui ne porte *explicitement* l'estampille de l'observateur', Tonnelat, 'La notion de réalité physique', George (ed.), *Louis de Broglie*, 150.

142 Salvador Dalí, *Le Mythe tragique de l'Angelus de Millet*, Paris: Jean-Jacques Pauvert, 1963.

143 'Comment, je me le demande, ne pas discerner la présence du mécanisme paranoïaque dans le phénomène extraordinairement déterminatif du choix 'experimental' qui prélude aux investigations des sciences naturelles? Comment, je le répète, un tel mécanisme ne serail-il pas là agissant, alors que l'étude des particularités délirantes paranoïaques nous fait assister à un *changement essentiel du monde objectif, changement qui se présente comme subit, et qui absorbe, par son pouvoir associatif instantané, toute notre attention et notre affectivité, lesquelles restent irrésistiblement fixées à un 'certain' nombre de faits et objets, au détriment et à l'exclusion de toute le reste*? Il me semble que de tels phénomènes, comportent d'aussi violentes déterminations de choix, ne peuvent être en vain, ne peuvent, à un degré plus ou moins accentué, marquer d'intervenir dans ce facteur, à mon avis, conditionnel de l'experimentation scientifique, qui, pour employer les mots mêmes de E. Schrödinger consiste dans "*la disposition momentanée de notre intérêt et de son influence déterminante dans la direction du travail postérieur*,"' Dalí, *Le Mythe tragique*, 91.

144 Also in his book on Millet's *Angelus*, Dalí turns to Einstein's Relativity, contending (in a manner that recalls Breton's 1921 introduction to Ernst's collages) that his definition of paranoia-criticism as the reconciliation of distant entities, or, as he calls it, 'the sum of concrete dialectics' ('somme de la "dialectique concréte"'), enables him to offer his method as a '"special theory of relativity"' ('"théorie restrainte de la relativité"'), Dalí, *Le Mythe tragique*, 91.

6 Relativity and the Fourth Dimension: *Salvador Dalí and Roger Caillois*

1 Laurent Jenny, 'Système et "délire" de Dalí à Caillois', *Europe*, year 78, no. 859–60, November–December 2000, 60–71.

2 One of the very first of this kind to appear was J. Eisenmann and Y. Rocard, 'Le principe de correspondance et l'état actuel de la théorie des Quanta', *Revue de Métaphysique et de Morale*, year 33, no. 2, April–June 1926, 183–200.

3 Arthur Eddington, *Space Time and Gravitation: An Outline of the General Relativity Theory*, Cambridge: Cambridge University Press, 1920, 22.

4 Eddington, *Space Time and Gravitation*, 23. Eddington probably means to refer here to the Warden's inquiry in *Sylvie and Bruno* as to the 'the length,

breadth, and thickness' of the impending Revolution, Lewis Carroll, *Sylvie and Bruno*, London and New York: Macmillan and Co., 1889, 30.

5 Arthur Eddington, 'Gravitation and the Principle of Relativity', *Nature*, vol. 101, no. 2523, 7 March 1918, 15–17, 15. Much of the material in Eddington's article, which Michael Whitworth brought to my attention, reappeared in *Space Time and Gravitation* and *The Nature of the Physical World*, Cambridge: Cambridge University Press, 1928.

6 Thomas F. Glick, 'Relativity in Spain', Thomas F. Glick (ed.), *The Comparative Reception of Relativity*, Dordrecht: D. Reidel, 1987, 231–63, 241. Elsewhere, Glick isolates Dalí as 'the prototypical consumer of science in twentieth-century Spain', asserting his incomprehension of scientific themes while characterising him as 'an example of a nonscientist whose personal cosmology was changed through the popularization of Special Relativity', Thomas F. Glick, *Einstein in Spain: Relativity and the Recovery of Science*, Princeton, New Jersey: Princeton University Press, 1988, 277–82, 278, 281.

7 Glick gives a brief description of Einstein's lecture but does not seem to know that Dalí was a student at the Residencia, *Einstein in Spain*, 144. The Residencia helped pave the way for Relativity in Spain, hosting a series of lectures on electricity in 1917 by the country's leading theoretical physicist Blas Cabrera, which were fully conversant with Special Relativity, and two lectures on Relativity by the philosopher García Morente, probably in 1921 just before Dalí's arrival, Glick, *Einstein in Spain*, 38, 158. I am grateful to Jordana Mendelson for advising me on Dalí's time at the Residencia.

8 José Ortega y Gasset, *El tema del nuestro tiempo* [1923], Madrid: Revista de Occidente, 1934, 207–35; José Ortega y Gasset, 'The Historical Significance of the Theory of Einstein', *The Modern Theme*, trans. James Cleugh, London: The C. W. Daniel Company, 1931, 135–52. The historical importance of this essay in the interpretation of Relativity is demonstrated by its (abridged) republication as José Ortega y Gasset, 'The Historical Significance of the Theory of Einstein', L. Pearce Williams (ed.), *Relativity Theory: Its Origins and Impact on Modern Thought*, New York: John Wiley and Sons, 1968, 147–57.

9 Ian Gibson, *The Shameful Life of Salvador Dalí*, London: Faber and Faber, 1997, 115–16.

10 Gibson, *Shameful Life of Salvador Dalí*, 117.

11 Salvador Dalí, 'La chèvre sanitaire', *La Femme visible*, Paris: Éditions Surréalistes, 1930, 21–34; Dalí, 'The Sanitary Goat' [1930], *The Collected Writings of Salvador Dalí*, ed. and trans. Haim Finkelstein, Cambridge: Cambridge University Press, 1998, 226–31.

12 Dalí, *Collected Writings*, 229–30 (trans. modified); '[l]orsque Einstein constate les différences entre la mesure-mètre et le mètre fait d'une matière dont nous devons nous servir, en faisant naître la possibilité d'une géométrie physique, il réalise sur un plan très spécial d'interférences de la pensée la rupture brutale de constantes psychologiques créatrices dans ce cas de grandes symboles et simulacres abstraits, non-physiques . . . De même qu'il faudra compter sur la dilation physique des mesures usuelles ou non, il faudra compter, parallèlement ou non, sur la dilation psychique des idées, il n'y aura plus lieu de faire appel à des critériums abstraits . . . La nouvelle géométrie de la pensée poétique exige une révision et une accomodation physiques de l'ordre de celles que la physique einsteinienne fait subir à toute mesure', Dalí, *La Femme visible*, 31.

13 George Gamow, *Mr. Tompkins in Wonderland*, Cambridge: Cambridge University Press, 1940. Duplicating the cartoon Einstein of the French dailies that Aragon had lamented in *Treatise on Style*, Spanish newspapers and cartoonists, not to mention the general public, had a field day when Einstein brought Relativity to Spain: '[t]he press fills column after absurd column, horribly distorting the relativistic harmonies. Cartoonists and humourists squander their charm and good humour on Einsteinian curves and spaces,' *El Noticiero* (Barcelona) quoted in Glick, 'Relativity in Spain', 255. According to Glick, Eddington's line 'light has weight' was favoured grist to the cartoonists' mill, 255.

14 Dalí, *Collected Writings*, 230 (trans. modified); 'La physique dans ce cas est dans l'ordre des choses que l'on peut croire, – que l'on peut croire ou interpréter, même si c'est dans le délire: la physique qui doit créer la nouvelle géométrie de la pensée sera précisément . . . le délire d'interprétation paranoïaque', *La Femme visible*, 31–2. I translate 'délire' as delirium rather than delusion or delusional, here, but see the note on translation in David Lomas, *The Haunted Self: Surrealism, Psychoanalysis, Subjectivity*, New Haven and London: Yale University Press, 2000, 243, n. 39.

15 The first double image attempted by Dalí was in *The Invisible Man* of 1929–33 according to Dawn Ades, *Dalí*, London: Thames and Hudson, 1982, 121.

16 Gerald Holton attempts to exacerbate art-science skirmishing in his abbreviated condemnation of Relativity inspired interpretations of artists' work in Holton, *Einstein, History, and Other Passions: The Rebellion against Science at the End of the Twentieth Century*, New York: Addison-Wesley, 1996, 130–32. His point that concepts from Relativity could just as easily be read into *pre*-twentieth-century art is worth taking on, however, and my focus on Dalí's art *and* writings is meant to circumvent this criticism of my reading. Holton's target is a speculative interpretation of Cubism by Paul Laporte which Einstein himself felt was 'unsatisfactory' (though his response seems to have guaranteed rather than prevented the airing of the theory), Paul M. Laporte, 'Cubism and Relativity, with a Letter of Albert Einstein', *Art Journal*, vol. 25, no. 3, Spring 1966, 246–8, 246. The latest and best word on Einstein, Relativity and Cubism's perturbed commentators is Linda Dalrymple Henderson, 'Four-Dimensional Space or Space-Time?: The Emergence of the Cubism-Relativity Myth in New York in the 1940s',

Michele Emmer (ed.), *The Visual Mind II*, Cambridge, Mass: MIT, 2005, 349–97. Also see Henderson, 'Cubism, Futurism, and Ether Physics in the Early Twentieth Century', *Science in Context*, vol. 17, no. 4, December 2004, 445–66.

17 Arkady Plotnitsky, *Complementarity: Anti-Epistemology after Bohr and Derrida*, Durham and London: Duke University Press, 1994, 122.

18 An intriguing sidebar to this part of the story is Dalí's friendship with René Crevel who wrote the first monograph on the artist. Published shortly after *La Femme visible* and typically baroque in its expression, Crevel's defence of paranoia-criticism shows evidence of an exchange between the two as it suggests Dalí's budding method offers a new means of representing time and space: 'Desséchés les marais du relativisme, le temps et l'espace ne stagneront plus, ici engorgement des richesses pourries, là, flapissures par manque du strict nécessaire', René Crevel, *Dalí ou l'anti-obscurantisme*, Paris: Éditions Surréalistes, 1931, 23.

19 In 1922, Bergson himself quoted from Eddington's 'already classic volume', *Space Time and Gravitation*, in his controversial critique of Special Relativity, Henri Bergson, *Duration and Simultaneity: Bergson and the Einsteinian Universe*, trans. Leon Jacobson, Manchester: Clinamen Press, 1999, 114; Bergson, *Durée et simultanéité: A propos de la théorie d'Einstein*, Paris: Librairie Félix Alcan, 1922, 222–3.

20 For a history of Bergsonism and French avant-garde art, see Mark Antliff, *Inventing Bergson: Cultural Politics and the Parisian Avant-Garde*, Princeton: Princeton University Press, 1993. Bergson is, of course, absent from all histories of Surrealism. Breton gave him a paltry −15 score compared to the 15 out of 20 he awarded Einstein in the Dadas' 1921 'referendum' in *Littérature*, year 3, no. 18, March 1921, 1–7. The Surrealists' resentment towards Bergson stems mainly from their objection to his patriotism during the First World War, his 'establishment' status, and his increasing sympathy for Catholicism in the interwar years. For Breton's thoughts on Bergson, see Breton, 'Situation of Surrealism between the Two Wars', *Free Rein*, trans. Michel Parmentier and Jacqueline d'Amboise, Lincoln and London: University of Nebraska Press, 1995, 51–67, 57; Breton, 'Situation du surréalisme entre les deux guerres', *Oeuvres complètes* [OC], vol. 3, Paris: Gallimard, 1999, 709–25, 715, and Breton, *Conversations: The Autobiography of Surrealism*, trans. Mark Polizzotti, New York: Paragon House, 1993, 14; Breton, *OC*, vol. 3, 436. In view of Breton's interest in Stéphane Lupasco's physics-inflected philosophy from the 1940s, it is worth noting that Jean Wahl regarded them as, on one level, 'un prolongement de certaines idées bergsoniennes', Wahl, 'Stéphane Lupasco, *Du devenir logique et de l'affectivité*' (book review), *Recherches philosophiques*, no. 6, 1936–7, 394–5, 395. In my view, Surrealism's 'resistance' to the Bergsonian *durée* was bypassed by artists such as Domínguez, Matta and Paalen (whose work, more 'painterly' than Dalí's, suggests that category), by their recourse to the reconceptualisation of temporality by modern physics and its philosophy. For the intersection of Bergson's thought with modern physics, particularly in relation to de Broglie, see Louis de Broglie, 'Les Conceptions de la physique contemporaine et les idées de Bergson sur le temps et sur le mouvement', *Revue de Métaphysique et de Morale*, year 48, no. 4, October–December 1941, 241–57; Milic Capek, *Bergson and Modern Physics: A Reinterpretaion and Reevaluation*, Dordrecht: D. Reidel, 1971; and Pete A. Y. Gunter (ed.), *Bergson and the Evolution of Physics*, Knoxville: University of Tennessee Press, 1969.

21 Salvador Dalí, 'Aerodynamic Apparitions of "Beings-Objects"', *Collected Writings*, 207–11, 209; 'la dimension délirante et surréaliste par excellence', 'Apparitions aérodynamiques des "Être-Objets"', *Minotaure*, no. 6, Winter 1934–5, 33–4, 33–4.

22 Sigmund Freud, 'The Unconscious', *On Metapsychology*, Penguin Freud Library, vol. 11, Harmondsworth: Penguin, 1991, 159–222.

23 Freud, 'The Unconscious', 205.

24 Freud, 'The Unconscious', 205.

25 Dalí, *Collected Writings*, 207; 'acte intime'; 'veritable "cérémoniale névrotique"', Dalí, 'Apparitions aérodynamiques des "Être-Objets"', 33.

26 Dalí, *Collected Writings*, 208; 'de plus vital et de plus lyrique dans la pensée morale, scientifique et artistique, contemporaine', Dalí, 'Apparitions aérodynamique des "Être-Objets,"' 33.

27 Sigmund Freud, 'Infantile Sexuality', *On Sexuality*, Pelican Freud Library, vol. 7, Harmondsworth: Penguin, 1986, 88–126.

28 Freud, *On Sexuality*, 117. See Haim Finkelstein, 'The Incarnation of Desire: Dalí and the Surrealist Object', *Res: Anthropology and Aesthetics*, no. 23, Spring 1993, 114–37, 126. Unusually for studies of Surrealism, my own inquiry affords psychoanalysis only an attenuated role, of course. Coupled with Freud's mesmeric hold over the discipline of art history, Surrealism's early investment in psychoanalysis has served to corral most significant studies of the movement over the last twenty years into Freudian territory: Rosalind Krauss and Jane Livingstone. *L'Amour fou: Photography and Surrealism*, Washington: Corcoran Gallery of Art, 1985; Hal Foster, *Compulsive Beauty*, Cambridge, Mass. and London: MIT, 1993; Lomas, *The Haunted Self*; Harris, *Surrealist Art and Thought in the 1930s*.

29 Dalí, *Collected Writings*, 209; 'matérielle et vraie', Dalí, 'Apparitions aérodynamiques des "Être-Objets"', 33.

30 Dalí, *Collected Writings*, 209 (trans. slightly modified); 'quatre dimensions comprenant le temps qui est bien la dimension délirante et surréaliste par excellence', Dalí, 'Apparitions aérodynamique des "Être-Objets"', 33–4.

31 Dalí, *Collected Writings*, 209 (trans. modified); 'sanguin et parfaitement visqueux', Dalí, 'Apparitions aérodynamique des "Être-Objets"', 34.

32 Dalí, *Collected Writings*, 209; Dalí, 'Apparitions aérodynamique des "Être-Objets"', 34.

33 'dès ses débuts, avait vaincu le matérialisme mécaniste et s'en tenait à un idéalisme relativiste tout provisoire', Dalí, 'Interprétation Paranoïaque-Critique de l'Image Obsédante "L'Angelus' de Millet,"' *Minotaure*, no. 1, February 1933, 65–7, 66.

34 Caillois, 'Spécification de la poésie', *SASDLR*, no. 5, 15 May 1933, 30–31. For a more in-depth analysis than I can conduct here, see Harris, *Surrealist Art and Thought in the 1930s*, 132–3.

35 Gibson, *Shameful Life of Salvador Dalí*, 323–4. For a humourless account of Dalí's 'trial' which airbrushes out his own role, see Roger Caillois, 'Testimony (Paul Éluard)' [1973], *The Edge of Surrealism: A Roger Caillois Reader*, ed. Claudine Frank, trans. Claudine Frank and Camille Naish, Durham and London: Duke University Press, 2003, 60–65, 63–5.

36 Roger Caillois, 'Systematization and Determination', *The Necessity of the Mind*, trans. Michael Syrotinski, Venice, CA: The Lapis Press, 1990, 87–95; Caillois, 'Systématisation et determination', *Documents* 34, no. 1 (special issue, 'Intervention surréaliste'), June 1934, 69–72.

37 Caillois, 'Letter to André Breton' [1934], *Edge of Surrealism*, 84–6, 84–5; 'il révélait une position antilyrique', Caillois, 'Lettre á André Breton', *Approches de l'imaginaire*, Paris: Gallimard, 1974, 35–8, 36.

38 Breton gives his own account of the jumping bean incident in his 'Fronton-Virage', *Free Rein*, 175–95, 178–9; Breton, 'Fronton-virage', *OC*, vol. 3, 840–61, 843–4.

39 Caillois, *Edge of Surrealism*, 84; Caillois, *Approches*, 35.

40 Caillois, *Edge of Surrealism*, 85; 'il s'agit surtout d'une question de méthodologie', Caillois, *Approches*, 36.

41 Caillois, *Edge of Surrealism*, 85; 'Ainsi, dans l'atomistique moderne, c'est l'aventure dans le noir: la découverte éblouie des fougères par des enfants élevés das des cubes (le mot est d'un physicien). Il ne s'agit pas là du *trouble* ou de la *jouissance* que procure une *belle image*, mais d'un désarroi, désarroi devant ce que je me plais à nommer la *déroute des évidences*. Car il ne subsiste rien des anciennes intuitions et toute philosophie qui ne compose pas avec cette nouvelle science du *pourquoi pas* est ridicule', Caillois, *Approches*, 36.

42 'We shall make room for paralogy alongside analogy and show how the ancient philosophy of "as if" is superseded, in the philosophy of science, by the philosophy of the "why not?"' Gaston Bachelard, *The New Scientific Spirit*, trans. Arthur Goldhammer, Boston: Beacon Press, 1984, 7; '[n]ous ferons place à la paralogie à côté de l'analogie et nous montrerons qu'à l'ancienne philosophie du *comme si* succède, en philosophie scientifique, la philosophie du *pourquoi pas*', Bachelard, *Le Nouvel Esprit scientifique*, Paris: Alcan, 1934, 6–7.

43 'je pense de plus en plus sérieusement que le surréalisme doit évoluer vers les bases d'une nouvelle religion anti-mystique, matérialiste, base sur le progrès des connaissances scientifiques (spécialement les nouvelles idées de l'espace, lesquelles étaient inaccessibles non seulement aux grecs mais aussi au christianisme) religion qui occuperait le vide imaginatif affectif, que l'écroulement des idées métaphysiques a produit à notre époque. J'ai des idées très concrètes à ce sujet . . .', Paris: Centre Georges Pompidou, Musée National d'Art Moderne, *Salvador Dalí, rétrospective* 1920–1980, 1979–80, 300.

44 Caillois could not possibly have overlooked Dalí's unique lucubrations even if he wanted to, because he published in *Minotaure* after his departure from the Surrealist group. The seventh number carried both his 'Mimétisme et psychanasthénie légendaire' and Dalí's 'Psychologie non-euclidienne d'une photographie', *Minotaure*, no. 7, June 1935, 5–10, 56–7.

45 Roger Caillois, *Procès intellectuel de l'art*, Marseille: Cahiers du Sud, 1935. Page references cited are from *Approches de l'imaginaire*, where the whole text is reprinted.

46 Caillois, 'Décision préliminaire sur la métaphysique', *Approches*, 38–41, 38.

47 'on ne peut ni de la terre déceler le mouvement de la terre, ni de la conscience le mouvement de la conscience', Caillois, *Approches*, 39. The Michelson–Morley experiment of 1881 attempted to detect the motion of the earth relative to the 'ether'. The experiment, which should have shown that different observers moving relative to the ether would see light coming towards them at different speeds, actually produced results showing light travelling at the same speed. By asserting all observers would register the same light speed no matter what their motion, Einstein's Special Theory of Relativity explained the results. Louis Aragon toyed with the Michelson–Morley experiment in 1928 to torment Paul Valéry, and satirise his devotion to Relativity, Louis Aragon, *Treatise on Style*, trans. Alyson Waters, Lincoln and London: University of Nebraska Press, 1991, 77; Aragon, *Traité du style*, Paris: Gallimard, 1928, 155. See Gerald Holton, 'Einstein, Michelson, and the "Crucial Experiment"', *Thematic Origins of Scientific Thought: Kepler to Einstein*, Cambridge, Mass.: Harvard University Press, 1974, 261–352.

48 'la situation d'êtres courbes infiniment plats vivant à la surface d'une sphère gigantesque', Caillois, *Approches*, 40. Albert Einstein, *Relativity: The Special and the General Theory* [1920], London: Methuen, 1960, 108–12. The other possible source for Caillois's analogy is, of course, E. A. Abbott, *Flatland: A Romance of Many Dimensions* [1884], London: Penguin, 1998.

49 Caillois, *Approches*, 41.

50 'ne peut être obtenu que par la patience ingrate du savant et non par des intuitions d'artiste', Caillois, *Approches*, 40.

51 Caillois, 'Notice sur l'impureté dans l'art', *Approches*, 41–52, 44.

52 Caillois, *Approches*, 44. Moreau, of course, was one of Breton's favourite pre-Surrealist painters.

53 Caillois, *Approches*, 46.

54 Caillois, *Approches*, 44–5.

55 'délirante fabulation', Caillois, *Approches*, 47.

56 'les résultats de la réflexion sur la science (l'épistémologie) sont singulièrement plus féconds et plus solides que ceux de la réflexion sur l'art'; 'nouveaux modes de pensée exigés par le développement des sciences physiques', Caillois, *Approches*, 49–50, n. 1.

57 For two refutations of Caillois's thesis from within Surrealism, see Nicolas Calas, *Foyers d'incendie*, Paris: Denoël, 1938, 22, and, less explicit, Salvador Dalí, 'The Spectral Surrealism of the Pre-Raphaelite Eternal Feminine', *Collected Writings*, 310–14; Dalí, 'Le Surréalisme spectrale de l'éternel féminin préraphaélite', *Minotaure*, no. 8, 1936, 46–9.

58 Caillois, 'Literature in Crisis', *Edge of Surrealism*, 86–8, 87; 'appliquer au domaine de l'art la méthode de la science, dont la rigueur nous semble, ici plus encore qu'ailleurs, nécessaire', Caillois, 'Crise de la littérature', *Approches*, 52–4, 53.

59 Caillois, *Edge of Surrealism*, 88 (trans. modified); 'il est risible de vouloir fonder par exemple la liberté psychologique sur la relation d'indétermination du monde intra-atomique'; 'constructions épistémologiques nécessitées par les problèmes de la méthodologie scientifique contemporaine', Caillois, *Approches*, 54.

60 Salvador Dalí, *La Conquête de l'irrationnel*, Paris: Éditions Surréalistes, 1935; Dalí, *Conquest of the Irrational*, trans. David Gascoyne, Paris: Éditions Surréalistes, 1935.

61 'le discrédit accablant de "l'intuition logique"', Dalí, *Conquête*, 7.

62 'une table à manger culturelle'; 'les restes refroidis et insubstantiels de l'art et de la littérature, et . . . les brulantes précisions analytiques des sciences particulières, inaccessibles, pour le moment, à une synthèse nutritive à cause de leurs démesurées extension et spécialisation, et, en tout cas, totalement inassimilables, en dehors des cannibalismes spéculatifs', Dalí, *Conquête*, 9.

63 'car nous, surréalistes . . . nous ne sommes pas exactement des artistes, et nous ne sommes pas non plus exactement des vraies hommes de science'; 'nous sommes des poissons carnassiers, qui, comme je l'ai déjà insinué, nageons entre deux eaux, l'eau froide de l'art, l'eau chaude de la science . . .', Dalí, *Conquête*, 10.

64 'modèle de débilité mentale qui s'appelle art abstrait', Dalí, *Conquête*, 19.

65 For some remarks on Dalí, Kant and photography which touch on Relativity, see John Roberts, 'Surrealism, Photography and the Everyday', *The Art of Interruption: Realism, Photography and the Everyday*, Manchester and New York: Manchester University Press, 1998, 98–113, 107–8.

66 An account of the implications for Kantian causality of Relativity and quantum mechanical indeterminacy (seemingly expressed from the position of Socrates) is given in Werner Heisenberg, 'Quantum Mechanics and Kantian Philosophy (1930–1934)', *Physics and Beyond: Encounters and Conversations*, trans. Arnold J. Pomerans, London: George Allen & Unwin, 1971, 116–24.

67 'les ambitions expérimentales / de la physique mathématique / moderne', Dalí, *Conquête*, 21.

68 'il n'existe pas / de nos jours / que je sache / une physique / de la psycho-pathologie / une physique de la paranoia / ce qui ne pourrait être considère / que / comme / la base expérimentale / de la prochaine / philosophie / de la / psycho-pathologie / de la prochaine / philosophie de l'activité "paranoïaque-critique" / laquelle un jour / je tenterai d'envisager polémiquement / si j'en ai les temps / et l'humeur', Dalí, *Conquête*, 23–4.

69 'cette angoissante et colossale question qui est celle de l'espace-temps einsteinien', Dalí, *Conquête*, 25.

70 'camembert paranoïaque-critique tendre, extravagant, solitaire, du temps et de l'espace', Dalí, *Conquête*, 25.

71 After completing this reading, I found that the connection between Dalí's *Persistence of Memory* and Einstein's time dilation had been made by Leonard Shlain, *Art & Physics: Parallel Visions in Space, Time & Light*, New York: Quill/William Morrow, 1991, 228–9. However, Shlain's is an intuitive response to the image, and, as the title of his book suggests, he sees it as a fortuitous illustration of Einstein's idea, and is unaware that Dalí and the Surrealists undertook a thorough engagement with modern physics during the period.

72 Writings by Dalí of the 1930s touching upon Relativity and/or quantum physics which I have been unable to consider here include 'Objets psycho-atmosphériques-anamorphiques', *SASDLR*, no. 5/6, 15 May 1933, 45–8; 'Psychologie non-euclidienne d'une photographie', *Minotaure*, no. 7, June 1935, 56–7; 'Le Surréalisme spectrale de l'éternel féminin préraphaélite', *Minotaure*, no. 8, 1936, 46–9; and *Le Mythe tragique de l'Angelus de Millet*, Paris: Jean-Jacques Pauvert, 1963, 91. I look at some of this material in my *Physique du Surréalisme: Surrealism, Modern Physics, and Epistemology*, dissertation submitted for the degree of Ph.D., Courtauld Institute of Art, University of London, 2000. I leave to future scholarship interpretation of the evolution of the imagery and spaces of Dalí's paintings of the mid-1930s along the lines I have laid down here, as there is much potential for this. In *White Calm* (1936), for instance, the expectation of temporal instantaneity is breached by the appearance in the foreground of a Greek urn (denoting a moment from the distant past) and the two similarly posed figures. Comparable, in their collapse of different events along the temporal continuum into the single three-dimensional space in which they occurred, are *Mediumistic-Paranoiac Image* (1935) and other canvases completed by Dalí around the time he was writing *Conquest of the Irrational*.

73 Roger Caillois, 'La Mante religieuse: De la Biologie à la Psychanalyse', *Minotaure*, no. 5, February 1934, 23–6.

74 Roger Caillois, 'L'alternative (Naturphilosophie ou Wissenschaftlehre)', *Cahiers du Sud*, year 24, May–June 1937, 111–21. Page references cited are from *Approches de l'imaginaire*, 25–34. The Surrealists were absent, keeping their usual factional distance, but contributors who were or had been close to Surrealism included Jules Monnerot, Denis de Rougemont, Jean Wahl, and the first historian of Surrealism, Walter Benjamin.

75 'uniquement un système de rapports esthétiques, sinon un réservoir d'expressions', Caillois, *Approches*, 27.

76 André Breton, 'Introduction to Achim von Arnim's *Strange Tales*', *Break of Day*, trans. Mark Polizzotti and Mary Ann Caws, Lincoln and London: University of Nebraska Press, 1999, 88–110; Breton, 'Introduction aux "Contes Bizarres" d'Achim d'Arnim', *OC*, vol. 2, Paris: Gallimard, 1992, 341–60.

77 Breton, *Break of Day*, 91; 'la méthode expérimentale et la méthode spéculative', Breton, *OC*, vol. 2, 344.

78 Breton, *Break of Day*, 92; '"surréaliste" avant la lettre', Breton, *OC*, vol. 2, 345. 'Il était, raconte-t-il lui-même dans l'introduction à ses Fragments, affecté d'un tic bizarre, qui prenait l'apparence d'un "esprit" lutin et rappelle, à s'y méprendre, "l'écriture automatique" des médiums spirites. Ce tic obligeait, à tout instant, à s'interrompre dans le feu même de la composition et à écrire en marge de son manuscript les inventions les plus burlesques', Édouard Spenlé, *Novalis: Essai sur l'idéalisme romantique en allemagne*, Paris: Librairie Hachette & Cie, 1904, 200. Footnoted by both Breton and Caillois, Spenlé's book gleans this information from Johann-Wilhelm Ritter, *Nachlass aus den Papieren eines jungen Physikers* [*Posthumous Fragments of a Young Physicist*], Heidelberg: Mohr und Zimmer, 1810. Although Breton footnotes Ritter's book too, we can safely assume he knew of it only second hand from Spenlé, especially given Breton's ignorance of German and his extensive use of, and close adherence to Spenlé's text. Walter Benjamin gave pride of place in his library to a first edition of Ritter's book, by the way, calling it 'the most important sample of personal prose of German Romanticism', Benjamin, 'Unpacking my Library: A Talk about Book Collecting' [1931], *Illuminations*, trans. Harry Zorn, London: Pimlico, 1990, 61–69, 67.

79 André Breton, 'What is Surrealism?' [1934], *What is Surrealism? Selected Writings*, ed. Franklin Rosemont, London: Pluto, 1978, 112–41, 122; 'surréaliste dans entre tous et en tout dans l'espace et le temps', Breton, 'Qu'est-ce que le surréalisme?' [1934], *OC*, vol. 2, 223–62, 239.

80 Caillois, *Approches*, 33.

81 Breton, *Break of Day*, 95; 'Une coalition frappante . . . tend à retenir dans le même camp les poètes, les artistes et les savants originaux', *OC*, vol. 2, 347. Breton uses again, here, his favourite Hegel quotation that we saw in 'Crisis of the Object', '"Ce qui est rationnel est réel et ce qui est réel est rationnel"', 350.

82 Breton refers in passing to the *Cahiers du Sud* collection years later in his *L'Art Magique* [1957], Paris: Éditions Phébus and Éditions Adam Biro, 1991, 50. Caillois actually mentions Breton's Achim von Arnim essay in a footnote, registering his disagreement with Breton over his interpretation of Fichte, *Approches*, 30, n. 3. Both drew from the same three-volume study by editor of the *Revue de Métaphysique et de Morale*, Xavier Léon, *Fichte et son temps*, Paris: Librairie Armand Colin, 1922–7.

83 'une situation analogue paraissant se dessiner aujourd'hui', Caillois, *Approches*, 25.

84 'cerveaux originaux qui pour le moment s'en tiennent encore au symbolisme des rêves . . .', Caillois, *Approches*, 32.

85 'Des sots se chargeront de tout expliquer par la quatrième dimension', Caillois, *Approches*, 32. Caillois gave a brief, more balanced assessment of Dalí's Freudian symbolism in his 'Déterminations inconscientes en peinture', *Documents* 35, year 3, no. 6, November–December 1935, 5–6. My thanks to Steven Harris.

86 André Breton, 'Des tendances les plus récentes de la peinture surréaliste', *Minotaure*, no. 12/13, May 1939, 16–17. Page references are from André Breton, 'The Most Recent Tendencies in Surrealist Painting', *Surrealism and Painting*, trans. Simon Watson Taylor, New York: Harper and Row, 1972, 145–50; Breton, 'Des tendances les plus récentes de la peinture surréaliste', *Le Surréalisme et la peinture*, Paris: Gallimard, 1965, 145–50.

87 Breton, *Surrealism and Painting*, 148; 'le domaine scientifique', Breton, *Le Surréalisme et la peinture*, 148.

88 Breton, *Surrealism and Painting*, 148; 'de passer outre à l'univers à trois dimensions', Breton, *Le Surréalisme et la peinture*, 148.

89 Breton, *Surrealism and Painting*, 148; 'l'introduction en physique de la conception de l'*espace-temps* par Einstein', Breton, *Le Surréalisme et la peinture*, 14–9.

90 Breton and these young recruits to Surrealism were reading the mystical writings of P. D. Ouspensky at the time, notably his *Tertium Organum* (1912), which includes discussion of Einstein's fourth dimension alongside the more mystical versions of the concept. Further research on Ouspensky's impact on Surrealist theory of this period is long overdue.

91 Breton, *Surrealism and Painting*, 148–9; 'besoin d'une représentation suggestive de l'univers quadridimensionnel', Breton, *Le Surréalisme et la peinture*, 149.

92 Breton, *Surrealism and Painting*, 146; 'tend à décrire un monde où se disjoignent les derniers angles clairs du cubisme', Breton, *Le Surréalisme et la peinture*, 146.

93 Breton, *Surrealism and Painting*, 148; 'à sa période héroïque, un des leitmotive du cubisme', Breton, *Le Surréalisme et la peinture*, 148.

94 'By misunderstanding, in its limitations, this essential liberty, Symbolism and Cubism failed', André Breton, 'Max Ernst', *The Lost Steps*, trans. Mark Polizzotti,

Lincoln and London: University of Nebraska Press, 1996, 60–61, 60; 'C'est pour avoir méconnu, dans ses bornes, cette liberté essentielle que le symbolisme et le cubisme ont échoué', Breton, 'Max Ernst', *OC*, vol. 1, Paris: Gallimard, 1988, 245–6, 245.

95 Breton, *Surrealism and Painting*, 149; 'le passage à la quatrième dimension tend à s'opérer sur le plan non plus physique, mais psychique', Breton, *Le Surréalisme et la peinture*, 149.

96 Oscar Domínguez and Ernesto Sábato quoted in Breton, *Surrealism and Painting*, 149, n. 1; 'ouvrent une fenêtre sur le monde étrange de la quatrième dimension, constituant une espèce de solidification du temps', Breton, *Le Surréalisme et la peinture*, 149, n. 1. For variations on Breton's favourite metaphor, see his early, 1925 statement, made as an innovative commentator on painting: 'it is impossible for me to envisage a picture as being other than a window . . .', Breton, *Surrealism and Painting*, 2; 'il m'est impossible de considérer un tableau autrement que comme une fenêtre . . .', Breton, *Le Surréalisme et la Peinture*, 2, and following Dalí's arrival on the scene in 1929, 'this is the first time that the mental windows can open wide . . .', Breton, 'The First Dalí Exhibit', *Break of Day*, 51–3, 52; 'C'est . . . la première fois que s'ouvrent toutes grandes les fenêtres mentales . . .', Breton, 'Première exposition Dalí', *OC*, vol. 2, 307–9, 308.

Coda: Physics and the Cold War: *Surrealism and Salvador Dalí*

1 André Breton, 'Interview with *Indice*', *What is Surrealism? Selected Writings*, ed. Franklin Rosemont, London: Pluto, 1978, 144–7, 147; 'Gardons-nous de contribuer à la formation d'une nouvelle religion qui soit, paradoxalement, la religion de la science', Breton, 'Interview d'*Indice*', *Oeuvres completes* [*OC*], vol. 2, Paris: Gallimard, 1992, 445–50, 450.

2 'Le surréalisme doit évoluer vers les bases d'une nouvelle religion anti-mystique, matérialiste, base sur le progrès des connaissances scientifiques (spécialement les nouvelles idées de l'espace . . .)', Paris: Centre Georges Pompidou, Musée National d'Art Moderne, *Salvador Dalí, rétrospective* 1920–1980, 1979–80, 300.

3 Dalí's Museum of Modern Art travelling retrospective had closed in May 1942, and since then Dalí's image, work and opinions had been uncommonly prevalent in the American popular press, as he promoted, with his usual flair for publicity, his 'autobiography', *The Secret Life of Salvador Dalí*, which appeared in October.

4 André Breton, 'Situation of Surrealism between the Two Wars', *Free Rein*, trans. Michel Parmentier and Jacqueline d'Amboise, Lincoln and London: University of Nebraska Press, 1995, 51–67, 60; 'découverte spectaculaire, dont certains signes avant-coureurs, qui en général ne trompent pas, ont déjà pu donner à penser qu'elle se produira sur le plan de la *physique*', Breton, 'Situation du surréalisme entre les deux guerres', *OC*, vol. 3, Paris: Gallimard, 1999, 709–25, 718.

5 André Breton, 'Situation du surréalisme entre les deux guerres', *VVV*, no. 2/3, March 1943, 44–53.

6 André Breton, 'Devant le rideau', *Le Surréalisme en 1947*, Paris: Éditions Pierre à Feu, 1947, 13–19, 14, n. 1.

7 Bertrand Goldschmidt, *The Atomic Complex: A Worldwide Political History of Nuclear Energy*, trans. Bruce M. Adkins, Illinois: American Nuclear Society, 1982, 8.

8 Goldschmidt, *The Atomic Complex*, 28. They did not know each other, but Goldschmidt recognised Masson as they left Marseilles on 1 April 1941 aboard the SS Carimaré (the Masson scholarship has it a day earlier), later recording his observations in Goldschmidt, *Atomic Rivals*, trans. Georges M. Tenner, New Brunswick and London: Rutgers University Press, 1990, 96–8. Substantially treated in the literature, Breton's own perilous departure from Marseilles on board the cramped freighter SS Capitaine Paul Lemerle on 25 March with his family, the artist Wifredo Lam, Claude Lévi-Strauss, Victor Serge and family and three hundred others, was the beginning of a sordid two-month journey to New York via Martinique.

9 Goldschmidt, *Atomic Rivals*, 153. Employed in Chicago for three months from 21 July 1942, Goldschmidt himself made an important contribution to the first nuclear chain reaction.

10 For a comprehensive history and analytic record, see Richard Rhodes, *The Making of the Atomic Bomb*, New York: Simon and Schuster, 1986. Also see the classic Robert Jungk, *Brighter than a Thousand Suns: The Moral and Political History of the Atomic Scientists*, trans. James Cleugh, London: Victor Gollancz/Rupert Hart-Davis, 1958, and for a history of nuclear energy in the twentieth century, Spencer R. Weart, *Nuclear Fear: A History of Images*, Cambridge, Mass. and London: Harvard University Press, 1988. For a history of nuclear energy in France, see Spencer R. Weart, *Scientists in Power*, Cambridge, Mass. and London: Harvard University Press, 1979.

11 I am grateful to Robert McNab, who placed the date of this photograph of Breton around 25 August 1945 in response to an inquiry from me.

12 J. Samuel Walker, *Prompt and Utter Destruction: Truman and the Use of Atomic Bombs Against Japan*, Chapel Hill and London: The University of North Carolina Press, 1997, 100–101.

13 Space limitations preclude discussion of Einstein's deep involvement in pacifism and anti-nuclear activism, which are relevant to the larger issues I am raising here, and in my opinion should not be thought subservient to his contributions to physics. See Albrecht Fölsing, *Albert Einstein: A Biography*, trans. Ewald Osers, Harmondsworth: Penguin, 1998, 659–741 which can act as an introduction to Einstein's two volumes of writings, *Einstein on Peace*, ed. Otto Nathan and Heinz Norden, New York:

Simon and Schuster, 1960, and *Ideas and Opinions* [1954], ed. Carl Seelig, trans. Sonja Bargmann, New York: The Modern Library, 1994. For some Surrealist context, see the criticism aimed at the then-recently translated Einstein–Freud exchange *Why War?* from the perspectives of dialectical materialism and revolutionary action in the article I touch on in chapter three by René Crevel, 'Notes en vue d'une psycho-dialectique', *Le Surréalisme au service de la Révolution*, no. 5/6, 15 May 1933, 48–52. One of Einstein's final acts was to help draft with Bertrand Russell and to sign the anti-nuclear Russell–Einstein Manifesto, which in the centenary year for Einstein the physicist fittingly received a fiftieth anniversary commentary in homage to Einstein the pacifist, *The Spokesman*, no. 85 (special issue, 'Albert Einstein, Bertrand Russell, Manifesto 50'), 2005.

14 Fölsing, *Albert Einstein*, 720.

15 'le choc a été rude', Breton, 'Interview de Jean Duché', *OC*, vol. 3, 588–99, 588.

16 André Breton, 'Interview with Jean Duché', *Conversations: The Autobiography of Surrealism*, trans. Mark Polizzotti, New York: Paragon House, 1993, 196–208, 196; 'la désintégration atomique était tenue pour la découverte la plus importante depuis celle du feu . . .', Breton, *OC*, vol. 3, 588.

17 Breton, *Conversations*, 198; Breton, *OC*, vol. 3, 589, 590. Breton's biographer Mark Polizzotti claims that all of his 'interviews' were probably scripted in advance, and the structure of the exchange here strongly suggests that Breton requested Duché mention the 'découverte spectaculaire' of the Yale lecture, Breton, *Conversations*, xi.

18 Salvador Dalí, *The Unspeakable Confessions of Salvador Dalí*, trans. Harold J. Salemson, London: W. H. Allen, 1976, 216.

19 David Pace, 'Old Wine – New Bottles: Atomic Energy and the Ideology of Science in Postwar France', *French Historical Studies*, vol. 17, no. 1, Spring 1991, 38–61, 38, n. 1.

20 Pace, 'Old Wine – New Bottles', 49.

21 Quoted in Pace, 'Old Wine – New Bottles', 42–3.

22 *Le Surréalisme en* 1947, Paris: Éditions Pierre à Feu, 1947.

23 The relevant texts in *Le Surréalisme en* 1947 are Arpad Mezei, 'Liberté du langage', 59–61; Henri Pastoureau, 'Pour une offensive de grand style contre la civilisation chrétienne', 78–83; Jean Brun, 'Le Problème de la sensation et le surréalisme', 87–91; Émile Malespine, 'Integralité Picurale', 112–14; Marcel Jean and Arpad Mezei, 'Les chants de Maldoror: Le VIe Chant et le monde occulte', 115–18. I gloss each of these in my *Physique du Surréalisme: Surrealism, Modern Physics, and Epistemology*, dissertation submitted for the degree of Ph.D., Courtauld Institute of Art, University of London, 2000, 323–5.

24 Pace, 'Old Wine – New Bottles', 41.

25 Immediately after Hiroshima, *Scientific American* cobbled together highlights from a set of articles it had published over the preceding years on atomic energy. One of them, on French experiments at Joliot-Curie's laboratory, made the biggest blunder in the history of journalism, by referring in its immortal title to the *impossibility* of an explosive chain reaction in uranium: '[r]eaders made insomnious by "newspaper talk" of terrific atomic war weapons held in reserve by dictators may now get sleep', Jean Harrington, 'Don't Worry – It Can't Happen', *Scientific American*, vol. 162, no. 5, May 1940, 268.

26 For a first-hand account of the tests, see Goldschmidt, *Atomic Rivals*, 300–308.

27 Quoted in Pace, 'Old Wine – New Bottles', 41. One would feel slightly better disposed toward this painting if it could be shown that Dalí was punning on an alternative title, *The Tree Sphinxes of Bikini*, but that interpretation does not fit with his usual practice and must be discarded.

28 Surrealism's methodological interpretation of cloud formations and the patterns created on damp walls (often associated with the writings of Leonardo da Vinci) is well-known, and is eulogised in André Breton, *Mad Love* [1937], trans. Mary Ann Caws, Lincoln and London: University of Nebraska Press, 1987, 84–94; Breton, L'amour fou, *OC*, vol. 2, 673–785, 751–61. In these pages, Breton mentions the best-known example in Western literature, which is the exchange between Hamlet and Polonius in Shakespeare's Hamlet, Act 3, Scene 2.

29 Dalí, *Unspeakable Confessions*, 216.

30 Salvador Dalí, 'Mystical Manifesto', *The Collected Writings of Salvador Dalí*, ed. and trans. Haim Finkelstein, Cambridge: Cambridge University Press, 1998, 363–6. Dalí's feverish account of his conversion to mysticism is given in Dalí, *Unspeakable Confessions*, 216–19.

31 Dalí, 'Mystical Manifesto', *Collected Writings*, 363.

32 'What distinguishes our age from the Renaissance is that now for the first time we realize that matter, instead of being something continuous, is discontinuous. As Eddington pointed out, if one wanted to give an accurate representation of a table, instead of being compact the table should resemble something like a swarm of flies', Dalí quoted in Carlton Lake, *In Quest for Dalí*, New York: G. P. Putnams Sons, 1969, 48. Thanks to Elliott King for directing me to Dalí's remark. For the two tables, see Arthur Eddington, *The Nature of the Physical World*, Cambridge: Cambridge University Press, 1928, xi–xix, reprised as a 'two desks' scenario in Eddington, *New Pathways in Science*, Cambridge: Cambridge University Press, 1935, 1. Eddington's idea of solid objects as 'swarms' of particles is returned to consistently by Dalí in the 1950s and 1960s.

33 'La théorie des quantums a déchiré la continuité unitaire de l'univers . . .', Carl Einstein, *Georges Braque*, trans. E. Zipruth, Paris: Chroniques du jour, 1934, 63.

34 Dalí, 'Mystical Manifesto', *Collected Writings*, 364.

35 Dalí quoted in Elliott H. King, '*Raphaelesque Head Exploding*, 1951', Venice: Palazzo Grassi, *Dalí*, 2004, 358–9, 358.

36 Michael R. Taylor, 'On the Road with Salvador Dalí: The 1952 Nuclear Mysticism Lecture Tour', Michael R. Taylor (ed.), *The Dalí Renaissance*, New Haven and London: Yale University Press, forthcoming

2008.

37 André Breton, 'The Lamp in the Clock', *Free Rein*, 108–121, 109; '[S]ur le plan politico-philosophique . . . dont précisément les dernières découvertes physiques devraient avoir suffi à démontrer l'opposition toute formelle . . .', Breton, 'La Lampe dans l'horlage', *OC*, vol. 3, 770–85, 771.

38 Breton, *Free Rein*, 111; '[n]ous avons beau nous interroger sur ce qui peut couver dans les boucles du professeur Einstein ou prospérer derrière la brosse dure de l'étrange camarade Staline, non: ce n'est vraiment pas de ce suprême tableau de chasse qu'il s'agissait', Breton, 'La Lampe dans l'horlage', *OC*, vol. 3, 773. In fact, there was plenty of shared ground between Breton and Einstein, chiefly their detestation of nationalism, leading both to lend their support to the Citizens of the World movement towards the end of 1948, formed following former American fighter pilot Garry Davis's renunciation of his citizenship and protest sit-ins at the United Nations.

39 As Dalí tells it, his 'conversion' coincided with his artistic practice as follows: 'In an ebullition of ideas of genius, I decided to undertake the plastic solution of the quantum theory of energy and invented 'quantified realism' so as to be master of gravity. I started to paint the *Leda atomica* . . .' [first unfinished version, 1948], Dalí, *Unspeakable Confessions*, 217–18.

40 Jolyon Howorth, *France: The Politics of Peace*, London: END/Merlin Press, 1984, 25, 38.

41 Colin Sweet, *A Study of Nuclear Power in France*, London: Polytechnic of the South Bank, 1981, 32.

42 Marcel Duchamp, Letter to André Breton of 10 January 1950, *Affect' Marcel._ The Selected Correspondence of Marcel Duchamp*, ed. Francis M. Naumann and Hector Obalk, trans. Jill Taylor, London: Thames and Hudson, 2000, 286–7, 286.

43 I am grateful to Linda Dalrymple Henderson for sharing her knowledge of *Trans/formation* with me in an email exchange.

44 Breton, Interview with André Parinaud, *Arts*, no. 349, 7 March 1952, reprinted in Breton, *Conversations*, 251; 'les poètes et les artistes que j'ai connus me semblent avoir adopté envers [la science] une attitude protestaire', Breton, *OC*, vol. 3, 634–9, 635.

45 Breton, *Conversations*, 252–3; 'unanimes, ou peu s'en faut, à se désintéresser de la théorie des quanta et de la mécanique de Heisenberg', Breton, *OC*, vol. 3, 637.

46 Breton, *Conversations*, 253; 'une cause d'enrichissement artistique: c'est plutôt le contraire', Breton, *OC*, vol. 3, 638.

47 Breton, *Conversations*, 253; 'a plus que jamais le droit de regarder de travers l'homme de science', Breton, *OC*, vol. 3, 637.

48 Breton, *Conversations*, 253; 'les têtes les plus encombrées de fatras', '[q]ue dire de plus des *savants atomiques*', Breton, *OC*, vol. 3, 637. Compare Pierre Mabille's remarks given at almost exactly the same time in a 1952 interview with Parinaud in *Arts* (to which Breton might have been responding), declaring the aims of artist and scientist are the same and they should therefore seek to fight on a common front, André Parinaud, 'Aux Frontières de l'Art et de la Science moderne: Le Docteur Mabille', *Arts*, no. 348, 29 February 1952, 1, 7. Elsewhere, shortly after, Mabille expressed his hope for a future reconciliation between the aims of art and science, Pierre Mabille, 'L'Art et la Science', *Education et Théâtre*, no. 12, April 1952, 1–8.

49 The Surrealist Group, 'Démasquez les physiciens, Videz les laboratoires', José Pierre (ed.), *Tracts surréalistes et déclarations collectives, 1922–1969*, Tome II: 1940–1969, Paris: Le Terrain Vague, 1982, 172–3, 377–8. Translations are from The Surrealist Group, 'Expose the Physicists, Empty the Laboratories', Michael Richardson and Krzysztof Fijalkowski (ed. and trans.), *Surrealism Against the Current: Tracts and Declarations*, London: Pluto Press, 2001, 164–5. The Surrealists' intention in echoing the title of the better-known 1925 declaration 'Ouvrez les prisons, Licenciez l'armée' ('Open the Prisons, Disband the Army') might have been to evoke the militaristic aspect of research into atomic energy, while displaying a continuity of linked anti-army and libertarian sentiment running through the history of the movement.

50 Richardson and Fijalkowski (eds), *Surrealism Against the Current*, 164; '[l]a physique moderne a pourtant promis, elle a tenu, et elle promet encore des résultats tangibles, sous forme de monceaux de cadavres', Pierre (ed.), *Tracts surréalistes*, Tome II, 172.

51 Richardson and Fijalkowski (eds), *Surrealism Against the Current*, 165; ' . . .l'opium du peuple', 'la théologie de la Bombe', Pierre (ed.), *Tracts surréalistes*, Tome II, 172, 173.

52 Dalí, 'Anti-Matter Manifesto', *Collected Writings*, 366–7.

53 Dalí, 'Anti-Matter Manifesto', *Collected Writings*, 366. A brief glance at Dalí's career alone puts paid to the erroneous remark, '[n]either quantum mechanics at the one extreme, nor psycho-analysis at the other, can be of serious use to a painter . . .', pronounced with typical confidence, only a few years after the publication of Dalí's 'Anti-Matter Manifesto', by Edgar Wind, *Art and Anarchy* [1963], London: Duckworth, 1985, 130, n. 105.

54 Richardson and Fijalkowski (eds), *Surrealism Against the Current*, 165; 'boycottons les conférences vouées à l'exaltation de l'atome', Pierre (ed.), *Tracts surréalistes*, Tome II, 173.

55 Richardson and Fijalkowski (eds), *Surrealism Against the Current*, 165; 'protestations contre la course aux armements, que certains physiciens affectent de signer aujourd'hui', Pierre (ed.), *Tracts surréalistes*, Tome II, 172. In an anti-clerical broadside of the previous year, the Surrealists described how they watched in amazement as 'Louis de Broglie, a militant catholic, return[ed] to a determinist interpretation of wave form phenomena', The Surrealist Group, 'Warning Shot', Richardson and Fijalkowski (eds), *Surrealism Against the Current*, 159–63, 161; 'Louis de Broglie, catholique militant, revenir à une

interprétation déterministe des phénomènes ondulatoires', 'Coup de semonce', Pierre (ed.), *Tracts surréalistes*, Tome II, 164–9, 167. This is the about-face and return to the determinism of *clarté cartésienne* with which de Broglie marked his sixtieth year in 1953, much to the relief of the French academy, see Milic Capek, 'Bergson and Louis de Broglie', *Bergson and Modern Physics: A Reinterpretaion and Reevaluation*, Dordrecht: D. Reidel, 1971, 292–301. Also in 'Warning Shot', Breton's former favourite Stéphane Lupasco and Dalí's friend the abstract painter Georges Mathieu come in for a bruising for alleged religious and nationalistic transgressions.

56 Marc Cioc, *Pax Atomica: The Nuclear Defense Debate in West Germany During the Adenauer Era*, New York: Columbia University Press, 1988, 72–91, 79. Naturally, there is much self-justification in Heisenberg's recollection of this episode, Werner Heisenberg, *Physics and Beyond: Encounters and Conversations*, trans. Arnold J. Pomerans, London: George Allen & Unwin, 1971, 218–19.

57 Heisenbergian readings of Dalí's postwar imagery, which the artist's comments authorise, are beginning to surface. Indeed, judging from its uncredited appearance in a wall text at the Venice-Philadelphia Dalí centenary exhibition in 2004–5, Elliott King's recent, technically proficient Heisenbergian interpretation of Dalí's *Still Life – Fast Moving* (1956) has even become the orthodoxy. The source of that anonymous wall text is King, 'Werner Karl Heisenberg (1901–76)', Palazzo Grassi, *Dalí*, 439.

58 Breton, 'Allocution au "Gala du Secours aux objecteurs de conscience,"' *Perspective cavalière*, Paris: Gallimard, 1970, 164–7; '[i]l me paraît pour le moins inutile qu'une telle conscience se réclame de préceptes religieux: elle est de nature à s'affirmer librement, hors de tout espoir parasite en une récompense au-delà de la vie. . . . Il n'est aucun besoin de faire appel à des instances surnaturelles pour se convaincre que le duel atomique est un monstre que nous alimentons de notre patience, ne serait-ce qu'à l'égard des physiciens ou que la guerre d'Algerie, suintante de pétrole, est une débauche de crimes', 166.

59 The Surrealist Group, 'We Don't Ear it that Way', Pierre (ed.), *Tracts surréalistes*, Tome II, 210. The Surrealists objected specifically to the inclusion of this recent painting by Dalí and to his attempt to gain publicity by showing up at the event. Indeed, his earlier, safely Surrealist work was allowed into the exhibition presumably by Breton himself, even though the 1939 model-and-lobster photograph reproduced in the catalogue is decidedly late- (or even post-) Surrealist, low quality Dalí, perhaps included mischievously by the Surrealists for that very reason, New York: d'Arcy Galleries, *Surrealist Intrusion in the Enchanters Domain*, 1960, 54–5. Duchamp did not bother explaining himself to Breton, only regretting he ever got involved in the show's organisation, Duchamp, Letter to André Breton of 11 December 1960, *Affect' Marcel._ The Selected Correspondence of Marcel Duchamp*, 371–2.

60 Pierre (ed.), *Tracts surréalistes*, Tome II, 210. Taunting the Surrealists up to the end, in the year of Breton's death, Dalí's mutation following his departure from the Surrealist group gave him the material for a book consisting of an exchange of letters between his two selves – Salvador Dalí and 'Avidadollars' – which contained some new writings alongside texts that first appeared in *Minotaure* in the 1930s, Dalí, *Lettre ouverte à Salvador Dalí*, Paris: Éditions Albin Michel, 1966. I am grateful to Elliott King for pointing this book out to me.

61 The best analyses of the painting are Robert Cozzolino, 'Why Are Salvador Dalí's "Late Works" His Most Contentious?', St. Petersburg, Florida: The Salvador Dalí Museum, *Persistence and Memory*, 2004, 145–8 and Michael R. Taylor, '*The Sistine Madonna*, 1958', Palazzo Grassi, *Dalí*, 2004, 386–9.

BIBLIOGRAPHY

Aaserud, Finn. *Redirecting Science: Niels Bohr, Philanthropy, and the Rise of Nuclear Physics*, Cambridge: Cambridge University Press, 1990.

Abbott, Edwin A. *Flatland. A Romance of Many Dimensions* [1884], London: Penguin, 1998.

Ades, Dawn. *Dada and Surrealism Reviewed*, London: Arts Council of Great Britain, 1978.

Ades, Dawn. *Dalí*, London: Thames and Hudson, 1982.

Ades, Dawn. 'Morphologies of Desire', Michael Raeburn (ed.), *Salvador Dalí: The Early Years*, London: South Bank Centre, 1994, 129–60.

Ades, Dawn (ed.). *Dalí's Optical Illusions*, New Haven and London: Yale University Press, 2000.

Albert, David Z. *Quantum Mechanics and Experience*, Cambridge, Mass.: Harvard University Press, 1992.

Alexandre, Maxime. 'A propos de morale', *La Révolution surréaliste*, no. 12, 15 December 1925, 49–50.

Alix, Josefina. 'Transparent Mountains, Luminous Lands', *Gordon Onslow Ford: Mirando en lo profundo/Seeing in Depth*, Santiago de Compostela: Fundación Eugenio Granell, 1998, 199–210.

Allendy, René. 'La Constitution de la matière', *L'Esprit Nouveau*, no. 14, (undated), 1579–85.

Allendy, René. 'La Constitution de la matière', *L'Esprit Nouveau*, no. 18, (undated), n.p.

Althusser, Louis. *For Marx*, trans. Ben Brewster, London: Penguin, 1969.

Amaldi, E. Review of Niels Bohr, *La théorie atomique et la description des phénomènes*, *Scientia*, vol. 54, no. CCLIX, November 1933, 359–60.

Ambrosino, Georges. 'La Structure de l'atome: Le noyau selon L. de Broglie', *Critique*, no. 1, June 1946, 91–6.

Anonymous. Review of H. Thirring, L'idée de la théorie de la Relativité, *Revue de Métaphysique et de Morale*, year 31, no. 4, October–December 1924, supplement, 3–5.

Anonymous. Review of Arthur Eddington, *Étoiles et Atomes*, *Revue de Métaphysique et de Morale*, year 38, no. 3, July–September 1931, supplement, 5–6.

Anonymous. Review of Hans Reichenbach, *La Philosophie Scientifique, vues nouvelles sur ses buts et ses méthodes*, *Revue de Métaphysique et de Morale*, year 40, no. 2, April–June 1933, supplement, 3–5.

Anonymous. Review of Gaston Bachelard, *Le Nouvel Esprit scientifique*, *Revue de Métaphysique et de Morale*, year 42, no. 1, January 1935, supplement, 7–9.

Anonymous. Review of Arthur Eddington, *L'univers en expansion*, *Revue de Métaphysique et de Morale*, year 42, no. 1, January 1935, supplement, 1–3.

Anonymous. Review of Hans Reichenbach, *Atome et Cosmos*, *Revue de Métaphysique et de Morale*, year 42, no. 2, April 1935, supplement, 1–2.

Anonymous. Review of Rudolph Carnap, *La Science et la Métaphysique*, *Revue de Métaphysique et de Morale*, year 42, no. 2, April 1935, supplement, 2–3.

Anonymous. 'Les Livres', *Cahiers d'Art*, year XI, May 1936, 67–8.

Antliff, Mark. 'Bergson and Cubism: A Reassessment', *Art Journal*, vol. 47, no. 4, Revising Cubism, Winter 1988, 341–9.

Antliff, Mark. *Inventing Bergson: Cultural Politics and the Parisian Avant-Garde*, Princeton, New Jersey: Princeton University Press, 1993.

Aragon, Louis. *Traité du style*, Paris: Gallimard, 1928.

Aragon, Louis. *Treatise on Style*, trans. Alyson Waters, Lincoln and London: University of Nebraska Press, 1991.

Armellini, G. Review of Arthur Eddington, *The Expanding Universe*, *Scientia*, vol. 57, no. CCLXXV, May 1935, 234.

Aron, Raymond. Reviews of *Recherches philosophiques* and Gaston Bachelard, *La dialectique de la durée*, *Zeitschrift für Sozialforschung*, year 6, no. 2, 1937, 417–19.

Aron, Robert and Arnaud Dandieu, *La Révolution nécessaire*, Paris: Bernard Grasset, 1933

Arp, Jean/Hans. *Jours effeuillés: Poèmes, essais, souvenirs*, 1920–1965, Paris: Gallimard, 1966.

Auffret, Dominique. *Alexandre Kojève: La Philosophie, l'État, la fin de l'Histoire*, Paris: Grasset & Fasquelle, 1990.

Babini, José. 'Los problemas de la ciencia', *SUR*, no. 33, June 1937, 84–7.

Bachelard, Gaston. 'La Richesse d'inférence de la physique mathématique', *Scientia*, vol. 50, no. CCXXXII, August 1931, 79–88.

Bachelard, Gaston. 'Noumène et microphysique',

Recherches philosophiques, no. 1, 1931–2, 55–66.
Bachelard, Gaston. 'Le Monde comme caprice et miniature', *Recherches philosophiques*, no. 3, 1933–4, 306–20.
Bachelard, Gaston. *Le Nouvel Esprit scientifique*, Paris: Alcan, 1934.
Bachelard, Gaston. 'Idéalisme discursif', *Recherches philosophiques*, no. 4, 1934–5, 21–9.
Bachelard, Gaston. 'Le Surrationalisme', *Inquisitions*, no. 1, June 1936, 1–6.
Bachelard, Gaston. '*Imagination et réalisation*, par Armand Petitjean' (book review), NRF, year 25, no. 282, 1 March 1937, 455–6.
Bachelard, Gaston. *La Formation de l'Esprit Scientifique*, Paris: Librairie Philosophique J. Vrin, 1938.
Bachelard, Gaston. *Lautréamont* [1939], Paris: Librairie José Corti, 1956.
Bachelard, Gaston. *La Philosophie du non: Essai d'une philosophie du nouvel esprit scientifique*, Paris: PUF, 1940.
Bachelard, Gaston. *L'eau et les rêves: Essai sur l'imagination de la matière*, Paris: José Corti, 1942.
Bachelard, Gaston. *La Poétique de l'espace*, Paris: PUF, 1957.
Bachelard, Gaston. *La Poétique de la rêverie*, Paris: PUF, 1960.
Bachelard, Gaston. *The Philosophy of No: A Philosophy of the New Scientific Mind*, trans. G. C. Waterston, New York: The Orion Press, 1968.
Bachelard, Gaston. *The Poetics of Space*, trans. Maria Jolas, Boston: Beacon Press, 1969.
Bachelard, Gaston. *L'Engagement rationaliste*, Paris: PUF, 1972.
Bachelard, Gaston. *The New Scientific Spirit*, trans. Arthur Goldhammer, Boston: Beacon Press, 1984.
Bachelard, Gaston. *Water and Dreams: An Essay on the Imagination of Matter*, trans. Edith R. Farrell, Dallas: Pegasus, 1999.
Bachelard, Gaston. *The Formation of the Scientific Mind: A Contribution to a Psychoanalysis of Objective Knowledge*, trans. Mary McAllester Jones, Manchester: Clinamen, 2002.
Balakian, Anna. *Surrealism: The Road to the Absolute* (revised edition), London: Unwin, 1970.
Bataille, Georges. 'Fatrasies', *La Révolution surréaliste*, no. 6, 1 March 1926, 2–3.
Bataille. Georges. 'Celestial Bodies' [1938], *October*, no. 36, Spring 1986, 75–8.
Bataille, Georges. *L'Expérience intérieure*, Paris: Gallimard, 1943.
Bataille, Georges. *La Part maudite: Essai d'économie générale. La Consomation*, Paris: Minuit, 1949.
Bataille, Georges. *Oeuvres complètes*, vol. 1, Paris: Gallimard, 1970.
Bataille, Georges. *Oeuvres complètes*, vol. 2, Paris: Gallimard, 1970.
Bataille, Georges. *Oeuvres complètes*, vol. 5, Paris: Gallimard, 1973.
Bataille, Georges. *Oeuvres complètes*, vol. 6, Paris: Gallimard, 1973.
Bataille, Georges. *Oeuvres complètes*, vol. 7, Paris: Gallimard, 1976.
Bataille, Georges. *Visions of Excess: Selected Writings, 1927–1939*, trans. and ed. Alan Stoekl, Minneapolis: University of Minnesota Press, 1985.
Bataille, Georges. 'Autobiographical Note' [1958?], *October*, no. 36, Spring 1986, 107–110.
Bataille, Georges. 'Un-Knowing: Laughter and Tears' [1953], *October*, no. 36, Spring 1986, 89–102.
Bataille, Georges. *Oeuvres complètes*, vol. 12, Paris: Gallimard, 1988.
Bataille, Georges. *Inner Experience*, trans. Leslie Anne Boldt, New York: SUNY, 1988.
Bataille, Georges (ed.). *Encyclopaedia Acephalica*, trans. Iain White, London: Atlas Press, 1995.
Bataille, Georges. *Choix de lettres, 1917–1962*, Paris: Gallimard, 1997.
Baugh, Bruce. *French Hegel: From Surrealism to Postmodernism*, New York and London: Routledge, 2003.
Bayet, Albert. *Qu'est-ce que le rationalisme?* Paris: Les Éditions Rationalistes, 1939.
Beer, Gillian. 'Eddington and the Idiom of Modernism', Henry Krips, J. E. McGuire, Trevor Melia (eds), *Science, Reason, and Rhetoric*, Pittsburgh, P.A.: University of Pittsburgh, 1995, 295–315.
Béhar, Henri and Michel Carassou. *Le Surréalisme: Textes et débats*, Paris: Librairie Générale Française, 1984.
Béhar, Henri (ed.). *Inquisitions: Du surréalisme au Front populaire*, Paris: Centre national de la recherche scientifique, 1990.
Beller, Mara, Robert S. Cohen, and Jürgen Renn (eds). *Einstein in Context*, Cambridge: Cambridge University Press, 1993.
Beller, Mara. *Quantum Dialogue: The Making of a Revolution*, Chicago and London: University of Chicago Press, 1999.
Belot, Émile. 'Le rôle capital de l'Astrophysique dans la cosmogonie', *Scientia*, vol. 62, no. CCIV, July 1937, 74–82.
Benayoun, Robert. 'Le Chandail d'Einstein', *Médium: Communication Surréaliste*, Series 2, no. 4, January 1955, 17.
Benjamin, Walter. Review of *Recherches philosophiques*, no. 4, 1934–5, *Zeitschrift für Sozialforschung*, year 6, no. 1, 1937, 173–4.
Benjamin, Walter. *Illuminations*, trans. Harry Zohn, London: Pimlico, 1990.
Benjamin, Walter. 'Surrealism: The Last Snapshot of the European Intelligentsia' [1929], *One Way Street and Other Writings*, trans. Edmund Jephcott and Kingsley Shorter, London and New York: Verso, 2000, 225–39.
Bergson, Henri. *Time and Free Will*, trans. F. L. Pogson, London: George Allen and Unwin, 1910.
Bergson, Henri. *Creative Evolution*, trans. Arthur Mitchell, London: MacMillan and Co., 1911.
Bergson, Henri. *Durée et simultanéité: A propos de la théorie d'Einstein*, Paris: Librairie Félix Alcan, 1922.
Bernstein, Jeremy. *Einstein*, Bungay, Suffolk: Fontana/Collins, 1973.
Biezunski, Michel. 'Popularisation and Scientific Contro-

versy: The Case of the Theory of Relativity in France', Terry Shinn and Richard Whitley (eds), *Expository Science: Forms and Functions of Popularisation*, Dordrecht: D. Reidel, 1985, 183–94.

Biezunski, Michel. 'Einstein's Reception in Paris in 1922', Thomas F. Glick (ed.), *The Comparative Reception of Relativity*, Dordrecht: D. Reidel, 1987, 169–88.

Blackmore, John T. *Ernst Mach: His Work, Life, and Influence*, Berkeley: University of California Press, 1972.

Bohr, Niels. 'Das Quantenpostulat und die neuere Entwicklung der Atomistik', *Die Naturwissenschaften*, vol. 16, 13 April 1928, 245–7.

Bohr, Niels. 'The Quantum Postulate and the Recent Development of Atomic Theory', *Nature*, vol. 121, no. 3050, 14 April 1928, 580–90.

Bohr, Niels. 'Wirkungsquantum und Naturbeschreibung', *Die Naturwissenschaften*, vol. 17, 28 June 1929, 483–6.

Bohr, Niels. 'Licht und Leben', *Die Naturwissenschaften*, vol. 21, 31 March 1933, 245–50.

Bohr, Niels. 'Light and Life', *Nature*, vol. 131, no. 3308, 25 March 1933, 421–3, and *Nature*, vol. 131, no. 3309, 1 April 1933, 457–9.

Bohr, Niels. *Atomic Theory and the Description of Nature*, Cambridge: Cambridge University Press, 1934.

Bohr, Niels. 'Kausalität und Komplementarität', *Erkenntnis*, no. 6, April 1937, 293–303.

Bohr, Niels. 'Causality and Complementarity', *Philosophy of Science*, no. 4, July 1937, 289–98.

Bohr, Niels. *On the Constitution of Atoms and Molecules*, Copenhagen: Munksgaard, 1963.

Boldt-Irons, Leslie Anne (ed.). *On Bataille: Critical Essays*, New York: SUNY, 1995.

Bonnet, Marguerite. *André Breton: Naissance de l'aventure surréaliste*, Paris: José Corti, 1975.

Born, Max and Pascual Jordan. 'Zur Quantenmechanik', *Zeitschrift für Physik*, vol. 34, 1925, 858–88.

Born, Max, Werner Heisenberg and Pascual Jordan. 'Zur Quantenmechanik. II', *Zeitschrift für Physik*, vol. 35, 4 February 1926, 557–615.

Bose, Jagadis Chunder. *Plant Response*, London: Longmans, Green, and Co., 1906.

Bose, Jagadis Chunder. *Life Movements in Plants*, London: Longmans, Green, and Co., 1918–31.

Brenkman, John. 'Introductsion to Bataille', *New German Critique*, no. 16, Winter 1979, 59–63.

Breton, André. *Les Pas perdus*, Paris: Gallimard, 1924.

Breton, André. 'Léon Trotsky: *Lénine*', *La Révolution surréaliste*, no. 5, 15 October 1925, 29.

Breton, André. 'Cabezas de tormenta', *SUR*, no. 32, May 1937, 7–40.

Breton, André. 'Des tendances les plus récentes de la peinture surréaliste', *Minotaure*, no. 12/13, May 1939, 16–17.

Breton, André. 'Situation du surréalisme entre les deux guerres', *VVV*, no. 2/3, March 1943, 44–53.

Breton, André and Marcel Duchamp. *Le Surréalisme en 1947*, Paris: Éditions Pierre à Feu, 1947.

Breton, André. *Anthologie de l'humour noir*, Paris: Sagittaire, 1950.

Breton, André. *Entretiens 1913–52*, Paris: Gallimard, 1952.

Breton, André et al. *Farouche à quatre feuilles*, Paris: Éditions Bernard Grasset, 1954.

Breton, André. 'Le Pont Suspendu', *Médium: Communication Surréaliste*, Series 2, no. 4, January 1955, 37.

Breton, André. *L'Art Magique* [1957], Paris: Éditions Phébus and Éditions Adam Biro, 1991.

Breton, André. *Le Surréalisme et la peinture*, Paris: Gallimard, 1965.

Breton, André. *La Clé des champs*, Paris: Jean Jacques Pauvert, 1967.

Breton, André. *Perspective cavalière*, Paris: Gallimard, 1970.

Breton, André. *Surrealism and Painting*, trans. Simon Watson Taylor, London: Macdonald and Co., 1972.

Breton, André. *Manifestoes of Surrealism*, trans. Richard Seaver and Helen R. Lane, Michigan: University of Michigan Press, 1972.

Breton, André. *L'Amour fou*, Paris: Gallimard, 1977.

Breton, André. *What is Surrealism? Selected Writings*, ed. Franklin Rosemont, London: Pluto, 1978.

Breton, André and Philippe Soupault. *The Magnetic Fields*, trans. David Gascoyne, London: Atlas, 1985.

Breton, André. *Oeuvres complètes*, vol. 1, Paris: Gallimard, 1988.

Breton, André. *Oeuvres complètes*, vol. 2, Paris: Gallimard, 1992.

Breton, André. *Conversations: The Autobiography of Surrealism*, trans. Mark Polizzotti, New York: Paragon House, 1993.

Breton, André. *Free Rein*, trans. Michel Parmentier and Jacqueline d'Amboise, Lincoln and London: University of Nebraska Press, 1995.

Breton, André. *The Lost Steps*, trans. Mark Polizzotti, Lincoln and London: University of Nebraska Press, 1996.

Breton, André. *Break of Day*, trans. Mark Polizzotti and Mary Ann Caws, Lincoln and London: University of Nebraska Press, 1999.

Breton, André. *Oeuvres complètes*, vol. 3, Paris: Gallimard, 1999.

Briot, Marie-Odile. '*L'Esprit Nouveau*: son regard sur les sciences', *Léger et l'esprit moderne*, 1918–1931, Paris: Musée d'art moderne de la ville de Paris, 1982, 27–59.

Broglie, Louis de. 'Sur les interférences et la théorie des quanta de lumière', *Comptes Rendus*, vol. 175, 1922, 811–13.

Broglie, Louis de. 'Rayonnement noir et quanta de lumière', *Journal de Physique*, vol. 3, November 1922, 422–8.

Broglie, Louis de. 'Ondes et quanta', *Comptes Rendus*, vol. 177, 1923, 507–10.

Broglie, Louis de. 'Quanta de lumière, diffraction et interférences', *Comptes Rendus*, vol. 177, 1923, 548–50.

Broglie, Louis de. 'Les quanta, la théorie cinétique des gaz et le principe de Fermat', *Comptes Rendus*, vol. 177, 1923, 630–32.

Broglie, Louis de. 'A Tentative Theory of Light Quanta',

The Philosophical Magazine and Journal of Science, vol. 47, no. 278, February 1924, 446–58.
Broglie, Louis de. 'Recherches sur la théorie des quanta', *Annales de Physique*, vol. 3/4, January–February 1925, 22–128.
Broglie, Louis de. 'La physique moderne et l'oeuvre de Fresnel', *Revue de Métaphysique et de Morale*, year 34, no. 4, October–December 1927, 421–40.
Broglie, Louis de. 'Déterminisme et causalité dans la physique contemporaine', *Bulletin de la Société française de Philosophie*, year 29, no. 5, October–December 1929.
Broglie, Louis de. *Matière et lumière*, Paris: Éditions Albin Michel, 1937.
Broglie, Louis de. *Matter and Light: The New Physics*, trans. W. H. Johnstone, London: George Allen and Unwin Ltd, 1939.
Broglie, Louis de. 'Les Conceptions de la physique contemporaine et les idées de Bergson sur le temps et sur le mouvement', *Revue de Métaphysique et de Morale*, year 48, no. 4, October–December 1941, 241–57.
Broglie, Louis de. 'Le Penchant de Paul Valéry pour la physique', *Cahiers du Sud*, 'Paul Valéry vivant', Paris, 1946, 109–10.
Brunet, Lucien. '*Le Nouvel Esprit scientifique* d'après G. Bachelard' (book review), *Revue de Synthèse*, Tome VIII, no. 2, October 1934, 250–2.
Brunet, Lucien. 'La Logique de la science et l'école de Vienne' (book review), *Revue de Synthèse*, Tome XII, no. 1, April 1936, 113–15.
Brunschvicg, Léon. 'Le Temps et la causalité', *Revue de Métaphysique et de Morale*, year 29, no. 1, January–March 1922, 1–33.
Brunschvicg, Léon. *L'Expérience humaine et la causalité physique*, Paris: Alcan, 1922.
Brunschvicg, Léon. 'La philosophie d'Émile Meyerson', *Revue de Métaphysique et de Morale*, year 33, no. 1, January–March 1926, 39–63.
Brunschvicg, Léon. 'Le déterminisme et la causalité dans la physique contemporaine', *Bulletin de la Société française de Philosophie*, year 30, no. 2, April–June 1930.
Brunschvicg, Léon. 'Science et prise de conscience', *Scientia*, vol. 55, no. CCLXV, May 1934, 329–40.
Brunschvicg, Léon. *La Physique du vingtième siècle et la philosophie*, Paris: Hermann, 1936.
Burnier, André. 'Néo-réalisme et idéalisme contemporain', *Revue de Métaphysique et de Morale*, year 46, no. 3, July 1939, 425–41.
Busi, Frederick. 'Hegel and the Origins of Surrealism and the New Left', *Contemporary French Civilization*, vol. II, no. 3, Spring 1978, 379–403.

Cabanne, Pierre. *Dialogues with Marcel Duchamp*, trans. Ron Padgett, London: Thames and Hudson, 1971.
Caillois, Roger. 'Spécification de la poésie', *SASDLR*, no. 5, May 1933, 30–31.
Caillois, Roger. 'Systématisation et determination', *Documents* 34, no. 1 (special issue, 'Intervention surréaliste'), June 1934, 69–72.
Caillois, Roger. 'Mimétisme et psychanasthénie légendaire', *Minotaure*, no. 7, June 1935, 5–10.
Caillois, Roger. *Procès intellectuel de l'art*, Marseille: Cahiers du Sud, 1935.
Caillois, Roger. 'Déterminations inconscientes en peinture', *Documents* 35, year 3, no. 6, November–December 1935, 5–6.
Caillois, Roger. 'Pour une orthodoxie militante: les taches immédiates de la pensée moderne', *Inquisitions*, no. 1, June 1936, 6–14.
Caillois, Roger. *Approches de l'imaginaire*, Paris: Gallimard, 1974.
Caillois, Roger. *The Necessity of the Mind*, trans. Michael Syrotinski, Venice, CA: The Lapis Press, 1990.
Caillois, Roger. *The Edge of Surrealism: A Roger Caillois Reader*, ed. Claudine Frank, trans. Claudine Frank and Camille Naish, Durham and London: Duke University Press, 2003.
Calas, Nicolas. *Foyers d'incendie*, Paris: Denoël, 1938.
Calas, Nicolas. *Confound the Wise*, New York: Arrow Editions, 1942.
Calas, Nicolas. *Matta: A Totemic World*, New York: Andrew Crispo Gallery, 1975.
Campbell, Joseph (ed.). *Man and Time*, Papers from the Eranos Yearbooks, vol. 3, New York: Princeton University Press, 1957.
Canguilhem, Georges. *The Normal and the Pathological* [1966], trans. Carolyn R. Fawcett with Robert S. Cohen, New York: Zone, 1991.
Capek, Milic. *Bergson and Modern Physics: A Reinterpretaion and Reevaluation*, Dordrecht: D. Reidel, 1971.
Capra, Fritjof. *The Tao of Physics: An Exploration of the Parallels between Modern Physics and Eastern Mysticism* [1975], London: Flamingo, 1983.
Carnap, Rudolph. 'Über Windung der Metaphysik durch logische Analyse der Sprache', *Erkenntnis*, no. 2, 1931, 219–41.
Carnap, Rudolph. *The Unity of Science*, trans. Max Black, London: Kegan Paul, 1934.
Carnap, Rudolph. *La Science et la Métaphysique devant l'analyse logique du langage*, Paris: Hermann, 1934.
Carnap Rudolph. *Foundations of Logic and Mathematics*, Chicago: University of Chicago Press, 1937.
Carrouges, Michel. *André Breton et les données fondamentales du surréalisme*, Paris: Gallimard, 1950.
Carrouges, Michel. *André Breton and the Basic Concepts of Surrealism*, trans. Maura Prendergast, Alabama: University of Alabama Press, 1974.
Cassidy, David C. *Uncertainty: The Life and Science of Werner Heisenberg*, New York: W. H. Freeman and Company, 1992.
Castelao, Teresa. 'Gaston Bachelard et le milieu scientifique et intellectuel français', Pascal Nouvel (ed.), *Actualité et Postérités de Gaston Bachelard*, Paris: PUF, 1997, 101–15.
Castelnuovo, G. 'Il principio di causalità', *Scientia*, vol. 60, no. CCXCII, August 1936, 61–8.
Cavaillès, Jean. 'L'École de Vienne au Congrès de Prague', *Revue de Métaphysique et de Morale*, year 42, no. 1, January 1935, 137–49.
Chandrasekhar, S. *Eddington: The Most Distinguished*

Astrophysicist of his Time, Cambridge: Cambridge University Press, 1983.
Chevalley, Claude and Alexandre Marc. 'Dialectique de l'action (En marge de la pensée d'Arnaud Dandieu)', *Recherches philosophiques*, no. 6, 1936–7, 373–83.
Chimisso, Cristina. *Gaston Bachelard: Critic of Science and the Imagination*, London and New York: Routledge, 2001.
Cioc, Marc. *Pax Atomica: The Nuclear Defense Debate in West Germany During the Adenauer Era*, New York: Columbia University Press, 1988.
Cohen, Robert S. and Raymond J. Seeger (eds), *Ernst Mach: Physicist and Philosopher*, Dordrecht: D. Reidel, 1970.
Compton, A. H. 'A Quantum Theory of the Scattering of X-Rays by Light Elements', *The Physical Review*, vol. 21, Series 2, no. 5, May 1923, 483–502.
Connor, Peter Tracey. *Georges Bataille and the Mysticism of Sin*, Baltimore and London: John Hopkins University Press, 2000.
Cozzolino, Robert. 'Why Are Salvador Dalí's "Late Works" His Most Contentious?', St. Petersburg, Florida: The Salvador Dalí Museum, *Persistence and Memory*, 2004, 145–8
Crevel, René. *Êtes-vous fous?* Paris: Éditions NRF, 1929.
Crevel, René. *Dalí ou l'anti-obscurantisme*, Paris: Éditions Surréalistes, 1931.
Crevel, René. 'Notes en vue d'une psycho-dialectique', *SASDLR*, no. 5/6, 15 May 1933, 48–52.
Crevel, René. *L'Esprit contre la raison*, Paris: Tchou, 1969.
Crevel, René. *Les Pieds dans le plat*, Paris: Jean-Jacques Pauvert, 1974.
Crevel, René. *Putting My Foot in It*, trans. Thomas Buckley, Illinois: Dalkey Archive Press, 1992.
Crow, Christine M. *Paul Valéry and Maxwell's Demon: Natural Order and Human Possibility*, Hull: University of Hull, 1972.

Dalí, Salvador. 'L'âne pourri', *SASDLR*, no. 1, July 1930, 9–12.
Dalí, Salvador. *La Femme visible*, Paris: Éditions Surréalistes, 1930.
Dalí, Salvador. 'Interprétation Paranoïaque-Critique de l'Image Obsédante 'L'Angelus' de Millet', *Minotaure*, no. 1, February 1933, 65–7.
Dalí, Salvador. 'Objets psycho-atmosphériques-anamorphiques', *SASDLR*, no. 5/6, 15 May 1933, 45–8.
Dalí, Salvador. 'De la beauté terrifiante et comestible, de l'architecture modern style', *Minotaure*, no. 3/4, December 1933, 69–76.
Dalí, Salvador. 'Apparitions aérodynamiques des "Être-Objets"', *Minotaure*, no. 6, Winter 1934–5, 33–4.
Dalí, Salvador. *La Conquête de l'irrationnel*, Paris: Éditions Surréalistes, 1935.
Dalí, Salvador. 'Psychologie non-euclidienne d'une photographie', *Minotaure*, no. 7, June 1935, 56–7.
Dalí, Salvador. 'Le Surréalisme spectrale de l'éternel féminin préraphaélite', *Minotaure*, no. 8, 1936, 46–9.
Dalí, Salvador. *Le Mythe tragique de l'Angelus de Millet*, Paris: Jean-Jacques Pauvert, 1963.
Dalí, Salvador. *Lettre ouverte à Salvador Dalí*, Paris: Éditions Albin Michel, 1966.
Dalí, Salvador. *The Unspeakable Confessions of Salvador Dalí*, trans. Harold J. Salemson, London: W. H. Allen, 1976.
Dalí, Salvador. *The Collected Writings of Salvador Dalí*, ed. and trans. Haim Finkelstein, Cambridge: Cambridge University Press, 1998.
Dandieu, Arnaud. 'Espace', *Documents*, year 2, no. 1, 1930, 41–4.
Dandieu, Arnaud. 'Émile Meyerson, '"Le Physicien et le Primitif"' (article review), *Documents*, year 2, no. 5, 1930, 312.
Dandieu, Arnaud. 'Le conflit du réel et du rationnel dans la psychologie du temps et de l'espace', *Revue philosophique*, year 55, no. 11/12, November–December 1930, 448–61.
Dandieu, Arnaud and Denis de Rougemont. 'L'acte comme point de départ', *Recherches philosophiques*, no. 6, 1936–7, 55–64.
Daumal, René. 'Les limites du langage philosophique et les savoirs traditionels', *Recherches philosophiques*, no. 4, 1934–5, 209–31.
Daumal, René. *Le Mont Analogue*, Paris: Gallimard, 1952.
Daumal, René. *You've Always Been Wrong*, trans. Thomas Vosteen, Lincoln and London: University of Nebraska Press, 1995.
De Regt, Henk W. 'Erwin Schrödinger', J. Blackmore, R. Itagaki and S. Tanaka (eds), *Ernst Mach's Vienna, 1895–1930, or Phenomenalism as Philosophy of Science*, Dordrecht: Kluwer Academic Publishers, 2001, 85–104.
De Rougemont, Denis. 'Marcel Duchamp mine de rien', *Preuves*, year 18, no. 204, February 1968, 43–7.
Derrida, Jacques. 'From Restricted to General Economy: A Hegelianism without Reserve' [1967], *Writing and Difference*, trans. Alan Bass, Chicago: University of Chicago Press, 1978, 251–77.
Derrida, Jacques. *Marges de la philosophie*, Paris: Minuit, 1972.
Derrida, Jacques. *Margins of Philosophy*, trans. Alan Bass, London: Harvester Wheatsheaf, 1982.
Descombes, Vincent. *Modern French Philosophy*, trans. L. Scott-Fox and J. M. Harding, Cambridge: Cambridge University Press, 1979.
Dirac, P. A. M. 'The Fundamental Equations of Quantum Mechanics', *Proceedings of the Royal Society of London*, vol. 109, no. A752, 1 December 1925, 642–53, 642.
Dirac, P. A. M. 'Quantum Mechanics and a Preliminary Investigation of the Hydrogen Atom', *Proceedings of the Royal Society of London*, vol. 110, no. A755, 1 March 1926, 561–79.
Dirac, P. A. M. 'The Elimination of the Nodes in Quantum Mechanics', *Proceedings of the Royal Society of London*, vol. 111, no. A757, 1 May 1926, 281–305.
Dirac, P. A. M. 'Relativity Quantum Mechanics with an Application to Compton Scattering', *Proceedings of the Royal Society of London*, vol. 111, no. A758, 2 June 1926, 405–23.
Dirac, P. A. M. 'On Quantum Algebra', *Proceedings of*

the Cambridge Philosophical Society, vol. 23, no. 4, October 1926, 412–18.
Dirac, P. A. M. *The Development of Quantum Theory*, New York: Gordon and Breach, 1971.
Dirac, P. A. M. *Directions in Physics*, New York: John Wiley & Sons, 1978.
Domínguez, Oscar. 'Le Pétrification du Temps', *La Conquête du Monde par l'Image*, 1942, 27.
Douglas, A. Vibert. *Arthur Stanley Eddington*, London: Thomas Nelson and Sons Ltd, 1957.
Duchamp, Marcel. *Marchand du sel, écrits de Marcel Duchamp*, ed. Michel Sanouillet, Paris: Le Terrain Vague, 1958.
Duchamp, Marcel. 'I Propose to Strain the Laws of Physics', interview with Francis Roberts, *Art News*, vol. 67, no. 8, December 1968, 46–7, 62–4.
Duchamp, Marcel. *Duchamp du signe: Écrits*, ed. Michel Sanouillet and Elmer Peterson, Paris: Flammarion, 1975.
Duchamp, Marcel. *Salt Seller: The Essential Writings of Marcel Duchamp*, ed. Michel Sanouillet and Elmer Peterson, London: Thames and Hudson, 1975.
Marcel Duchamp, *Affect' Marcel._ The Selected Correspondence of Marcel Duchamp*, ed. Francis M. Naumann and Hector Obalk, trans. Jill Taylor, London: Thames and Hudson, 2000.
Dunne, J. W. *An Experiment with Time*, London: A. & C. Black, 1927.
Dunne, J. W. *Le Temps et le Rêve*, trans. Eugène de Veauce, Paris: Éditions du Seuil, 1948.
Duplessis, Yves. *Le Surréalisme*, Paris: PUF, 1950.
Durkheim, Émile. *Les Formes élémentaires de la vie religieuse*, Paris: Alcan, 1912.
Durozoi, Gérard. *History of the Surrealist Movement* [1997], trans. Alison Anderson, Chicago and London: University of Chicago Press, 2002.
Dyck, Walther (ed.). *Katalog mathematischer und mathematisch-physikalischer Modelle, Apparate und Instrumente*, Munich: K. Hof.- u. Universitätsbuchdruckerei von Dr. C. Wolf & Sohn, 1892.

Earman, John and Clark Glymour. 'Relativity and Eclipses: The British Expeditions of 1919 and Their Predecessors', *Historical Studies in the Physical Sciences*, vol. 11, 1980, 49–85.
Eco, Umberto. *The Role of the Reader: Explorations in the Semiotics of Texts*, Bloomington and London: Indiana University Press, 1979.
Eddington, Arthur. 'Gravitation and the Principle of Relativity', *Nature*, vol. 101, no. 2523, 7 March 1918, 15–17.
Eddington, Arthur. 'The Internal Constitution of the Stars', *The Observatory*, vol. XLIII, October 1920, no. 557, 341–58.
Eddington, Arthur. *Space Time and Gravitation: An Outline of the General Relativity Theory*, Cambridge: Cambridge University Press, 1920.
Eddington, Arthur. *Stars and Atoms*, Oxford: Clarendon Press, 1927.
Eddington, Arthur. *The Nature of the Physical World*, Cambridge: Cambridge University Press, 1928.
Eddington, Arthur Stanley. *Science and the Unseen World*, London: George Allen & Unwin, 1929.
Eddington, Arthur. *The Rotation of the Galaxy*, Oxford: Clarendon Press, 1930.
Eddington, Arthur. *Science and Religion*, London: The Friends Home Service Committee, 1931.
Eddington, Arthur. *The Expanding Universe*, Cambridge: Cambridge University Press, 1933.
Eddington, Arthur. *New Pathways in Science*, Cambridge: Cambridge University Press, 1935.
Einstein, Albert. 'Über einen die Erzeugung und Verwandlung des Lichtes betreffenden heuristischen Gesichtspunkt', *Annalen der Physik*, vol. 17, no. 6, 9 June 1905, 132–48.
Einstein, Albert. 'Über die von der molekularkinetischen Theorie der Wärme geforderte Bewegung von in ruhenden Flüssigkeiten suspendierten Teilchen', *Annalen der Physik*, vol. 17, no. 7, 18 July 1905, 549–60.
Einstein, Albert. 'Zur Elektrodynamik bewegter Körper', *Annalen der Physik*, vol. 17, no. 10, 26 September 1905, 891–921.
Einstein, Albert. 'Über den Einfluß der Schwerkraft auf die Ausbreitung des Lichtes', *Annalen der Physik*, vol. 35, no. 10, 1 September 1911, 898–908.
Einstein, Albert. 'Die Grundlage der allgemeinen Relativitätstheorie', *Annalen der Physik*, vol. 49, no. 7, 11 May 1916, 769–822.
Einstein, Albert. *Relativity: The Special and the General Theory* [1920], London: Methuen, 1960.
Einstein, Albert. *La théorie de la Relativité: Restreinte et Généralisée*, trans. J. Rouvière, Paris: Gauthier-Villars et Cie, 1921.
Einstein, Albert, et al. *The Principle of Relativity: A Collection of Original Papers on the Special and General Theory of Relativity*, trans. W. Perrett and G.B. Jeffery [1923], New York: Dover, 1952.
Einstein, Albert. *Einstein on Peace*, ed. Otto Nathan and Heinz Norden, New York: Simon and Schuster, 1960.
Einstein, Albert. *Ideas and Opinions* [1954], ed. Carl Seelig, trans. Sonja Bargmann, New York: The Modern Library, 1994.
Einstein, Carl. 'André Masson, étude ethnologique', *Documents*, year 1, no. 2, May 1929, 93–102.
Einstein, Carl. *Georges Braque*, trans. E. Zipruth, Paris: Chroniques du jour, 1934.
Eisenmann, J. and Y. Rocard, 'Le principe de correspondance et l'état actuel de la théorie des Quanta', *Revue de Métaphysique et de Morale*, year 33, no. 2, April–June 1926, 183–200.
Elton, Lewis. 'Einstein, General Relativity, and the German Press, 1919–1920', *ISIS*, vol. 77, no. 1, March 1986, 95–103.
Éluard, Paul. *La Rose publique*, Paris: Gallimard, 1934.
Éluard, Paul. 'Physique de la poésie', *Minotaure*, no. 6, Winter 1934–5, 6–12.
Éluard, Paul. 'Premières vues anciennes', *Minotaure*, no. 10, Winter 1937, 49–50.
Éluard, Paul. *Oeuvres complètes*, vol. 1, Paris: Gallimard, 1968.
Éluard, Paul. *Lettres à Gala*, 1924–1948, Paris: Gallimard, 1984.
Enriques, Frederigo. 'Philipp Frank, *Théorie de la con-*

naissance et physique moderne' (book review), *Scientia*, vol. 57, no. CCLXXV, March 1935, 227–9.
Ernst, Max et al. *Max Ernst: Beyond Painting*, ed. Robert Motherwell, New York: Wittenborn, Schultz, Inc., 1948.
Ernst, Max. 'What is Surrealism?', trans. Werner Spies, London: Tate Gallery, *Max Ernst: A Retrospective*, 1991, 309–10.
Ey, Henri. Review of Jacques Lacan, *De la psychose paranoïaque dans ses rapports avec la personalité*, *Recherches philosophiques*, no. 3, 1933–4, 457–60.

Fabre, Lucien. *Les Théories d'Einstein*, Paris: Payot & Cie, 1921.
Fer, Briony, David Batchelor, Paul Wood, *Realism, Rationalism, Surrealism: Art between the Wars*, New Haven and London: Yale University Press and the Open University, 1993.
Ferrari, Germana (ed.). *Matta: Entretiens Morphologiques: Notebook no. 1, 1936–44*, London: Sistan, 1987.
Feuer, Lewis S. *Einstein and the Generations of Science* (second ed.), New Brunswick and London: Transaction, 1982.
Feyerabend, Paul. *Against Method*, London: Verso, 1978.
Feyerabend, Paul. *Farewell to Reason*, London and New York: Verso, 1987.
Feynman, Richard P. *Six Easy Pieces: The Fundamentals of Physics Explained*, Harmondsworth: Penguin, 1998.
Finkelstein, Haim. 'The Incarnation of Desire: Dalí and the Surrealist Object', *Res: Anthropology and Aesthetics*, no. 23, Spring 1993, 114–37.
Fischer, Gerd. *Mathematische Modelle/Mathematical Models* (2 vols), Braunschweig/Wiesbaden: Friedr. Vieweg and Sohn, 1986.
Fletcher, Valerie (ed.). *Crosscurrents of Modernism: Four Latin American Painters*, Washington DC: Smithsonian Institution Press, 1992.
Fölsing, Albrecht. *Albert Einstein*, trans. Ewald Osers, London: Penguin, 1997.
Fort, Charles. 'Les Livres des damnés', *Médium: Communication Surréaliste*, Series 2, no. 4, January 1955, 18–19.
Fortuné, Isabelle. 'Man Ray et les objets mathématiques', *Études photographiques*, no. 6, May 1999, 100–113.
Foster, Hal. *Compulsive Beauty*, Cambridge, Mass. and London: MIT, 1993.
Foucault, Michel. *The Archaeology of Knowledge*, trans. A. M. Sheridan Smith, London and New York: Routledge, 1989.
Frank, Philipp. *Einstein: His Life and Times*, London: Jonathan Cape, 1948
Frank, Philipp. 'Ernst Mach and the Unity of Science', Robert S. Cohen and Raymond J. Seeger (eds), *Ernst Mach: Physicist and Philosopher*, Dordrecht: D. Reidel, 1970, 235–44.
Frank-Elster, Claudine. 'Les Retournements Petitjeans: The Return of Armand Petitjean', *MODERNISM/modernity*, vol. 3, no. 1, 2001, 87–116.
Freud, Sigmund. *The Interpretation of Dreams*, Pelican Freud Library vol. 4, Harmondsworth: Penguin, 1978.
Freud, Sigmund. *Historical and Expository Works on Psychoanalysis*, Pelican Freud Library vol. 15, London: Penguin, 1986.
Freud, Sigmund. *On Sexuality*, Pelican Freud Library vol. 7, London: Penguin, 1986.
Freud, Sigmund. *On Metapsychology*, Penguin Freud Library vol. 11, London: Penguin, 1991.
Freud, Sigmund. *New Introductory Lectures in Psychoanalysis*, Penguin Freud Library vol. 2, London: Penguin, 1991.
Frost, Rosamund. 'Matta's Third Surrealist Manifesto', *ARTNews*, vol. XLIII, no. 1, 15–29 February 1944, 18.
Fruteau de Laclos, Frédéric. *La philosophie de l'intellect d'Émile Meyerson. De l'épistémologie à la psychologie*, unpublished dissertation submitted for the degree of Ph.D., Université de Paris X–Nanterre, 2004.

Galison, Peter. *Einstein's Clocks, Poincaré's Time: Empires of Time*, London: Sceptre, 2003.
Galison, Peter, Michael Gordin, David Kaiser (eds). *Science and Society: The History of Modern Physical Science in the Twentieth Century*, vol. 2, 'Making General Relativity', New York and London: Routledge, 2001.
Gamow, George. *Mr. Tompkins in Wonderland*, Cambridge: Cambridge University Press, 1940.
Gamow, George. *Thirty Years that Shook Physics*, New York: Doubleday, 1966.
Gibson, Ian. *The Shameful Life of Salvador Dalí*, London: Faber and Faber, 1997.
Gide, André. 'Dada', Justin O'Brien (ed.), *The Most Significant Writings from the* Nouvelle Revue française, 1919–1940, London: Eyre & Spottiswoode, 1958, 12–15.
Gleizes, Albert. 'Art and Science' [1931], *Art and Religion, Art and Science, Art and Production*, trans. Peter Brooke, London: Francis Boutle, 1999, 60–95.
Glick, Thomas F. (ed.). *The Comparative Reception of Relativity*, Dordrecht: D. Reidel, 1987.
Glick, Thomas F. *Einstein in Spain: Relativity and the Recovery of Science*, Princeton, New Jersey: Princeton University Press, 1988.
Golan, Romy. 'R. E. Matta, 1937–1947. The Last Phase of Surrealism', M.A. report, Courtauld Institute of Art, 1982.
Goldberg, Stanley. 'Henri Poincaré and Einstein's Theory of Relativity', *American Journal of Physics*, vol. 35 (ii), no. 10, October 1967, 934–44.
Goldberg, Stanley. 'Poincaré's Silence and Einstein's Relativity', *The British Journal for the History of Science*, vol. 5, part 1, no. 17, June 1970, 73–84.
Goldberg, Stanley. *Understanding Relativity: Origin and Impact of a Scientific Revolution*, Oxford: Clarendon Press, 1984.
Goldschmidt, Bertrand. *The Atomic Complex: A Worldwide Political History of Nuclear Energy*, trans. Bruce M. Adkins, Illinois: American Nuclear Society, 1982.

Goldschmidt, Bertrand. *Atomic Rivals*, trans. Georges M. Tenner, New Brunswick and London: Rutgers University Press, 1990.
Gracq, Julien. *André Breton: Quelques aspects de l'écrivain*, Paris: Librairie José Corti, 1948.
Greenberg, Clement. *The Collected Essays and Criticism*, vol. II: *Arrogant Purpose, 1945–1949*, Chicago and London: University of Chicago Press, 1986.
Gribbin, John. *In Search of Schrödinger's Cat: Quantum Physics and Reality*, London: Black Swan Press, 1998.
Grogin, R. C. *The Bergson Controversy in France 1900–1914*, Calgary: University of Calgary Press, 1988.
'G.S.' 'Gaston Bachelard, *Le Nouvel Esprit scientifique*' (book review), *Scientia*, vol. 57, no. CCLXXVI, July 1935, 318–19.
Guicharnaud, Jacques. *Raymond Queneau*, trans. June Guicharnaud, New York and London: Columbia University Press, 1965.
Gunter, Pete A. Y. (ed.). *Bergson and the Evolution of Physics*, Knoxville: University of Tennessee Press, 1969.
Gunter, Pete A. Y. 'Henri Bergson', David Ray Griffin et al. (eds), *Founders of Constructive Postmodern Philosophy*, New York: State University of New York Press, 1993, 133–63.

Harrington, Jean. 'Don't Worry – It Can't Happen', *Scientific American*, vol. 162, no. 5, May 1940, 268.
Harris, Steven. *Surrealist Art and Thought in the 1930s: Art, Politics, and the Psyche*, Cambridge: Cambridge University Press, 2004.
Heidegger, Martin. 'Qu'est-ce que la métaphysique?' *Bifur*, no. 8, 10 June 1931, 5–27.
Heidegger, Martin. 'De la nature de la cause', *Recherches philosophiques*, no. 1, 1931–2, 83–124.
Heilbron, John L. 'The Earliest Missionaries of the Copenhagen Spirit', Edna Ullman-Margalit (ed.), *Science in Reflection*, Dordrecht: Kluwer Academic Publishers, 1988, 201–33.
Heisenberg, Werner. 'Über quantentheoretische Umdeutung kinematischer und mechanischer Beziehungen', *Zeitschrift für Physik*, vol. 33, 1925, 879–93.
Heisenberg, Werner. 'Über den anschaulichen Inhalt der quantentheoretischen Kinematik und Mechanik', *Zeitschrift für Physik*, vol. 43, 31 May 1927, 172–98.
Heisenberg, Werner. *The Physical Principles of the Quantum Theory*, trans. Carl Eckart and Frank C. Hoyt, Chicago: University of Chicago Press, 1930.
Heisenberg, Werner. 'Zur Geschichte der physikalischen Naturerklärung', *Wandlungen in den Grundlagen der Naturwissenschaft*, Leipzig: Verlag von S. Hirzel, 1935, 27–45.
Heisenberg, Werner. 'Die Gothesche und die Newtonsche Farbenlehre im Lichte der modernen Physik', *Geist der Zeit*, 19, 1941, 261–75.
Heisenberg, Werner. *Philosophic Problems of Nuclear Science*, trans. F. C. Hayes, London: Faber and Faber, 1952.
Heisenberg, Werner. *Physics and Beyond: Encounters and Conversations*, trans. Arnold J. Pomerans, London: George Allen & Unwin, 1971.
Heisenberg, Werner. *Encounters with Einstein and Other Essays on People, Particles, and Places* [1983], Princeton, New Jersey: Princeton University Press, 1989.
Heisenberg, Werner. *Physics and Philosophy: The Revolution in Modern Science*, London: Penguin, 1990.
Hellman, John. *The Communitarian Third Way: Alexandre Marc's Ordre Nouveau, 1930–2000*, Montreal and Kingston: McGill-Queen's University Press, 2002.
Henderson, Linda Dalrymple. *The Fourth Dimension and Non-Euclidean Geometry in Modern Art*, Princeton: Princeton University Press, 1983.
Henderson, Linda Dalrymple. *Duchamp in Context: Science and Technology in the* Large Glass *and Related Works*, Princeton: Princeton University Press, 1998.
Henderson, Linda Dalrymple. Editor's Introduction: I. 'Writing Modern Art and Science – An Overview'; II. 'Cubism, Futurism, and Ether Physics in the Early Twentieth Century', *Science in Context* (special issue, 'Modern Art and Science'), vol. 17, no. 4, December 2004, 423–66.
Henderson, Linda Dalrymple. 'Four-Dimensional Space or Space–Time?: The Emergence of the Cubism-Relativity Myth in New York in the 1940s', Michele Emmer (ed.), *The Visual Mind II*, Cambridge, Mass.: MIT Press, 2005, 349–97.
Hermann, Armin. *The Genesis of Quantum Theory (1899–1913)*, trans. Claude W. Nash, Cambridge, Mass. and London: MIT, 1971.
Hesnard, Angelo and Édouard Pichon. 'Aperçu historique du mouvement psychanalytique français', *Revue de Psychologie Concrète*, no. 1, 1 February 1929, 102–20.
Hollier, Denis (ed.). *Le Collège de Sociologie (1937–1939)*, Paris: Gallimard, 1979.
Hollier, Denis (ed.). *The College of Sociology (1937–39)*, trans. Betsy Wing, Minneapolis: University of Minnesota Press, 1988.
Hollier, Denis. *Against Architecture: The Writings of Georges Bataille*, trans. Betsy Wing, Cambridge, Mass. and London: MIT, 1992.
Hollier, Denis. *Absent Without Leave*, trans. Catherine Porter, Cambridge, Mass. and London: Harvard University Press, 1997.
Hollier, Denis. 'The Dualist Materialism of Georges Bataille', Fred Botting and Scott Wilson (eds), *Bataille: A Critical Reader*, Oxford: Blackwell, 1998, 59–73.
Holton, Gerald. *Thematic Origins of Scientific Thought: Kepler to Einstein*, Cambridge, Mass.: Harvard University Press, 1974.
Holton, Gerald. *Einstein, History, and Other Passions: The Rebellion against Science at the End of the Twentieth Century*, New York: Addison-Wesley, 1996.
Horton, Robin. 'Lévy-Bruhl, Durkheim and the Scientific Revolution', Robin Horton and Ruth Finnegan (eds), *Modes of Thought: Essays on Thinking in Western and Non-Western Societies*, London: Faber & Faber,

1973, 249–305.
Howard, Don and John Stachel (eds). *Einstein and the History of General Relativity*, Einstein Studies, vol. 1, Boston, Basel, Berlin: Birkhäuser: 1989.
Howorth, Jolyon. *France: The Politics of Peace*, London: END/Merlin Press, 1984.
Hussey, Andrew. *The Inner Scar: The Mysticism of Georges Bataille*, Amsterdam: Editions Rodopi, 2000.

Isidore Isou. 'Qu'est-ce que le Lettrisme?' ,*Fontaine*, vol. 11, no. 62, October 1947, 529–50.

Jacob, Max. *Les Pénitents en maillots roses*, Paris: Éditions du Sagittaire, 1925.
Jaffard, Paul. 'Philosophie, science et psychanalyse', *Critique*, vol. 7, no. 46, 15 March 1951, 219–32.
Jammer, Max. *The Conceptual Development of Quantum Mechanics*, New York: McGraw-Hill, 1966.
Jammer, Max. *The Philosophy of Quantum Mechanics: The Interpretations of Quantum Mechanics in Historical Perspective*, New York: John Wiley and Sons, 1974.
Janet, Maurice. 'La finalité en mathématiques et en physique', *Recherches philosophiques*, no. 2, 1932–3, 1–17.
Jarry, Alfred. 'Gestes et Opinions du Docteur Faustroll' [1898], *Oeuvres complètes*, vol. 1, Monte Carlo: Éditions du Livre, 1948, 195–320.
Jay, Martin. *Downcast Eyes: The Denigration of Vision in Twentieth-Century French Thought*, California and London: University of California Press, 1993.
Jean, Marcel. *The History of Surrealist Painting*, trans. Simon Watson Taylor, New York: Grove Press, 1960.
Jeans, James. *Physics and Philosophy*, Cambridge: Cambridge University Press, 1943.
Jenny, Laurent. 'Système et "délire" de Dalí à Caillois', *Europe*, year 78, no. 859–60, November–December 2000, 60–71.
Johan, R. 'La raison et l'irrationnel chez M. Meyerson', *Recherches philosophiques*, no. 1, 1931–2, 138–65.
'Joint Eclipse Meeting of the Royal Society and the Royal Astronomical Society', *The Observatory*, vol. XLII, no. 545, November 1919, 389–98.
Joliot-Curie, Frédéric et al. 'Linéaments d'un deuxième univers', *NRF*, vol. 41, year 21, no. 238, 1 July 1933, 1–32.
Jones, Henri. *Le Surréalisme ignoré*, Paris: Naaman Dilif, 1969.
Joravsky, David. *Soviet Marxism and Natural Science, 1917–1932*, London: Routledge and Kegan Paul, 1961.
Jordan, H. J. 'L'indeterminisme vital et le dynamisme des structures causales', *Recherches philosophiques*, no. 2, 1932–3, 18–47.
Jordan, H. J. 'Esprit et matière', *Recherches philosophiques*, no. 5, 1935–6, 299–320.
Jordan, Pascual. 'Die Quantenmechanik und die Grundprobleme der Biologie und Psychologie', *Die Naturwissenschaften*, vol. 20, 4 November 1932, 815–21.
Jordan, Pascual. 'Quantenphysikalische Bemerkungen zur Biologie und Psychologie', *Erkenntnis*, no. 4, 1934, 215–52.
Jordan, Pascual. 'Ergänzende Bemerkungen über Biologie und Quantenmechanik', *Erkenntnis*, no. 5, 1935, 348–52.
Jordan, Pascual. 'Die Verstärkertheorie der Organismen in ihrem gegenwärtigen Stand', *Die Naturwissenschaften*, vol. 26, 19 August 1938, 537–45.
Jordan, Pascual. *Verdrängung und Komplementarität: Eine philosophische Untersuchung*, Hamburg-Bergedorf, 1947.
Jungk, Robert. *Brighter than a Thousand Suns: The Moral and Political History of the Atomic Scientists*, trans. James Cleugh, London: Victor Gollancz/Rupert Hart-Davis, 1958.
Juvet, Gustave. *La Structure des nouvelles théories physiques*, Paris: Alcan, 1933.

Keller, Thomas. *Deutsch-französische Dritte-Weg-Diskurse: Personalistische Intellektuellendebatten der Zwischenkriegszeit*, Munich: Wilhelm Fink, 2001.
Kelly, Michael. *Hegel in France*, Birmingham: University of Birmingham, 1992.
Kern, Stephen. *The Culture of Time and Space, 1880–1918* [1983], Cambridge, Mass. and London: Harvard University Press, 2003.
King, John. Sur*: A Study of the Argentine Literary Journal and its Role in the Development of Culture, 1931–1970*, Cambridge: Cambridge University Press, 1986.
Kojève, Alexandre. *Introduction à la lecture de Hegel*, ed. Raymond Queneau, Paris: Gallimard, 1947.
Kojève, Alexandre. *Introduction to the Reading of Hegel: Lectures on the Phenomenology of Spirit*, ed. Raymond Queneau and Allan Bloom, trans. James H. Nichols, Jr., Ithaca and London: Cornell University Press, 1980.
Kojève, Alexandre. *L'Idée du déterminisme dans la physique classique et dans la physique moderne*, Paris: Librairie Générale Française, 1990.
Kojevnikoff, Alexandre. Book reviews, *Recherches philosophiques*, no. 3, 1933–4, 464–6.
Koyré, Alexandre. Review of Paul Langevin, *La notion de corpuscules et d'atomes*, *Recherches philosophiques*, no. 4, 1934–5, 436–8.
Koyré, Alexandre. *Études galiléennes*, Paris: Hermann, 1940.
Koyré, Alexandre. 'Rapport sur l'état des études hégéliennes en France' [1930–31], *Études d'histoire de la pensée philosophique*, Paris: Librairie Armand Colin, 1961, 205–27.
Kozloff, Max. 'An Interview with Matta', *Artforum*, vol. IV, no. 1, September 1965, 23–6.
Kraft, Victor. *The Vienna Circle: The Origin of Neo-Positivism*, trans. Arthur Pap, New York: Philosophical Library, 1953.
Kragh, Helge. *Quantum Generations: A History of Physics in the Twentieth Century*, Princeton: Princeton University Press, 1999.
Krauss, Rosalind. '"Michel, Bataille et moi"', *October*, no. 68, Spring 1994, 3–20.
Kuhn, Thomas S. *The Structure of Scientific Revolutions* [1962], Chicago: University of Chicago Press, 1970.

Labadié, Jean. 'La Mathématique de la matière (d'Einstein à Bergson)', *L'Esprit Nouveau*, no. 16, March 1922, 1865–70.
Lacan, Jacques. 'Le problème du style et les formes paranoïaque de l'expérience', *Minotaure*, no. 1, February 1933, 68–9.
Lacan, Jacques. Review of Eugène Minkowski, *Le temps vécu*, *Recherches philosophiques*, no. 5, 1935–6, 424–31.
Lake, Carlton. *In Quest for Dalí*, New York: G. P. Putnams Sons, 1969.
Lalo, Charles. Obituary of Albert Spaïer, *Recherches philosophiques*, no. 3, 1933–4, vii–viii.
Langevin, Paul. 'L'évolution de l'espace et du temps', *Scientia*, vol. 10, no. XIX, 1911, 31–54.
Langevin, Paul. *La notion de corpuscules et d'atomes*, Paris: Hermann, 1934.
Laporte, Paul M. 'Cubism and Relativity, with a Letter of Albert Einstein', *Art Journal*, vol. 25, no. 3, Spring 1966, 246–8.
Las Palmas de Gran Canaria: Centro Atlántico de Arte Moderno. *Oscar Domínguez, 1926–1957*, 1995.
Laude, Jean. 'Un portrait de Carl Einstein', *Cahiers du Musée National d'Art Moderne*, no. 1, July–September 1979, 10–13.
Lavelle, Louis. *La Philosophie française entre les deux guerres*, Paris: Éditions Montaigne, 1942.
Laville, Remy. *Pierre Mabille: Un Compagnon du Surréalisme*, Faculté des Lettres et Sciences humaines de l'Université de Clermont-Ferrand II, 1983.
Leavis, F. R. 'Two Cultures? The Significance of Lord Snow', *Nor Shall My Sword: Discourses on Pluralism, Compassion and Social Hope*, London: Chatto and Windus, 1972, 39–74.
Le Becq, Paul. 'À propos des Théories d'Einstein', *L'Esprit Nouveau*, no. 7, 719–27.
Lecourt, Dominique. *L'épistémologie historique de Gaston Bachelard*, Paris: Vrin, 1974.
Lecourt, Dominique. *Marxism and Epistemology: Bachelard, Canguilhem and Foucault*, trans. Ben Brewster, London: NLB, 1975.
Legrand, Gérard. *André Breton et son temps*, Paris: Le Soleil Noir, 1976.
Legrand, Gérard. 'Breton et l'inauguration philosophique du surréalisme', Christian Deschamps (ed.), *Surréalisme et philosophie*, Paris: Centre Georges Pompidou, 1992, 13–19.
Leiris, Michel. 'Notes sur deux figures microcosmiques des XIV et XV siècles', *Documents*, no. 1, April 1929, 48–52.
Leiris, Michel. 'Civilisation', *Documents*, year 1, no. 4, September 1929, 221–2.
Léon, Xavier. *Fichte et son temps*, 3 vols., Paris: Librairie Armand Colin, 1922–7.
Leroux, Jean. 'Bachelard et le cercle de Vienne', *Cahiers Gaston Bachelard*, no. 5, 2003, 107–27.
Levy, Julien (ed.). *Surrealism*, New York: The Black Sun Press, 1936.
Lewis, Wyndham. *Time and Western Man*, London: Chatto and Windus, 1927.
Lingua Franca (eds). *The Sokal Hoax: The Sham that Shook the Academy*, Lincoln: University of Nebraska Press, 2000.
Lippard, Lucy R. (ed.). *Surrealists on Art*, New Jersey: Spectrum, 1975.
Lomas, David. *The Haunted Self: Surrealism, Psychoanalysis, Subjectivity*, New Haven and London: Yale University Press, 2000.
London: Hayward Gallery. *Matta: Coïgitum*, 1977.
Loubet del Bayle, Jean-Louis. *Les non-conformistes des années 30. Une tentative de renouvellement de la pensée politique française*, Paris: Éditions du Seuil, 1969.
Lupasco, Stéphane. *Logique et Contradiction*, Paris: PUF, 1947.

Mabille, Pierre. *La Construction de l'homme*, Paris: Jean Flory, 1936.
Mabille, Pierre. 'L'oeil du peintre', *Minotaure*, no. 12/13, May 1939, 53–6.
Mabille, Pierre. *Le Miroir du Merveilleux*, Paris: Sagittaire, 1940.
Mabille, Pierre. 'Del Nuevo Mundo', *Cuadernos Americanos*, vol. XI, no. 5, September–October 1943, 94–110.
Mabille, Pierre. 'L'Art et la Science', *Education et Théâtre*, no. 12, April 1952, 1–8.
Mabille, Pierre. *Traversées de Nuit*, Paris: Plasma, 1981.
Mabille, Pierre. *Messages de l'étranger*, Paris: Plasma, 1981.
Mabille, Pierre. *Mirror of the Marvelous*, trans. Jody Gladding, Rochester, Vermont: Inner Traditions, 1998.
MacKinnon, Edward M. *Scientific Explanation and Atomic Physics*, Chicago and London: University of Chicago Press, 1982.
Madrid: Palacio de Velázquez. *Henry Moore: Sculptures, Drawings, Graphics 1921–1981*, 1981.
Mallarmé, Stéphane. 'Crise de vers', *Divagations*, Paris: Charpentier, 1897, 235–51.
Marmande, Francis. *Georges Bataille politique*, Lyon: Presses Universitaires de Lyon, 1985.
Masson, André. *Anatomy of My Universe*, New York: Curt Valentin, 1943.
Matthews, Eric. *Twentieth Century French Philosophy*, Oxford: Oxford University Press, 1996.
Matthews, J. H. *Eight Painters: The Surrealist Context*, New York: Syracuse University Press, 1982.
Matthews, J. H. *Languages of Surrealism*, Columbia: University of Missouri Press, 1986.
Maudlin, Tim. *Quantum Non-Locality and Relativity: Metaphysical Intimations of Modern Physics*, Oxford: Blackwell, 1994.
Mauss, Marcel. 'Essai sur le Don: Forme et raison de l'échange dans les sociétés archaïques', *L'Année Sociologique*, New Series, vol. 1, 1923–1924, 30–186.
Mauss, Marcel. *The Gift: The Form and Reason for Exchange in Archaic Societies* [1925], trans. W. D. Halls, London: Routledge, 1990.
Meffre, Liliane. 'Aspects de la théorie de l'art de Carl Einstein', *Cahiers du Musée National d'Art Moderne*, no. 1, July–September 1979, 14–39.
Meffre, Liliane. 'Georges Bataille et Carl Einstein',

Dominique Lecoq (ed.), *Georges Bataille et la pensée allemande*, Paris: Les Amis de Georges Bataille, 1986, 95–9.

Mehra, Jagdish and Helmut Rechenberg, *The Historical Development of Quantum Theory: The Quantum Theory of Planck, Einstein, Bohr and Sommerfeld: Its Foundations and the Rise of its Difficulties 1900–1925*, vol. 1, part 1, New York, Heidelberg, Berlin: Springer-Verlag, 1982.

Merleau-Ponty, Maurice. *The Primacy of Perception*, trans. James M. Edie, Evanston: Northwestern University Press, 1964.

Merrifield, C. W. *A Catalogue of a Collection of Models of Ruled Surfaces*, London: George E. Eyre and William Spottiswoode, 1872.

Metz, André. Review of Arthur Eddington, *La nature du Monde physique*, *Revue Philosophique*, year 55, no. 11/12, November–December 1930, 418–47.

Meyerson, Émile. 'Le relativisme, théorie du réel', *Revue de Métaphysique et de Morale*, year 31, no. 1, January–March 1924, 29–48.

Meyerson, Émile. *La Déduction relativiste*, Paris: Payot, 1925.

Meyerson, Émile. 'Le physicien et le primitif', *Revue philosophique*, year 55, no. 5/6, May–June 1930, 319–58.

Meyerson, Émile. *Réel et déterminisme dans la physique quantique*, Paris: Hermann, 1933.

Meyerson, Émile. 'La notion de l'identique', *Recherches philosophiques*, no. 3, 1933–4, 1–17.

Miller, Arthur I. *Insights of Genius: Imagery and Creativity in Science and Art*, New York: Copernicus, 1996.

Miller, Arthur I. 'Erotica, Aesthetics and Schrödinger's Equation', Graham Farmelo (ed.), *It Must be Beautiful: Great Equations of Modern Science*, London and New York: Granta, 2002, 81–101.

Minkowski, Eugène. 'Le problème du temps en psychopathologie', *Recherches philosophiques*, no. 2, 1932–3, 231–56.

Monnerot, Jules. *La Poésie moderne et le sacré*, Paris: Gallimard, 1945.

Monod-Herzen, Édouard. *Principes de Morphologie générale*, (2 vols), Paris: Gauthier-Villars et Cie, 1927.

Monod-Herzen, Édouard. 'Science et Intuition' [1928], Marc Thivolet (ed.), *Le Grand Jeu: les textes parus dans les trois numéros publiées de la revue*, Paris: Éditions de l'Herne, 1968, 77.

Moore, Walter. *A Life of Erwin Schrödinger*, Cambridge: Cambridge University Press, 1994.

Mukherji, Visvapriya. *Jagadis Chandra Bose*, New Delhi: Ministry of Information and Broadcasting, Government of India, 1983.

Müller-Freienfels, Richard. 'Rationalisme et irrationalisme', *Recherches philosophiques*, no. 1, 1931–2, 125–37.

Neuhuys, Paul. 'Quelques Poètes', *Ça Ira!* no. 14, 1921, 53–67.

New York: d'Arcy Galleries. *Surrealist Intrusion in the Enchanters Domain*, 1960.

Nietzsche, Friedrich. *Twilight of the Idols and The Anti-Christ*, trans. R. J. Hollingdale, London: Penguin, 1968.

Nietzsche, Friedrich. *The Gay Science*, trans. Walter Kaufmann, New York: Vintage Books, 1974.

Nietzsche, Friedrich. *Ecce Homo*, trans. R. J. Hollingdale, London: Penguin, 1980.

Nye, Mary Jo. 'Science and Socialism: The Case of Jean Perrin in the Third Republic', *French Historical Studies*, vol. IX, no. 1, Spring 1975, 141–69.

Oberhelman, Harley D. *Ernesto Sábato*, New York: Twayne, 1970.

Okun, Henry. *The Surrealist Object*, dissertation submitted for the degree of Ph.D., New York University, 1981.

Onslow Ford, Gordon. 'The Wooden Giantess of Henry Moore', *London Bulletin*, no. 18/19/20, June 1940, 10.

Ortega y Gasset, José. *El tema del nuestro tiempo* [1923], Madrid: Revista de Occidente, 1934.

Ortega y Gasset, José. 'The Historical Significance of the Theory of Einstein', *The Modern Theme*, trans., James Cleugh, London: The C. W. Daniel Company, 1931, 135–52.

Ory, Pascal. *La Belle Illusion: Culture et politique sous le signe du Front populaire*, 1935–38, Paris: Plon, 1994.

Paalen, Wolfgang. *Dyn*, nos. 1–6, 1942–4.

Paalen, Wolfgang. *Form and Sense*, New York: Wittenborn and Co., 1945.

Pace, David. 'Old Wine – New Bottles: Atomic Energy and the Ideology of Science in Postwar France', *French Historical Studies*, vol. 17, no. 1, Spring 1991, 38–61.

Pais, Abraham. *'Subtle is the Lord . . .' The Science and the Life of Albert Einstein*, Oxford: Oxford University Press, 1982.

Pansaers, Clément. 'Dada et moi', *Ça Ira!* no. 16 (special issue, 'Dada: sa naissance, sa vie, sa mort'), November 1921, 111–15.

Paris: Centre Georges Pompidou, Musée National d'Art Moderne. *Salvador Dalí, rétrospective 1920–1980*, 1979–80.

Paris: Centre Georges Pompidou, Musée National d'Art Moderne. *Matta*, 1985.

Paris: Centre Georges Pompidou, Musée National d'Art Moderne. *André Breton: La Beauté convulsive*, 1991.

Paris: Fondation Cartier pour l'art contemporain. *Hiroshi Sugimoto: Conceptual Forms*, 2004.

Parkinson, Gavin. *Physique du Surréalisme: Surrealism, Modern Physics, and Epistemology*, dissertation submitted for the degree of Ph.D., Courtauld Institute of Art, University of London, 2000.

Parodi, D. '"De l'explication dans les sciences" par É. Meyerson', *Revue de Métaphysique et de Morale*, year 31, no. 4, October–December 1924, 585–97.

Parodi, D. 'En marge de M. Brunschvicg', *Revue de Métaphysique et de Morale*, year 35, no. 3, July–September 1928, 297–310.

Paul, Harry W. *From Knowledge to Power: The Rise of

the Science Empire in France, 1860–1939, Cambridge: Cambridge University Press, 1985.
Pauli, Wolfgang. 'Über den Einfluß der Geschwindigkeitsabhängigkeit der Elektronenmasse auf den Zeemaneffekt', *Zeitschrift für Physik*, vol. 31, 1925, 373–85.
Pauli, Wolfgang. 'Über den Zusammenhang des Abschlusses der Elektronengruppen im Atom mit der Komplexstruktur der Spektren', *Zeitschrift für Physik*, vol. 31, 1925, 765–83.
Pauli, Wolfgang. *Exclusion Principle and Quantum Mechanics*, Neuchatel: Griffon, 1947.
Péret, Benjamin. *Death to the Pigs: Selected Writings*, trans. and ed. Rachel Stella, London: Atlas, 1988.
Perks, Simone. *The Dice-Box of Chance: Chance and Determinism in Surrealism, Science, and the Occult*, unpublished dissertation submitted for the degree of Ph.D., University of Manchester, 2007.
Pestre, Dominique. *Physique et physiciens en France,* 1918–1940, Paris: Archives Contemporaines, 1984.
Peterson, Elmer Theodore. *Tristan Tzara: Dada and Surrational Theorist*, New Brunswick, New Jersey: Rutgers University Press, 1971.
Petitjean, Armand. Reply to 'L'enquête sur la rencontre', *Minotaure*, no. 3/4, December 1933, 112.
Petitjean, Armand. 'Analyse spectrale du singe', *Minotaure*, no. 7, June 1935, 58–9.
Petitjean, Armand. *Imagination et réalisation*, Paris: Denoël et Steele, 1936.
Petitjean, Armand. 'L'oeil', *Mesures*, year 2, no. 4, 15 October 1936, 55–65.
Pettigrew, J. Bell. *Design in Nature* (3 vols), London: Longmans and Co., 1908.
Piaget, Jean. 'La philosophie de Gustave Juvet', *A la mémoire de Gustave Juvet,* 1896–1936, Lausanne: Université de Lausanne, 1937, 37–52.
Pierre, José. *Domaine de Paalen*, Paris: Éditions Galanis, 1970.
Pierre, José (ed.). *Tracts surréalistes et déclarations collectives,* 1922–1969, Tome II: 1940–1969, Paris: Le Terrain Vague, 1982.
Pilkington, A. E. *Bergson and His Influence: A Reassessment*, Cambridge: Cambridge University Press, 1976.
Planck, Max. 'Zur Theorie des Gesetzes der Energieverteilung im Normalspectrum', *Verhandlungen der Deutschen Physikalischen Gesellschaft*, vol. 2, no. 17, 14 December 1900, 237–45.
Planck, Max. 'Ueber das Gesetz der Energieverteilung im Normalspektrum', *Annalen der Physik*, vol. 4, no. 3, 1 March 1901, 553–63.
Plotnitsky, Arkady. *Reconfigurations: Critical Theory and General Economy*, Florida: University Press of Florida, 1993.
Plotnitsky, Arkady. *In the Shadow of Hegel: Complementarity, History, and the Unconscious*, Florida: University Press of Florida, 1993.
Plotnitsky, Arkady. *Complementarity: Anti-Epistemology after Bohr and Derrida*, Durham and London: Duke University Press, 1994.
Plymouth: City Museum and Art Gallery. *Matta: The Logic of Hallucination*, 1984.
Podin, G. 'Inquiétudes actuelles des sciences physiques et métaphysiques', *L'Esprit Nouveau*, no. 20 (undated), n.p.
Poincaré, Henri. *The Value of Science*, trans. George Bruce Halsted, New York: The Science Press, 1907.
Poirier, René. 'Temps spirituels et temps matériel', *Recherches philosophiques*, no. 5, 1935–6, 1–40.
Polizzotti, Mark. *Revolution of the Mind: The Life of André Breton*, New York: Farrar, Straus and Giroux, 1995.
Prévost, Pierre. *Rencontre Georges Bataille*, Paris: Jean-Michel Place, 1987.
Price, Kate. *Poetry and Physics: Interchange within the Writings of Arthur Eddington, I.A. Richards, William Empson and Laura Riding*, unpublished dissertation submitted for the degree of Ph.D., Cambridge University, 2002.
Privitera, Walter. *Problems of Style: Michel Foucault's Epistemology* [1990], trans. Jean Keller, Albany, New York: State University of New York Press, 1995.
Puech, Henri-Charles. 'Picasso et la représentation', *Documents*, no. 3 (special number, 'Hommage à Picasso'), 1930, 118–22.
Puech, Henri-Charles. 'Les "Prisons" de Jean-Baptiste Piranèse', *Documents*, no. 4, 1930, 199–204.
Puech, Henri-Charles. 'Le dieu Bésa et la magie hellénistique', *Documents*, no. 7, 1930, 415–25.
Puech, Henri-Charles. 'Signification et représentation', *Minotaure*, no. 6, Winter 1934–5, 52.

Queneau, Raymond. Review of V. Vernadsky, 'L'étude de la vie et la nouvelle physique', *La Critique sociale*, no. 3, October 1931, 133–4.
Queneau, Raymond. Review of René Allendy, La Justice intérieure, *La Critique sociale*, no. 3, October 1931, 122–3.
Queneau, Raymond. *Le Chiendent*, Paris: Gallimard, 1933.
Queneau, Raymond. *Bâtons, chiffres et lettres*, Paris: Gallimard, 1950.
Queneau, Raymond. *The Bark Tree*, trans. Barbara Wright, London: Calder and Boyers, 1968.
Queneau, Raymond. *Le Voyage en Grèce*, Paris: Gallimard, 1973.
Queneau, Raymond. *Journaux* 1914–1965, Paris: Gallimard, 1996.
Queneau, Raymond. *Oeuvres complètes*, vol. 2, Paris: Gallimard, 2002.

Rae, Alastair I. M. *Quantum Physics: Illusion or Reality?* Cambridge: Cambridge University Press, 1986.
Ray, Man. *Self Portrait*, London: Andre Deutsch, 1963.
Regler, Gustav. *Wolfgang Paalen*, New York: Nierendorf Editions, 1946.
Reich, Zdenko. 'Propos sur une prétendue orthodoxie militante', *Europe*, no. 171, 15 March 1937, 385–98.
Reichenbach, Hans. 'Crise de la causalité', *Documents*, year 1, no. 2, May 1929, 105–8.
Reichenbach, Hans. 'Das Kausalproblem in der Physik', *Die Naturwissenschaften*, vol. 19, 21 Aug 1931, 713–22.
Renouvier, Charles. *Les dilemmes de la Métaphysique*

pure, Paris: Alcan, 1901.
Rhodes, Richard. *The Making of the Atomic Bomb*, New York: Simon and Schuster, 1986.
Richman, Michèle H. *Reading Georges Bataille: Beyond the Gift*, Baltimore and London: The John Hopkins University Press, 1982.
Richman, Michèle. 'Anthropology and Modernism in France: From Durkheim to the Collège de sociologie', Marc Manganaro (ed.), *Modernist Anthropology: From Fieldwork to Text*, Princeton, New Jersey: Princeton University Press, 1990, 183–214.
Richardson, Michael and Krzysztof Fijalkowski (ed. and trans.). *Surrealism Against the Current: Tracts and Declarations*, London: Pluto Press, 2001, 164–5.
Rignano, Eugenio. 'La psychologie dans ses rapports avec la philosophie et avec la science', *Revue de Métaphysique et de Morale*, year 33, no. 4, October–December 1926, 455–67.
Ritter, Johann-Wilhelm. *Nachlass aus den Papieren eines jungen Physikers*, Heidelberg: Mohr und Zimmer, 1810.
Rivière, Jacques. 'Reconnaisance à dada', *NRF*, year 7, no. 83, 1 August 1920, 216–37.
Roberts, John. 'Surrealism, Photography and the Everyday', *The Art of Interruption: Realism, Photography and the Everyday*, Manchester and New York: Manchester University Press, 1998, 98–113.
Robinson-Valéry, Judith (ed.). *Fonctions de l'esprit: Treize savants redécouvrent Paul Valéry*, Paris: Hermann, 1983.
Robinson-Valéry, Judith. 'The Fascination of Science', Paul Gifford and Brian Stimpson (eds), *Reading Paul Valéry: Universe in Mind*, Cambridge: Cambridge University Press, 1998, 77–84.
Roche, Anne and Christian Tarting (eds). *Des Années trente: groupes et ruptures*, Paris: Centre national de la recherche scientifique, 1985.
Romains, Jules. 'A Sketch of Psychoanalysis', trans. Blanche A. Price, Justin O'Brien (ed.), *The Most Significant Writings from the* Nouvelle Revue française, 1919–1940, London: Eyre & Spottiswoode, 1958, 240–55.
Roth, Michael S. *Knowing and History: Appropriations of Hegel in Twentieth-Century France*, Ithaca: Cornell University Press, 1988.
Roudinesco, Elisabeth. *A History of Psychoanalysis in France*, 1925–1985, trans. Jeffrey Mehlman, London: Free Association Books, 1990.
Roudinesco, Elisabeth. *Jacques Lacan*, trans. Barbara Bray, Cambridge: Polity Press, 1997.
Roy, Christian. *Alexandre Marc et la Jeune Europe*, 1904–1934*: L'Ordre Nouveau aux origins de personnalisme*, unpublished dissertation submitted for the degree of Ph.D., McGill University, 1993.
Roy, Christian. 'Arnaud Dandieu's Epistemological Contributions to *Documents*: Evidence of Common Ground between Surrealism and "Nietzschean" Personalism', paper delivered at Fourth Annual Ph.D. Symposium, Tate Modern, London, 2 June 2006.
Rubin, William. *Matta*, New York: MoMA, 1957.
Rucker, Rudolf v. B. *Geometry, Relativity and the Fourth Dimension*, New York: Dover, 1977.
Rusk, Rogers D. *Les Atomes, les hommes et les étoiles*, trans. L. Hervé and Armand Petitjean, Paris: Gallimard, 1939.
Ruyer, Raymond. 'Sur quelques arguments nouveaux contre le réalisme', *Recherches philosophiques*, no. 4, 1934–5, 30–50.
Ruyer, Raymond. 'Micro-physique et Micro-spiritualisme', *Revue de Métaphysique et de Morale*, year 45, no. 4, October 1938, 527–41.

Sageret, Jules. *Le Système du Monde de Pythagore à Eddington*, Paris: Payot, 1931.
San Francisco: San Francisco Museum of Art. *Dynaton*, 1951.
Sapper, K. 'Die Erschütterung des Kausalitätsprinzips in der Neuesten Physik', *Scientia*, vol. 58, no. CCLXXI, September 1935, 149–56.
Sartre, Jean-Paul. 'La transcendance de l'Ego. Esquisse d'une description phénoménologique', *Recherches philosophiques*, no. 6, 1936–7, 85–123.
Sartre, Jean-Paul. *Situations*, vol. 1, Paris: Gallimard, 1947.
Sartre, Jean-Paul. *The Transcendence of the Ego: An Existentialist Theory of Consciousness*, trans. Forrest Williams and Robert Kirkpatrick, New York: Noonday, 1957.
Sawin, Martica. *Surrealism in Exile and the Beginning of the New York School*, Cambridge, Mass.: MIT, 1995.
Sawin, Martica. 'Onslow Ford in New York, June 1940 to June 1941', *Gordon Onslow Ford: Mirando en lo profundo/Seeing in Depth*, Santiago de Compostela: Fundación Eugenio Granell, 1998, 193–7.
Schlipp, Paul Arthur (ed.). *Albert Einstein: Philosopher-Scientist*, Illinois: Open Court, 1949.
Schrödinger, Erwin. *Science and the Human Temperament*, trans. James Murphy, London: George Allen and Unwin Ltd, 1935.
Schrödinger, Erwin. *Collected Papers on Wave Mechanics* [1927], trans. J. F. Shearer, New York: Chelsea Publishing Co., 1982.
Schrödinger, Erwin. *Nature and the Greeks and Science and Humanism*, Cambridge: Cambridge University Press, 1996.
Schuster, Jean. 'Interview de Max Ernst sur l'allemagne', *Médium: Communication Surréaliste*, Series 2, no. 2, February 1954, 27–9.
Sheppard, Richard. *Modernism–Dada–Postmodernism*, Evanston, Illinois: Northwestern University Press, 2000.
Shlain, Leonard. *Art & Physics: Parallel Visions in Space, Time & Light*, New York: Quill/William Morrow, 1991.
Short, Robert. *The Political History of the Surrealist Movement*, unpublished dissertation submitted for the degree of Ph.D., University of Essex, 1965.
Short, Robert Stuart. 'Contre-attaque', Ferdinand Alquié (ed.), *Entretiens sur le surréalisme*, Paris: Mouton, 1968, 144–76.
Simon, Sidney. 'Concerning the Beginnings of the New York School: 1939–1943. An Interview with Peter Busa and Matta Conducted by Sidney Simon in Minneapolis in December 1966', *Art International*, vol.

xi, no. 6, Summer 1967, 17–20.
Simon, Sidney. 'An Interview with Robert Motherwell Conducted by Sidney Simon in New York in January 1967', *Art International*, vol. xi, no. 6, Summer 1967, 20–23.
Snow, C. P. *The Two Cultures*, Cambridge: Cambridge University Press, 1993.
Soby, James Thrall. *Contemporary Painters*, New York: Museum of Modern Art, 1948.
Sokal, Alan and Jean Bricmont. *Intellectual Impostures*, London: Profile, 1998.
Solomon, Jacques. 'À propos des entretiens de physique à Varsovie', *Cahiers du Bolchevisme*, year 15, no. 7, July 1938, 322–5.
Souriau, Michel. 'La Matière et le concret', *Recherches philosophiques*, no. 2, 1932–3, 81–111.
Souriau, Michel. 'Matière, Lettre et Verbe', *Recherches philosophiques*, no. 3, 1933–4, 89–124.
Souriau, Michel. 'G. Bachelard, *Le Nouvel Esprit scientifique*' (book review), *Recherches philosophiques*, no. 4, 1934–5, 440–42.
Souriau, Michel. 'J. W. Dunne, *An Experiment with Time*' (book review), *Recherches philosophiques*, no. 6, 1936–7, 399–400.
Spaïer, Albert. *La Pensée concrète: Essai sur le symbolisme intellectuel*, Paris: Alcan, 1927.
Spaïer, Alfred. 'Pensée et étendue', *Recherches philosophiques*, no. 1, 1931–2, 21–54.
Spenlé, Édouard. *Novalis: Essai sur l'idéalisme romantique en allemagne*, Paris: Librairie Hachette et Cie, 1904.
Spies, Werner (ed.). *Max Ernst, Oeuvre-Katalog: Werke 1939–1953*, Houston, Texas: Menil Foundation, 1987.
Spies, Werner. *Max Ernst, Collages: The Invention of the Surrealist Universe*, trans. John William Gabriel, New York: Harry N. Williams, 1991.
Spitz, Jacques. '*Lautréamont*, par Léon-Pierre Quint' (book review), NRF, year 18, no. 203, 1 August 1930, 273–4.
Spitz, Jacques. 'La Théorie Quantique et le problème de la connaissance', *Inquisitions*, no. 1, June 1936, 20–28.
Spitz, Jacques. 'Les Quanta et l'individu', NRF, year 26, no. 303, 1 December 1938, 998–1007.
The Spokesman, no. 85 (special issue, 'Albert Einstein, Bertrand Russell, Manifesto 50'), 2005.
Stadler, Friedrich. *The Vienna Circle: Studies in the Origins, Development, and Influence of Logical Empiricism* [1997], trans. Camilla Nielson, Joel Golb, Sabine Schmidt, Thomas Ernst, New York: Springer, 2001.
Stallabrass, Julian. *The Mechanical Arts: Science, Mind and Mechanism in Art, France 1918–1931*, unpublished dissertation submitted for the degree of Ph.D., Courtauld Institute, University of London, 1992.
Stanley, Matthew. '"An Expedition to Heal the Wounds of War": The 1919 Eclipse and Eddington as Quaker Adventurer', *ISIS*, no. 94, 2003, 57–89.
Stockholm: Moderna Museet. *Sebastian Matta: 15 former av trivel/15 forms of doubting*, 1959.
Strauss, Jonathan. 'The Inverted Icarus', *Yale French Studies*, no. 78 ('On Bataille'), 1990, 106–23.
Surya, Michel. *Georges Bataille: An Intellectual Biography*, trans. Krzysztof Fijalkowski and Michael Richardson, London and New York: Verso, 2002.
Sweet, Colin. *A Study of Nuclear Power in France*, London: Polytechnic of the South Bank, 1981.

Taylor, Michael R. 'On the Road with Salvador Dalí: The 1952 Nuclear Mysticism Lecture Tour', Michael R. Taylor (ed.), *The Dalí Renaissance*, New Haven and London: Yale University Press, forthcoming 2008.
Therrien, Vincent. *La Révolution de Gaston Bachelard en critique littéraire*, Paris: Klincksieck, 1970.
Thirion, André. *Révolutionnaires sans Révolution*, Paris: Éditions Robert Laffont, 1972.
Thivolet, Marc (ed.). *Le Grand Jeu: les textes parus dans les trois numéros publiées de la revue*, Paris: Éditions de l'Herne, 1968.
Thompson, D'Arcy Wentworth. *On Growth and Form*, Cambridge: Cambridge University Press, 1942.
Thomson, William (Lord Kelvin). *Popular Lectures and Addresses* (3 vols), London: Macmillan and Co., 1889.
Tiryakian, E. A. 'L'École Durkheimienne à la recherche de la société perdue: La sociologie naissante et son milieu culturel', *Cahiers Internationaux de sociologie*, vol. LXVI, year 26, January–June 1979, 97–114.
Tomkins, Calvin. *The Bride and the Bachelors*, London: Weidenfeld and Nicolson, 1965.
Tomkins, Calvin. *Duchamp: A Biography*, London: Pimlico, 1998.
Tonnelat, Marie-Antoinette. 'La notion de réalité physique et l'œuvre de M. Louis de Broglie', André George (ed.), *Louis de Broglie: Physicien et penseur*, Paris: Éditions Albin Michel, 1953, 145–52.
Tzara, Tristan. Letter to the editor, *Cahiers du Sud*, year 22, no. 170, March 1935, 231–2.
Tzara, Tristan. *Le Surréalisme et l'Après-Guerre*, Paris: Les Éditions Nagel, 1948.
Tzara, Tristan. *Oeuvres complètes*, vol. 3, (1934–6), Paris: Flammarion, 1979.
Tzara, Tristan. *Grains et issues: rêve experimental*, Paris: Garnier Flammarion, 1981.

Uhlenbeck, George and Samuel Goudsmit. 'Ersetzung der Hypothese vom unmechanischen Zwang durch eine Forderung bezüglich des inneren Verhaltens jedes einzelnen Elektrons', *Die Naturwissenschaften*, vol. 13, 20 November 1925, 953–4.
Uhlenbeck, George and Samuel Goudsmit. 'Spinning Electrons and the Structure of Spectra', *Nature*, vol. 117, no. 2938, 20 February 1926, 264–5.
Ullmo, Jean. 'L'évolution de la notion de corpuscule d'après M. Langevin', *Scientia*, vol. 55, no. CCLXII, February 1934, 103–17.
Ushenko, Andrew Paul. *Power and Events: An Essay on Dynamics in Philosophy*, Princeton: Princeton University Press, 1946.

Valéry, Paul. 'Un article de *L'Athenaeum*', NRF, vol. 18, year 7, no. 75, 1 December 1919, 1118–22.

Valéry, Paul. 'Les Sciences de l'esprit sont-elles essentiellement différentes des Sciences de la nature?', *Revue de Synthèse*, no. 3, October 1931, 9–11.

Valéry, Paul. *Cahiers* (facsimile edition of 29 vols.), Paris: CNRS, 1957–61.

Valéry, Paul. *Ideé fixe*, trans. David Paul, London: Routledge and Kegan Paul, 1965.

Valéry, Paul. *Cahiers* (2 vols), Paris: Éditions Gallimard, 1973.

Venice: Palazzo Grassi. *Dalí*, 2004.

Vierling, Angela G. http://math.harvard.edu/~angelavc/models.

View, vol. 1, no. 7/8, (October–November 1941).

Virtanen, Reino. *L'Imagerie scientifique de Paul Valéry*, Paris: Librairie Philosophique J. Vrin, 1975.

Wahl, Jean. 'Hegel et Kierkegaard', *Revue philosophique*, year 56, no. 11/12, November–December 1931, 321–80.

Wahl, Jean. 'Vers le concret', *Recherches Philosophiques*, no. 1, 1931–2, 1–20.

Wahl, Jean. *Vers le concret. Études d'histoire de la philosophie contemporaine*, Paris: Librairie Philosophique J. Vrin, 1932.

Wahl, Jean. 'Heidegger et Kierkegaard', *Recherches philosophiques*, no. 2, 1932–3, 349–70.

Wahl, Jean. Reviews of *Inquisitions* and *Acéphale*, NRF, year 24, no. 275, 1 August 1936, 402–3.

Wahl, Jean. 'Stéphane Lupasco, *Du devenir logique et de l'affectivité*' (book review), *Recherches philosophiques*, no. 6, 1936–7, 394–5.

Wahl, Jean. 'La Philosophie française en 1939', *Renaissance*, vol. 2/3, 1944–5, 336–9.

Walker, J. Samuel. *Prompt and Utter Destruction: Truman and the use of Atomic Bombs Against Japan*, Chapel Hill and London: The University of North Carolina Press, 1997.

Wallon, Henri (ed.). *À la lumière du Marxisme*, Paris: Ési, 1935–7.

Warnod, Jeanine (ed.). *Focus on Minotaure. The Animal-Headed Review*, Geneva: Musée d'Art et d'Histoire, 1987.

Watts, Harriett Ann. *Chance: A Perspective on Dada*, Ann Arbor, Michigan: UMI Research Press, 1980.

Weart, Spencer R. *Scientists in Power*, Cambridge, Mass. and London: Harvard University Press, 1979.

Weart, Spencer R. *Nuclear Fear: A History of Images*, Cambridge, Mass. and London: Harvard University Press, 1988.

Weber, Louis. '"L'expérience humaine et la causalité physique" par Léon Brunschvicg', *Revue de Métaphysique et de Morale*, year 30, no. 1, January–March 1923, 59–95.

Weber, Robert L. *Pioneers of Science: Nobel Prize Winners in Physics*, Bristol and London: The Institute of Physics, 1980.

Werner, Gabriele. *Mathematik im Surrealismus: Man Ray – Max Ernst – Dorothea Tanning*, Marburg: Jonas Verlag, 2002.

Whitehead, Alfred North. *Process and Reality*, Cambridge: Cambridge University Press, 1929.

Whitworth, Michael. 'The Clothbound Universe: Popular Physics Books, 1919–1939', *Publishing History*, vol. 40, 1996, 53–82.

Whitworth, Michael H. *Einstein's Wake: Relativity, Metaphor, and Modernist Literature*, Oxford: Oxford University Press, 2001.

Williams, L. Pearce (ed.). *Relativity Theory: Its Origins and Impact on Modern Thought*, New York: John Wiley and Sons, 1968.

Williams, Virginia Parrott. *Surrealism, Quantum Philosophy, and World War I*, New York and London: Garland Publishing Inc., 1987.

Winter, Amy. *Wolfgang Paalen*, Dyn, *and the American Avant-Garde of the 1940s*, dissertation submitted for the degree of Ph.D., City University of New York, 1995.

Winter, Amy. *Wolfgang Paalen: Artist and Theorist of the Avant-Garde*, Westport, Connecticut: Praeger, 2003.

'X' (Bertrand Russell). 'Einstein's Theory of Gravitation', *The Athenaeum*, no. 4672, 14 November 1919, 1189.

Zervos, Christian. 'ENQUÊTE', *Cahiers d'Art*, year 10, 1935, 5–72.

Zervos, Christian. 'Mathématiques et art abstrait', *Cahiers d'Art*, year 11, May 1936, 4–10.

Zukav, Gary. *The Dancing Wu Li Masters: An Overview of the New Physics* [1979], New York: Bantam Books, 1986.

PHOTOGRAPH CREDITS

INDEX

Rien, plus rien aujourd'hui ne distingue la Science (généralisée : la querelle est close, de savoir si elle devait hommes, tant il est évident qu'elle a cessé d'être un moy moderne a pourtant promis, elle a tenu, et elle promet forme de monceaux de cadavres. Jusqu'alors, en présenc possible anéantissement d'une civilisation, nous réagissi moraux habituels. Mais voici l'espèce humaine promise à par l'emploi cynique des bombes nucléaires, fussent-elles dus aux déchets qui, *en attendant*, polluent de manière in phérique et biologique de l'espèce, puisqu'une surenchère rimentales » continue sous le couvert des « fins pacifiq les conditions élémentaires de son activité réduites à une à ses sources de révolte, et, en deçà d'un monde qui ne sai retrouver les chances inconnues de la *fureur*.

Ce n'est donc pas à une attitude humaniste que no longtemps l'opium du peuple, la Science est en bonne pla tations contre la course aux armements, que certains physi nous éclairent au plus sur leur complexe de culpabilité, des vices les plus infâmes de l'homme. La poitrine qu'on s aux mornes bêlements du troupeau par la même main q cette antienne. Le christianisme, et ses miroirs grossissan nous y ont habitués.